T0300940

STATISTICAL CHALLENGES IN 21ST CENTURY COSMOLOGY

IAU SYMPOSIUM No. 306

COVER ILLUSTRATION:

Mean of the posterior pdf of filaments in the matter density field, given the observed SDSS galaxy positions. For further details on this Bayesian method, see the article by Leclercq and Wandelt in this volume.

IAU SYMPOSIUM PROCEEDINGS SERIES

INTERNATIONAL ASTRONOMICAL UNION
UNION ASTRONOMIQUE INTERNATIONALE

STATISTICAL CHALLENGES IN 21ST CENTURY COSMOLOGY

PROCEEDINGS OF THE 306th SYMPOSIUM OF THE INTERNATIONAL ASTRONOMICAL UNION HELD IN LISBON, PORTUGAL MAY 25–29, 2014

Edited by

ALAN HEAVENS
Imperial College London, UK

JEAN-LUC STARCK
CEA Saclay, France

and

ALBERTO KRONE-MARTINS
Universidade de Lisboa, Portugal

CAMBRIDGE
UNIVERSITY PRESS

University Printing House, Cambridge CB2 8BS, United Kingdom

One Liberty Plaza, 20th Floor, New York, NY 10006, USA

477 Williamstown Road, Port Melbourne, VIC 3207, Australia

314-321, 3rd Floor, Plot 3, Splendor Forum, Jasola District Centre, New Delhi - 110025, India

79 Anson Road, #06-04/06, Singapore 079906

Cambridge University Press is part of the University of Cambridge.

It furthers the University's mission by disseminating knowledge in the pursuit of education, learning and research at the highest international levels of excellence.

www.cambridge.org
Information on this title: www.cambridge.org/9781107078567

First published 2015

A catalogue record for this publication is available from the British Library

ISBN 978-1-107-07856-7 Hardback

Table of Contents

CMB section

Radio section

Joint probes section

LSS section

Data mining section

Supernovae section

Survey section

Future project

Other

Conclusions

Preface

On behalf of the International Astronomical Union, I am very happy to welcome the speakers, participants and readers of the IAU Symposium 306, 'Statistical Challenges in 21st Century Cosmology' conference in Lisbon Portugal. In the 19th century, when confronting observational data with predictions of Newtonian celestial mechanics, mathematicians like Legendre, Laplace and Gauss laid the foundations for modern statistics with least squares theory. During the 20th century, the astronomical and statistical communities drifted apart, one towards other branches of astrophysics and the other towards terrestrial affairs.

But, with the advent of advanced instruments and huge surveys, the 21st century is witnessing a resurgence of astrostatistical science. Interpretation of the cosmic microwave background, weak and strong gravitational lensing, galaxy clustering and other signatures of the early Universe all require advanced statistical techniques. This was the motivation for the IAU's Working Group in Astrostatistics and Astroinformatics (recently formed in 2012) to encourage first-class leaders – Alan Heavens, Jean-Luc Starck, Alberto Krone-Martins and their colleagues – to organize this conference. Unlike most cosmology research conferences, the emphasis here is not on the latest survey findings or their astrophysical implications, but rather on the intricate mathematical methods needed for the extraction of scientific insights from the large and complicated datasets.

The conference auditorium was full and the invited presentations were excellent. Together with poster papers, informal discussions and social events, the entire experience was very exciting. The IAU Working Group could not be more pleased with this inaugural astrostatistical Symposium, and we hope that it stimulates the larger community of astronomers and statisticians to further development of methodology for cosmology.

Eric D. Feigelson[1,2]
[1] *Dept. of Astronomy and Astrophysics, Pennsylvania State University, 525 Davey Laboratory, University Park PA 16802 USA*
[2] *Chair, Working Group in Astrostatistics and Astroinformatics, International Astronomical Union*

THE ORGANIZING COMMITTEE

Scientific

Alan Heavens (UK; co-chair)
Raul Abramo (Brazil)
Bruce Bassett (South Africa)
Joseph Hilbe (USA)
Vicent Martinez (Spain)
Enn Saar (Estonia)
Qingjuan Yu (China)
Jean-Luc Starck (France; co-chair)
Adrian Baddeley (Australia)
Dalia Chakrabarty (UK)
Alberto Krone-Martins (Portugal)
Thanu Padmanabhan (India)
Ralf Siebenmorgen (ESO)
Yanxia Zhang (China)

Local

Alberto Krone-Martins (chair)
Sónia Antón
Ana Mourão
André Moitinho (co-chair)
Carlos Martins
António da Silva

Acknowledgements

The symposium is sponsored and supported by IAU Division B (with its Commission 5 and Working Group in Astrostatistics and Astroinformatics) and Division J, Commissariat à l'Energie Atomique, Service d'Astrophysique, France, European Research Council (ERC), International Astrostatistics Association, Academia das Ciências de Lisboa and Systems, Instrumentation and Modeling Laboratory (SIM), Universidade de Lisboa, Portugal, and the Imperial Centre for Inference and Cosmology (ICIC), Imperial College, London.

The Local Organizing Committee operated under the auspices of the Portuguese Astronomical Society.

Funding by the
International Astronomical Union,
European Research Council,
Imperial Centre for Inference and Cosmology (ICIC),
is gratefully acknowledged.

CONFERENCE PHOTOGRAPH

Participants

Filipe **Abdalla**, University College London, United Kingdom, fba@star.ucl.ac.uk
Viviana **Acquaviva**, New York City College of Technology, United States of America, vacquaviva@citytech.cuny.edu
Simone **Aiola**, University of Pittsburgh, United States of America, sia21@pitt.edu
Sonia **Antón**, Universidade de Lisboa - SIM & IAA-CSIC, Portugal, santon@fc.ul.pt
Frederico **Arez**, Universidade de Lisboa, Portugal, frederico.arez@gmail.com
Pablo **Arnalte-Mur**, Institute for Comp. Cosmology, Durham University, United Kingdom, pablo.arnalte-mur@durham.ac.uk
Vicken **Asadourian**, University of Houston, United States of America, vickenasadourian@gmail.com
Metin **Ata**, Leibniz-Institut fuer Astrophysik Potsdam AIP, Germany, mata@aip.de
Jonathan **Aumont**, Institut d'Astrophysique Spatiale, France, jonathan.aumont@ias.u-psud.fr
Adrija **Banerjee**, Indian Statistical Institute, India, adrija@isichennai.res.in
Robert **Beck**, Eotvos Lorand University, Hungary, robert.beck23@gmail.com
Assaf **Ben-David**, The Niels Bohr Institute, Copenhagen, Denmark, bendavid@nbi.dk
Narciso **Benítez**, Instituto de Astrofísica de Andalucía, Spain, txitxo@pha.jhu.edu
Joel **Bergé**, ONERA, France, joel.berge@onera.fr
Federico **Bianchini**, SISSA, Italy, fbianchini@sissa.it
Jérome **Bobin**, CEA-Saclay, France, jbobin@cea.fr
Vanessa **Böhm**, Max-Planck-Institut für Astrophysik, Germany, vboehm@mpa-garching.mpg.de
Christopher **Bonnett**, IFAE, Spain, c.bonnett@gmail.com
Marta **Bruno Silva**, University of Lisbon - CENTRA/IST, Portugal, marta.bruno.silva@ist.utl.pt
Filomena **Bufano**, Universidad Andrés Bello, Chile, milena.bufano@gmail.com
Emory **Bunn**, University of Richmond, United States of America, ebunn@richmond.edu
Sander **Bus**, Kapteyn Astronomical Institute, Netherlands, s.bus@astro.rug.nl
Stefano **Camera**, CENTRA, Instituto Superior Técnico, Universidade de Lisboa, Portugal, stefano.camera@tecnico.ulisboa.pt
Ewan **Cameron**, Oxford Zoology, United Kingdom, dr.ewan.cameron@gmail.com
Julien **Carron**, IfA Hawaii, United States of America, carron_julien@hotmail.com
Carla Sofia **Carvalho**, CAAUL, University of Lisbon, Portugal, cscarvalho@oal.ul.pt
Stefano **Cavuoti**, INAF - Osservatorio Astronomico di Capodimonte, Italy, stefano.cavuoti@gmail.com
Dalia **Chakrabarty**, University of Leicester & Warwick, United Kingdom, d.chakrabarty@warwick.ac.uk
Emma **Chapman**, University College London, United Kingdom, eow@star.ucl.ac.uk
Yen-Chi **Chen**, Carnegie Mellon University, United States of America, ga014528@gmail.com
Vinicius **Consolini Busti**, University of Cape Town, South Africa, vinicius.busti@iag.usp.br
Dagoberto **Contreras**, University of British Columbia, Canada, dagocont@phas.ubc.ca
Manuel **da Silva**, Universidade de Lisboa - SIM, Portugal, madusilva@gmail.com
Aurore **Delaigle**, University of Melbourne, Australia, A.Delaigle@ms.unimelb.edu.au
Barkats **Denis**, ALMA - ESO, Chile, dbarkats@alma.cl
Sandro **Dias Pinto Vitenti**, Institut d'Astrophysique de Paris, France, dias@iap.fr
Sebastian **Dorn**, Max-Planck-Institut for Astrophysics, Germany, sdorn@mpa-garching.mpg.de
Lise **du Buisson**, University of Cape Town / AIMS, South Africa, lisedubuisson@gmail.com
Filipe **Duarte dos Santos**, Universidade de Lisboa - SIM, Portugal, fdsantos@siam.fis.fc.ul.pt
Xavier **Dupac**, ESA, Spain, xdupac@sciops.esa.int
Claudio **Durastanti**, University of Tor Vergata, Roma, Italy, durastan@mat.uniroma2.it
Franz **Elsner**, University College London, United Kingdom, f.elsner@ucl.ac.uk
Martin Borsta **Eriksen**, Leiden University, Netherlands, martin.b.eriksen@gmail.com
August **Evrard**, University of Michigan, United States of America, gus.evrard@gmail.com
Ophélia **Fabre**, Institut d'Astrophysique de Paris, France, fabre@iap.fr
Jalal **Fadili**, Ecole Nationale Supérieure d'Ingénieurs de Caen, France, Jalal.Fadili@greyc.ensicaen.fr
Yabebal **Fantaye**, University of Rome2 Tor Vergata, Italy, fantaye@mat.uniroma2.it
Stephen **Feeney**, Imperial College London, United Kingdom, s.feeney@imperial.ac.uk
Eric **Feigelson**, Pennsylvania State University, United States of America, e5f@psu.edu
José **Fonseca**, University of the Western Cape, South Africa, josecarlos.s.fonseca@gmail.com
Francisco **Forster**, Universidad de Chile, Center for Mathematical Modelling, Chile, francisco.forster@gmail.com
Lluis **Galbany**, DAS - Universidad de Chile, Chile, lluisgalbany@gmail.com
Lin **Ganghua**, National Astronomical Observatories, Chinese Academy of Sciences, China, lgh@bao.ac.cn
Santiago **Gonzalez-Gaitan**, Universidad de Chile, Chile, gongsale@gmail.com
Ben **Granett**, INAF Brera Observatory, Italy, ben.granett@brera.inaf.it
Alejandro **Guarnizo Trilleras**, Ruprecht-Karls-Universität Heidelberg, Germany, a.guarnizo@thphys.uni-heidelberg.de
Mario **Hamuy**, Universidad de Chile, Chile, mhamuy@das.uchile.cl
Shan **Hao**, Xinjiang Observatory, China, shanhao@xao.ac.cn
Alan **Heavens**, Imperial College London, United Kingdom, a.heavens@imperial.ac.uk
Caroline **Heneka**, DARK Cosmology Center, University of Copenhagen, Denmark, caroline@dark-cosmology.dk
Joseph **Hilbe**, Arizona State University, United States of America, hilbe@asu.edu
Mike **Hobson**, University of Cambridge, United Kingdom, mph@mrao.cam.ac.uk
Hector Javier **Hortua Orjuela**, Institucion los Libertadores / O.A.N-Universidad Nacional, Colombia, hjhortuao@unal.edu.co
Zhuoxi **Huo**, Tsinghua University, China, huozx@tsinghua.edu.cn
Lluís **Hurtado-Gil**, Observatori Astronòmic de la Universitat de València, Spain, lluis.hurtado@uv.es
Stéphane **Ilic**, Institut de Recherche en Astrophysique et Planétologie, France, stephane.ilic@irap.omp.eu
Emille **Ishida**, Max Planck Institute for Astrophysics, Germany, emille@mpa-garching.mpg.de
Andrew **Jaffe**, Imperial College London, United Kingdom, a.jaffe@imperial.ac.uk
Benjamin **Joachimi**, University College London, United Kingdom, b.joachimi@ucl.ac.uk
Natallia **Karpenka**, Stockholm University, Sweden, nkarp@fysik.su.se
Eyal **Kazin**, Swinburne University of Technology, Australia, eyalkazin@gmail.com
Madhura **Killedar**, Ludwig-Maximilians Universität, Germany, killedar@usm.lmu.de
Francisco **Kitaura**, Leibniz Institute for Astrophysics, Germany, kitaura@aip.de
Michelle **Knights**, University of Cape Town/AIMS, South Africa, michelle.knights@gmail.com
Andras **Kovacs**, Eotvos University, Hungary, andraspankasz@gmail.com
Alberto **Krone-Martins**, Universidade de Lisboa - SIM, Portugal, algol@sim.ul.pt
Konrad **Kuijken**, Leiden Observatory, Netherlands, kuijken@strw.leidenuniv.nl
Noah **Kurinsky**, Tufts University, United States of America, Noah.Kurinsky@tufts.edu

Fabien **Lacasa**, ICTP - SAIFR, Brazil, fabien@ift.unesp.br
François **Lanusse**, CEA Saclay, France, francois.lanusse@cea.fr
Florent **Leclercq**, Institut d'Astrophysique de Paris, France, florent.leclercq@polytechnique.org
Yajuan **Lei**, National Astronomical Observatory, Chinese Academy of Sciences, China, leiyj@lamost.org
Boris **Leistedt**, University College London, United Kingdom, boris.leistedt.11@ucl.ac.uk
Adrienne **Leonard**, University College London, United Kingdom, adrienne.leonard@ucl.ac.uk
Maggie **Lieu**, University of Birmingham, United Kingdom, mlieu@star.sr.bham.ac.uk
Lauri Juhan **Liivamägi**, Tartu Observatory, Estonia, wire@ut.ee
Chieh-An **Lin**, CEA Saclay, France, chieh-an.lin@cea.fr
Jon **Loveday**, University of Sussex, United Kingdom, J.Loveday@sussex.ac.uk
Jufu **Lu**, Xiamen University, China, lujf@xmu.edu.cn
Marc **Manera**, University College London, United Kingdom, m.miret@ucl.ac.uk
Anna **Mangilli**, Insitut d'Astrophysique de Paris, France, mangilli@iap.fr
Vicent J. **Martínez**, Astronomical Observatory - University of Valencia, Spain, martinez@uv.es
Carlos **Martins**, Centro de Astrofísica da Universidade do Porto, Portugal, Carlos.Martins@astro.up.pt
Sabino **Matarrese**, Università degli Studi di Padova, Italy, sabino.matarrese@pd.infn.it
Jason **McEwen**, University College London, United Kingdom, jason.mcewen@ucl.ac.uk
Josh **Meyers**, Stanford University, United States of America, jmeyers3@stanford.edu
José Pedro **Mimoso**, Universidade de Lisboa, Portugal, jpmimoso@fc.ul.pt
André **Moitinho**, Universidade de Lisboa - SIM, Portugal, andre@sim.ul.pt
Daniel **Mortlock**, Imperial College London, United Kingdom, mortlock@ic.ac.uk
Ana **Mourão**, Universidade de Lisboa - CENTRA/IST, Portugal, amourao@ist.utl.pt
Volker **Mueller**, Leibniz-Institut fuer Astrophysik Potsdam AIP, Germany, vmueller@aip.de
Steven **Murray**, ICRAR UWA, Australia, steven.murray@uwa.edu.au
Mark **Neyrinck**, Johns Hopkins University, United States of America, neyrinck@pha.jhu.edu
Hyerim **Noh**, KASI - Korea Astronomy and Space Science Institute, South Korea, hr@kasi.re.kr
Minji **Oh**, UST/KASI(Korea Astronomy and Space Science Institute), South Korea, minjioh@kasi.re.kr
Niels **Oppermann**, Canadian Institute for Theoretical Astrophysics, Canada, niels@cita.utoronto.ca
René A. **Ortega-Minakata**, Universidad de Guanajuato, Mexico, rene@astro.ugto.mx
Arnab **Pal**, Indian Statistical Institute, India, arnabandstats@gmail.com
Prina **Patel**, UWC, Cape Town, South Africa, prina83@gmail.com
Ajinkya **Patil**, Kapteyn Astronomical Institute, Netherlands, patil@astro.rug.nl
Paniez **Paykari**, CEA-Saclay, France, paniez.paykari@cea.fr
Hiranya **Peiris**, University College London, United Kingdom, h.peiris@ucl.ac.uk
Vincent **Pelgrims**, IFPA, AGO dept., University of Liège, Belgium, pelgrims@astro.ulg.ac.be
Mariana **Penna Lima Vitenti**, National Institute for Space Research - INPE, Brazil, pennalima@gmail.com
Giuliano **Pignata**, Universidad Andrés Bello, Chile, pignago@gmail.com
Athina **Pouri**, University of Athens - RCAAM, Greece, ath.pouri@gmail.com
Chris **Pritchet**, University of Victoria, Canada, pritchet@uvic.ca
Anais **Rassat**, École Polytechnique Fédérale de Lausanne, Switzerland, anais.rassat@epfl.ch
Marcelo **Reboucas**, Centro Brasileiro de Pesquias Fisicas, Brazil, reboucas.marcelo@gmail.com
Eniko **Regos Dr**, Eotvos University, Hungary, Enikoe.Regoes@cern.ch
Graça **Rocha**, Jet Propulsion Laboratory, Caltech, United States of America, graca@caltech.edu
Ana **Rodrigues**, Universidade de Lisboa, Portugal, aparodrigues.27@gmail.com
Nina **Roth**, University College London, United Kingdom, n.roth@ucl.ac.uk
Matthias **Rubart**, Bielefeld University, Germany, matthiasr@physik.uni-bielefeld.de
Rafael **S. de Souza**, Korea Astronomy and Space Science Institute, Korea, rafael.2706@gmail.com
Enn **Saar**, Tartu Observatory, Estonia, saar@to.ee
Iftach **Sadeh**, University College London, United Kingdom, i.sadeh@ucl.ac.uk
Mário **Santos**, University of the Western Cape, South Africa, mgrsantos@ist.utl.pt
Tomás **Santos**, Universidade de Lisboa, Portugal, tomassantos95@gmail.com
Roberto **Scaramella**, Osservatorio Astronomico di Roma, Italy, kosmobob@oa-roma.inaf.it
Robert **Schuhmann**, University College London, United Kingdom, robert.schuhmann.13@ucl.ac.uk
Sebastian **Seehars**, ETH Zurich, Switzerland, seehars@phys.ethz.ch
Elena **Sellentin**, ITP, Universität Heidelberg, Germany, sellentin@stud.uni-heidelberg.de
Paolo **Serra**, IAS-Orsay, France, paoloserr@gmail.com
Ralf **Siebenmorgen**, ESO, Germany, ralf.siebenmorgen@eso.org
António **Silva**, Universidade de Lisboa - CAAUL, Portugal, asilva@astro.up.pt
Fergus **Simpson**, University of Barcelona, Spain, fergus2@icc.ub.edu
Navin **Sivanandam**, African Institute for Mathematical Sciences, South Africa, navin.sivanandam@gmail.com
Jéssica **Soares**, Universidade de Lisboa, Portugal, jessica.soares@sapo.pt
Lara **Sousa**, Centro de Astrofísica da Universidade do Porto, Portugal, Lara.Sousa@astro.up.pt
Vallery **Stanishev**, University of Lisbon - CENTRA/IST, Portugal, vallery.stanishev@ist.utl.pt
Jean-Luc **Starck**, CEA, Service d'Astrophysique, France, jstarck@cea.fr
Radu **Stoica**, Université Lille 1 & IMCCE Paris, France, radu.stoica@univ-lille1.fr
Alex **Szalay**, Johns Hopkins University, United States of America, szalay@jhu.edu
Masahiro **Takada**, University of Tokyo - IPMU / TODIAS, Japan, masahiro.takada@ipmu.jp
Elmo **Tempel**, Tartu Observatory, KBFI, Estonia, elmo.tempel@to.ee
Ismael **Tereno**, CAAUL - Universidade de Lisboa, Portugal, tereno@fc.ul.pt
Haijun **Tian**, China Three Gorges University, China, hjtian2000@gmail.com
Jesus **Torrado Cacho**, Instituut-Lorentz, Leiden University, Netherlands, torradocacho@lorentz.leidenuniv.nl
Arlindo **Trindade**, Centro de Astrofísica da Universidade do Porto, Portugal, Arlindo.Trindade@astro.up.pt
Antonino **Troja**, University of Milan - Physics department, Italy, antonino.troja@unimi.it
David **van Dyk**, Imperial College London, United Kingdom, dvandyk@imperial.ac.uk
Jesús **Varela**, Centro de Estudios de Física del Cosmos de Aragón, Spain, jvarela@cefca.es
Licia **Verde**, Universidad de Barcelona, Spain, liciaverde@icc.ub.edu
Michael **Vespe**, Carnegie Mellon University, United States of America, mvespe@andrew.cmu.edu
Pedro **Viana**, Centro de Astrofísica da Universidade do Porto, Portugal, viana@astro.up.pt
Giuseppe **Vinci**, Carnegie Mellon University, United States of America, giuseppevinci88@gmail.com

Vladimir **Vinnikov**, Dorodnicyn Computing Centre of Russian Academy of Sciences, Russia, vvinnikov@list.ru
Massimo **Viola**, Leiden University, Netherlands, viola@strw.leidenuniv.nl
Vincenzo **Vitagliano**, University of Lisbon - CENTRA/IST, Portugal, vitaglia@gmail.com
Jasper **Wall**, University of British Columbia, Canada, jvw@phas.ubc.ca
Ben **Wandelt**, Institut d'Astrophysique de Paris, France, bwandelt@illinois.edu
Catherine **Watkinson**, Imperial College London, United Kingdom, catherine.watkinson@gmail.com
Yves **Wiaux**, Heriot-Watt University, United Kingdom, y.wiaux@hw.ac.uk
Godlowski **Wlodzimierz**, Opole University, Poland, godlowski@uni.opole.pl
Yue **Wu**, National Astronomical Observatories, Chinese Academy of Sciences, China, wuyue@bao.ac.cn
Yanxia **Zhang**, National Astronomical Observatory, Chinese Academy of Sciences, China, zyxsunny@gmail.com
Jianfeng **Zhou**, Center for Astrophysics, Tsinghua UniversityChina, zhoujf@tsinghua.edu.cn

Address by the Scientific and Local Organizing Committees

Dear Colleagues,

In the era of large astronomical surveys that are grappling with unsolved methodological and data challenges, transforming Data into Science is a huge, and exciting, problem.

With surveys and instruments such as Planck, Pan-STARRS1, DES, VST KiDS, LSST, Gaia, Euclid, JPAS, SKA, wide-field spectroscopic surveys and the large and interconnected databases of archival material coming online, a special scientific focus on cosmological inference is of great interest. Without this focus there is no guarantee that the best possible science will be the outcome of this flood of data. However, transforming data of the size and complexity of current and future surveys into knowledge presents considerable challenges, and is a problem that can benefit from cross-disciplinary efforts.

During recent years there has been a resumption of the dialogue between astronomers and statisticians, led by the Penn State conferences organized by Professors Jogesh Babu and Eric Feigelson. This dialogue has been fruitful and has been at the origin of a new Astrostatistics discipline. Since 2009 this discipline has a committee in the International Statistical Institute, (roughly) the counterpart of the IAU, and later, in 2012 an IAU Working Group in Astrostatistics and Astroinformatics was created.

It was thus timely to hold the first IAU Symposium devoted to Astrostatistics, which was also the first IAU Symposium in Portugal. This was a symposium with a cross-disciplinary nature, reflected in the inclusion of themes and speakers from the Astronomy and from the Statistics community, and covering topics from data processing to model selection. The aim of the meeting was to explore methods at a more technical level than would be normal at a typical cosmology conference, providing exposure to new ideas and techniques, to establish fruitful collaborations and to provide a natural discussion forum for both communities. We were delighted at the standard and technical level of the presentations and posters, and hope that these proceedings act as a useful starting point for researchers in search of the latest thinking in this exciting area of research.

Alan Heavens and Jean-Luc Starck (co-chairs SOC), Alberto Krone-Martins (chair LOC)
London, Paris and Lisbon, 27 August 2014

Statistical Challenges in 21st Century Cosmology
Proceedings IAU Symposium No. 306, 2014
A. F. Heavens, J.-L. Starck & A. Krone-Martins, eds.

doi:10.1017/S1743921314011120

Bayesian large-scale structure inference: initial conditions and the cosmic web

Florent Leclercq[1,2,3] **and Benjamin Wandelt**[1,2,4]

[1]Institut d'Astrophysique de Paris (IAP), UMR 7095, CNRS - UPMC Université Paris 6, 98bis boulevard Arago, F-75014 Paris, France

[2]Institut Lagrange de Paris (ILP), Sorbonne Universités, 98bis boulevard Arago, F-75014 Paris, France

[3]École polytechnique ParisTech, Route de Saclay, F-91128 Palaiseau, France

[4]Departments of Physics and Astronomy, University of Illinois at Urbana-Champaign, Urbana, IL 61801

emails: `florent.leclercq@polytechnique.org`, `wandelt@iap.fr`

Abstract. We describe an innovative statistical approach for the *ab initio* simultaneous analysis of the formation history and morphology of the large-scale structure of the inhomogeneous Universe. Our algorithm explores the joint posterior distribution of the many millions of parameters involved via efficient Hamiltonian Markov Chain Monte Carlo sampling. We describe its application to the Sloan Digital Sky Survey data release 7 and an additional non-linear filtering step. We illustrate the use of our findings for cosmic web analysis: identification of structures via tidal shear analysis and inference of dark matter voids.

Keywords. large-scale structure of universe, methods: statistical

1. Introduction

How did the Universe begin? How do we understand the shape of the present-day cosmic web? Within standard cosmology, we have an observationally well-supported model for the initial conditions (ICs) – a Gaussian random field – and the evolution and growth of cosmic structures is well-understood in principle. It is therefore natural to analyze large-scale structure (LSS) surveys in terms of the simultaneous constraints they place on the statistical properties of the initial conditions of the Universe and on the shape of the cosmic web. Due to the computational challenge and to the lack of detailed physical understanding of the non-Gaussian and non-linear processes that link galaxy formation to the large-scale dark matter distribution, the current state of the art of statistical analyses of LSS surveys is far from this ideal and these problems are addressed in isolation. Here, we describe progress towards the full reconstruction of four-dimensional states and illustrate the use of these results for cosmic web classification.

2. Statistical approach

2.1. *Why Bayesian inference?*

Cosmological observations are subject to a variety of intrinsic and experimental uncertainties (incomplete observations – survey geometry and selection effects –, cosmic variance, noise, biases, systematic effects), which make the inference of signals a fundamentally ill-posed problem. For this reason, no unique recovery of the initial conditions and of the shape of the present-day cosmic web is possible; it is more relevant to quantify a probability distribution for such signals, given the observations. Adopting this point of view for large-scale structure surveys, Bayesian forward modeling (gravitational structure formation is the generative model for the complex final state, starting from a simple initial state – Gaussian or nearly-Gaussian ICs) offers a conceptual basis for dealing with the problem of inference in presence of uncertainty (e.g. Jasche & Wandelt 2013a; Kitaura 2013; Wang *et al.* 2013).

2.2. *High-dimensionality*

Statistical analysis of LSS surveys requires to go from the few parameters describing the homogeneous Universe to a point-by-point characterization of the inhomogeneous Universe. The latter description typically involves tens of millions of parameters: the density in each voxel of the survey volume. No obvious reduction of the problem size exists. "Curse of dimensionality" phenomena (Bellman 1961) are therefore the significant obstacle in this high-dimensional data analysis problem. They refer to the problems caused by the exponential increase in volume associated with adding extra dimensions to a mathematical space, and therefore in sparsity given a fixed amount of sampling points. Numerical representations of high-dimensional probability distribution functions (pdfs) will tend to have very peaked features and narrow support, which means that traditional sampling methods will fail. However, gradients of these functions carry capital information, as they indicate the direction to high-density regions, permitting fast travel through a very large volume in parameter space.

2.3. *Hamiltonian Monte Carlo*

The Hamiltonian Monte Carlo algorithm (Duane *et al.* 1987) is an algorithm for exploring parameter spaces with particles (samples). The general idea is to use classical mechanics to solve statistical problems. The algorithm interprets the negative logarithm of the pdf to sample from, $\mathcal{P}(\mathbf{x})$, as a potential, $\psi \equiv -\ln(\mathcal{P}(\mathbf{x}))$, and integrates Hamilton's equation in parameter space. Due to the conservation of energy in classical mechanics, the theoretical acceptance rate is always unity. Therefore, HMC beats the "curse of dimensionality" by exploiting gradients ($\partial\psi(\mathbf{x})/\partial\mathbf{x}$ in Hamilton's equations) and using conserved quantities.

3. Physical reconstructions

3.1. *Bayesian large-scale structure inference in the SDSS DR7*

The full-scale Bayesian inference code BORG (Bayesian Origin Reconstruction from Galaxies, Jasche & Wandelt 2013a) uses HMC for four-dimensional inference of density fields in the linear and mildly non-linear regime. The (approximate) physical model for gravitational dynamics included in the likelihood is second-order Lagrangian perturbation theory (2LPT), linking initial density fields (at a scale factor $a = 10^{-3}$) to the presently observed large-scale structure (at $a = 1$). The galaxy distribution is modeled as a Poisson sample from these evolved density fields. The algorithm also accounts for luminosity dependent galaxy biases (Jasche & Wandelt 2013b). In Jasche *et al.* (2015), we apply the BORG code to 463,230 galaxies from the `Sample dr72` of the New York University Value Added Catalogue (NYU-VAGC, Blanton *et al.* 2005), based ot the final data release (DR7) of the Sloan Digital Sky Survey (SDSS, Adelman-McCarthy *et al.* 2008; Padmanabhan *et al.* 2008).

Each inferred sample (Fig. 1, left) is a "possible version of the truth" in the form of a full physical realization of dark matter particles. The variation between samples (Fig. 1, right) quantifies joint and correlated uncertainties inherent to any cosmological observation and accounts for all non-linearities and non-Gaussianities involved.

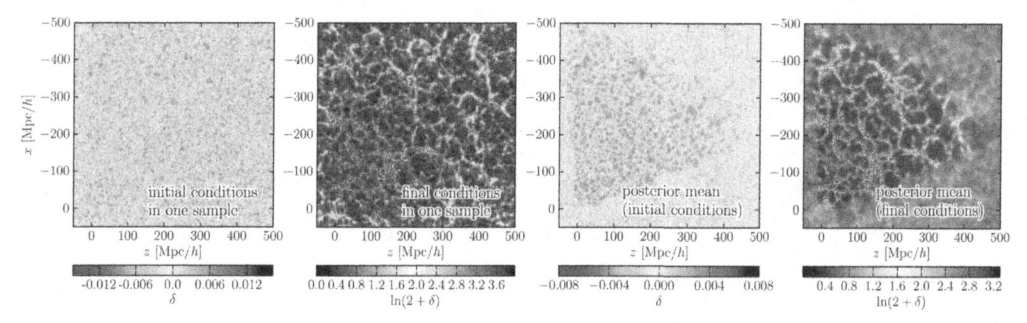

Figure 1. Bayesian LSS inference with BORG in the SDSS DR7. Slices through one sample of the posterior for the initial and final density fields (left) and posterior mean in the initial and final conditions (right). The input galaxies are overplotted on the final conditions as red dots.

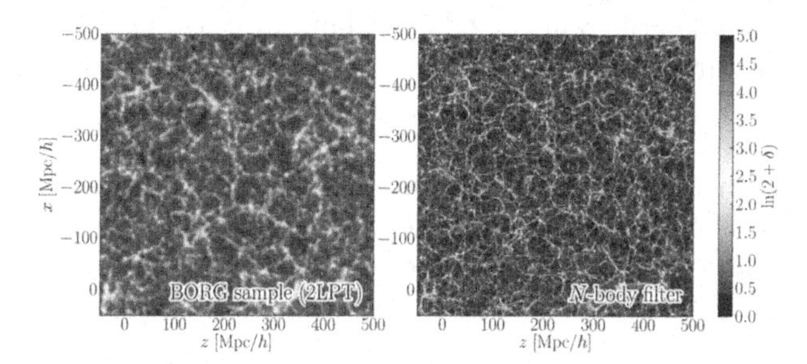

Figure 2. N-body filtering of a BORG sample (left), to produce a non-linear data-constrained realization of the redshift-zero large-scale structure (right).

3.2. *Non-linear filtering*

Building upon these results, it is possible to post-process the samples using fully non-linear dynamics as an additional filtering step (Leclercq *et al.* 2015a). We generate a set of data-constrained realizations of the present large-scale structure: some samples of inferred initial conditions are evolved with 2LPT to $z = 69$, then with a fully non-linear cosmological simulation (using GADGET-2) from $z = 69$ to $z = 0$. This filtering step yields a much more precise view of the deeply non-linear regime of cosmic structure formation, sharpening overdense, virialized structures and resolving more finely the substructure of voids (Fig. 2).

4. Cosmic web analysis

4.1. *Tidal shear classification*

The results presented in § 3.1 form the basis of the analysis of Leclercq *et al.* (2015b), where we classify the cosmic large scale structure into four distinct web-types (voids, sheets, filaments and clusters) and quantify corresponding uncertainties. We follow the dynamic cosmic web classification procedure proposed by Hahn *et al.* (2007), based on the eigenvalues $\lambda_1 < \lambda_2 < \lambda_3$ of the tidal tensor T_{ij}, Hessian of the rescaled gravitational potential: $T_{ij} \equiv \partial^2\Phi/\partial\mathbf{x}_i\,\partial\mathbf{x}_j$, where Φ follows the Poisson equation ($\nabla^2\Phi = \delta$). A voxel is in a cluster (resp. in a filament, in a sheet, in a void) if three (resp. two, one, zero) of the λs are positive.

By applying this classification procedure to all density samples, we are able to estimate the posterior of the four different web-types, conditional on the observations. The means of these pdfs are represented in Fig. 3.

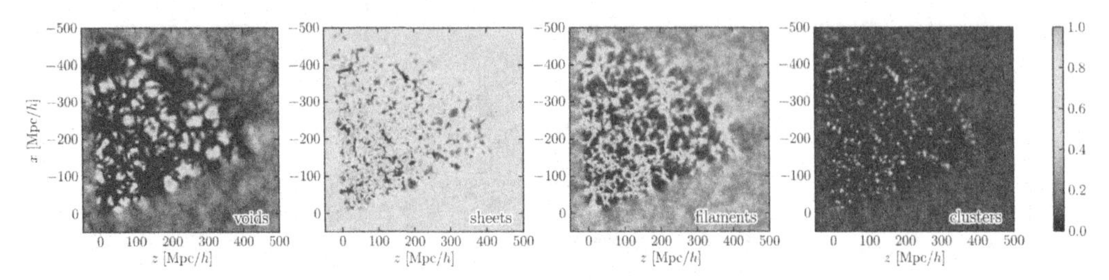

Figure 3. Mean of the posterior pdf for the four different web-types in the SDSS DR7.

4.2. *Dark matter voids*

In Leclercq *et al.* (2015a), we apply computational geometry tools (VIDE: the Void IDentification and Examination pipeline, Sutter *et al.* 2015) to the constrained parts of the non-linear realizations described in § 3.2. We find physical cosmic voids in the field traced by the dark matter particles, probing a level deeper in the mass distribution hierarchy than galaxies. Due to the high density of tracers, we find about an order of magnitude more voids at all scales than the voids directly traced by the SDSS galaxies. In this fashion, we circumvent the issues due to the conjugate and intricate effects of sparsity and biasing on galaxy void catalogs (Sutter *et al.* 2014) and drastically reduce the statistical uncertainty. For usual void statistics (number count, radial density profiles, ellipticities), all the results we obtain are consistent with N-body simulations prepared with the same setup.

Acknowledgements

We thank Jacopo Chevallard, Jens Jasche, Nico Hamaus, Guilhem Lavaux, Emilio Romano-Díaz, Paul Sutter and Alice Pisani for a fruitful collaboration on the projects presented here. FL acknowledges funding from an AMX grant (École polytechnique) and BW from a senior Excellence Chair by the Agence Nationale de la Recherche (ANR-10-CEXC-004-01). This work made in the ILP LABEX (ANR-10-LABX-63) was supported by French state funds managed by the ANR within the Investissements d'Avenir programme (ANR-11-IDEX-0004-02).

References

Adelman-McCarthy, J. K., Agüeros, M. A., *et al.* 2008, *Astrophys. J. Supp.*, 175, 297
Blanton, M. R., Schlegel, D. J., Strauss, M. A., *et al.* 2005, *AJ*, 129, 2562
Bellman, R. E. 1961, *Adaptive Control Processes: A Guided Tour* (Princeton University Press)
Duane, S., Kennedy, A. D., Pendleton, B. J., & Roweth, D. 1987, *Physics Letters B*, 195, 216
Hahn, O., Porciani, C., Carollo, C. M., & Dekel, A. 2007, *Mon. Not. R. Astron. Soc.*, 375, 489
Jasche, J. & Wandelt, B. D. 2013, *Mon. Not. R. Astron. Soc.*, 432, 894
Jasche, J. & Wandelt, B. D. 2013, *ApJ*, 779, 15
Jasche, J., Leclercq, F., & Wandelt, B. D. 2015, JCAP, 1, 036
Kitaura, F.-S. 2013, *Mon. Not. R. Astron. Soc.*, 429, L84
Leclercq, F., Jasche, J., Sutter, P. M., Hamaus, N., & Wandelt, B. 2015, *JCAP*, 3, 047
Leclercq, F., Jasche, J., & Wandelt, B. 2015, arXiv:1502.02690
Sutter, P. M., Lavaux, G., Hamaus, N., et al. 2014, *Mon. Not. R. Astron. Soc.*, 442, 462
Sutter, P. M., Lavaux, G., Hamaus, N., et al. 2015, *Astronomy and Computing*, 9, 1
Padmanabhan, N., Schlegel, D. J., Finkbeiner, D. P., *et al.* 2008, *ApJ*, 674, 1217
Wang, H., Mo, H. J., Yang, X., & van den Bosch, F. C. 2013, *ApJ*, 772, 63

Statistical Challenges in 21st Century Cosmology
Proceedings IAU Symposium No. 306, 2014
A. F. Heavens, J.-L. Starck & A. Krone-Martins, eds.

doi:10.1017/S1743921314013908

Bayesian model comparison in cosmology

Daniel J. Mortlock
Astrophysics Group, Blackett Laboratory, and Department of Mathematics
Imperial College London, London SW7 2AZ, United Kingdom
email: mortlock@ic.ac.uk

Abstract. The standard Bayesian model formalism comparison cannot be applied to most cosmological models as they lack well-motivated parameter priors. However, if the data-set being used is separable, then it is possible to use some of the data to obtain the necessary parameter distributions, the rest of the data being retained for model comparison. While such methods are not fully prescriptive, they provide a route to applying Bayesian model comparison in cosmological situations where it could not otherwise be used.

Keywords. methods: statistical, cosmology: cosmological parameters

---

1. Introduction

Much of observational cosmology can be thought of as an attempt to use astronomical data to discriminate between the different cosmological models under consideration. Given both the inevitably imperfect data and the intrinsically stochastic nature of many cosmological measurements (*i.e.*, cosmic variance), it is generally impossible to come to absolute conclusions about the various candidate models; the best that can be hoped for is to evaluate the probabilities, conditional on the the available data, that each of the candidate models is the correct description of the Universe. The fact that there is, as far as is known, just a single observable Universe (*i.e.*, there is no ensemble from which it has been drawn), means that such probabilities cannot be frequency-based, and must instead must represent a degree of implication. Self-consistency arguments then require (Cox 1946) that these probabilities be manipulated and inverted using Bayes's theorem.

Taken together, the above facts imply that Bayesian model comparison (Section 2) should be used to assess how well different cosmological models explain the available data, although the fact that most such models have unspecified parameters is a significant difficulty for this approach (Section 3). This problem can be solved for separable data-sets as it is possible to use a two-step method of model comparison (Section 4), illustrated here with high-redshift supernova (SN) data (Section 5).

2. Bayesian model comparison

Given that one of a set of N models, $\{M_1, M_2, \ldots, M_N\}$, is assumed to be true, the state of knowledge conditional on all the available (and relevant) information, I, is fully summarised by the probabilities $\Pr(M_1|I), \Pr(M_2|I), \ldots, \Pr(M_N|I)$, where $\Pr(M_i|I)$ is the probability that the i'th model is correct (and $i \in \{1, 2, \ldots, N\}$). In the light of some new data, d, that has not already been included in the above probabilities, Bayes's theorem gives the updated probability that model i is correct as

$$\Pr(M_i|d, I) = \frac{\Pr(M_i|I)\Pr(d|M_i, I)}{\sum_{j=1}^{N} \Pr(M_j|I)\Pr(d|M_j, I)}, \tag{2.1}$$

where $\Pr(d|M_i, I)$ is the marginal likelihood under model M_i.

If model M_i has N_i unspecified parameters $\{\theta_i\} = \{\theta_{i,1}, \theta_{i,2}, \ldots, \theta_{i,N_i}\}$ then the model-averaged likelihood is obtained by marginalising over these parameters to give

$$\Pr(d|M_i, I) = \int \Pr(\{\theta_i\}|M_i, I) \Pr(d|\{\theta_i\}, M_i, I) \, \mathrm{d}\theta_{i,1} \, \mathrm{d}\theta_{i,2} \ldots \mathrm{d}\theta_{i,N_i}, \quad (2.2)$$

where $\Pr(\{\theta_i\}|M_i, I)$ is the prior distribution of the parameter values in this model. This expression demonstrates that the full specification of a model requires not just an explicit parameterisation, but a distribution for those parameters as well; two mathematically identical descriptions with different parameter priors are, in fact, different models.

3. Comparison of models without parameter priors

Equations 2.1 and 2.2 together summarise a self-consistent method for assessing which of a set of models is better supported by the available information, provided that the parameter priors for all the models are explicitly defined and unit-normalised. In particular, while it is often possible to obtain sensible parameter constraints based on an improper prior, such as $\Pr(\{\theta_i\}|M_i, I)$ constant for all $\{\theta_i\}$, the resultant marginal likelihood is meaningless (Dickey 1961). Unfortunately, it is commonly the case in astronomy and cosmology that there is no compelling form for the models' parameter priors and, further, that the natural uninformative prior distributions are improper and cannot be normalised. The apparent implication is that Bayesian model comparison, at least in the form described in Section 2, cannot be used in cosmology, an idea that has been explored previously by, *e.g.*, Efstathiou (2008) and Jenkins & Peacock (2011). The disturbing corollary would be that there is no self-consistent method to choose between the available cosmological models, even if they are completely quantitative and mathematically well-defined.

4. Model comparison with separable data

The idea that the relative degree of support for models with unspecified parameters is undefined is at odds with the marked – and data-driven – progress that has been made in cosmology over the last century. Clearly it *is* possible to use data to choose sensibly between models even if they do not have well-motivated parameter priors; but can this be formalised in a way that satisfies Bayes's theorem and is hence logically self-consistent?

One possibility is, for separable data-sets (such as those which consist of measurements of many astronomical sources), to use some of the available data to obtain the necessary parameter priors and to then use the remaining data for model comparison. This is an old concept, dating back at least to Lempers (1971) and explored subsequently by, *e.g.*, Spiegelhalter & Smith (1982) and O'Hagan (1995). The central idea is to partition the data as $d = (d_1, d_2)$, with the first partition of training data used to obtain the (partial) posterior distribution for the parameters of i'th model as

$$\Pr(\{\theta_i\}|d_1, M_i, I) = \frac{\Pr(\{\theta_i\}|M_i, I) \Pr(d_1|\{\theta_i\}, M_i, I)}{\int \Pr(\{\theta'_i\}|M_i, I) \Pr(d_1|\{\theta'_i\}, M_i, I) \, \mathrm{d}\theta'_{i,1} \, \mathrm{d}\theta'_{i,2} \ldots \mathrm{d}\theta'_{i,N_i}}, \quad (4.1)$$

where $\Pr(\{\theta_i\}|M_i, I)$, which need *not* be normaliseable, should be a highly uninformative prior. This posterior distribution can then be used as the prior needed to obtain a meaningful marginal likelihood, which can then be evaluated for the testing data as

$$\Pr(d_2|d_1, M_i, I) = \int \Pr(\{\theta_i\}|d_1, M_i, I) \Pr(d_2|\{\theta_i\}, M_i, I) \, \mathrm{d}\theta_{i,1} \, \mathrm{d}\theta_{i,2} \ldots \mathrm{d}\theta_{i,N_i}. \quad (4.2)$$

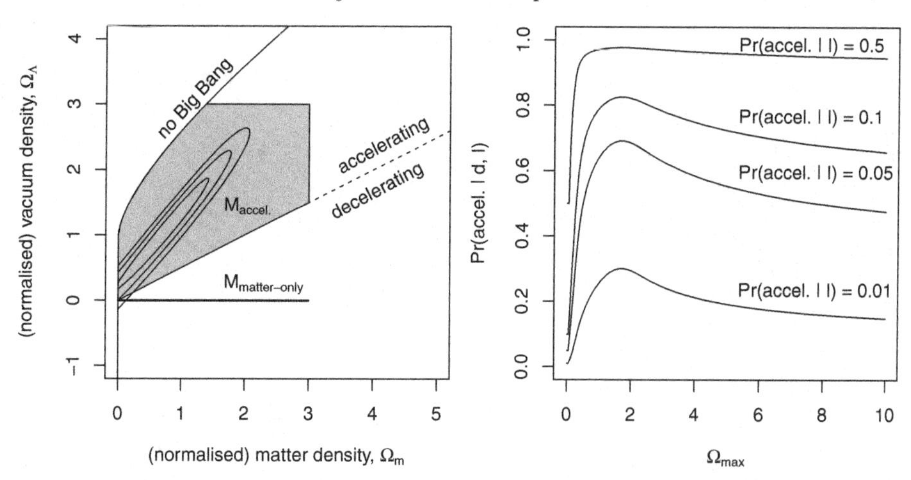

Figure 1. (left) The posterior distribution of Ω_m and Ω_Λ implied by from the Perlmutter *et al.* (1999) SCP SN data and a uniform prior with $\Omega_m \geqslant 0$. Highest posterior density contours enclosing 68.3%, 95.4% and 99.7% of the posterior probability are shown. Also shown are the prior distributions of the accelerating model and matter only model for $\Omega_{max} = 3$. (right) The dependendence of $\Pr(\text{accel.}|d, I)$ on Ω_{max}, shown for different prior probabilities, $\Pr(\text{accel.}|I)$.

This marginal likelihood is coherent, in the sense that it provides self-consistent updated posterior probabilities when inserted into Equation 2.1, but there is also ambiguity: there is no compelling scheme for partitioning the data. It is tempting to average over the possible partitions, but this approach does not have a rigorous motivation. Despite these ambiguities, this two-step method of Bayesian model comparison for separable data does satisfy the Cox (1946) self-consistency requirements and so provide a means of calculating posterior probabilities for cosmological models with unspecified parameter priors.

5. Example: late-time acceleration and supernovae

One of the most significant recent cosmological discoveries was that the Universe's expansion rate is increasing, a result which is often linked most strongly to the observations of distant SNe made by Riess *et al.* (1998) and Perlmutter *et al.* (1999). The comparative faintness of the SNe, given their redshifts and light-curve decay timescales, indicated that the (normalised) cosmological constant, Ω_Λ, is sufficiently large to override the deceleration caused by the (normalised) matter density, Ω_m. Riess *et al.* (1998) and Perlmutter *et al.* (1999) used their SNe measurements, d, to obtain posterior distributions of the form $\Pr(\Omega_\Lambda, \Omega_m|d, I)$, under the assumption of unimformative (and improper) uniform priors of the form $\Pr(\Omega_m, \Omega_\Lambda) \propto \Theta(\Omega_m)$, where $\Theta(x)$ is the Heaviside step function. The posterior distribution for the 42 SCP SNe from Perlmutter *et al.* (1999), reproduced in Fig. 1, reveals that most of the models that are consistent with the data correspond to an accelerating universe (*i.e.*, $\Omega_\Lambda > \Omega_m/2$).

But do these data provide *quantitive* evidence of cosmological acceleration? Riess *et al.* (1998) approached this question by calculating the fraction of the posterior with $\Omega_\Lambda > \Omega_m/2$, which is an apparently compelling 0.997 for the case shown in Fig. 1. The relevant Bayesian calculation (*c.f.* Drell *et al.* 2000) should, however, be based on the marginal likelihoods of an accelerating model (for which the prior is non-zero only for $\Omega_\Lambda > \Omega_m/2$) and a decelerating model (for which the obvious option is a matter-only model with $\Omega_\Lambda = 0$). Such models can be fully specified (in the sense defined in Section 2) by adding the restrictions that $0 \leqslant \Omega_m \leqslant \Omega_{max}$ and $0 \leqslant \Omega_\Lambda \leqslant \min[\Omega_{max}, \Omega_{\Lambda,BB}(\Omega_m)]$ (defined to

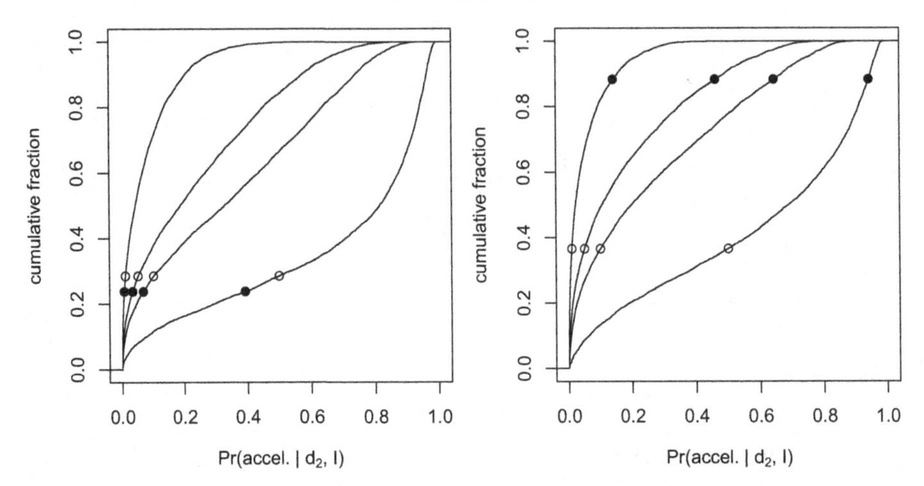

Figure 2. The distribution of $\Pr(\text{accel.}|d_2, I)$ obtained from different partitions of the Perlmutter *et al.* (1999) SN data set with training sets of 10 (left) and 21 (right) SNe. The open symbols indicate the prior values (of, from left to right, 0.01, 0.05, 0.1 and 0.5) and the solid symbols show the posterior values given by training and testing samples that alternate in redshift.

reject models that did not begin with a Big Bang), where $\Omega_{\rm max} \geqslant 0$ is an unspecified "hyper-parameter". Figure 1 shows the dependence of the posterior probability of the accelerating model, $\Pr(\text{accel.}|d, I)$, on $\Omega_{\rm max}$. Even the peak values of $\Pr(\text{accel.}|d, I)$ are considerably lower than the posterior fraction quoted above, and the dependence on the unknown value of $\Omega_{\rm max}$ is significant as well.

Rather than introducing an arbitrary new parameter, another option is to adopt the two-step method described in Section 4, using some of the SN data to obtain a partial posterior in $\Omega_{\rm m}$ and Ω_Λ for both the accelerating and matter-only models and then using the remainder to perform model comparison. The results of doing so are shown in Fig. 2 for several different partitioning options. These results again illustrate the standard Bayesian result that the better-fitting accelerating model is not favoured so decisively over the more predictive (*i.e.*, "simpler") matter-only model, a result that robust to prior choice.

This two-step approach to model comparison could be applied to a variety of problems in astrophysics and cosmology (*e.g.*, Bailer-Jones 2012, Khanin & Mortlock 2014).

References

Bailer-Jones, C. 2012 *Astronomy & Astrophysics*, 546, A89

Cox, R. T. 1946, *American Journal of Physics*, 14, 1

Drell, P. S., Loredo, T. J., & Wasserman, I. 2000, *The Astrophysical Journal*, 530, 593

Efstathiou, G. 2008, *Monthly Notices of the Royal Astronomical Society*, 388, 1314

Jenkins, C. R. & Peacock, J. A. 2011, *Monthly Notices of the Royal Astronomical Society*, 413, 2895

Khanin, A. & Mortlock, D. J. 2014, *Monthly Notices of the Royal Astronomical Society*, 444, 1591

Lempers, F. B. 1971, *Posterior Probabilities of Alternative Linear Models*, Rotterdam: University Press

O'Hagan, A. 1995, *Journal of the Royal Statistical Society B*, 30, 490

Perlmutter, S., *et al.* 1999, *The Astrophysical Journal*, 517, 565

Spiegelhalter, D. J. & Smith, A. F. M., 1982, *Journal of the Royal Statistical Society B*, 44, 377

Riess, A. G., *et al.* 1998, *The Astronomical Journal*, 116, 1009

Statistical Challenges in 21st Century Cosmology
Proceedings IAU Symposium No. 306, 2014
A. F. Heavens, J.-L. Starck & A. Krone-Martins, eds.

doi:10.1017/S1743921314010722

What we talk about when we talk about fields

Ewan Cameron

Department of Zoology, University of Oxford, Tinbergen Building, South Parks Road, Oxford, OX1 3PS, United Kingdom
email: dr.ewan.cameron@gmail.com
website: astrostatistics.wordpress.com

Abstract. In astronomical and cosmological studies one often wishes to infer some properties of an infinite-dimensional field indexed within a finite-dimensional metric space given only a finite collection of noisy observational data. Bayesian inference offers an increasingly-popular strategy to overcome the inherent ill-posedness of this signal reconstruction challenge. However, there remains a great deal of confusion within the astronomical community regarding the appropriate mathematical devices for framing such analyses and the diversity of available computational procedures for recovering posterior functionals. In this brief research note I will attempt to clarify both these issues from an "applied statistics" perpective, with insights garnered from my post-astronomy experiences as a computational Bayesian / epidemiological geostatistician.

Keywords. Methods: data analysis–methods: statistical

1. Introduction

The potential afforded by Bayesian techniques for inferring the properties of infinite-dimensional mathematical structures, such as random fields (to be understood here as random functions defined at each point of some finite-dimensional metric space), has long been recognised by both probability theorists, e.g. O'Hagan (1978), and practitioners: with the first wave of practical applications in geoscience (e.g. Omre (1987), Handcock & Stein (1993)) and machine learning (e.g. Rasmussen & Williams (1996), Neal (1997)) contemporaneous with the advent of sufficiently powerful desktop computers. Cosmologists were at this time notable as *'early adopters'* and pioneers of the new techniques for field inference. Indeed, the Monte Carlo methods for constrained simulation from Gaussian random fields developed by Bertschinger (1987) and Hoffman & Ribak (1991) remain key tools for efficient conditional simulation, cf. Doucet (2010).

However, over the past decade the sophistication of statistical analysis techniques brought to bear on the study of cosmological fields has not kept pace with progress outside of astronomy. With modern tools such as the Integrated Nested Laplace Approximation (INLA; Rue *et al.* (2009)), 'variational inference' (Hensman *et al.* (2013)), particle filtering (Del Moral *et al.* (2007)), and Approximate Bayesian Computation (ABC; Marjoram *et al.* (2003)) almost entirely ignored to-date by the cosmological community we have, in my opinion, become the *'laggards'* of the technology adoption lifecycle.

There are multiple factors seemingly to blame for this divergence: (i) the emergence of an isolationist attitude to the practice of cosmological statistics; (ii) an over-emphasis on the path integration-based conceptulisation of random fields, rather than the measure-theory-based mathematics of mainstream statistics; and (iii) an under-appreciation of the potential for stochastic process priors (including, *but not limited to*, the Gaussian process) as flexible modelling components within the hierarchical Bayesian framework. With the first already being well fought back against by inter-disciplinary programming in conferences such as the IAUS306 and the SCMA series I will therefore focus in this proceedings (as in my contributed talk) on the latter two. In particular, I aim to clarify a number of mathematical concepts crucial to a high-level understanding of Bayesian inference over random fields and measures (Section 2), and then to highlight just a few of the exciting techniques to have recently emerged in this area (Section 3).

2. The Mathematics of Bayesian Field Inference

Cosmologists and astronomers already well-versed in the practice and theory of Bayesian statistics in the finite-dimensional setting will typically have one of two contrasting experiences upon first attempting to extend these ideas to infinite-dimensional inference problems. The pragmatist will happily observe that the mechanics of computation are little changed (e.g. the Gaussian random field at finite sample points is distributed just as the familiar multivariate Gaussian), while the cautious theorist will more likely be overwhelmed by a first acquaintance with measure-theoretic probability (i.e., probability triples and the algebra of sets). But ideally one will have both experiences, since each offers an equally important perspective, as I discuss in this Section.

2.1. *Distributions over Infinite-Dimensional Space*

In formal statistics the core of probabilistic computation is framed within the language of measure theory: the key object being the *'probability triple'* of (i) a sample space, Ω, i.e., some non-empty set; (ii) a σ-algebra, Σ, i.e., a collection of subsets of Ω with $\emptyset, \Omega \in \Sigma$, closed under the formation of complements, countable unions and intersections; and (iii) a probability measure, P, i.e., a countably additive set function from Σ to [0,1] for which $P(\emptyset) = 0$ and $P(\Omega) = 1$. In this context Carathéodory's Extension Theorem provides the theorist with the machinary to build complex probability triples and forge a rigorous notion of random variables as measures on the pre-images of Borel σ-algebra sets of the real numbers; and from this to the familiar mechanics of probability densities defined with respect to the Lebesgue measure (e.g. the standard Normal with $f(x) = \frac{1}{\sqrt{2\pi}} \exp -\frac{x^2}{2} dx$). Nevertheless, with the Lebesgue measure behaving intuitively as a product measure in $\mathcal{R}^n$, and with Lebesgue and Riemann integration interchangable in practice for all but a few rare cases, the pragmatist can safely ignore these theoretical foundations in the study of 'real-world' problems in finite-dimensional settings.

In the context of probabilistic inference over *fields*, however, one must proceed with care as there exists no equivalent to the Lebesgue measure to serve as a natural reference for defining densities in an infinite-dimensional Banach space (e.g. the L^p function spaces). Hence, for Bayesian analysis in infinite-dimensional space we must be deliberate in our choice of reference measure, which we encode into the prior. Typically we will do this indirectly by assigning as prior the implicit measure (or 'law') belonging to a given *stochastic process* (e.g. a Gaussian process, or Poisson process) having sample paths within the field space under study. Although quite technical the distinction between this formal statistical approach and the path integration-based language of cosmological papers, e.g. Enßlin *et al.* (2009), Kitching & Taylor (2011), is important if we are to connect with, *and thereby benefit from*, the rich body of applied statistics literature on infinite-dimensional inference. Worth noting also is that the measure-theoretic equivalent of the probability density is the 'Radon-Nikodym (R-N) derivative', with a trivial but illustrative example being that of the R-N derivative of posterior against prior given by the likelhood function divided by the marginal likelihood, c.f. Cotter *et al.* (2009).

Finally, the measure theoretic definition of a stochastic process is a collection of random variables *indexed* by a set; here all points of the physical metric space over which the field problem is to be studied. A key theoretical tool for the construction of stochastic processes, which is greatly illustrative of their behaviour, is Kolmogorov's Extension Theorem. The theorem gives conditions under which a rule for assigning the distributions of the finite-dimensional projections of an infinite-dimensional indexing set can be considered sufficient to define a proper stochastic process. The key condition here is one of *mutual consistency* of 'coarser-binned' finite-dimensional projected distributions with respect to 'finer-binned' ones; this idea was well-understood by the MaxEnt pioneers, cf. Skilling (1998): "the prior must depend ... on the pixel size h in such a way that subsidiary pixelisation is immaterial".

2.2. *Hierarchical Bayesian Models with Stochastic Process Priors*

As mentioned earlier for the pragmatist the mechanics of Bayesian field analysis need differ little from those of finite-dimensional inference; especially when we are able to write our prior–likelihood pairing via a stochastic process embedded within a *hierarchical Bayesian model.* A typical hierarchical model for Bayesian field inference is the following from Gething *et al.* (2010) in the context of epidemiological geostatistics,

$$\begin{aligned} N_i^+ &\sim \mathrm{Bin}(N_i, p(x)) \\ p(x) &= g^{-1}(f(x)) \\ f|\phi &\sim \mathrm{GP}(M_\phi, C_\phi). \end{aligned}$$

We can read this model from the top down to understand the generative process for the data: binomial sampling from a population of N_i at each site with underlying prevalence (probability), $p(x)$, depending on the location of the site, x. The prevalence field, $p(x)$, is in turn the realisation of a Gaussian random field, GP, transformed to the range, (0,1), by the *link function*, $g^{-1}(\cdot)$. A prior on the parameters of the mean and/or correlation function of the GP would be a natural extension of this particular model; while natural extensions to other problems using the GP could include, e.g., exponentation of f via $g^{-1}(\cdot)$, use of a Poisson process likelihood function, or some penalisation of the GP sample paths. An even greater diversity of hierarchical forms can then be built with the addition of *non*-Gaussian processes: cf. systematic error analysis (Burr & Doss (2005)) and *non*-proportional hazards modelling (De Iorio *et al.* (2009)) with Dirichlet processes, online Bayesian classification with Mondrian processes (Lakshminarayanan *et al.* (2013)), and so on (Poisson processes, Gamma processes, Negative binomial processes, etc.).

For complicated real-world problems there is rarely an analytical solution to the resulting posterior so, as in ordinary Bayesian analysis, one will almost always turn to computational methods. Some of these are well-known to cosmologists already, such as Gibbs sampling (e.g. Wandelt *et al.* (2004)) and Hamiltonian Monte Carlo (e.g. Hajian (2007), Jasche & Kitaura (2010)); however, a much greater number of specialised techniques for Bayesian field inference remain largely undiscovered. I highlight just a few of these briefly in the following Section.

3. Some Computational Techniques for Bayesian Field Inference

From a geostatistical perspective it is difficult to overstate the revolutionary impact lately effected by the emergence of INLA (Rue *et al.* (2009)) and the stochastic partial differential equation (SPDE) approach to Gaussian processes (Lindgren *et al.* (2011)). Using the machinery of the finite-element method already familiar to astronomers, the SPDE approach aims to identify discretely-indexed Gaussian *Markov* random fields providing weak approximations to their continuously-indexed counterparts. The result is a system for approximate Bayesian inference over random fields amenable to *fast* computation via sparse matrix operations; whereas Hamiltonian MCMC over field posteriors is typically the reserve of cluster-computing, the INLA method enables approximate field inference on desktop computers with run-times small enough to allow for the important follow-up inference steps of model testing and refinement. A first glimpse of the potential for INLA in cosmological applications was provided in a 2010 study of CMB reconstruction by Wilson & Yoon (2010); but as yet this pioneering work remains unappreciated (or, at least, uncited).

Another seminal technique for 'Big Data' Bayesian inference over random fields is that of 'variational inference' (Hensman *et al.* (2013)), in which approximation to the full field posterior is made using sets of data-driven 'inducing points'. Ongoing research efforts into the application of Approximate Bayesian Computation (Soubeyrand *et al.* (2013)) and Kalman filtering (Särkkä *et al.* (2014)) in the spatial domain are also well worth keeping an eye on.

References

Bertschinger, E. 1987, *ApJ*, 323, L103–L106

Burr, D. & Doss, H. 2005, *J. Am. Statist. Assoc.*, 100, 242–251

Cotter, S. L., Dashti, M., Robinson, J. C., & Stuart, A. M. 2009, *Inverse Probl.*, 25, 115008

De Iorio, M., Johnson, W. O., Müller, P., & Rosner, G. L. 2009, *Biometrics*, 65, 762–771

Del Moral, P., Doucet, A., & Jasra, A. 2007, 'Sequential Monte Carlo for Bayesian Computation' in *Bayesian Statistics 8*, J.M. Bernardo, M.J. Bayarri, J.O. Berger, A.P. Dawid, A.F.M. Smith & M. West, eds., OUP, 1–34

Doucet, A. 2010, 'A Note on Efficient Conditional Simulation of Gaussian Distributions', Departments of Computer Science and Statistics, University of British Columbia

Enßlin, T. A., Frommert, M., & Kitaura, F. S. 2009, *Phys. Rev. D*, 80, 105005

Handcock, M. S. & Stein, M. L. 1993, *Technometrics*, 35, 403–410

Gething, P. W., Patil, A. P., & Hay, S. I. 2010, *PLOS Computat. Biol.*, 6, e1000724

Hajian, A. 2007, *Phys. Rev. D.*, 75, 083525

Hensman, J., Fusi, N., & Lawrence, N. D. 2013, 'Gaussian Processes for Big Data' in *Association for Uncertainty in Artificial Intelligence*, UAI2013, 244

Hoffman, Y. & Ribak, E. 1991, *ApJ*, 380, L5–L8

Jasche, J. & Kitaura, F. S. 2010, *MNRAS*, 407, 29–42

Kitching, T. D. & Taylor, A. N. 2011, *MNRAS*, 410, 1677–1686

Lakshminarayanan, B., Roy, D. M., & Teh, Y. W. 2013, 'Top-down Particle Filtering for Bayesian Decision Trees' in *International Conference on Machine Learning*, ICML2013

Lingren, F., Rue, H., & Lindström, J. 2011, *J. R. Statist. Soc. B*, 73, 423–498

Marjoram, P., Molitor, J., Plagnol, V., & Tavaré, S. 2003, *PNAS*, 100, 15324–15328

Neal, R. M. 1997, 'Monte Carlo Implementation of Gaussian Process Models for Bayesian Regression and Classification', *Tech. Rep. 9702*, Department of Statistics, University of Toronto

O'Hagan, A. 1978, *J. R. Statist. Soc. B*, 40, 1–42

Omre, H. 1987, *Math. Geol.*, 19, 25–39

Rasmussen, C. E. & Williams, C. K. I. 1996, 'Gaussian Processes for Regression' in *Advances in Neural Information Processing Systems 8*, eds. D.S. Touretzky, M.C. Mozer, M.E. Hasselmo, MIT Press, 514–520

Rue, Y., Martino, S., & Chopin, N. 2009, *J. R. Statist. Soc. B*, 71, 319–392

Särkkä, S., Solin, A., & Hartikainen, J. 2014, 'Spatio-Temporal Learning via Infinite-Dimensional Bayesian Filtering and Smoothing', to appear in The IEEE Signal Processing Magazine

Skilling, J. 1998, *J. Microscop.*, 190, 28–36

Soubeyrand, S., Carpentier, F., Guiton, F., & Klein, E. K. 2013, *Stat. Appl. Genet. Mol. Biol.*, 12, 17–37

Wandelt, B. D., Larson, D. L., & Lakshminarayanan, A. 2009, *Phys. Rev. D*, 70, 083511

Wilson, S. P. & Yoon, J. 2010, preprint(arXiv:1011.4018)

Statistical Challenges in 21st Century Cosmology
Proceedings IAU Symposium No. 306, 2014
A. F. Heavens, J.-L. Starck & A. Krone-Martins, eds.

doi:10.1017/S1743921314013854

Back to Normal! Gaussianizing posterior distributions for cosmological probes

Robert L. Schuhmann[1], Benjamin Joachimi[1] and Hiranya V. Peiris[1]

[1]Dept. of Physics and Astronomy
University College London
Gower Place
London, UK WCE1 6BT
email: robert.schuhmann.13@ucl.ac.uk

Abstract. We present a method to map multivariate non-Gaussian posterior probability densities into Gaussian ones via nonlinear Box-Cox transformations, and generalizations thereof. This is analogous to the search for normal parameters in the CMB, but can in principle be applied to any probability density that is continuous and unimodal. The search for the optimally Gaussianizing transformation amongst the Box-Cox family is performed via a maximum likelihood formalism. We can judge the quality of the found transformation *a posteriori*: qualitatively via statistical tests of Gaussianity, and more illustratively by how well it reproduces the credible regions. The method permits an analytical reconstruction of the posterior from a sample, *e.g.* a Markov chain, and simplifies the subsequent joint analysis with other experiments. Furthermore, it permits the characterization of a non-Gaussian posterior in a compact and efficient way. The expression for the non-Gaussian posterior can be employed to find analytic formulae for the Bayesian evidence, and consequently be used for model comparison.

Keywords. methods: data analysis, methods: numerical, cosmology: observations

---

1. The Idea

The quest for normal parameters, where the cosmological likelihood is of approximately multivariate Gaussian form, has proven to be highly advantageous for rapid computation of CMB likelihoods, *e.g.* CMBfit, CMBwarp, PICO (Kosowsky *et al.* 2002, Chu *et al.* 2003, Jimenez *et al.* 2004, Sandvik *et al.* 2004, Fendt & Wandelt 2007). To achieve the same for systems in which this is not feasible analytically, we computationally search for a parameter transformation which renders the transformed probability density approximately Gaussian. This allows for a reliable reconstruction of the posterior distribution, *e.g.* from an MCMC sample, and enormously simplifies the reporting and further usage of the multivariate posterior constraints, via data compression.

Our proposed method can be used to characterize and to reproduce high-dimensional probability densities (see Fig. 1). Instead of publishing an MCMC sample ($\gtrsim 10^5$ numbers), the posterior can be reconstructed simply from the first and second moments of the Gaussianized distribution, and the optimal transformation parameters. This provides substantial data compression, while simultaneously providing versatility and accuracy. Further, there is no need for smoothing of contours, in contrast with density estimates based on MCMC samples.

2. The Tools: Box-Cox Transformations and their Kin

We employ families of one-dimensional transformations mapping cosmological parameters to normal parameters. The transformations are defined by a set of parameters for

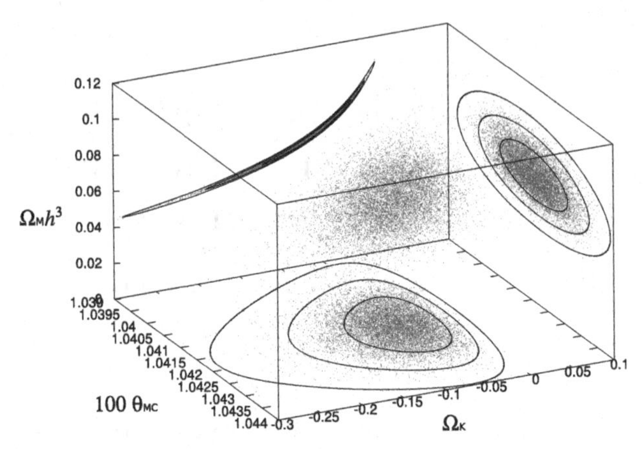

Figure 1. *3D marginalized MCMC sample from Planck XVI (2013) (red), the marginal 2D sample (blue), and 2D contours of the full 3D analytic posterior (black), as reconstructed via Box-Cox transformations.*

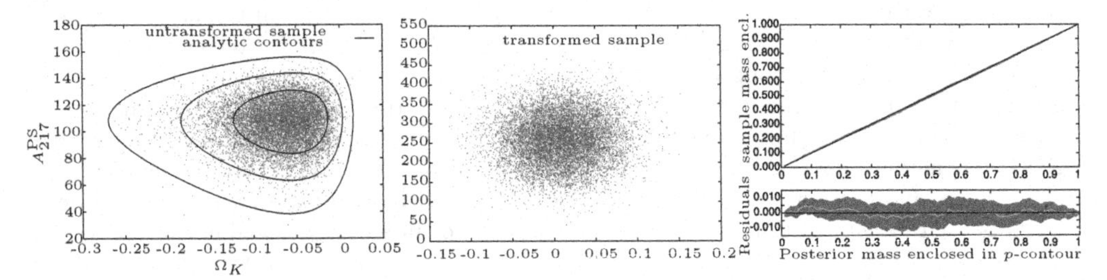

Figure 2. *left to right: 1) 2D MCMC sample from Planck XVI (2013) (red) and reconstructed analytic posterior (black). 2) Sample, as Gaussianized with ABC transformation (blue). 3) CC plot showing the validity of the reconstructed black contours. "Residuals" shows the deviation of the point sample probability mass in a given contour from the probability mass in the analytic posterior.*

each dimension. One particularly useful family was introduced by Box & Cox (1964):

$$x \mapsto B_{a,\lambda}(x) = \begin{cases} [\lambda^{-1}(x+a)^{\lambda} - 1] & (\lambda \neq 0) \\ \log(x+a) & (\lambda = 0). \end{cases} \tag{2.1}$$

Notice that this family is continuous in λ. As a useful generalization for removing residual kurtosis we propose the ABC transformation,

$$x \mapsto F_{a,\lambda,t}(x) = \begin{cases} t^{-1}\sinh[t\,B_{a,\lambda}(x)] & (t > 0) \\ B_{a,\lambda}(x) & (t = 0) \\ t^{-1}\operatorname{arsinh}[t\,B_{a,\lambda}(x)] & (t < 0). \end{cases} \tag{2.2}$$

Our algorithm can be applied to arbitrary parameterized transformations, suitable for various forms of non-Gaussianity.

3. The Method: Optimization and Verification

Given a sample from a non-Gaussian posterior distribution, we find the optimal transformation parameters by maximizing the log-likelihood of the transformed sample after a given transformation (Joachimi & Taylor 2011). Typically, degeneracies appear in the transformation parameters. To avoid these, we penalize transformations too far from the identity.

The quality of the reproduced contours can be evaluated via cross-contour (CC) plots: given a certain posterior contour region of enclosed probability mass m, compute the

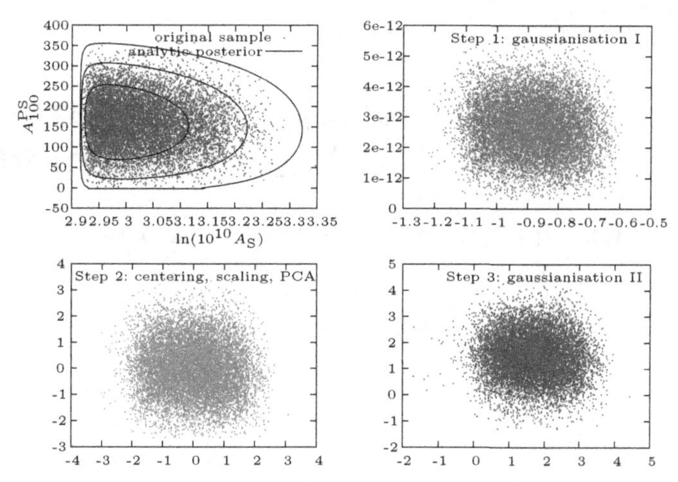

Figure 3. *From top left to bottom right: 1) Two-pass optimization of 2D sample from Planck XVI (2013), showing A_{100}^{PS} vs. $\ln(10^{10}A_s)$, and reconstructed posterior (black contours). 2) After first Gaussianization transformation (all y values plus 0.91585740528). 3) After linear reshaping. 4) After second Gaussianization transformation.*

fraction f of points inside. Plotting f over m, we can determine whether every credible region of the analytic posterior encloses the correct probability mass of the point sample, *i.e.* whether $m \simeq f$ within the margin of noise. To quantify the sampling noise, we perform a non-parametric bootstrap to find the 95% confidence interval for every f (see Fig. 2).

4. Two-pass transformations

If the transformed posterior is not sufficiently Gaussian (as, *e.g.*, determined via a CC plot), the process can be repeated. We have implemented the following protocol for the two-pass transformation (see Fig. 3):

Step 1: Optimize the parameters of the first transformation.

Step 2: Linear reshaping: Center the sample, scale every direction to unit variance, then rotate into the PCA eigenbasis.

Step 3: Optimize the parameters of the second transformation.

Note that this two-pass protocol is more effective for Gaussianization than one single transformation, because the transformation families employed (Box-Cox, ABC) do not form groups – in particular, this means that two subsequent transformations do not result in a transformation of the same family.

5. Outlook

Transforming the unnormalized posterior distribution to a Gaussian form can allow for a simple and fast computation of the marginal likelihood, and hence used for model comparison (Schuhmann *et al.*, in preparation).

References

Box, G. E. P. & Cox, D. R. 1964, *J. Roy. Stat. Soc.* B 26/2, 211-252

Chu, M., Kaplinghat, M., & Knox, L. 2003, *ApJ*, 596, 725

Fendt, W. A. & Wandelt, B. D. 2007, *ApJ*, 654, 2 (astro-ph/0606709)

Jimenez, R., Verde, L., & Peiris, H. 2004, *Phys. Rev. D*, 70, 023005 (astro-ph/0404237)

Joachimi, B. & Taylor, A. N. 2011, *MNRAS*, 416(2), 1010-1022, (arXiv:1103.3370)

Kosowsky, A., Milosavljevic, M., & Jimenez, R. 2002, *Phys. Rev. D*, 66, 063007

Planck Collaboration (Ade *et al.*) 2013, *A&A*, in press (arXiv:1311.1657)

Sandvik, H. B., Tegmark, M., Wang, X., & Zaldarriaga, M. 2004, *Phys. Rev. D*, 69, 063005 (astro-ph/0311544v2)

Statistical Challenges in 21st Century Cosmology
Proceedings IAU Symposium No. 306, 2014
A. F. Heavens, J.-L. Starck & A. Krone-Martins, eds.

doi:10.1017/S1743921314013532

Bayesian CMB foreground separation with a correlated log-normal model

Niels Oppermann[1] and Torsten A. Enßlin[2]

[1]Canadian Instittute for Theoretical Astrophysics, University of Toronto,
60 St. George Street, Toronto, ON, M5S 3H8, Canada
email: niels@cita.utoronto.ca

[2]Max Planck Institute for Astrophysics,
Karl-Schwarzschild-Straße 1, 85748 Garching, Germany
email: ensslin@mpa-garching.mpg.de

Abstract. The extraction of foreground and CMB maps from multi-frequency observations relies mostly on the different frequency behavior of the different components. Existing Bayesian methods additionally make use of a Gaussian prior for the CMB whose correlation structure is described by an unknown angular power spectrum. We argue for the natural extension of this by using non-trivial priors also for the foreground components. Focusing on diffuse Galactic foregrounds, we propose a log-normal model including unknown spatial correlations within each component and cross-correlations between the different foreground components. We present case studies at low resolution that demonstrate the superior performance of this model when compared to an analysis with flat priors for all components.

Keywords. Methods: data analysis, methods: statistical, ISM: general, cosmic microwave background

1. Introduction

The separation of the CMB and the various diffuse foreground components relies mostly on their different frequency dependence. Bayesian methods are especially useful for this, since all their assumptions are made explicitly and the remaining uncertainties are easily quantifiable (see, e.g., Planck Collaboration *et al.* 2013). Established Bayesian methods infer a set of parameters explicitly describing the frequency spectra of a set of physical foreground components under the assumption of non-trivial priors for these parameters (Eriksen *et al.* 2006). On the other hand, the priors for the parameters describing the spatial dependence of the emission components, i.e., the pixel values, are usually assumed to be flat (Planck Collaboration *et al.* 2013), with the exception of the CMB component, for which an isotropic Gaussian prior is sometimes assumed (Eriksen *et al.* 2008). Here, we propose a natural extension, namely the inclusion of non-trivial spatial priors for the foreground components. This will enable us to allow for spatial correlations both within each component and across components. Specifically, we suggest the use of isotropic log-normal priors for the diffuse Galactic components. For simplicity, we will assume in the following that the frequency dependence of every physical component is known.

2. Data and signal model

We can relate the observed frequency maps d to maps of the emission components ϕ via the linear model

$$d = R\phi + n. \tag{2.1}$$

Here, the linear operator R mixes the components into several frequency maps according to their frequency spectra. It can in principle also mix the information from different points in the component maps into different pixels of the observed maps, e.g., via a beam convolution. The last term in Eq. (2.1) is an additive noise term, for which we assume Gaussian statistics with zero mean and a known covariance matrix. Thus, the likelihood $\mathcal{P}(d|\phi)$ is fully determined.

To find the posterior, $\mathcal{P}(\phi|d)$, which is the probability distribution that we are interested in, we need to augment the likelihood with a suitable prior distribution $\mathcal{P}(\phi)$. For this we suggest to model the CMB as an isotropic Gaussian random field and the foreground components as isotropic log-normal fields, i.e., we define $s_{\hat{n}}^{(0)} = \phi_{\hat{n}}^{(0)}/\bar{\phi}^{(0)}$ for the CMB component and $s_{\hat{n}}^{(\alpha)} = \log\left(\phi_{\hat{n}}^{(\alpha)}/\bar{\phi}^{(\alpha)}\right)$ for each of the foreground components, where the barred quantities are suitable dimensional normalization constants. We then use an isotropic Gaussian prior for the transformed components, s, that is described by a zero mean and an isotropic covariance. The latter can be written in the spherical harmonic basis as

$$S_{(\ell,m),(\ell',m')}^{(\alpha,\alpha')} = \left\langle s_{(\ell,m)}^{(\alpha)}\, s_{(\ell',m')}^{(\alpha')*}\right\rangle_{(s)} = \delta_{\ell\ell'}\,\delta_{mm'}\,C_\ell^{(\alpha,\alpha')}, \tag{2.2}$$

where the asterisk denotes complex conjugation. Here, the auto-spectra, $C_\ell^{(\alpha,\alpha)}$, describe the angular correlation structure within the components and the cross-spectra, $C_\ell^{(\alpha,\alpha')}$ for $\alpha \neq \alpha'$, describe the correlations between different components. Both types of quantities are clearly non-zero in nature and accounting for them will allow us to transfer some of the information from one pixel into the reconstruction of other pixels and from one component into the reconstruction of other components.

To complete the model, we need to specify a prior for the unknown auto- and cross-spectra. For this we use an inverse-Wishart distribution as the natural generalization of the inverse-Gamma prior commonly used for unknown angular power spectra (O'Hagan 1994). With this, the angular spectra can be marginalized and the posterior $\mathcal{P}(s|d)$ calculated.

A simpler way to account for correlations would be to model all emission components as Gaussian random fields. However, the log-normal model has three advantages that make it more suited to describe the Galactic foregrounds: The foreground intensities will automatically be positive, their fluctuations will easily be able to vary over orders of magnitude, such as observed between the Galactic plane and the Galactic halo, and the logarithmic model will suppress the anisotropies that are definitely present in the sky to some degree.

3. Reconstruction algorithm and test cases

The posterior probability distribution $\mathcal{P}(s|d)$ can be calculated analytically. However, it is non-Gaussian. One way to explore this probability distribution would be to draw samples from it using a Monte Carlo algorithm. To minimize computational costs, we pursue a different route here.

Following Oppermann *et al.* (2013), we instead apply an iterative scheme that finds the Gaussian distribution that best approximates (in a minimum Kullback-Leibler distance sense) the true posterior, described by a mean and a covariance. The details of the resulting algorithm are beyond the scope of these proceedings and will be described in a forthcoming publication.

In Fig. 1, we show a simplistic low-resolution simulation of a CMB sky and three foregrounds meant to mimic synchrotron radiation, free-free radiation, and thermal dust

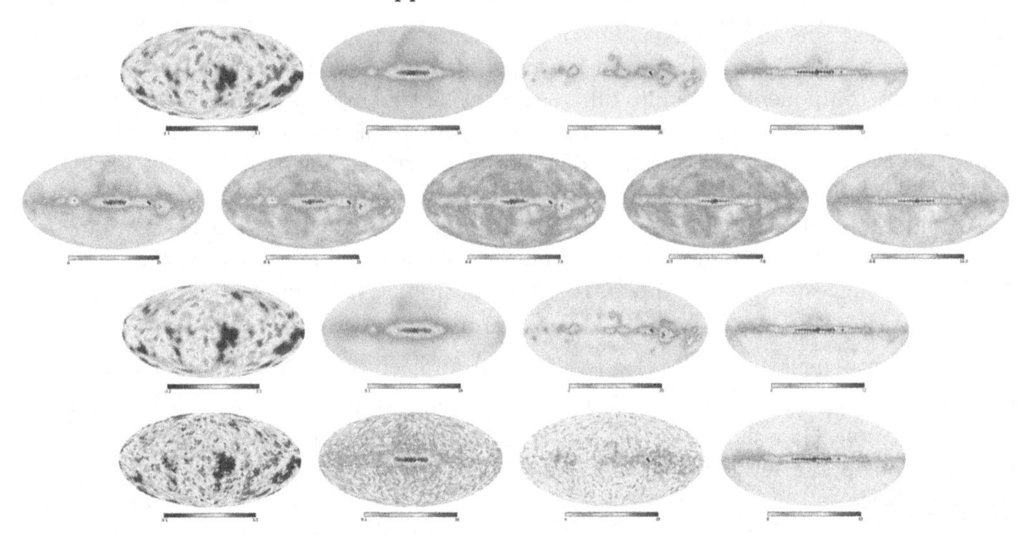

Figure 1. Simulated sky and its reconstructions. The top row shows, from left to right, a CMB realization (based on the `Commander-Ruler` samples of Planck Collaboration *et al.* 2013), a synchrotron map (based on Haslam *et al.* 1982), a free-free map (based on Bennett *et al.* 2013), and a dust map (taken from the `Commander-Ruler` results of Planck Collaboration *et al.* 2013). The second row shows simulated frequency maps, containing an isotropic noise contribution. Typical frequency spectra for the different components have been assumed. The third and fourth rows show the reconstructions of the four simulated components using the algorithm presented here and a maximum likelihood calculation, respectively.

radiation, along with a simulated data set generated from these maps. The third and fourth rows of the figure show the posterior mean reconstruction using our approximative algorithm and a maximum likelihood reconstruction, which corresponds to the posterior mean under the assumption of flat priors for the fields ϕ, respectively. This implementation has made use of the `NIFTy` package (Selig *et al.* 2013) and the `HEALPix` package (Górski *et al.* 2005).

4. Conclusion

Our test case demonstrates that the correlated log-normal prior model is not only well motivated as a good approximation of reality, but also produces superior results in practice, compared to an analysis using flat priors. This is true even if reality, like the test case presented here, is not precisely described by isotropic log-normal distributions.

References

Bennett, C. L., Larson, D., Weiland, J. L., Jarosik, N., Hinshaw, G., Odegard, N., Smith, K. M., *et al.* 2013. *ApJS* 208, 20

Eriksen, H. K., Dickinson, C., Lawrence, *et al.* 2006. *ApJ* 641, 665

Eriksen, H. K., Jewell, J. B., Dickinson, *et al.* 2008. *ApJ* 676, 10

Górski, K. M., Hivon, E., Banday, A. J., *et al.* 2005. *ApJ* 622, 759

Haslam, C. G. T., Salter, C. J., Stoffel, H., *et al.* 1982. *A&AS* 47, 1

O'Hagan, A. 1994. *Kendall's Advanced Theory of Statistics, Bayesian Inference.* John Wiley & Sons

Oppermann, N., Selig, M., Bell, M. R., *et al.* 2013. *Phys. Rev. E* 87 032136

Planck Collaboration, Ade, P. A. R, Aghanim, N., *et al.* 2013. *arXiv eprint* 1303.5072

Selig, M., Bell, M. R., Junklewitz, H., *et al.* 2013. *A&A* 554, A26

Statistical Challenges in 21st Century Cosmology
Proceedings IAU Symposium No. 306, 2014
A. F. Heavens, J.-L. Starck & A. Krone-Martins, eds.

doi:10.1017/S1743921315000010

Searching for bias and correlations in a Bayesian way - Example: SN Ia data

Caroline Heneka[1], Alexandre Posada[2], Valerio Marra[3] and Luca Amendola[4]

[1]Dark Cosmology Centre, Niels Bohr Institute, University of Copenhagen
Juliane Maries Vej 30, DK-2100 Copenhagen, Denmark
email: caroline@dark-cosmology.dk

[2]Centre de Physique théorique, Université d'Aix-Marseille
Campus de Luminy Case 907, 13288 Marseille cedex 9, France

[3]Instituto de Física, Universidade Federal do Rio de Janeiro
CEP 21941-972, Rio de Janeiro, RJ, Brazil

[4]Institut für Theoretische Physik, Universität Heidelberg
Philosophenweg 16, D-69120 Heidelberg, Germany

Abstract. A range of Bayesian tools has become widely used in cosmological data treatment and parameter inference (see Kunz *et al.* 2007, Trotta 2008, Amendola *et al.* 2013). With increasingly big datasets and higher precision, tools that enable us to further enhance the accuracy of our measurements gain importance. Here we present an approach based on internal robustness, introduced in Amendola *et al.* (2013) and adopted in Heneka *et al.* (2014), to identify biased subsets of data and hidden correlation in a model independent way.

Keywords. methods: statistical, cosmology: cosmological parameters, stars: supernovae: general

1. Introduction and method

Our objective is the identification of subsets that differ from the overall data set in having a deviating underlying model. This deviation becomes evident in form of a shift and change in size of likelihood contours (see 'biased' subset d_1 in blue as opposed to overall set d in green in Fig.1, left). Our method is useful for identifying deviating populations otherwise not distinguishable 'by eye' (see blue data points of lowest robustness in Fig.1, right). We apply the formalism on supernova Ia data, the Union2.1 compilation (Suzuki *et al.* 2012) of 580 supernovae from z=0.015 to z=1.414. Observables are apparent magnitudes, stretch and colour corrected, as well as apparent magnitude errors.

Internal Robustness Formalism. We employ the Bayes' ratio to assess the compatibility between subsets statistically, making use of the full likelihood information. The hypothesis of having one model set of parameters M_C to describe the overall dataset d is compared with the hypothesis of having two independent distributions, i.e. parameter sets M_C and M_S for subsets d_1 and d_2. The corresponding Bayes' ratio of the evidences states, where we dub the logarithm of this ratio *internal robustness R*,

$$\mathcal{B}_{\rm tot,ind} = \frac{\mathcal{E}\,(d; M_C)}{\mathcal{E}\,(d_1; M_S)\,\mathcal{E}\,(d_2; M_C)} \quad and \quad R \equiv \log \mathcal{B}_{\rm tot,ind} \tag{1.1}$$

Internal Robustness probability distribution function (iR-PDF). As an analytical form for the distribution of robustness values is not available, unbiased synthetic catalogues are necessary to test for the significance of the robustness values of the real data. They were created by randomising the best-fit function of the observable. In practice we start by partitioning the data into subsets and choosing a parametrisation, followed by the

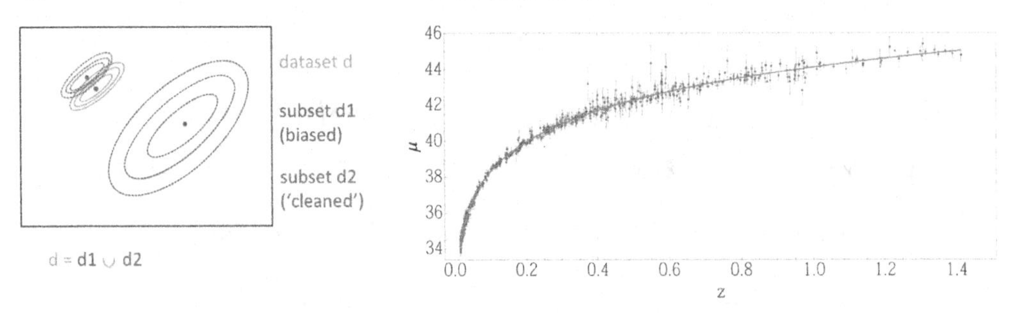

Figure 1. Left: Sketch showing shift and change of size for likelihood contours when removing a biased subset (d_1) from the overall set (d). Right: Hubble diagram for the 580 SN Ia of the Union2.1 compilation (Suzuki *et al.* 2012), best-fit cosmology in green, distance moduli with errors of subset of minimised robustness (R ≈ -280) in blue, complementary set in red. Note that the otherwise indistinguishable biased set d_1 is identified.

evaluation of the robustness value for each partition. Finally, robustness values for real and synthetic catalogues can be compared to detect deviations, see Fig.2.
Genetic Algorithm (GA). We employ a genetic algorithm in order to find subsets of minimal robustness. Again, the parametrisation and initial subsets are chosen and their robustness values evaluated, followed by an iteration cycle of selection (in favour of subsets of lower robustness), reproduction (replacement of disfavoured with favoured subsets) and mutation (random data points are replaced) till convergence.

2. Results

We employ the internal robustness formalism to search for statistically significant bias or correlations present in SN Ia data. The applicability to detect biased subsets, i.e. to identify subsets of deviating underlying best-fit parametrisation, is demonstrated. There are two ways to treat the data to form the iR-PDF: by randomly partitioning it into subsets to test in an unprejudiced way or by sorting the data after specific criteria to test prejudice on the occurrence of bias (for example angular separation, redshift, survey or hemisphere). Observables are both supernova Ia distance moduli and distance modulus errors. The tests of subsets partitioned due to certain prejudices showed no statistically significant deviation between real data and unbiased synthetic catalogues. This result demonstrates a successful removal of systematics for these cases and possible non-standard signals of anisotropy or inhomogeneity at only low level of significance. Fig.2 compares the iR-PDF of unbiased synthetic catalogues in grey with the real catalogue robustness value in red for anisotropies as reported by Planck (Ade *et al.* 2013). For random partitioning subsets of low robustness can be identified. We show in Fig.3, left panel, the occurrence of distance modulus errors for the least robust set found by random selection.

The genetic algorithm (GA) randomly selects subsets for robustness analyses and transforms them due to selection rules in order to find subsets of minimal robustness. Seeking for the detection of systematics, distance modulus errors are analysed. Subsets of minimal robustness are found at low values of $R \leqslant -280$. Fig.3, right panel, shows the occurrence of distance modulus errors with redshift for a subset of lowest robustness found via genetic algorithm minimisation. Remarkably, most SNe found in these subsets occupy a confined region in distance modulus error - redshift - space and belong to distinct surveys of the overall compilation.

3. Conclusion

The applicability of the internal robustness formalism to detect subsets of data whose underlying model deviates significantly from the overall best-fit model is demonstrated.

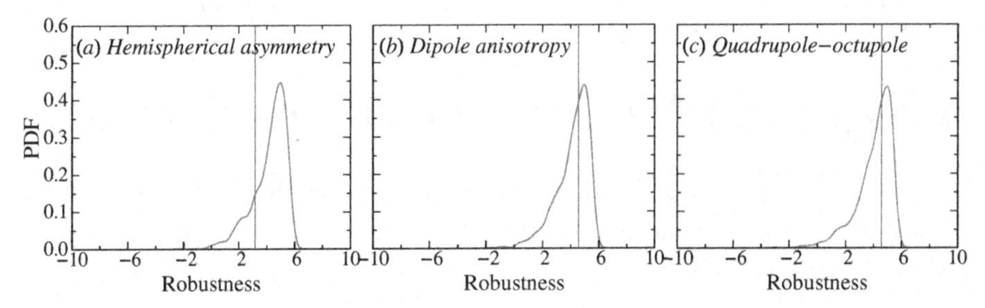

Figure 2. Robustness test for three anisotropies reported by Planck: hemispherical asymmetry (left), dipole anisotropy (centre) and quadrupole-octupole alignment (right). The red vertical lines are robustness values of the Union2.1 Compilation, the distribution of the 1000 unbiased synthetic catalogues is shown in grey; taken from Heneka *et al.* 2014.

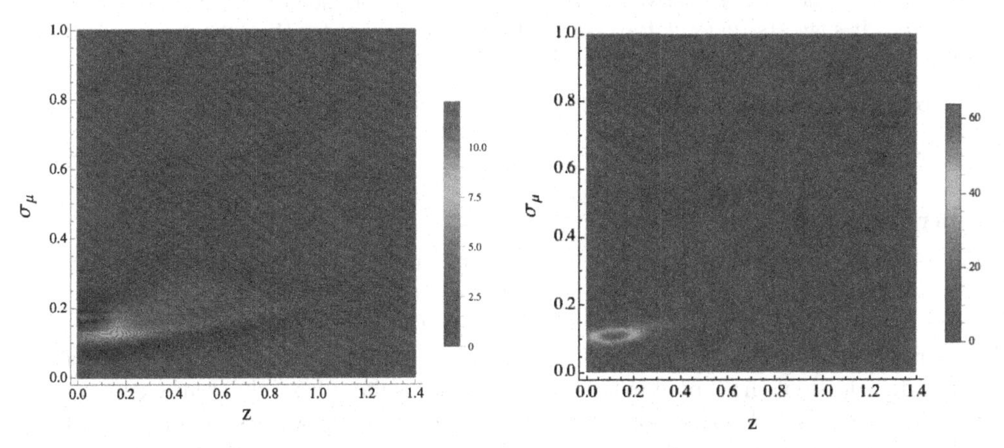

Figure 3. Colour-coded contour plots for the occurrence of SN Ia in distance modulus error-red-shift-space. Left: Contour plot for a subset of $R \approx$-31, the subset of lowest robustness found for random 10^5 subsets. Right: Contour-plot for the SN subset of minimal $R \approx$-283 found via GA.

Subsets of lowest robustness for further investigation are identified, having higher probabilities of being biased. Both the degree to which systematics or cosmological signals unaccounted for are present can be quantified in an unprejudiced and model-independent way. This is crucial in order to detect contaminants or signals in cosmological or any astronomical data, especially with upcoming surveys rendering a hunt for bias by-hand more and more problematic.

Acknowledgements

The Dark Cosmology Centre is funded by the DNRF. V.M. is supported by a Science Without Borders fellowship from the Brazilian Foundation for the Coordination of Improvement of Higher Education Personnel (CAPES). L.A. acknowledges support from DFG through the project TRR33 "The Dark Universe".

References

Ade, P., *et al.* (Planck Collaboration) 2013, 1303.5083
Amendola, L., Marra, V., & Quartin, M. 2013, *MNRAS*, 430, 1867
Heneka, C., Marra, V., & Amendola, L. 2014, *MNRAS*, 439, 1855
Kunz, M., Bassett, B. A., & Hlozek, R. 2007, *Phys. Rev. D*, 75, 103508
Suzuki, N. 2012, *ApJ*, 746, 85
Trotta, R. 2008, *Contemp. Phys.*, 49, 71

Statistical Challenges in 21st Century Cosmology
Proceedings IAU Symposium No. 306, 2014
A. F. Heavens, J.-L. Starck & A. Krone-Martins, eds.

doi:10.1017/S1743921314010692

A Bayesian Method for the Extinction

Hai-Jun Tian[1,2], Chao Liu[1], Jing-Yao Hu[1], Yang Xu[1], Xue-Lei Chen[1]

[1]National Astronomical Observatories, Chinese Academy of Sciences, Beijing 100012
[2]China Three Gorges University, Yichang, 443002. Email: hjtian@lamost.org

Abstract. We propose a Bayesian method to measure the total Galactic extinction parameters, R_V and A_V. Validation tests based on the simulated data indicate that the method can achieve the accuracy of around 0.01 mag. We apply this method to the SDSS BHB stars in the northern Galactic cap and find that the derived extinctions are highly consistent with those from Schlegel *et al.* (1998). It suggests that the Bayesian method is promising for the extinction estimation, even the reddening values are close to the observational errors.

Keywords. dust, extinction stars: horizontal-branch methods: statistical

1. Introduction

The Galactic interstellar extinction is attributed to the absorption and scattering of the interstellar medium, such as gas and dust grains(Draine 2003). The extinction, as a function of the wavelength, is related to the size distribution and abundances of the grains. Therefore, it plays an important role in understanding the nature of the interstellar medium. In addition, the flux of extragalactic objects suffers from different extinction in different bands, which leads to some bias on the extragalactic studies (Tian *et al.* 2011). Hence, understanding the total interstellar extinction in each line of sight (hereafter *los*) is crucial for accurate flux measurements.

The all-sky dust map can either be constrained by measuring interstellar extinction, or by employing a tracer of ISM, e.g., HI. One of the most broadly used dust maps was published by Schlegel *et al.* (1998) (hereafter SFD), who derived it from the dust emission at 100 μm and 240 μm. Since then, many other works have claimed discrepancy with their results(Arce & Goodman 1999; Guy *et al.* 2010).

This paper propose an effective method to examine the extinction values of SFD using the BHB stars as tracer in the northern Galactic cap. BHB stars are luminous and far behind the dusty disk, which contribute to most of the interstellar extinction.

2. Methods

Bayesian Method for Color Excess. The total Galactic extinction in a given *los* is measured from the offset of the observed color indexes of the BHB stars from their intrinsic values. A set of BHB stars, which dereddened color indexes, $\{\mathbf{c_k}\}$ (where $k = 1, 2, \ldots, N_{BHB}$), are known, are selected as template stars. The reddening of a field BHB stars can then be estimated by comparing their observed colors with the templates. Given a *los* i with N_i field BHB stars, the posterior probability of the reddening $\mathbf{E_i}$ is denoted as $p(\mathbf{E_i}|\{\mathbf{\hat{c}_{ij}}\}, \{\mathbf{c_k}\})$, where $\mathbf{\hat{c}_{ij}}$ is the observed color index vector of the BHB star j in the *los* i, and $\mathbf{c_k}$ the intrinsic color index vector of the template BHB star k. According to the Bayes theorem, this probability can be written as

$$p(\mathbf{E_i}|\{\mathbf{\hat{c}_{ij}}\}, \{\mathbf{c_k}\}) = \mathbf{p}(\{\mathbf{\hat{c}_{ij}}\}|\mathbf{E_i}, \{\mathbf{c_k}\})\mathbf{P}(\mathbf{E_i}|\{\mathbf{c_k}\}). \tag{2.1}$$

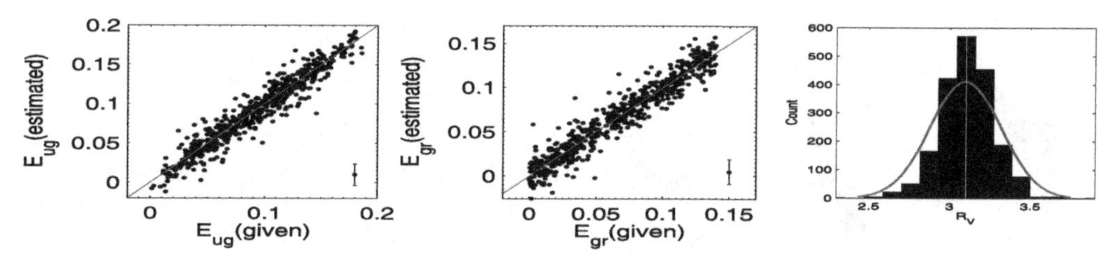

Figure 1. Comparison between the estimated and true values, and the histogram of R_V.

The right-hand side can now be rewritten as

$$p(\mathbf{E_i}|\{\mathbf{\hat{c}_{ij}}\}, \{\mathbf{c_k}\}) = \prod_{j=1}^{N_i} \sum_{k=1}^{N_{BHB}} \mathbf{p}(\mathbf{\hat{c}_{ij}}|\mathbf{E_i}, \mathbf{c_k})\mathbf{p}(\mathbf{E_i}). \tag{2.2}$$

We assume that the likelihood $p(\mathbf{\hat{c}_{ij}}|(\mathbf{E_i}, \mathbf{c_k}))$ is a multivariate Gaussian:

$$p(\mathbf{\hat{c}}_{ij}|\mathbf{E}_i, \mathbf{c_k}) = \frac{1}{(2\pi|\mathbf{\Sigma}|)^{m/2}} \exp(-\mathbf{x}^T \mathbf{\Sigma}^{-1} \mathbf{x}), \tag{2.3}$$

where $\mathbf{x} = \mathbf{E} + \mathbf{c}_k - \mathbf{\hat{c}}_{ij}$, and $\mathbf{\Sigma}$ is the m×m covariance matrix of the measurement of the color indexes of the star j,

$$\mathbf{\Sigma} = \begin{bmatrix} \sigma_u^2 + \sigma_g^2 & -\sigma_g^2 & 0 & 0 \\ -\sigma_g^2 & \sigma_g^2 + \sigma_r^2 & -\sigma_r^2 & 0 \\ 0 & -\sigma_r^2 & \sigma_r^2 + \sigma_i^2 & -\sigma_i^2 \\ 0 & 0 & -\sigma_i^2 & \sigma_i^2 + \sigma_z^2 \end{bmatrix}, \tag{2.4}$$

where the σ_u,σ_g, σ_r, σ_i, and σ_z are the measurement uncertainties.

Least-squares Method for R_V and A_V. After deriving the probability of the reddening in a *los*, the most likely reddening values,

$$E_i = (E(u-g), E(g-r), E(r-i), E(i-z)), \tag{2.5}$$

can be obtained from the probability density function (PDF). They can then be used to derive the R_V and A_V given an extinction model, such as Cardelli *et al.* (1989) (hereafter CCM), from the following equations,

$$E(u-g) = ((a_u + \frac{b_u}{R_V}) - (a_g + \frac{b_g}{R_V})) * A_V, \quad E(g-r) = ((a_g + \frac{b_g}{R_V}) - (a_r + \frac{b_r}{R_V})) * A_V$$
$$E(r-i) = ((a_r + \frac{b_r}{R_V}) - (a_i + \frac{b_i}{R_V})) * A_V, \quad E(i-z) = ((a_i + \frac{b_i}{R_V}) - (a_z + \frac{b_z}{R_V})) * A_V. \tag{2.6}$$

These are linear equations for A_V and A_V/R_V and can be easily solved with a least-squares or χ^2 method to find the best fit A_V and R_V for each BHB star. The averaged A_V and R_V in each *los* are obtained from the median values of all the stars located in the *los*; and the uncertainties can be estimated from the median absolute deviation. The terms a_x and b_x are given from CCM.

Validation of the Methods. We used 900 Monte Carlo simulations to validate the Bayesian method, the left and middle panels in Fig. 1 show the comparison between the estimated and the true extinction values (given in the simulations) in two colors. The mean 1σ error bars (less than 0.01 mag) are marked at the bottom, which suggests that the Bayesian method we employed in this work is robust.

To validate the least-squares method, we solve Eq. 2.6 for the $E(B-V)$ data looked up from the SFD extinction maps for each *los* and the fixed value of $R_V = 3.1$. The right panel in Fig. 1 presents the histogram of R_V in the simulation. The red curve is the Gaussian fit profile with the mean value $< R_V > \simeq 3.1$ and $\sigma \simeq 0.16$, the yellow line marks the location of $R_V = 3.1$.

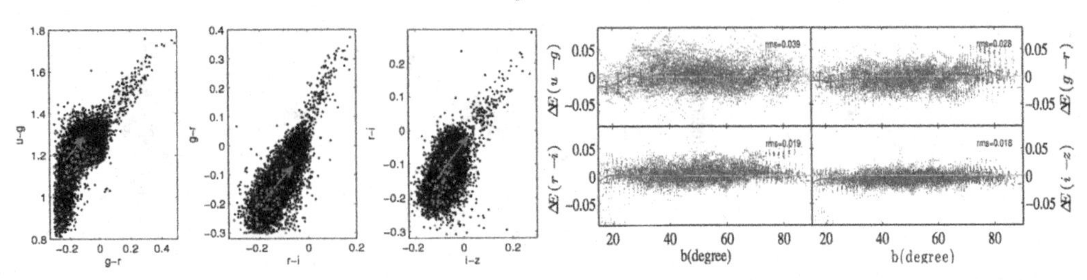

Figure 2. Sample distribution in the 2-color space, and the reddening contrasts with SFD.

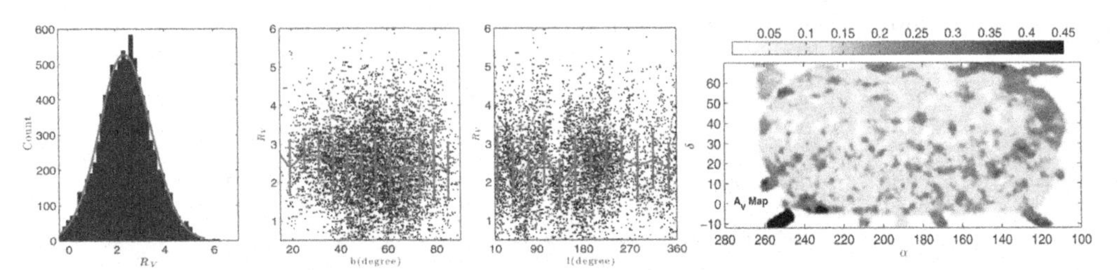

Figure 3. Distributions of the measured R_V (the first three subplots), and A_V map.

3. Application to the BHB stars

Data Selection. A total of 12 530 field BHB stars are selected from Smith *et al.* (2010), as shown with the black points in the left panel of Fig. 2. The red points are the 94 zero-reddened template BHB stars selected from seven known globular clusters. The magenta arrows show the reddening direction.

Reddening Values. The reddening values estimated by Bayesian method are compared with SFD in the right panel of Fig. 2. They are well in agreement with each other.

R_V and A_V Values. The left panel in Fig. 3 shows the histogram distribution of the measured R_V, best-fitted by a Gaussian with $\mu \simeq 2.4$ and $\sigma \simeq 1.05$. The middle two are the measured R_V as a function of the Galactic latitude (the second panel) and longitude (the right panel), respectively. The red curves show $< R_V >$, which keeps constant at ~ 2.5 over all latitudes and longitudes. The right is the estimated A_V map.

4. Conclusions

To measure the extinction, we propose a Bayesian method, and validated the method with simulations, which indicates accuracy is around 0.01 mag. It is robust even in the case that the reddening values are close to the observational errors. The extinctions derived from the SDSS BHB stars with this method are high consistent with SFD.

Acknowledgments. We thank the grants (No. U1231123, U1331202, U1231119, 11073024, 11103027, U1331113, 11303020) from NSFC and the support from LAMOST fellowship.

References

Arce, H. G. & Goodman, A. A. 1999, *ApJ*, 512, L135
Cardelli, J. A., Clayton, G. C., & Mathis, J. S. 1989, *ApJ*, 345, 245
Draine, B. T. 2003,*ARA&A*, 41, 241
Guy, J., Sullivan, M., Conley, *et al.* 2010, *A&A*, 537, A7
Schlegel, D. J., Finkbeiner, D. P., & Davis, M. 1998, *ApJ*, 500, 525
Smith, K. W., Bailer-Jones, C. A. L., Klement, R. J., & Xue, X. X., 2010, *A&A*, 522, A88
Tian, H. J., Neyrinck, M. C., Budavári, T., & Szalay, A. S., 2011, *ApJ*, 728, 34T

Statistical Challenges in 21st Century Cosmology
Proceedings IAU Symposium No. 306, 2014
A. F. Heavens, J. -L. Starck & A. Krone-Martins, eds.

doi:10.1017/S1743921314013751

The Value of H_0 from Gaussian Processes

Vinicius C. Busti[1], Chris Clarkson[1] and Marina Seikel[2,1]

[1]Astrophysics, Cosmology & Gravity Centre (ACGC), and Department of Mathematics and Applied Mathematics,
University of Cape Town, Rondebosch 7701, Cape Town,
South Africa
email: `vinicius.busti@iag.usp.br`

[2] Physics Department, University of Western Cape, Cape Town 7535, South Africa

Abstract. A new non-parametric method based on Gaussian Processes (GP) was proposed recently to measure the Hubble constant H_0. The freedom in this approach comes in the chosen covariance function, which determines how smooth the process is and how nearby points are correlated. We perform coverage tests with a thousand mock samples within the ΛCDM model in order to determine what covariance function provides the least biased results. The function Matérn(5/2) is the best with sligthly higher errors than other covariance functions, although much more stable when compared to standard parametric analyses.

Keywords. methods: statistical, (cosmology:) cosmological parameters, (cosmology:) large-scale structure of universe, cosmology: observations, cosmology: theory, (cosmology:) distance scale

1. Introduction

The difference between the value of the Hubble constant H_0 determined by *Planck* (Planck Collaboration 2014) and by local measurements (Riess *et al.* 2011) shows a 2.3σ tension. In order to understand what could generate such a discrepancy, many attempts in the literature were done searching for new physics or systematic errors (e.g. Marra *et al.* 2013, Spergel *et al.* 2015, Efstathiou 2014, Wyman *et al.* 2014, Holanda *et al.* 2014, Clarkson *et al.* (2014)).

Recently, we proposed a new method to determine H_0 by applying *Gaussian Processes* (GP), which is a non-parametric procedure, to reconstruct $H(z)$ data and extrapolating to redshift zero (Busti *et al.* 2014). We selected 19 $H(z)$ measurements (Simon *et al.* 2005, Stern *et al.* 2010, Moresco *et al.* 2012) based on cosmic chronometers and obtained $H_0 = 64.9 \pm 4.2$ km s^{-1} Mpc^{-1}, which is compatible with *Planck* but shows a tension with local measurements.

Here, we use mock samples in order to test our method. Basically, we are interested to see which covariance function adopted in the GP analysis provides the best match with the fiducial cosmological model. As we shall see, the Matérn(5/2) performs better, with the standard squared exponential covariance function underestimating the errors. In the next section GP will be briefly described, showing our results in Sec. 3. We draw our conclusions and discuss future improvements in Sec. 4.

2. Gaussian Processes (GP)

A gaussian process allows one to reconstruct a function from data without assuming a parametrisation for it. While a gaussian distribution is a distribution over random variables, a gaussian process is a distribution over functions. We use GaPP (Gaussian Processes in Python)† (Seikel *et al.* 2012)) in order to reconstruct the Hubble parameter as a function of the redshift.

† http://www.acgc.uct.ac.za/~seikel/GAPP/index.html

Table 1. H_0 constraints from 19 $H(z)$ measurements.

Method	$H_0 \pm 1\sigma$ (km s^{-1} Mpc^{-1})	Coverage 1σ	Coverage 2σ
Sq. Exp.	$64.9 \pm 4.2(5.9)$	0.527	0.905
Matérn(9/2)	$65.9 \pm 4.5(5.6)$	0.594	0.939
Matérn(7/2)	$66.4 \pm 4.7(5.7)$	0.610	0.946
Matérn(5/2)	$67.4 \pm 5.2(5.5)$	0.665	0.959
ΛCDM	68.9 ± 2.8	0.676	0.938
XCDM	69.0 ± 6.7	0.685	0.939

Basically, the reconstruction is given by a mean function with gaussian error bands, where the function values at different points z and $\tilde{z}$ are connected through a covariance function $k(z, \tilde{z})$. This covariance function depends on a set of hyperparameters. For example, the general purpose squared exponential (Sq. Exp.) covariance function is given by

$$k(z, \tilde{z}) = \sigma_f^2 \exp\left\{-\frac{(z-\tilde{z})^2}{2l^2}\right\}. \tag{2.1}$$

In the above equation we have two hyperparameters, the first σ_f is related to typical changes in the function value while the second l is related to the distance one needs to move in input space before the function value changes significantly. We follow the steps of Seikel *et al.* (2012) and determine the maximum likelihood value for σ_f and l in order to obtain the value of the function. In this way, we are able to reconstruct the Hubble parameter as a function of the redshift from $H(z)$ measurements. Many choices of covariance function are possible, and we consider a variety below.

3. Results

The freedom in the GP approach comes in the covariance function. While in traditional parametric analyses we choose a model to characterise what is our prior belief about the function in which we are interested, with GP we ascribe in the covariance function our priors about the expected function properties (e.g. smoothness, correlation scales etc.).

We consider the Sq. Exp. covariance function and three examples from the Matérn family:

$$k(z, \tilde{z}) = \sigma_f^2 \frac{2^{1-\nu}}{\Gamma(\nu)} \left(\frac{\sqrt{2\nu(z-\tilde{z})^2}}{l}\right)^{\nu} K_\nu \left(\frac{\sqrt{2\nu(z-\tilde{z})^2}}{l}\right), \tag{3.1}$$

where K_ν is a modified Bessel function and we choose $\nu = 5/2$, $7/2$ and $9/2$ (see Seikel & Clarkson 2013 for more discussions). Writing $\nu = p + 1/2$, each Matérn function is p times differentiable as are functions drawn from it, and the squared exponential is recovered for $\nu \to \infty$. Increasing ν increases the width of the covariance function near the peak implying stronger correlations from nearby points for a fixed correlation length ℓ. For comparison purposes, we also consider two standard parametric models: a flat ΛCDM model and a flat XCDM model.

The results are shown in Table 1 together with the constraints from the 19 $H(z)$ data. The coverage test of each covariance function and parametric model was performed by creating 1000 mock catalogues of 19 data points with the same redshifts and error-bars of the measured points in a fiducial ΛCDM model. For each model realisation a value of H_0

was derived. The third and fourth columns of Table 1 show the frequency the true value for H_0 was recovered inside the 1σ and 2σ regions. So, for example, the Matérn(9/2) covariance function captures the true value at $1(2)\sigma$ about 60%(94%) of the time – alternatively, the $1(2)\sigma$ region should be interpreted as a 60%(94%) confidence interval. This provides a way to re-normalise the $n\sigma$ intervals for a given covariance function and a prior model assumption, which we show between parentheses for 1σ(68%) errors in Table 1. Therefore, this is an attempt to quantify a possible systematic error from the covariance functions *assuming the true model is* Λ*CDM*. We also considered some different fiducial models with a time-varying dark energy equation of state, 64 data points in the redshift range $0.1 < z < 1.8$, with coverages showing the same pattern as depicted in Table 1. It is important to note this is a model-dependent comparison which relies on the knowledge of the true model in advance, which is never the case, and changes with the quality of the data. The coverage can change with a different underlying model as well – but note that the errors are actually much more stable than switching from ΛCDM to XCDM.

4. Conclusions

We have applied GP to reconstruct $H(z)$ data and from it extrapolate to redshift zero to obtain H_0 (Busti *et al.* 2014). Based on a set of 1000 mock samples, we have tested the method assuming a fiducial flat ΛCDM model by considering four different covariance functions and applying a coverage test. We have shown Matérn(5/2) represents better the errors, with errors slightly higher than the other covariance functions. A heuristic method to recalibrate the errors for different covariance functions was also provided within the ΛCDM model.

Possible improvements can be achieved by marginalizing over the hyperparameters and comparing the results using Bayesian model comparison tools, which will allow a direct assessment of performance with no need to rely on a fiducial model.

References

Busti, V. C., Clarkson, C., & Seikel, M. 2014, *MNRAS* (Letters) 441, L11 [preprint(arXiv:1402.5429)]

Clarkson, C., Umeh, O., Maartens, R., & Durrer, R. 2014, *J. Cosmol. Astropart. Phys.*, 11, 36

Efstathiou, G. 2014, *MNRAS*, 440, 1138

Holanda, R. F. L., Busti, V. C., & Pordeus da Silva, G. 2014, *MNRAS* (Letters), 443, L74

Marra, V., Amendola, L., Sawicky, I., & Walkenburg, W. 2013, *Phys. Rev. Lett.*, 110, 241305

Moresco, M. *et al.* 2012, *J. Cosmol. Astropart. Phys.*, 8, 6

Planck Collaboration, Ade, P. A. R. *et al.* 2014, *A&A* 571, A16

Riess, A. G. *et al.* 2011, *ApJ*, 730, 119

Seikel, M., Clarkson, C., & Smith, M. 2012, *J. Cosmol. Astropart. Phys.*, 6, 36

Seikel, M. & Clarkson, C. 2013, preprint(arXiv:1311.6678)

Simon, J., Verde, L., & Jimenez, R. 2005, *Phys. Rev. D*, 71, 123001

Spergel, D., Flauger, R., & Hlozek, R., 2015, *Phys. Rev. D*, 91, 023518

Stern, D., Jimenez, R., Verde, L., Kamionkowski, M., & Stanford, S. A. 2010, *J. Cosmol. Astropart. Phys.*, 2, 8

Wyman, M., Rudd, D. H., Vanderveld, A., & Hu, W. 2014, *Phys. Rev. Lett.*, 112, 051302

Statistical Challenges in 21st Century Cosmology
Proceedings IAU Symposium No. 306, 2014
A. F. Heavens, J.-L. Starck & A. Krone-Martins, eds.

doi:10.1017/S1743921314013489

Nonparametric kernel methods for curve estimation and measurement errors

Aurore Delaigle

School of Mathematics and Statistics, University of Melbourne,
Parkville, VIC 3010, Australia
email: A.Delaigle@ms.unimelb.edu.au

Abstract. We consider the problem of estimating an unknown density or regression curve from data. In the parametric setting, the curve to estimate is modelled by a function which is known up to the value of a finite number of parameters. We consider the nonparametric setting, where the curve is not modelled a priori. We focus on kernel methods, which are popular nonparametric techniques that can be used for both density and regression estimation. While these methods are appropriate when the data are observed accurately, they cannot be directly applied to astronomical data, which are often measured with a certain degree of error. It is well known in the statistics literature that when the observations are measured with errors, nonparametric procedures become biased, and need to be adjusted for the errors. Correction techniques have been developed, and are often referred to as deconvolution methods. We introduce those methods, in both the homoscedastic and heteroscedastic error cases, and discuss their practical implementation.

Keywords. methods: data analysis, methods: statistical, stars: statistics, galaxies: statistics.

---

1. Nonparametric curve estimation

1.1. *Regression estimation*

In the regression problem, we are interested in modelling the unknown relationship between two random variables X and Y. Specifically, we wish to estimate the unknown regression curve $m(x) = E(Y|X = x)$ from a sample of independent and identically distributed (i.i.d.) data $(X_1, Y_1), \ldots, (X_n, Y_n)$ satisfying

$$Y_i = m(X_i) + \epsilon_i, \tag{1.1}$$

where, for all x, $E(\epsilon_i|X_i = x) = 0$ and $\text{var}(\epsilon_i|X_i = x) < \infty$.

There are many examples in astronomy where one is interested in modelling the relationship between two variables X and Y. For example, in the Hubble diagram, $(X, Y)=$ (redshift, distance modulus). In Fig. 1, we show a scatterplot of the (X_i, Y_i)'s for $n = 557$ SNe from the Union2 compilation; see Amanullah *et al.* (2010).

In the parametric estimation setting, we assume that we know the shape of m up to the value of a finite number, d say, of parameters. Then, estimating the regression curve m reduces to the estimation of these unknown parameters. For example, if we assume that the regression curve is a quadratic curve, then $m(x) = \theta_0 + \theta_1 x + \theta_2 x^2$, where θ_0, θ_1 and θ_2 are unknown parameters that need to be estimated from the data. Several approaches are possible for computing these estimators, such as the least squares and maximum likelihood procedures.

Parametric estimators can have excellent properties, such as fast convergence rates, but this is only true when the parametric model (i.e. the assumed shape for m) is correct, at least approximately. In particular, if we do not have sufficient information about m, and

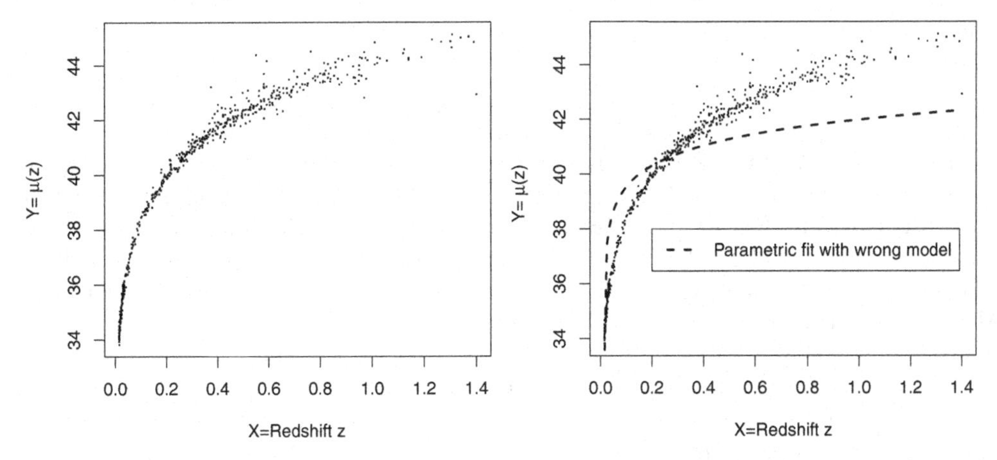

Figure 1. Left: distance moduls versus redshift for 557 SNe from the Union2 compilation. Right: parametric estiimator of $E(Y|X = x)$ using a wrong parametric model.

use a parametric model that is far from the truth, then parametric estimators produce biased, inconsistent estimators of m. See for example Fig. 1, where we depict a parametric estimator of m obtained when assuming that $m(x) = \log(\theta_0 + \theta_1 x)$.

One way to overcome this difficulty is to estimate m using a nonparametric estimator, that is, an estimator that does not require to formulate a parametric model. For excellent introductions to techniques of nonparametric estimation of regression curves, see for example Wand & Jones (1994) and Fan & Gijbels (1996). To understand how such an estimator may be constructed, recall that the regression curve $m(x)$ is the expectation of Y, conditional on $X = x$. Motivated by this, we could think of constructing an estimator of m by taking $\hat{m}_0(x) = \bar{Y} = n^{-1} \sum_{i=1}^{n} Y_i$, the empirical mean of the Y_i's. Clearly, this estimator is too naive since it estimates m by a constant. A more sophisticated approach for estimating m at a point x could be to take the average of only the Y_i's whose X_i is relatively close to x. Letting I denote the indicator function, with this approach we would estimate m by

$$\hat{m}_1(x) = \sum_{i=1}^{n} Y_i \cdot I(X_i \text{ close to } x) \Big/ \sum_{i=1}^{n} I(X_i \text{ close to } x).$$

As long as we define "close to x" properly, this estimator produces a reasonable estimator of m, but is not sufficiently smooth to be attractive. See Fig. 2 below for an illustration of the non smoothness of the estimator in the closely related density estimation problem introduced in section 1.2.

A more sophisticated approach consists in using all the data (X_i, Y_i), assigning to each pair (X_i, Y_i) a weight $w(X_i)$ which is small if X_i is far from x, and large if X_i is close to x. This leads to the estimator

$$\hat{m}_2(x) = \sum_{i=1}^{n} Y_i\, w(X_i) \Big/ \sum_{i=1}^{n} w(X_i).$$

It remains to define the weights $w(X_i)$. In Statistics, a very popular way to choose the weights is to take $w(X_i) = K\{(x - X_i)/h\}$, where $h > 0$ is a smoothing parameter called the bandwidth and K is a weight function called the kernel, wich is usually smooth and symmetric. Computed with these particular weights, the estimator $\hat{m}_2$ of m is called the

Nadaraya-Watson estimator. It is defined by

$$\hat{m}_{\mathrm{NW}}(x) = \frac{\sum_{i=1}^{n} Y_i K\{(X_i - x)/h\}}{\sum_{i=1}^{n} K\{(X_i - x)/h\}}. \tag{1.2}$$

The Nadaraya-Watson estimator is a particular case of a more general class of nonparametric estimators of m called local polynomial estimators. Local polynomial estimators are constructed in a very intuitive way, as follows: while many curves can not be expressed as a polynomial, locally around each point x, if they are smooth, they can be well approximated by a polynomial (one way to understand this is through Taylor's expansion). Motivated by this, the local polynomial estimator of $m(x)$, of order p, is obtained by fitting a polynomial of order p locally around x (that is, using mostly the data (X_i, Y_i) for which X_i is close to x). Formally, at each x, approximate the function $m(u)$ by a pth order polynomial $m_{\mathrm{pol},p}(u) = \beta_{0,x} + \beta_{1,x}(u - x) + \ldots + \beta_{p,x}(u - x)^p$, and estimate the parameters $\beta_{k,x}$ by $\hat{\beta}_{k,x}$, obtained by minimising a local least squares sum which puts more weight on observations whose X_i is close to x:

$$(\hat{\beta}_{0,x}, \ldots, \hat{\beta}_{p,x}) = \mathrm{argmin}_{(\beta_{0,x},\ldots,\beta_{p,x})} \sum_{i=1}^{n} \{Y_i - m_{\mathrm{pol},p}(X_i)\}^2 K\{(X_i - x)/h\}.$$

Then, estimate $m(x)$ by $\hat{m}(x) = \hat{m}_{\mathrm{pol},p}(x) = \hat{\beta}_{0,x} + \hat{\beta}_{1,x}(x-x) + \ldots + \hat{\beta}_{p,x}(x-x)^p = \hat{\beta}_{0,x}$.

It can be proved that the Nadaraya-Watson estimator is equal to the local polynomial estimator of order $p = 0$, which is also called the local constant estimator. In both theory and practice, the Nadaraya-Watson is known to suffer from boundary effects when the density f_X of the X_i's is compactly supported and is not continuous at the endpoints of its support. Specifically, the bias of the Nadaraya-Watson estimator near such endpoints tends to be larger (see below for an illustration). The local polynomial estimator of order $p = 1$, which is also called the local linear estimator, is less affected by this boundary effect, and tends to perform better than the local constant estimator. It is one of the most popular nonparametric regression estimators. It can be written as

$$\hat{m}_{\mathrm{LL}}(x) = \frac{S_2(x)T_0(x) - S_1(x)T_1(x)}{S_2(x)S_0(x) - S_1^2(x)},$$

where, for $k = 0, 1$, $T_k(x) = n^{-1}h^{-k-1}\sum_{i=1}^{n} Y_i (X_i - x)^k K\{(X_i - x)/h\}$ and for $k = 0, 1, 2$, $S_k(x) = n^{-1}h^{-k-1}\sum_{i=1}^{n}(X_i - x)^k K\{(X_i - x)/h\}$. The right panel of Fig. 3 compares the local constant and the local linear estimators computed from the data in Fig. 1. The boundary problem of the local constant estimator is apparent for x small.

1.2. *Density estimation*

Similar ideas can be used to estimate a density f_X from i.i.d. data $X_1, \ldots, X_n \sim f_X$. For excellent introductions to the problem, see Silverman (1986) and Wand & Jones (1994). There too, we can construct a nonparametric estimator of $f_X(x)$ using mainly the observations X_i that are close to x. To define this estimator, recall that $f_X(x) = F_X'(x)$, where $F_X(x) = \int_{-\infty}^{x} f_X(u)\,du$ is the cumulative distribution function. This implies that

$$f_X(x) = \lim_{h\to 0} \frac{F_X(x+h) - F_X(x-h)}{2h} = \lim_{h\to 0} \frac{P(x - h \leqslant X \leqslant x + h)}{2h},$$

which can be estimated by replacing the unknown probability by a proportion computed from the data:

$$\hat{f}_X(x) = \frac{\sum_{i=1}^{n} I(x - h \leqslant X_i \leqslant x + h)}{2nh}, \tag{1.3}$$

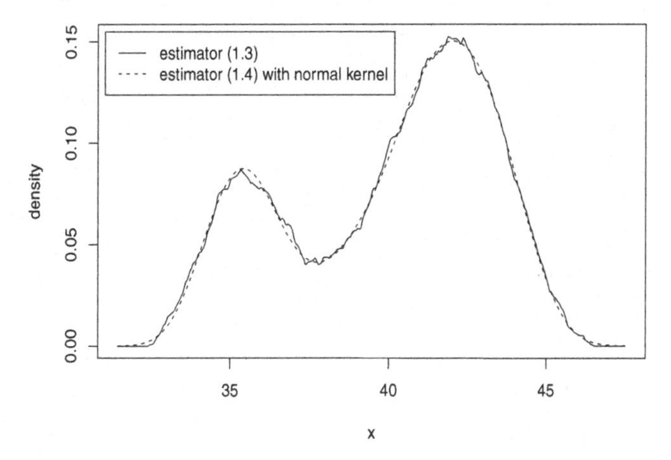

Figure 2. Estimator of the density of distance modulus constructed from the data shown in Fig. 1, using the non smooth estimator at (1.3) or the kernel density estimator at (1.4) with the standard normal kernel.

where h is a small positive number. As in the regression case, while this estimator is reasonable, its lack of smoothness make is relatively unattractive. See for example Fig. 2, where we show this estimator in the case where the X_i's are the distance moduli from $n = 557$ SNe from the Union2 compilation and $h = 0.8$.

A more sophisticated version of this naive estimator is the kernel density estimator defined by

$$\hat{f}_X(x) = (nh)^{-1} \sum_{i=1}^{n} K\{(X_i - x)/h\}, \tag{1.4}$$

where the kernel K is a smooth and symmetric density and $h > 0$ is a bandwidth. In Fig. 2, we depict this estimator in the case where the X_i's are the distance moduli from 557 SNe from the Union2 compilation, taking K to be the density of a standard normal random variable, and $h = 0.8$. It is clear from this example that this estimator is very similar to one at (1.3). However, the fact that it is nice and smooth makes it more attractive. Such density estimators are useful to understand properties of a population. For example in this case the two modes of the density suggest two groups or clusters. See Sun *et al.* (2002) for interesting aspects of the detection of bumps using nonparametric density estimators in astronomy.

1.3. *Choosing the bandwidth h and the kernel K*

As long as it is smooth, the choice of the kernel K is not very important and does not play a major role in the quality of the estimator. It is usually chosen to be a smooth and symmetric density, such as the density of a standard normal random variable.

The role of the bandwidth h is much more important. It dictates the closeness of an observation X_i to x. For example, in the regression case, if h is too small, most observations will be deemed far from x, and the estimator $\hat{m}_{\mathrm{NW}}(x)$ will essentially be based on the few observations (X_i, Y_i) for which X_i is very close to x. As a result it will tend be too wiggly. On the other hand, if h is too large, most observations will be considered to be close to x, and the estimator $\hat{m}_{\mathrm{NW}}(x)$ will be quite similar to the naive estimator $\hat{m}_0(x)$ introduced above. The bandwidth plays the same role for the more general local polynomial estimators. For example in Fig. 3 we show the local linear estimator of $m(x) = E(Y|X = x)$ computed from the data plotted in Fig. 1, using a standard normal kernel and three different bandwidths: a too small bandwidth, which

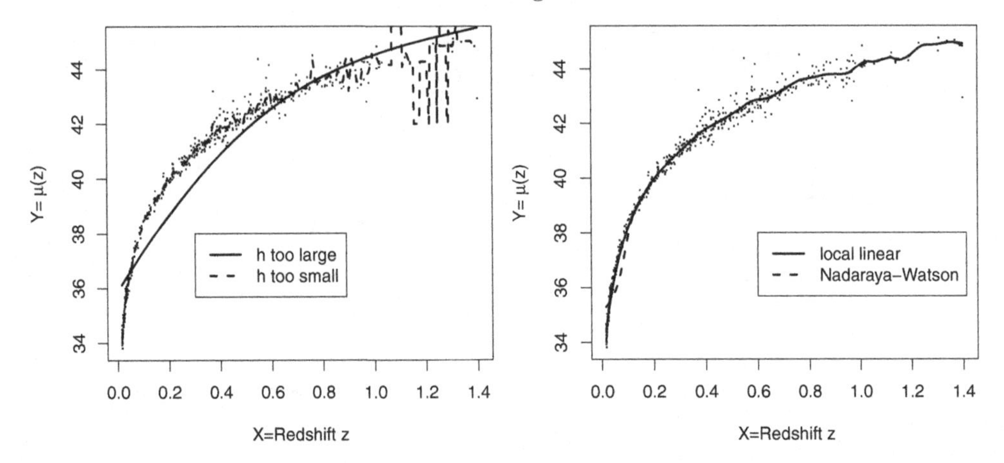

Figure 3. Local linear estimator constructed from the data shown in Fig. 1, using three bandwidths: a too large or too small bandwidth (left graph) or a good bandwidth (right graph). The right graph depicts the local linear and the Nadaraya-Watson estimators, both computed with a standard normal kernel and a bandwidth chosen by an automatic procedure.

produces an estimator which almost interpolates the data and causes numerical difficulty for x large, a too large bandwidth, which oversmoothes the data, and a good bandwidth, computed using one of the automatic procedures described below.

In practice, h should preferably be chosen by a fully automatic data-driven procedure and not by eye (the user may find that the estimator with a given bandwidth looks more attractive than one with a different bandwidth, but in the absence of detailed information about the true curve, the user's impression does not necessarily reflect the reality). Let $\hat{m}$ denote one of the regression estimators introduced above, which depend on a bandwidth h. Ideally, if we knew the curve m, we would choose h to minimise the error committed by estimating m by $\hat{m}$. This error is not unique. It could be the L_2 distance between m and $\hat{m}$, the L_1 distance, or any other sensible criterion. In nonparametric regression, we often employ an L_2 distance called the conditional mean integrated squared error, defined by $\text{MISE}(h) = \text{E}\left[\int\{\hat{m}(x) - m(x)\}^2\, dx \middle| X_1, \ldots, X_n\right]$. With the latter, the ideal bandwidth is defined by $h_{\text{opt}} = \text{argmin}_h \text{MISE}(h)$.

Of course we cannot compute this bandwidth in practice, since it depends on the unknown m. However, a large statistics literature has been devoted to developing estimators of the MISE which can be computed from the data. Once a good estimator of the MISE has been computed in this way, the bandwidth can be chosen by minimising this MISE estimator. Perhaps the most popular data-driven bandwidth is the so-called plug-in bandwidth of Ruppert *et al.* (1995), which is obtained in such a way. There, in a first step the MISE is approximated by its asymptotic dominating part (i.e. the part that dominates the MISE as the sample size n increases) denoted by AMISE (asymptotic mean integrated squared error). In a second step, the unknown quantities of the AMISE are replaced by estimators computed from the data, producing an estimator $\widehat{\text{AMISE}}$ of AMISE. Finally, the plug-in bandwidth h_{PI} is defined by $h_{\text{PI}} = \text{argmin}_h \widehat{\text{AMISE}}(h)$, or, more commonly, by $h_{\text{PI}} = \text{argmin}_h \widehat{\text{AMISE}}_w(h)$, where $\widehat{\text{AMISE}}_w$ denotes a weighted version of the $\widehat{\text{AMISE}}$. See section 4 of Fan & Gijbels (1996) for a more detailed description. We refer to the R package `KernSmooth` of Wand R port by Ripley (2011) for R codes for computing this bandwidth for the local linear estimator (see `dpill`).

Another popular data-driven bandwidth for computing regression estimators is the cross-validation bandwidth h_{CV}. It is defined by

$$h_{\mathrm{CV}} = \mathrm{argmin}_h \sum_{i=1}^{n} \{Y_i - \hat{m}^{(-i)}(X_i)\}^2, \tag{1.5}$$

where $\hat{m}^{(-i)}$ denotes the version of the estimator $\hat{m}$ computed without using the ith observation. For example, in the case of the Nadaraya-Watson estimator,

$$\hat{m}_{\mathrm{NW}}^{(-i)}(x) = \frac{\sum_{j \neq i}^{n} Y_j K\{(X_j - x)/h\}}{\sum_{j \neq i}^{n} K\{(X_j - x)/h\}}.$$

This bandwidth is simple to define but it tends to be too small. Moreover, it is not always unique (the sum on the right hand side of (1.5) does not always have a unique minimum).

In the density case, typically the bandwidth is chosen to minimise an estimator of MISE $= \mathrm{E} \int \{\hat{f}_X(x) - f(x)\}^2 \, dx$ (e.g. the plug-in bandwidth), or of ISE $= \int \{\hat{f}_X(x) - f_X(x)\}^2 \, dx$ (e.g. the cross-validation bandwidth). The plug-in bandwidth (Sheather & Jones (1991)) is constructed using ideas similar to those explained above in the regression case. Like there, it usually performs very well in practice. We refer to the R function `bw.SJ` for computing this bandwidth in the case where K is the standard normal kernel. The cross-validation bandwidth, h_{CV}, suffers from the same difficulties as those mentioned in the regression case above, but it is simpler to define than the plug-in bandwidth. Specifically,

$$h_{\mathrm{CV}} = \mathrm{argmin}_h \left\{ \int \hat{f}_X^2(x) \, dx - \frac{2}{n} \sum_{i=1}^{n} \hat{f}_X^{(-i)}(X_i) \right\},$$

where $\hat{f}_X^{(-i)}(x) = \{(n-1)h\}^{-1} \sum_{j \neq i}^{n} K\{(X_j - x)/h\}$.

We conclude this section with three remarks on the bandwidth. First, it is important to note that different kernels usually require different bandwidths. This is easy to understand since $\hat{m}$ and $\hat{f}_X$ both depend on K. In particular, the MISE and thus the optimal bandwidth depend on K. Likewise, h_{PI} and h_{CV} both depend on K. What this means in practice is that if we have computed h_{PI} or h_{CV} for a given kernel, those bandwidths are generally not appropriate for kernel estimators computed with another kernel. Another important remark is that a good bandwidth should depend on the sample size n. Specifically, as n increases the optimal bandwidth decreases. What this means in practice is that a bandwidth computed for a sample of a given size n is generally not appropriate for a kernel estimator computed from a sample of a different size. Finally, good bandwidths are usually different for density and for regression estimators. For example, a bandwidth computed by `dpill` should not be used to compute a kernel density estimator.

2. Measurement errors

2.1. *Introduction*

In astronomy, data are rarely measured with perfect accuracy and the quantities we observe are often approximated versions of those we are interested in. When computed with data contaminated by measurement errors, the nonparametric procedures introduced in the previous section are not valid and need to be corrected for the measurement errors. In this section we consider the classical measurement error problem, where, instead of observing the variable X of interest, we only manage to observe $W = X + U$, where U represents a measurement error. Importantly, X and U are independent and $E(U) = 0$. See Carroll *et al.* (2006) for an introduction to measurement errors.

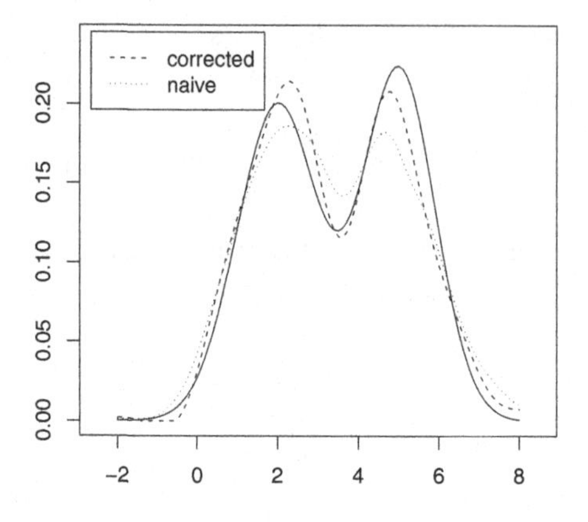

Figure 4. Kernel density estimator (naive) constructed from a sample $W_1, \ldots, W_n$ of size $n = 1000$ contaminated with normal errors U_i such that $\text{var}(U_i) = 0.2\,\text{var}(X_i)$, and modified kernel estimator (corrected) that takes measurement errors into account. The true density f_X is depicted by a continuous line.

Clearly, if we compute the kernel density estimator at (1.4) using data $W_1, \ldots, W_n$ having the distribution of W, instead of data $X_1, \ldots, X_n$ having the distribution of X, then instead of obtaining a consistent estimator of the density f_X, we will obtain a consistent estimator of the density f_W of W. For example, in Fig. 4, we show the kernel density estimator computed from a sample $W_1, \ldots, W_n$ of size $n = 1000$, where, for each i, $W_i = X_i + U_i$, the X_i's have a bimodal density shown in Fig. 4, and the U_i's are normally distributed, with $\text{var}(U_i) = 0.2\,\text{var}(X_i)$. This estimator, which ignores the presence of measurement errors, is often referred to as a naive estimator. It is a consistent estimator of f_W, but a non consistent, biased, estimator of f_X. In this example the bias is noticeable from the fact that the peaks and the valleys of the estimator are attenuated compared to those of f_X. In Fig. 4 we also show a corrected estimator that takes the measurement errors into account. It is a consistent estimator of f_X, and, for example, is able to better estimate the peaks and the valleys of f_X.

Likewise, if we compute one of the regression estimators of $m(x) = E(Y|X = x)$ defined in section 1.1 from data $(W_1, Y_1), \ldots, (W_n, Y_n)$ instead of data $(X_1, Y_1), \ldots, (X_n, Y_n)$, where $W_i = X_i + U_i$ with X_i and U_i independent as above, then instead of a consistent estimator of $m(x)$, we will obtain a consistent estimator of $E(Y|W = x)$.

While the measurement errors U_i are not observed, often in astronomy we can compute the distribution of U_i. Exploiting this fact, we shall assume throughout section 2 that the distribution of the U_i's in known, and under this assumption we shall see how to adapt the kernel density and regression estimators to this errors-in-variables context. In particular we shall see how to transform them into consistent estimators of f_X and m.

2.2. *Deconvolution kernel density estimator*

Suppose we observe i.i.d. data $W_1, \ldots, W_n$, where, for $i = 1, \ldots, n$, $W_i = X_i + U_i$ with $X_i \sim f_X$ and $U_i \sim f_U$ independent. The error density f_U is known, $E(U_i) = 0$ and the goal is to estimate the density f_X from the W_i's. For $V = X$, U and W, let $\phi_V(t) = \int e^{itx} f_V(x)\, dx$ denote the characteristic function of V. By the Fourier inversion theorem, we can write

$$f_X(x) = \frac{1}{2\pi} \int e^{-itx} \phi_X(t)\, dt.$$

Moreover, the independence of X_i and U_i implies that $\phi_W(t) = \phi_X(t)\,\phi_U(t)$. Therefore, assuming that $\phi_U(t) \neq 0$, we have $\phi_X(t) = \phi_W(t)/\phi_U(t)$. Since f_U is known, then ϕ_U is known too, and since we observe data $W_1, \ldots, W_n$, then we can estimate $\phi_W(t)$ by the empirical characteristic function $\hat{\phi}_W(t) = n^{-1}\sum_{j=1}^n e^{itW_j}$. Therefore, we can estimate $\phi_X(t)$ by $\hat{\phi}_X(t) = \hat{\phi}_W(t)/\phi_U(t)$. From there, it is tempting to define an estimator of $f_X(x)$ by $\hat{f}_X(x) = (2\pi)^{-1}\int e^{-itx}\hat{\phi}_X(t)\,dt$. However, $\hat{\phi}_X(t)$ is a very poor estimator of $\phi_X(t)$ for $|t|$ large, which makes the estimator $\hat{f}_X(x)$ inappropriate.

To overcome this difficulty, we need to modify $\hat{\phi}_X(t)$ so as to put less emphasis on it for $|t|$ large. One way to do this is to replace $\hat{\phi}_W(t)$ by the Fourier transform of the kernel estimator $\hat{f}_W(w) = (nh)^{-1}\sum_{j=1}^n K\{(W_i - w)/h\}$ of f_W. Indeed, it can be proved that the Fourier transform of $\hat{f}_W$ is given by $\tilde{\phi}_W(t) = \hat{\phi}_W(t)\phi_K(ht)$, where ϕ_K denotes the Fourier transform of the kernel K. Based on this, we can define a new estimator of $\phi_X(t)$ by $\tilde{\phi}_X(t) = \tilde{\phi}_W(t)/\phi_U(t) = \hat{\phi}_X(t)\phi_K(ht)$. Now the factor $\phi_K(ht)$ is small when $|t|$ is large, which reduces the impact of $\hat{\phi}_X(t)$ when the latter is a poor estimate of $\phi_X(t)$. Motivated by this, the deconvolution kernel estimator of Carroll & Hall (1988) and Stefanski & Carroll (1990) is defined by

$$\hat{f}_X(x) = \frac{1}{2\pi}\int e^{-itx}\hat{\phi}_W(t)\phi_K(ht)/\phi_U(t)\,dt.$$

It can be rewritten as

$$\hat{f}_X(x) = (nh)^{-1}\sum_{j=1}^n K_U\{(W_j - x)/h\}, \tag{2.1}$$

where

$$K_U(x) = (2\pi)^{-1}\int e^{-itx}\phi_K(t)/\phi_U(t/h)\,dt. \tag{2.2}$$

It is interesting to note that $\inf_x K_U(x) < 0$, even if $\inf_x K(x) \geqslant 0$. As a result, in finite samples, while $\hat{f}_X$ integrates to 1, it is often not positive everywhere, although in general, $\hat{f}_X(x)$ vanishes only at points x where $f_X(x)$ is rather small. Since a density is always positive, it is convenient to replace $\hat{f}_X(x)$ by $\tilde{f}_X(x) = \max\{0, \hat{f}_X(x)\}$. This is the estimator we used to construct the corrected estimator shown in Fig. 4. If needed, $\tilde{f}_X$ can also be rescaled so that it integrates to 1. See Hall & Murison (1993).

While the choice of the kernel K is usually not important in kernel estimation procedures from data measured without errors, one has to be more careful in this case since the kernel needs to be such that the integral in (2.2) exists. This is not trivially the case. For example if K is the standard normal kernel and $U \sim N(0, \sigma^2)$, then this integral only exists for sufficiently large values of h. However, when the sample size n is large, we should use a sufficiently small bandwidth (as already noticed earlier, h should decrease to zero as n increases, and this is true both in theory and in practice). To ensure that the integral at (2.2) exists, in the deconvolution literature it is standard to take a kernel whose characteristic function is compactly supported. Two such kernels are usually employed: the sinc kernel, denoted here by K_1, whose Fourier transform is defined by $\phi_{K_1}(t) = I(|t| \leqslant 1)$, and the kernel which we shall denote here by K_2, whose Fourier transform is defined by $\phi_{K_2}(t) = (1 - t^2)^3 \cdot I(|t| \leqslant 1)$. See, for example, Fan (1991).

2.3. *Errors-in-variables regression estimator*

In the errors-in-variables regression context, we observe i.i.d. data $(W_1, Y_1), \ldots, (W_n, Y_n)$, where, for $i = 1, \ldots, n$, $W_i = X_i + U_i$ with $X_i \sim f_X$ and $U_i \sim f_U$. Moreover, $E(U_i) = 0$

and $Y_i = m(X_i) + \epsilon_i$, where $E(\epsilon_i|X_i) = 0$, $\text{var}(\epsilon_i|X_i) < \infty$, and the U_i's are independent of the ϵ_i's, the Y_i's and the X_i's. Finally m and f_X are unknown but f_U is known, and the goal is to estimate m from the (W_i, Y_i)'s.

To construct a consistent estimator of m in this context, we start by comparing the standard kernel density estimator with the deconvolution kernel estimator introduced in the previous section. In particular, comparing (1.4) and (2.1), we can see that the deconvolution kernel density estimator takes the same form as the standard kernel density estimator, except that K is replaced by K_U and the X_i's are replaced by the W_i's. This motivates us to modify the Nadaraya-Watson estimator at (1.2) in a similar manner, and define a kernel estimator that takes measurement errors into account by

$$\hat{m}(x) = \frac{\sum_{i=1}^n Y_i K_U\{(W_i - x)/h\}}{\sum_{i=1}^n K_U\{(W_i - x)/h\}}. \tag{2.3}$$

This estimator was introduced by Fan & Truong (1993). Under appropriate regularity conditions, including the one that $|\phi_U(t)| > 0$ for all t, it can be proved that it is a consistent estimator of m.

It is also possible to define a version of local polynomial estimators which takes the measurement errors into account. These estimators are less easy to define, and we refer to Delaigle *et al.* (2009) for details. See also Delaigle (2014) for a more general description of how to construct consistent nonparametric estimators in errors-in-variables problems. In practice, unlike the error-free case, in the errors-in-variables context the local constant estimator, which corresponds to the estimator defined at (2.3), tends to perform better than the local linear estimator.

As indicated in section 2.2, the function K_U is not positive everywhere and, in practice, the denominator of the right hand side of (2.3) can vanish (or be very close to zero) at some points x, which creates numerical problems. The latter can be avoided by preventing the denominator from getting too small. One way to do this is to replace (2.3) by

$$\hat{m}(x) = \frac{\sum_{i=1}^n Y_i K_U\{(W_i - x)/h\}}{\max\left[\sum_{i=1}^n K_U\{(W_i - x)/h\}, \rho\right]},$$

where ρ is a positive number, sometimes referred to as a ridge parameter.

2.4. *Heteroscedastic errors*

As highlighted by Feigelson & Babu (2012), in astronomy, the measurement errors are often heteroscedastic. There, the independent contaminated data $W_1, \ldots, W_n$ are such that $W_i = X_i + U_i$, with X_i and U_i independent, $X_i \sim f_X$ and $U_i \sim f_{U_i}$. In particular, each observation may have its own error density f_{U_i}, which we assume to be known. As before, $E(U_i) = 0$. In this case, the observations W_i that are contaminated by "a lot of noise" contain less information than the observations that are contaminated by "less noise". Intuitively, when constructing estimators, the least contaminated observations should be given more emphasis than the most contaminated observations. It remains to define the notions of "a lot of noise", "less noise", etc, and to seek a way of giving more importance to more reliable observations.

First, contrary to what may be thought, "a lot of noise" does not always mean "a large error variance". In nonparametric errors-in-variables problems, the effect of measurement errors is also measured by the speed at which the characteristic function of the noise, $\phi_U(t)$, tends to zero as $|t|$ tends to infinity. Specifically, the faster $\phi_U(t)$ tends to zero as $|t|$ increases, the more the measurement errors affect the quality of the estimators. For example, if the errors U_i are normally distributed, then nonparametric estimators of f_X and m converge at a logarithmic rate, i.e. like $(\log n)^{-\alpha}$ for some $\alpha > 0$. By contrast, if the errors U_i have a Laplace distribution, then this rate is rather $n^{-\alpha}$ for some $\alpha > 0$. Of course the variance of the U_i's also plays a role in the quality of nonparametric estimators: the larger that variance, the more difficult it is to estimate f_X and m.

These considerations indicate that combining observations that are contaminated by errors which are not identically distributed is a rather subtle problem. For example, a naive construction could be as follows. Let ϕ_{U_i} and ϕ_{W_i} denote the characteristic functions of U_i and W_i, and assume that $|\phi_{U_i}(t)| > 0$ for all t. Then we have

$$\phi_X(t) = \phi_{W_j}(t)/\phi_{U_j}(t) = n^{-1}\sum_{j=1}^{n} \phi_{W_j}(t)/\phi_{U_j}(t).$$

Since $E(e^{itW_j}) = \phi_{W_j}(t)$, using the same ideas as in section 2.2, and in particular using the Fourier inversion theorem, we could define an estimator of $f_X(x)$ by

$$\hat{f}_X(x) = \frac{1}{2\pi n}\sum_{j=1}^{n}\int e^{-itx}\frac{e^{itW_j}}{\phi_{U_j}(t)}\phi_K(ht)\,dt.$$

As highlighted by Delaigle & Meister (2008), this would be a consistent estimator, but with rather poor properties. For example, suppose that half of the observations were observed with Laplace error, and the other half with normal errors. Then it can be proved that the converge rate of this estimator would be logarithmic. In other words, the estimator would inherit from the convergence rate induced by the least favourable errors U_i. In this example, we would do worse by using all the observations than by using only the observations contaminated by Laplace errors (the latter would lead to a convergence rate of order $n^{-\alpha}$ for some $\alpha > 0$). This indicates that the observations W_j were not combined in an adequate way since our estimator should improve as we use more data.

Delaigle & Meister (2008) proposed an estimator which does not suffer from this problem. They proceed as follows. To understand their estimator, note that $\phi_X(t) = \phi_{W_j}(t)/\phi_{U_j}(t)$ implies that $\phi_X(t) = \phi_{W_j}(t)\bar{\phi}_{U_j}(t)/|\phi_{U_j}(t)|^2$ or again that $\phi_X(t)|\phi_{U_j}(t)|^2 = \phi_{W_j}(t)\bar{\phi}_{U_j}(t)$ (here $\bar{a}$ denotes the conjugate of a complex number a). In turn this implies that $\phi_X(t)\sum_{j=1}^{n}|\phi_{U_j}(t)|^2 = \sum_{j=1}^{n}\phi_{W_j}(t)\bar{\phi}_{U_j}(t)$, so that

$$\phi_X(t) = \sum_{j=1}^{n}\phi_{W_j}(t)\bar{\phi}_{U_j}(t)\Big/\sum_{k=1}^{n}|\phi_{U_k}(t)|^2.$$

Motivated by the fact that $E(e^{itW_j}) = \phi_{W_j}(t)$, using arguments similar to those used above, Delaigle & Meister (2008) propose to estimate $f_X(x)$ by

$$\hat{f}_X(x) = \frac{1}{2\pi n}\sum_{j=1}^{n}\int e^{-itx}\frac{e^{itW_j}\bar{\phi}_{U_j}(t)}{\sum_{k=1}^{n}|\phi_{U_k}(t)|^2}\phi_K(ht)\,dt. \tag{2.4}$$

See Delaigle & Meister (2008) for properties of this estimator.

To illustrate this estimator, we used the data described by De Blok *et al.* (2001) and used by Wang & Wang (2011). They concern the velocity of $n = 318$ stars from 26 low surface brightness galaxies, and for these data the variance of each U_i is available. In Fig. 5 we show the estimator $\hat{f}_X(x)$ at (2.4) computed from these data, assuming that the U_i's are normally distributed with the known error variances, or Laplace distributed with those error variances. In this example the two estimators are so close that they can hardly be distinguished on the graph (this is not always the case!).

In the regression case, where we observe independent data $(W_1, Y_1), \ldots, (W_n, Y_n)$ with the W_i's as above and where $Y_i = m(X_i) + \epsilon_i$ as in section 2.3, Delaigle & Meister (2007)

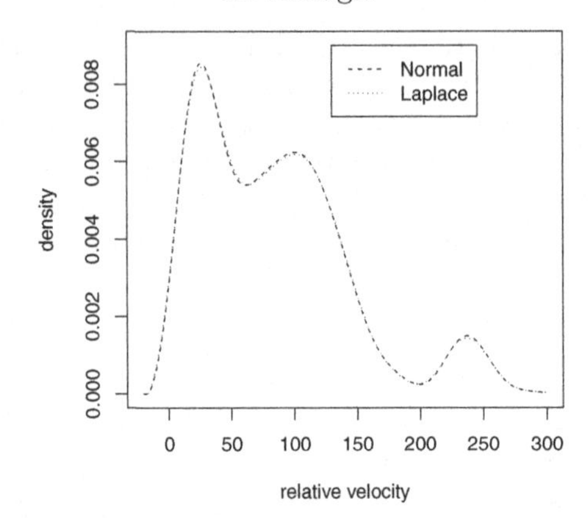

Figure 5. Deconvolution kernel estimator of the density of relative velocity for 318 stars from 26 low surface brightness galaxies, using the estimator of Delaigle & Meister (2008) for heteroscedastic errors and assuming that the errors have a Laplace distribution or a normal distribution.

suggest the following regression estimator:

$$\hat{m}(x) = \frac{\sum_{j=1}^n Y_j \int e^{-itx} \left\{\bar{\phi}_{U_j}(t)/\sum_{k=1}^n |\phi_{U_k}(t)|^2\right\} \phi_K(ht)\, dt}{\sum_{j=1}^n \int e^{-it(x-W_j)} \left\{\bar{\phi}_{U_j}(t)/\sum_{k=1}^n |\phi_{U_k}(t)|^2\right\} \phi_K(ht)\, dt}.$$

2.5. *Bandwidth choice and code for computing the estimators*

As in the case where the data are observed without measurement errors, in order for the estimators introduced above to work well, the bandwidth h needs to be chosen with a lot of care. In the density case, the plug-in techniques of Sheather & Jones (1991) can be adapted to the measurement error context. See Delaigle & Gijbels (2002) and Delaigle & Gijbels (2004). A cross-validation bandwidth can also be constructed; see Stefanski & Carroll (1990).

The situation is much more complex in the regression case. For example, there the plug-in techniques depend on many more unknown functions than in the standard error-free context, which makes them particularly unattractive. Moreover, in the measurement error context it is not possible to compute the cross-validation bandwidth defined at (1.5): even though we can compute the estimator $\hat{m}^{(-i)}$, we need to compute it at X_i, which we cannot do since we only observe the W_i's. Thus it does not seem that standard bandwidth selection techniques can be used in this context. Delaigle & Hall (2008) suggested a procedure based on Simulation Extrapolation (SIMEX) which can be applied to select the bandwidth of a variety of errors-in-variables problems, including the one of regression estimation.

Matlab codes for computing all the estimators described in section 2 are available on the author's webpage at `www.ms.unimelb.edu.au/~aurored`. On that webpage, code for computing the plug-in, cross-validation and SIMEX bandwidths in the errors-in-variables density and regression estimation problems are also available. Some limited R codes written by Achilleas Achilleos, and which focus on density estimation with a local bandwidth as described in Achilleos & Delaigle (2012), are also available there. An R package called `decon` written by Wang & Wang (2011) also exists. However, at the time of writing this paper, this package did not seem to compute the bandwidths in an appropriate manner. See Delaigle (2014) for a detailed description of the problems with this R package, and Delaigle & Gijbels (2007) for a description of numerical issues that can be encountered when computing deconvolution kernel estimators.

Acknowledgements

This work was supported by the Australian Research Council. The author thanks Véronique Delouille for useful discussion.

References

Amanullah, R., *et al.* (2010). Spectra and Hubble space telescope light curves of six type Ia supernovae at $0.511 < z < 1.12$ and the Union2 compilation. *ApJ*, **716**, 712.

Achilleos, A. & Delaigle, A. (2012). Local bandwidth selectors for deconvolution kernel density estimation. *Statistics and Computing*, **22**, 563–577.

Carroll, R. J. & Hall, P. (1988). Optimal rates of convergence for deconvolving a density. *J. Amer. Statist. Assoc.*, **83**, 1184–1186.

Carroll, R. J., Ruppert, D., Stefanski, L. A., & Crainiceanu, C. M. (2006). *Measurement Error in Nonlinear Models*, 2nd Edn. Chapman and Hall CRC Press, Boca Raton.

De Blok, W., McGaugh, S. & Rubin, V. (2001). High-resolution rotation curves of low surface brightness galaxies: Mass Models. *Astr J.*, **122**.

Delaigle, A. (2014). Kernel methods with errors-in-variables: constructing estimators, computing them, and avoiding common mistakes. *Australian and New Zealand J. Statist.*, **56**, 105–124.

Delaigle, A., Fan, J., & Carroll, R. J. (2009). A design-adaptive local polynomial estimator for the errors-in-variables problem. *J. Amer. Statist. Assoc.*, **104**, 348–359.

Delaigle, A. & Gijbels, I. (2002). Estimation of integrated squared density derivatives from a contaminated sample. *J. Roy. Statist. Soc.* Series B, **64**, 869–886.

Delaigle, A. & Gijbels, I. (2004). Comparison of data-driven bandwidth selection procedures in deconvolution kernel density estimation. *Comp. Statist. Data Anal.*, **45**, 249–267.

Delaigle, A. & Gijbels, I. (2007). Frequent problems in calculating integrals and optimizing objective functions: a case study in density deconvolution. *Statis. Comput.*, **17**, 349–355.

Delaigle, A. & Hall, P. (2008). Using SIMEX for smoothing-parameter choice in errors-in-variables problems. *J. Amer. Statist. Assoc.*, **103**, 280–287.

Delaigle, A. & Meister, A. (2007). Nonparametric regression estimation in the heteroscedastic errors-in-variables problem. *J. Amer. Statist. Assoc.*, **102**, 1416–1426.

Delaigle, A. & Meister, A. (2008). Density estimation with heteroscedastic error. *Bernoulli*, **14**, 562–579.

Fan, J., (1991). On the optimal rates of convergence for nonparametric deconvolution problems. *Ann. Statist.*, **19**, 1257–1272.

Fan, J. & Gijbels, I. (1996). Local polynomial modeling and its applications. *Chapman and Hall*, London.

Fan, J. & Truong, Y. K. (1993). Nonparametric regression with errors-in-variables. *Ann. Statist.* **21**, 1900–1925.

Feigelson, E. D. & Babu, G. J. (2012). Modern Statistical Methods for Astronomy. With R Applications. *Cambridge University Press.*

Hall, P. & Murison, R. D. (1993). Correcting the Negativity of High-Order Kernel Density Estimators. *J. Multivar. Anal.*, **47**, 103–122.

Matt Wand R port by Brian Ripley (2011). KernSmooth: Functions for kernel smoothing for Wand & Jones (1995). R package version 2.23-6.

Ruppert, D., Sheather, S. J., & Wand, M. P. (1995). An effective bandwidth selector for local least squares regression. *J. Amer. Statist. Assoc.*, **90**, 1257–1270.

Sheather, S. J. & Jones, M. C. (1991). A reliable data-based bandwidth selection method for kernel density estimation. *J. Roy. Statist. Soc.* Series B, **53**, 683–690.

Silverman, B., W. (1986). Density Estimation for Statistics and Data Analysis. *CRC Press.*

Stefanski, L. A. & Carroll, R. J., (1990). Deconvoluting kernel density estimators. *Statistics*, **21**, 169–184.

Sun, J., Morrisson, H., Harding, P., & Woodroofe (2002). Mixtures and bumps: errors in measurement. *Technical report.* Case Western Reserve University.

Wand, M. P., & Jones, M. C. (1994). Kernel smoothing. *CRC Press.*

Wang, X. F. & Wang, B. (2011). Deconvolution estimation in measurement error models: The R package decon. *J. Statist. Soft.*, **39**, 1–24.

Statistical Challenges in 21st Century Cosmology
Proceedings IAU Symposium No. 306, 2014
A. F. Heavens, J. -L. Starck & A. Krone-Martins, eds.

doi:10.1017/S174392131401357X

Galaxy and Mass Assembly (GAMA): luminosity function evolution

Jon Loveday and the GAMA team

Astronomy Centre, University of Sussex, Falmer, Brighton BN1 9QH, UK
email: J.Loveday@sussex.ac.uk

Abstract. We describe modifications to the joint stepwise maximum likelihood method of Cole (2011) in order to simultaneously fit the GAMA-II galaxy luminosity function (LF), corrected for radial density variations, and its evolution with redshift. The whole sample is reasonably well-fit with luminosity (Q) and density (P) evolution parameters $Q, P \approx 0.8, 1.7$. Red galaxies show larger luminosity but smaller density evolution than blue galaxies, as expected.

Keywords. galaxies: evolution, galaxies: luminosity function, mass function, galaxies: statistics

---

1. Introduction

The luminosity function (LF) is perhaps the most fundamental model-independent quantity that can be measured from a galaxy redshift survey. Reproducing the observed LF is the first requirement of a successful model of galaxy formation, and thus accurate measurements of the LF are important in constraining the physics of galaxy formation and evolution (e.g. Benson et al. 2003). In addition, accurate knowledge of the survey selection function (and hence LF) is required in order to determine the clustering of a flux-limited sample of galaxies (Cole 2011).

A standard $1/V_{\rm max}$ (Schmidt 1968) estimate of the LF is sensitive to radial density variations within the sample. This sensitivity can be largely mitigated by multiplying the maximum volume $V_{\rm max}$ in which each galaxy is visible by the integrated radial overdensity of a density-defining population (Baldry et al. 2006, 2012). Maximum-likelihood methods (Sandage et al. 1979; Efstathiou et al. 1988) are also unaffected by density fluctuations due to galaxy clustering. However, if the sample covers a significant redshift range, galaxy properties (such as luminosity) and number density are subject to systematic evolution with lookback time. All of the above methods must then either be applied to restricted redshift subsets of the data, or be modified to explicitly allow for evolution (e.g. Lin et al. 1999; Loveday et al. 2012).

Cole (2011) recently introduced a *joint stepwise maximum likelihood* (JSWML) method, which jointly fits non-parametric estimates of the LF and the galaxy overdensity in radial bins, along with an evolution model. In this paper we adapt Cole's JSWML method in order to determine the LF and its evolution from the Galaxy and Mass Assembly (GAMA) survey (Driver et al. 2011).

Throughout, we assume a Hubble constant of $H_0 = 100h$ km/s/Mpc and an $\Omega_M = 0.3, \Omega_\Lambda = 0.7$ cosmology in calculating distances, co-moving volumes and luminosities.

2. Parametrizing the evolution

We parametrize luminosity and density evolution using the parameters Q and P introduced by Lin et al. (1999). Evolution in absolute magnitude is assumed to be linear with redshift, $E(z) = Q(z - z_0)$, such that absolute magnitude M is determined from apparent magnitude m using

$$M = m - 5\log_{10} d_L(z) - 25 - K(z; z_0) + Q(z - z_0), \tag{2.1}$$

where $d_L(z)$ is the luminosity distance at redshift z and $K(z; z_0)$ is the K-correction, relative to a passband blueshifted by z_0, determined from fitting a spectral energy distribution to the model *ugriz* magnitudes for each galaxy using KCORRECT V4_2 (Blanton et al. 2003; Blanton & Roweis 2007). Luminosity evolution is determined relative to the same redshift z_0 as the K-correction. In order to minimize errors introduced by uncertainties in individual galaxy K-corrections and evolution histories, one should set z_0 close to the mean redshift of the sample. Here, we choose $z_0 = 0.1$ so that results may be directly compared with a previous estimate of the GAMA LF and its evolution (Loveday et al. 2012).

Evolution in number density $\bar{n}(z)$ is parametrized as

$$\bar{n}(z) = \bar{n}(z_0)10^{0.4P(z-z_0)} = \bar{n}(z=0)10^{0.4Pz}. \tag{2.2}$$

3. Maximum likelihood density-corrected $V_{\rm max}$ method

We have adapted the Cole (2011) method by (i) including incompleteness-correction weights and (ii) finding optimum evolution parameters by minimising the combined χ^2 from radial overdensities and the comparison of the evolution-corrected LF in redshift slices:

$$\chi^2 = \sum_i \frac{(\phi_i^{z_{\rm lo}} - \phi_i^{z_{\rm hi}})^2}{{\rm Var}(\phi_i^{z_{\rm lo}}) + {\rm Var}(\phi_i^{z_{\rm hi}})} + \sum_p \frac{(\Delta_p - 1)^2}{\sigma_p^2}, \tag{3.1}$$

where ϕ is estimated in two broad redshift ranges $z_{\rm lo}$ and $z_{\rm hi}$ split near the mean redshift of the sample, $\bar{z} \approx 0.2$; ${\rm Var}(\phi)$ is the corresponding variance, determined by jackknife resampling; Δ_p is the overdensity in radial bin p and σ_p^2 its variance, also determined by jackknife resampling.

We first evaluate χ^2 values on a rectangular grid of (P, Q), thus allowing one to visualise the correlations between the evolution parameters. The grid point with the smallest χ^2 value is then used as a starting point for a downhill simplex minimisation to refine the parameter values corresponding to minimum χ^2.

4. Results

Fig. 1 shows χ^2 contours in the space of the evolution parameters P, Q determined using equation (3.1) for the full GAMA-II sample and for blue and red galaxies separately. The multi-modal nature of the likelihood contours is most likely due to changes in the noise properties of individual LF bins as the varying value of Q moves galaxies from one luminosity bin to another.† We see a significant trend of increasing density evolution P from red to blue galaxies. This is to be expected, since many galaxies that were star-forming and hence blue at higher redshift have since ceased forming stars and now lie on the red sequence, e.g. Peng et al. (2010). Less significantly, one also observes a trend of decreasing luminosity evolution Q from red to blue, again as expected since the luminosity of the latter population is maintained by ongoing star formation. The differences between red and blue galaxies agree qualitatively with those of Loveday et al. (2012), although in

† We are currently investigating a method which fits a smooth kernel density estimate of the LF, thus avoiding these binning effects.

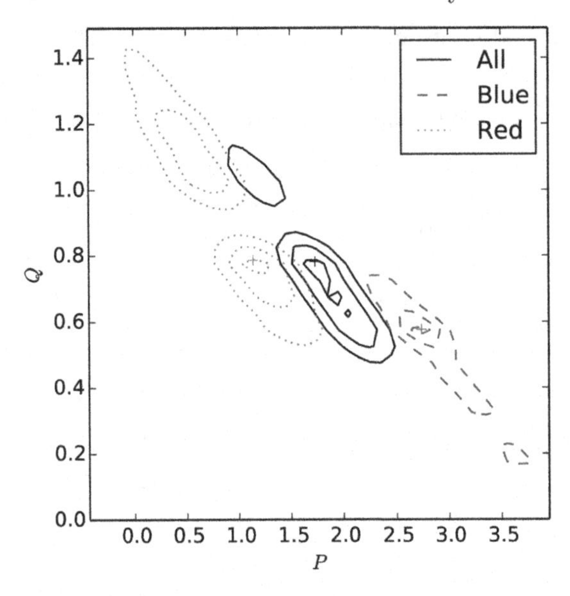

Figure 1. 1, 2, and 3σ χ^2 contours for GAMA-II evolution parameters for all, blue and red galaxies as labelled. The plus signs indicate the location of minimum χ^2.

Table 1. Best-fitting evolution and r-band LF parameters. $\chi^2_{\rm ev}/\nu$ is the reduced χ^2 from equation (3.1), χ^2_ϕ/ν is the reduced χ^2 from least-squares Schechter function fits to the LF estimates; none of the LFs are well-fit in detail by a Schechter function, particularly at the bright end. The uncertainties quoted on the LF parameters come from jackknife sampling, but do not explicitly include the large degeneracies between parameters.

Sample	Q	P	$\chi^2_{\rm ev}/\nu$	α	$M^* - 5\log h$	$\log \phi^*/h^3\rm Mpc^{-3}$	χ^2_Φ/ν
All	0.78	1.73	3.18	-1.28 ± 0.07	-20.76 ± 0.06	-2.06 ± 0.05	2.11
Blue	0.58	2.74	3.77	-1.47 ± 0.06	-20.53 ± 0.08	-2.41 ± 0.06	1.54
Red	0.79	1.14	3.09	-0.71 ± 0.14	-20.57 ± 0.07	-2.15 ± 0.05	2.78

the present analysis we no longer see any evidence for negative density evolution for red galaxies. See Table 1 for best-fitting evolution parameters.

Radial overdensities are shown in the left panel of Fig. 2; Petrosian r-band LFs are shown in the right panel. Surface brightness and redshift incompleteness have been taken into account by appropriately weighting each galaxy. We have fit a Schechter function to each binned LF using least squares, the parameters are tabulated in Table 1. Note that the Schechter fit for red galaxies grossly underestimates the faint end of the LF. It is likely that the faint-end upturn for red galaxies is at least partly due to the inclusion of dusty spirals; the stellar mass function of E–Sa galaxies of Kelvin *et al.* (in prep.) shows no indication of a low-mass upturn. In detail, none of the LFs are well-described by Schechter functions, with large reduced χ^2 values. This is due to the high statistical precision of the GAMA data (jackknife errors for most LF bins are smaller than the plotting symbols), resulting in the large statistical significance of apparently small deviations from a Schechter function. There are significantly more high-luminosity ($M_r - 5\lg h < -23$ mag) galaxies than predicted by the Schechter function fit. Since these very luminous galaxies lie at high redshift, this could be due to non-linearity in true luminosity evolution. These LFs are consistent with the r-band LFs determined from the GAMA-I sample by Loveday et al. (2012), using slightly different methods, and shown in the Figure as dotted lines. We also show the 'corrected' LF from the Blanton et al. (2005) low-redshift SDSS sample (without colour selection). Considering that we are comparing the LFs of SDSS galaxies within only $150h^{-1}$Mpc with GAMA galaxies out to $z \approx 0.65$, the agreement

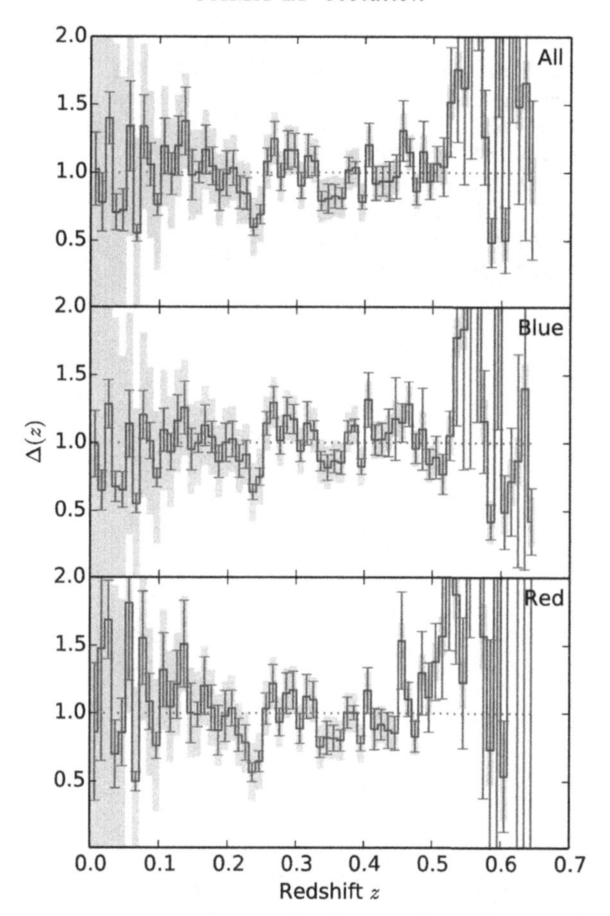

Figure 2. Left: radial overdensities determined from GAMA-II using the entire sample and blue and red subsets as labelled. The error bars show uncertainties estimated from jackknife sampling and the shaded regions show the expected variance assuming $J_3 = \int r^2 \xi(r) dr \approx 2,000 h^{-3}\text{Mpc}^3$. Right: GAMA-II evolution- and density-corrected Petrosian r-band LFs with best-fitting Schechter functions (solid lines) assuming evolution parameters for each sample as given in Table 1. The dotted lines show the best-fit r-band Schechter functions from Table 5 of Loveday et al. (2012). The open diamonds in the top panel show the 'corrected' LF from Fig. 7 of Blanton et al. (2005).

is remarkably good, and provides further evidence that our simple evolutionary model allows one to accurately recover the evolution-corrected LF.

The GAMA website is: `http://www.gama-survey.org/`.

References

Baldry I. K., Balogh M. L., Bower R. G., Glazebrook K., Nichol R. C., Bamford S. P., Budavari T., 2006, *MNRAS*, 373, 469

Baldry I. K. *et al.*, 2012, *MNRAS*, 421, 621

Benson A., Bower R., Frenk C., Lacey C., Baugh C., Cole S., 2003, *ApJ*, 599, 38

Blanton M. R. *et al.*, 2003, *AJ*, 125, 2348

Blanton M. R., Lupton R. H., Schlegel D. J., Strauss M. A., Brinkmann J., Fukugita M., Loveday J., 2005, *ApJ*, 631, 208

Blanton M. R., Roweis S., 2007, *AJ*, 133, 734

Cole S., 2011, *MNRAS*, 416, 739

Driver S. P. *et al.*, 2011, *MNRAS*, 413, 971
Efstathiou G., Ellis R. S., Peterson B. A., 1988, *MNRAS*, 232, 431
Lin H., Yee H. K. C., Carlberg R. G., Morris S. L., Sawicki M., Patton D. R., Wirth G., Shepherd C. W., 1999, *ApJ*, 518, 533
Loveday J. *et al.*, 2012, *MNRAS*, 420, 1239
Peng Y.-j. *et al.*, 2010, *ApJ*, 721, 193
Sandage A., Tammann G. A., Yahil A., 1979, *ApJ*, 232, 352
Schmidt M., 1968, *ApJ*, 151, 393
Taylor E. N. *et al.*, 2011, *MNRAS*, 418, 1587

Statistical Challenges in 21st Century Cosmology
Proceedings IAU Symposium No. 306, 2014
A. F. Heavens, J.-L. Starck & A. Krone-Martins, eds.

doi:10.1017/S1743921314010680

Detecting multi-scale filaments in galaxy distribution

Elmo Tempel[1,2]

[1]Tartu Observatory, Observatooriumi 1, 61602 Tõravere, Estonia
email: elmo.tempel@to.ee

[2]National Institute of Chemical Physics and Biophysics,
Rävala pst 10, 10143 Tallinn, Estonia

Abstract. The main feature of the spatial large-scale galaxy distribution is its intricate network of galaxy filaments. This network is spanned by the galaxy locations that can be interpreted as a three-dimensional point distribution. The global properties of the point process can be measured by different statistical methods, which, however, do not describe directly the structure elements. The morphology of the large-scale structure, on the other hand, is an important property of the galaxy distribution. Here, we apply an object point process with interactions (the Bisous model) to trace and extract the filamentary network in the presently largest galaxy redshift survey, the Sloan Digital Sky Survey (SDSS data release 10). We search for multi-scale filaments in the galaxy distribution that have a radius of about 0.5, 1.0, 2.0, and 4.0 h^{-1}Mpc. We extract the spines of the filamentary network and divide the detected network into single filaments.

Keywords. Methods: data analysis, methods: statistical, large-scale structure of Universe.

1. Introduction

Large galaxy redshift surveys reveal that the Universe has a salient weblike structure, called the cosmic web (Jõeveer & Einasto 1978; Bond *et al.* 1996). The cosmic web consists of highly complex geometrical patterns. Of these patterns, the most dominant ones are galaxy clusters at the intersections of filaments and filaments at the intersections of walls. These web elements display structures and substructures over a wide range of scales. Their hierarchical and interconnected nature is the defining characteristics of the web. However, identifying and describing the cosmic network is not a trivial task due to the overwhelming complexity of the structures, their connectivity and the intrinsic multi-scale nature.

Translating the visual impression of the cosmic web into an algorithm that classifies the local geometry into different environments is not a trivial task, and much work is being done in this direction. In this work, the detection of filaments is performed using a marked point process with interactions, called Bisous model (Stoica *et al.* 2005). This model approximates the filamentary network by a random configuration of small segments or cylinders that interact and connect while building the network. The model was already successfully applied to observational data and to mock catalogues (Stoica *et al.* 2007,2010). This approach has the advantage that it works directly with the original point process and does not require smoothing to create a continuous density field. Our method can be applied to relatively poorly sampled data sets, as the galaxy maps are; it can be applied both to observations and simulations.

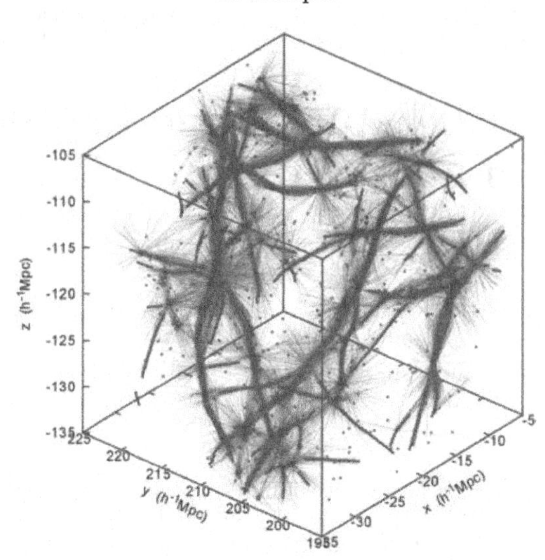

Figure 1. Detected filamentary pattern (cylinder axes) in a small sample volume within a pattern of galaxies (points). Galaxies in groups with 10 or more members are shown with red points; other galaxies are shown with grey points. Green lines show the cylinders from 1000 realisations (it corresponds to the visit map) used to extract the filament spines. The extracted filament spines are shown with blue lines.

2. Galaxy sample

The present work is based on the Sloan Digital Sky Survey (York *et al.* 2000) data release 10 (Ahn *et al.* 2014). We use the galaxy and group samples as compiled in Tempel *et al.* (2014b) that cover the main contiguous area of the survey. The flux-limited catalogue extends to 574 h^{-1}Mpc and includes 588193 galaxies and 82458 groups. In Tempel *et al.* (2014b) the finger-of-god effect is suppressed using the detected galaxy groups.

3. Bisous model

The catalogue of filaments is built by applying an object point process with interactions (the Bisous process; Stoica *et al.* 2005) to the distribution of galaxies. The method and parameters are exactly the same as in Tempel *et al.* (2014a), where the Bisous model was applied to the SDSS DR8 data. The assumed scale (radius) for the extracted filaments in the current study is roughly 0.5, 1.0, 2.0, and 4.0 h^{-1}Mpc.

A detailed description of the Bisous model is given in Stoica *et al.* (2007,2010) and Tempel *et al.* (2014b). In the Bisous algorithm, random segments (thin cylinders) based on the positions of galaxies are used to form the filamentary network according to the connection and alignment rules between these segments. The morphological and quantitative characteristics of these complex geometrical objects can be obtained by following a straightforward procedure: constructing a model, sampling the probability density describing the model, and, finally, applying the methods of statistical inference. In practice, after fixing the approximate scale of the filaments, the algorithm returns filament detection probability and filament orientation fields. The filament probability field is detected using a Markov-Chain Monte-Carlo scheme that effectively samples a large parameter space. A deterministic filamentary pattern spine can be extracted based on the detection probability and filament orientation fields. Using this method, we extract single filaments in the survey.

The Bisous model takes full advantage of the Bayesian data analysis, one of the most successful and increasingly popular tools in statistical analysis nowadays. The application of Bayesian analysis grants the Bisous model a big advantage over the other methods – the probabilistic nature. The Bisous model does not attempt to classify the web into strict components. Instead, it assigns a confidence estimate to each detected structure. The filamentary network is modelled as a whole and the connectivity between structures is intrinsically implemented to the model.

Figure 1 shows the result of Bisous model. The filament probability field is shown with green lines and the orientation of lines illustrates the filament orientation field. The detected filament spines are shown as blue lines.

4. Implications

The major driving force shaping the cosmic web is the large-scale tidal field. The gravitational collapse amplifies any initial anisotropies to give rise to highly asymmetrical structures, exhibiting strong planar or filamentary characteristics. Since filaments are the dominant features in the cosmic web, a better understanding of them could reveal the principal structure formation and evolution processes and the energy state of the Universe at different cosmological epochs. Since dark energy affects the speed of structure formation, the statistics of galaxy filaments would provide insight into the driving forces behind the web formation. Distinctive signatures of different dark energy models can be detected by analysing the statistics of cosmic structures.

Until now, the full research potential of galactic filaments has not yet been utilised. Comparing e.g. with galaxy clusters and cosmic voids, filaments are very rarely used as a probe of cosmology and also the role of filaments in galactic evolution is poorly known. In principle, statistics of galaxy filament properties, such as length, width and connectivity, can be used to measure the large-scale structure and to test cosmological as well as galaxy formation models.

Galaxy formation and evolution is largely affected by the environment where they reside. One of the unknowns of galaxy formation is the dominant scale (and structure) that determines the formation scenario of a given galaxy. The answer to this question is not straightforward, since the speed of structure evolution itself depends on the large-scale environment. Therefore it is not easy to disentangle general structure evolution from galaxy evolution.

Analysing the multi-scale nature of cosmic structures will potentially help to understand some of the unknown factors in structure and galaxy formation scenarios.

References

Ahn, C. P., Alexandroff, R., Allende Prieto, C., *et al.* 2014, *ApJS*, 211, 17

Bond, J. R., Kofman, L., & Pogosyan, D. 1996, *Nature* 380, 603

Jõeveer, M. & Einasto, J. 1978, in M. S. Longair, J. Einasto (eds.), *Large Scale Structures in the Universe*, Proc. IAU Symposium No. 79, 241

Stoica, R. S., Gregori, P., & Mateu, J. 2005, *Stochastic Processes and their Applications*, 115, 1860

Stoica, R. S., Martínez, V. J., & Saar, E. 2007, *Journal of the Royal Statistical Society Series C*, 56, 459

Stoica, R. S., Martínez, V. J., & Saar, E. 2010, *A&A*, 510, A38

Tempel, E., Stoica, R. S., Martínez, V. J., Liivamägi, L. J., Castellan, G., & Saar, E. 2014a, *MNRAS*, 438, 3465

Tempel, E., Tamm, A., Gramann, M., Tuvikene, T., Liivamägi, L. J., Suhhonenko, I., Kipper, R., Einasto, M., & Saar, E. 2014b, *A&A*, 566, A1

York, D. G., Adelman, J., Anderson, Jr., J. E., *et al.* 2000, *AJ*, 120, 1579

Statistical Challenges in 21st Century Cosmology
Proceedings IAU Symposium No. 306, 2014
A. F. Heavens, J.-L. Starck & A. Krone-Martins, eds.

doi:10.1017/S1743921314010813

The Needlet CMB Trispectrum

Antonino Troja[1,†], Simona Donzelli[2], Davide Maino[1] and Domenico Marinucci[3]

[1]Dipartimento di Fisica, Università degli Studi di Milano,
Via Celoria 16, 20133, Milano (MI), Italy
[†]email: antonino.troja@unimi.it

[2]INAF - Istituto di Astrofisica Spaziale e Fisica Cosmica, Milano,
Via E. Bassini 15, 20133, Milano (MI), Italy

[3]Dipartimento di Matematica, Università degli Studi di Roma Tor Vergata,
Via della Ricerca Scientifica, 00133, Roma (RM), Italy

Abstract. We propose a computationally feasible estimator for the needlet trispectrum, which develops earlier work on the bispectrum by Donzelli *et al.* (2012). Our proposal seems to enjoy a number of useful properties, in particular a) the construction exploits the localization properties of the needlet system, and hence it automatically handles masked regions; b) the procedure incorporates a quadratic correction term to correct for the presence of instrumental noise and sky-cuts; c) it is possible to provide analytic results on its statistical properties, which can serve as a guidance for simulations. The needlet trispectrum we present here provides the natural building blocks for the efficient estimation of nonlinearity parameters on CMB data, and in particular for the third order constants g_{NL} and τ_{NL}.

Keywords. cosmic microwave background, early universe

1. Introduction

The Inflationary models describe the dynamics of the first instants of the Universe. Each model introduces a characteristic signature of non-Gaussianity into the anisotropy distribution of the Cosmic Microwave Background (CMB) radiation. The level of non-Gaussianity is described by the so-called *Bardeen's Potential* (Bardeen (1980)):

$$\Phi(\mathbf{x}) = \Phi_L(\mathbf{x}) + f_{NL}[\Phi_L^2(\mathbf{x}) - \langle\Phi_L^2(\mathbf{x})\rangle] + g_{NL}[\Phi_L^3(\mathbf{x})] \tag{1.1}$$

where $\Phi(\mathbf{x})$ is the gravitational potential field of the Universe and $\Phi_L(\mathbf{x})$ is its Gaussian part. The constants f_{NL} and g_{NL}, which parametrize the non-Gaussian part, can be measured from the amplitude of the bispectrum and trispectrum of the field, i.e. the harmonic counterpart of the 3-point and 4-point correlation function of the field, respectively.

There exist several optimal bispectrum estimators that allow to evaluate f_{NL} with high confidence (see for instance Komatsu *et al.* (2005) or Planck Coll. (2013)). Optimal trispectrum estimators has been already applied to CMB data at WMAP resolution (e.g. Sekiguchi & Sugiyama (2013) in real space and Regan *et al.* (2013) in needlet space), but the lack of an optimal trispectrum estimator applied to high-resolution sky simulations (e.g. Planck resolution) has prevented the evaluation of strong constraints on g_{NL} so far.

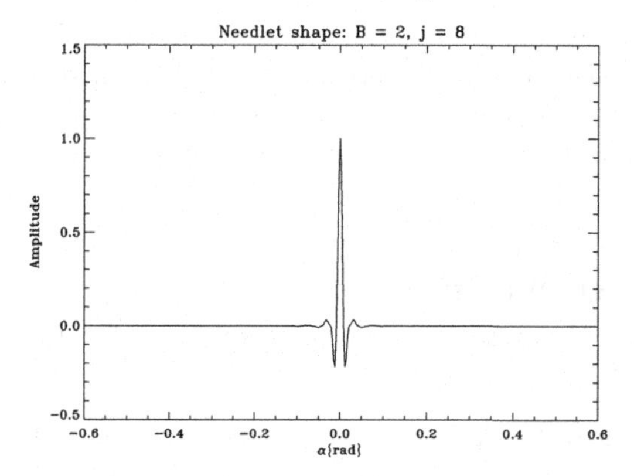

Figure 1. 1D-Spherical needlet. The figure shows the characteristic needle shape of the function, from which take its name (from Marinucci *et al.* (2008))

2. Spherical Needlet System

2.1. *Definition*

Spherical needlets (Narcowich *et al.* (2006), Baldi *et al.* (2009)) are a wavelet system on the sphere defined by setting:

$$\psi_{jk}(x) := \sqrt{\lambda_{jk}} \sum_l b\left(\frac{l}{B^j}\right) \sum_{m=-l}^{l} Y_{lm}(\xi_{jk})\overline{Y_{lm}}(x); \tag{2.1}$$

where $x \in S^2$, $\{\lambda_{jk}, \xi_{jk}\}$ are a set of cubature points and weights on the sphere, $B > 1$ is a constant related to the width of the needlet and $b(.)$ is a weight function satisfying the three following conditions: a) Compact Support, $b(\xi) > 0$ if $\xi \in (B^{-1}, B)$, 0 otherwise; b) Partition of Unity, for all $\zeta \geqslant 1$, $\sum_{j=0}^{\infty} b^2\left(\frac{\xi}{B^j}\right) = 1$; c) Smoothness, $b(.) \in C^M$, i.e., $b(.)$ is M times continuously differentiable, for some $M = 1, 2, \ldots$ or $M = \infty$.

2.2. *Reconstruction Formula*

The spherical needlet coefficients are provided by the analytical formula:

$$\beta_{jk} = \int_{S^2} T(x)\psi_{jk}(x)d\sigma(x) = \sqrt{\lambda_{jk}} \sum_l b\left(\frac{l}{B^j}\right) \sum_{m=-l}^{l} a_{lm} Y_{lm}(\xi_{jk}). \tag{2.2}$$

As a consequence of the partition of unity property, the following well-known reconstruction formula holds:

$$f(x) = \sum_{jk} \beta_{jk}\psi_{jk}(x) \equiv \sum_{lm} a_{lm} Y_{lm}(x). \tag{2.3}$$

2.3. *Localization and Uncorrelation Properties*

As argued earlier in Baldi *et al.* (2009), spherical needlets enjoy an excellent localization property in real domain (fig. 1):

$$|\psi_{jk}(x)| \leqslant \frac{c_M B^j}{(1 + B^j \arccos\langle \xi_{jk}, x\rangle)^M}. \tag{2.4}$$

More explicitly, spherical needlets are then quasi-exponentially localized around any cubature point ξ_{jk}. Moreover, as extensively argued in the literature, the needlet coefficients evaluated on isotropic random fields are asymptotically uncorrelated, under mild regularity conditions. It is then possible to derive analytically their statistical properties, and to understand the role and expressions for correction terms under realistic experimental conditions (noise and masks, see Donzelli *et al.* for the bispectrum case)

3. Optimal Trispectrum Estimator

The sample trispectrum estimator is usually written using the spherical harmonics coefficients as:

$$\sum_{m_1 m_2 m_3 m_4} \mathcal{G}(l_1, m_1, l_2, m_2, l_3, m_3, l_4, m_4) \times a_{l_1 m_1} a_{l_2 m_2} a_{l_3 m_3} a_{l_4 m_4}, \tag{3.1}$$

where $\mathcal{G}(l_1, m_1, l_2, m_2, l_3, m_3, l_4, m_4)$ is the Gaunt integral defined for instance in Marinucci & Peccati (2011). This estimator is clearly unfeasible under realistic experimental conditions, due to the presence of missing data and anisotropic noise. As a consequence of the previous discussion, it is possible to exploit the needlet coefficients to derive an alternative, computationally feasible and statistically sound estimators of the trispectrum. More precisely, exploiting Wick theorem on higher order moments of Gaussian variables we propose the following needlet trispectrum:

$$\begin{aligned} J^{j_1 j_2}_{j_3 j_4} =& \frac{1}{\sigma_{j_1}\sigma_{j_2}\sigma_{j_3}\sigma_{j_4}} \sum_k [\beta_{j_1 k}\beta_{j_2 k}\beta_{j_3 k}\beta_{j_4 k} \\ & - \{\langle \beta_{j_1 k}\beta_{j_2 k}\rangle \beta_{j_3 k}\beta_{j_4 k} + 5\, perms.\} + \{\langle \beta_{j_1 k}\beta_{j_2 k}\rangle \langle \beta_{j_3 k}\beta_{j_4 k}\rangle + 2\, perms.\}] \end{aligned} \tag{3.2}$$

Heuristically, the needlet trispectrum is constructed combining quadruples of coefficients, evaluated at the scales of interest, and subtracting linear and quadratic components in order to cancel bias and minimize the variance. Further details, and a software which exploits the needlet trispectrum to evaluate g_{NL} on CMB maps, will be provided in the forthcoming paper Troja *et al.* (2014) .

References

Baldi, P., Kerkyacharian, G. Marinucci, D., & Picard, D. 2009, *Annals of Statistics*, 37, 1150
Bardeen, J. M. 1980, *Phys. Rev. D*, 22, 1882
Donzelli, S., Hansen, F. K., Liguori, M., Marinucci, D., & Matarrese, S. *Apj*
Komatsu, E., Spergel, D. N., & Wandelt, B. D. 2005, *ApJ*, 634, 14
Marinucci, D. & Peccati, G. 2011, *Random Fields on the Sphere*, Cambridge University Press
Marinucci, D., Pietrobon, D., Balbi, A., Baldi, P., Cabella, P., Kerkyacharian, G., Natoli, P., Picard, D., & Vittorio, N. 2008, *MNRAS*, 383, 539
Narcowich, F. J., Petrushev, P., & Ward, J. D. 2006, *SIAM J. Math. Anal.*, 38, 574
PLANCK Collaboration 2013, *ArXiv e-prints* arxiv:1303.5084
Regan, D., Gosenca M., Seery D. 2013, *ArXiv e-prints* arxiv:1310.8617
Sekiguchi, T., Sugiyama, N. 2013, *JCAP*, 09, 2
Troja, A. *et al.* 2014, Manuscript in preparation
Wick, G. C. 1950, *Physical Review*, 80, 268

Statistical Challenges in 21st Century Cosmology
Proceedings IAU Symposium No. 306, 2014
A. F. Heavens, J. -L. Starck & A. Krone-Martins, eds.

doi:10.1017/S1743921314010667

Generic inference of inflation models by local non-Gaussianity

Sebastian Dorn[1], Erandy Ramirez[2], Kerstin E. Kunze[3], Stefan Hofmann[4], and Torsten A. Enßlin[1,5]

[1]Max-Planck-Institut für Astrophysik, Karl-Schwarzschild-Str. 1, D-85748 Garching, Germany
email: sdorn@mpa-garching.mpg.de

[2]Instituto de Ciencias Nucleares, UNAM A. Postal 70-543, Mexico D.F. 04510, Mexico

[3]Departamento de Física Fundamental and IUFFyM, Universidad de Salamanca, Plaza de la Merced s/n, 37008 Salamanca, Spain

[4]Arnold Sommerfeld Center for Theoretical Physics, Ludwigs-Maximilians-Universität München, Theresienstraße 37, D-80333 Munich, Germany

[5]Ludwigs-Maximilians-Universität München, Geschwister-Scholl-Platz 1, D-80539 Munich, Germany

Abstract. The presence of multiple fields during inflation might seed a detectable amount of non-Gaussianity in the curvature perturbations, which in turn becomes observable in present data sets like the cosmic microwave background (CMB) or the large scale structure (LSS). Within this proceeding we present a fully analytic method to infer inflationary parameters from observations by exploiting higher-order statistics of the curvature perturbations. To keep this analyticity, and thereby to dispense with numerically expensive sampling techniques, a saddle-point approximation is introduced whose precision has been validated for a numerical toy example. Applied to real data, this approach might enable to discriminate among the still viable models of inflation.

Keywords. cosmology: early universe, cosmology: cosmic microwave background, methods: data analysis, methods: statistical

1. Motivation & data model

Precision measurements of the CMB and the LSS have opened a window to the physics of the early universe (WMAP 2012, Planck 2013a). In particular it has become possible to measure the exact statistics of the primordial curvature perturbations on uniform density hypersurfaces, ζ, which turned out to be almost Gaussian (Planck 2013b). These small deviations from Gaussianity are commonly represented by the parameters f_{NL} and g_{NL} and allow to write

$$\zeta = \zeta_1 + \frac{3}{5} f_{\mathrm{NL}} \zeta_1^2 + \frac{9}{25} g_{\mathrm{NL}} \zeta_1^3 + \mathcal{O}(\zeta_1^4), \tag{1.1}$$

where ζ_1 denotes the Gaussian curvature perturbations.

On the other side, the exact statistics of ζ are predicted by inflation models. This suggests to relate current observations directly to inflation models, parametrized by inflationary parameters, p, by higher order statistics of the curvature perturbations (see Fig. 1). How to set up such an inference approach in the framework of information field theory (Enßlin *et al.* 2011) was originally developed in (Dorn *et al.* 2013a, 2014) and is addressed in the work at hand.

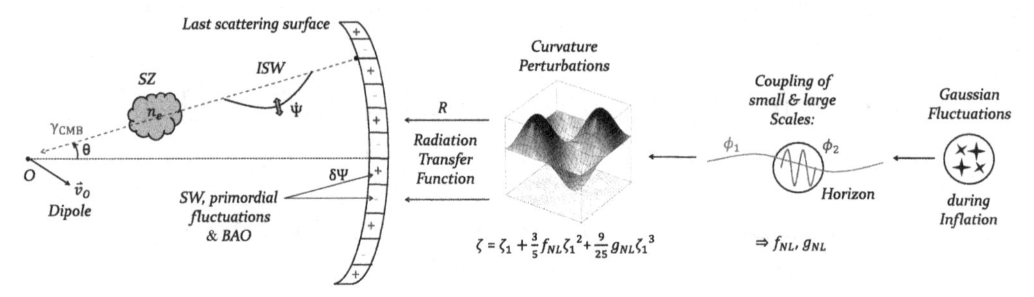

Figure 1. Schematic illustration of the generation of non-Gaussian curvature perturbations out of inflation and their translation to the observable CMB.

To do inference of inflationary parameter we consider the observation (CMB, LSS) to be a discrete data set, $d = (d_1, \dots, d_m)^T \in \mathbb{R}^m,\ m \in \mathbb{N}$, given by

$$d = \frac{\delta T}{T_{\text{CMB}}} = R\zeta + n = R\left(\zeta_1 + \frac{3}{5} f_{\text{NL}} \zeta_1^2 + \frac{9}{25} g_{\text{NL}} \zeta_1^3 + \mathcal{O}\left(\zeta_1^4\right)\right) + n, \tag{1.2}$$

where n denotes some Gaussian noise, $P(n) = \mathcal{G}(n, N) \equiv |2\pi N|^{-1/2} \exp(-n^\dagger N^{-1} n/2)$, and R the so-called response operator. If we consider the CMB, the latter is a linear operator that transfers the curvature perturbations into temperature anisotropies, i.e. the radiation transfer function, and contains all measurement and instrumental effects.

2. Posterior of inflationary parameters

Since the non-Gaussianity parameters depend on a particular inflation model, the calculation of their posterior,

$$\mathcal{P}(f_{\text{NL}}, g_{\text{NL}}|d) \propto \mathcal{P}(f_{\text{NL}}, g_{\text{NL}}) \int \mathcal{D}\zeta_1 \exp\left[-\mathcal{H}(\zeta_1, d|f_{\text{NL}}, g_{\text{NL}})\right], \tag{2.1}$$

is done first (Dorn *et al.* 2014). $\mathcal{H}$ denotes the information Hamiltonian, $\mathcal{H}(.) \equiv -\ln \mathcal{P}(.)$. A straightforward calculation of this Hamiltonian shows that it contains terms up to $\mathcal{O}(\zeta_1^4)$. Therefore the integral of Eq. (2.1) cannot be performed analytically. However, this obstacle can be circumvented by conducting a saddle-point approximation in ζ_1 around $\bar{\zeta_1} \equiv \arg\min\left[\mathcal{H}(\zeta_1, d|f_{\text{NL}}, g_{\text{NL}})\right]$ up to the second order in ζ_1 to be still able to perform the path-integration analytically. This means, we replace $\mathcal{P}(\zeta_1, d|f_{\text{NL}}, g_{\text{NL}})$ by the Gaussian $\mathcal{G}(\zeta_1 - \bar{\zeta_1}, D_{d,f_{\text{NL}},g_{\text{NL}}})$, with

$$D^{-1}_{d,f_{\text{NL}},g_{\text{NL}}} \equiv \left.\frac{\delta^2 \mathcal{H}(\zeta_1, d|f_{\text{NL}}, g_{\text{NL}})}{\delta \zeta_1^2}\right|_{\zeta_1 = \bar{\zeta_1}}, \quad 0 = \left.\frac{\delta \mathcal{H}(\zeta_1, d|f_{\text{NL}}, g_{\text{NL}})}{\delta \zeta_1}\right|_{\zeta_1 = \bar{\zeta_1}}. \tag{2.2}$$

Including the saddle-point approximation and performing the path integral in Eq. (2.1) yields the final expression of the posterior (Dorn *et al.* 2014),

$$\mathcal{P}(f_{\text{NL}}, g_{\text{NL}}|d) \propto \sqrt{|2\pi D_{d,f_{\text{NL}},g_{\text{NL}}}|} \exp\left[-\mathcal{H}(d, \bar{\zeta_1}|f_{\text{NL}}, g_{\text{NL}})\right] \mathcal{P}(f_{\text{NL}}, g_{\text{NL}}). \tag{2.3}$$

Eq. (2.3) enables to calculate the posterior of $f_{\text{NL}}, g_{\text{NL}}$ fully analytic without expensive Monte Carlo sampling techniques. This analyticity has been conserved by conducting a saddle-point approximation, whose sufficiency has been validated by the DIP test (Dorn *et al.* 2013b). Note that $\mathcal{H}(d, \bar{\zeta_1}|f_{\text{NL}}, g_{\text{NL}})$ as well as $D_{d,f_{\text{NL}},g_{\text{NL}}}$ denpends on the two point correlation function of ζ_1 and thus requires some a priori knowledge on the primordial power spectrum. One might use the currently measured and therefore well motivated primordial power spectrum (pure power law) with best fit parameters from Planck (2013c).

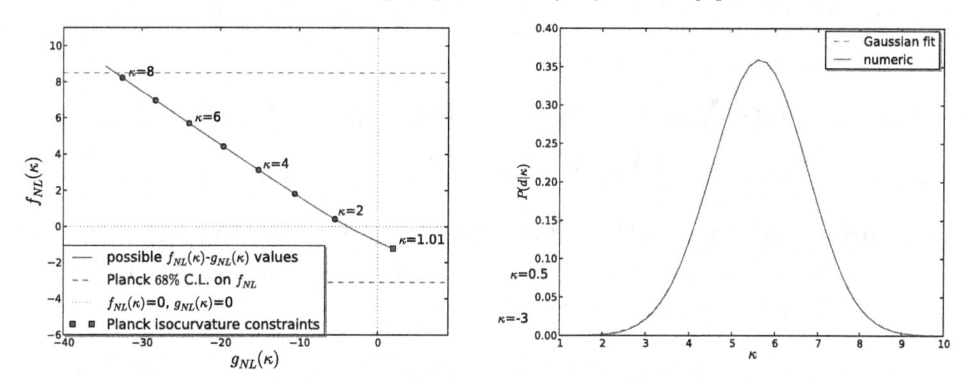

Figure 2. Simplest curvaton model: (Left) Possible values of $f_{\rm NL}$ and $g_{\rm NL}$ within current *Planck* constraints parametrized by the curvaton parameter κ. (Right) Normalized likelihood distributions for κ in a two-dimensional test case in the Sachs-Wolfe limit with data generated from $\kappa_{\rm gen} = 5$. Figures taken from (Dorn *et al.* 2014).

To obtain the posterior of inflationary parameters we replace $f_{\rm NL}(p), g_{\rm NL}(p)$ by their parameter dependent expressions, predicted by inflation models, e.g., for the simplest curvaton model with potential $V(\phi, \chi) = \frac{1}{2}m_\phi \phi^2 + \frac{1}{2}m_\chi \chi^2$ (Bartolo *et al.* 2002;Sasaki *et al.* 2006) one obtains

$$f_{\rm NL} = \frac{5}{4}\kappa - \frac{5}{3} - \frac{5}{6\kappa}, \qquad g_{\rm NL} = \frac{25}{54}\left(-9\kappa + \frac{1}{2} + \frac{10}{\kappa} + \frac{3}{\kappa^2}\right), \qquad \kappa \equiv \frac{4\bar{\rho}_r}{3\bar{\rho}_\chi} + 1. \quad (2.4)$$

The posterior for the curvaton parameter κ for a toy example in two dimensions as well as possible values of $f_{\rm NL}(\kappa), g_{\rm NL}(\kappa)$ are illustrated in Fig. 2.

3. Conclusions

We presented a novel and generic method to infer inflationary parameters from observations (CMB, LSS) by local non-Gaussianity. The method is fully analytic and thereby avoids expensive sampling techniques. The introduced approximation, necessary to conserve the analyticity, has been validated successfully.

References

WMAP Collaboration 2013a, *APJS* 208 20, arXiv:astro-ph/1212.5225

Planck Collaboration 2013b, arXiv:astro-ph/1303.5076

Planck Collaboration 2013, arXiv:astro-ph/13035084

Enßlin, T. A., Frommert, M., & Kitaura, F. S. 2011, *Phys. Rev. D*, 80, 105005, arXiv:astro-ph/0806.3474

Dorn, S., Oppermann, N., Khatri, R., & Enßlin, T. A. 2013a, *Phys. Rev. D*, 88, 103516, arXiv:astro-ph/1307.3884

Dorn, S., Ramirez, E., Kunze, K. E., Hofmann, S., & Enßlin, T. A. 2014, arXiv:astro-ph/1403.5067

Dorn, S., Oppermann, N., & Enßlin, T. A. 2013b, *Phys. Rev. E*, 88, 053303, arXiv:astro-ph/1307.3889

Planck Collaboration 2013c, arXiv:astro-ph/1303.5082

Bartolo, N. & Liddle, A. R. 2002, *Phys. Rev. D* 65 121301, arXiv:astro-ph/0203076

Sasaki, M., Väliviita, J., & Wands, D. 2006, *Phys. Rev. D*, 74, 103003, arXiv:astro-ph/0607627

Statistical Challenges in 21st Century Cosmology
Proceedings IAU Symposium No. 306, 2014
A. F. Heavens, J.-L. Starck & A. Krone-Martins, eds.

doi:10.1017/S1743921314013763

Extreme-Value Statistics for Testing Dark Energy

Simone Aiola[1], **Arthur Kosowsky and Bingjie Wang**

Department of Physics and Astronomy & Pitt-PACC, University of Pittsburgh, PA 15260
[1]email: sia21@pitt.edu

Abstract. The *integrated Sachs-Wolfe effect* was recently detected at a level of 4.4σ by Granett *et al.* (2008), by stacking compensated CMB temperature patches corresponding to superstructures in the universe. We test the reported signal using realistic gaussian random realizations of the CMB sky, based on the temperature power spectrum predicted by the concordance ΛCDM model. Such simulations provide a complementary approach to the largely used N-body simulations and allow to include the contaminant effects due to small-scale temperature fluctuations. We also apply our pipeline to foreground-cleaned CMB sky maps using the Granett *et al.* (2008) voids/clusters catalog. We confirm the detection of a signal, which depart from the null hypothesis by 3.5σ, and we report a tension with our theoretical estimates at a significance of about 2.5σ.

Keywords. cosmic microwave background, large-scale structure of universe, methods: numerical

---

1. Introduction

The current state of accelerated expansion of the universe, driven by Dark Energy (DE), leaves an imprint in the large-scale cosmic structure (at redshifts $z \lesssim 2$), as well as on the CMB temperature fluctuations. The time-evolving gravitational potential weakly redshifts/blueshifts microwave photons, traveling from the last scattering surface to us. This effect is known as *late-time integrated Sachs-Wolfe effect* (late-ISW, hereafter) (Sachs & Wolfe (1967)), and its detection represents an independent test for DE, and in principle a useful probe to characterize its *properties* and *dynamics*.

In 2008, Granett *et al.* (2008) (GNS08, hereafter) reported the strongest detection of the late-ISW, at a significance level of 4.4σ. This strong detection exploited a novel technique involving analysis of stacked CMB patches centered on 100 superstructures, which are detected in `SDSS DR6` in a redshift range $0.4 < z < 0.75$. The photometric signal averaged over a 4° compensated disk corresponds to $\langle T_m^{GNS08} \rangle = 9.6 \pm 2.2 \mu K$. This was confirmed by the Planck collaboration (PLK13, hereafter), which reports a value of $\langle T_m^{PLK13} \rangle = 8.7 \pm 2.2 \mu K$ (Planck Collaboration, XIX, 2013).

It has been argued by Flender *et al.* (2013) (FHN13, hereafter), that this strong signal is in tension with the underlying ΛCDM model. Analytical estimates of the stacked late-ISW signal (only), in a comoving volume that corresponds to what probed by GNS08 (ISW-SDSS, hereafter), predict an average signal of $\langle T_m^{FHN13} \rangle = 2.27 \pm 0.14 \mu K$. This estimate is in tension with the GNS08 results at a level bigger than 3.0σ. This discrepancy is confirmed in the same work using computationally intense N-body simulations, which may have problems related to the size of the simulations boxes.

The GNS08 technique relies on *extreme-value statistics*, which may be easily reveal small departures from the ΛCDM model. On the other hand, an efficient control of the systematics and careful consistency checks between data and simulations are required. In this work, we aim to test this prerequisites, which are fundamental to promote the

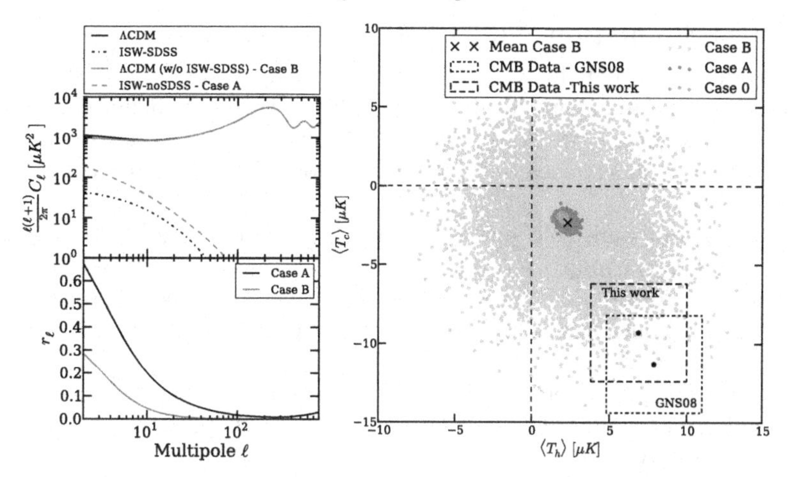

Figure 1. (*Left panel*) CMB temperature auto-spectra, assuming concordance ΛCDM model. The late-ISW is broken up into two different components, in order to show the relative amplitudes compared to the fluctuations originated at $z \sim 1100$. The correlation coefficients $r_\ell = C_\ell^{XY}/\sqrt{C_\ell^{XX}C_\ell^{YY}}$ are plotted in the bottom part, showing the importance of the cross-correlation between the probed ISW-SDSS and the other contributions to the CMB map. (*Right panel*) summary of the results from this analysis. From the simulations, we notice that the effect of the correlations slightly shifts the expectation values. However, the effect of the small-scale fluctuations and the cosmic variance drastically change the intrinsic signal-to-noise ratio of such a measurement. The results from the analysis of the CMB map with our pipeline are compared with the previous analysis of GNS08.

GNS08 measurement from a simple detection to a cosmological probe of the late-time universe.

2. A Simple Description of the CMB Sky

The large-scale CMB sky is well described by a gaussian random field, which statistical properties are fully specified by the usual power spectrum C_ℓ^{TT}. Specifically, different physical processes cause temperature fluctuations at different angular scales. Consequently, the CMB sky map can be considered as the sum of different *correlated gaussian fields*, each separately described by auto-spectra, and the cross-spectra define the correlations between them. In the left panel of Fig. 1, we show the auto-spectra of the ISW-SDSS (corresponding to the signal we aim to detect) and the auto-spectra of the late-ISW coming from neighboring regions of what it is probed by GNS08, as well as the temperature fluctuations generated at the last scattering surface. We also plot (in the bottom part of the same panel) the normalized correlation coefficients between two different components. For a 2-component sky, we need only two coefficients A_ℓ and B_ℓ:

$$A_\ell = |B_\ell||F_\ell|\sqrt{C_\ell^{1,1}}\left(1 + \frac{C_\ell^{1,2}}{C_\ell^{1,1}}\right), \quad B_\ell = |B_\ell||F_\ell|\left(C_\ell^{2,2} - \frac{\left(C_\ell^{1,2}\right)^2}{C_\ell^{1,1}}\right)^{1/2} \tag{2.1}$$

where F_ℓ is the Legendre transform of the 4° compensated TopHat filter used in the GNS08 analysis. From such coefficients, we can generate two temperature maps and simply co-add them. This algorithm gives a realistic realization of a filtered CMB sky, which is properly correlated with a *given realization* of the ISW-SDSS map (separately used in the analysis).

3. Numerical Simulations and CMB Data Analysis

From the assumption that the late-ISW is traced by superstructures in the universe, we can use the ISW-SDSS map (previously generated) as a tracer of clusters/voids (hot/cold spots) catalog. Specifically, we consider an ensemble of 5000 simulated skies, focusing only on 20% of the total sky area (GNS08 sky fraction). The pipeline we adopted can be described as follows:

(a) we identify and rank cold and hot spots in a randomly-generated *filtered* ISW-SDSS map (called tracer map);

(b) we generate *filtered* CMB temperature map, which contains different temperature components. These components are properly correlated with the previous tracer map (called temperature map);

(c) we average the pixels temperatures of the *temperature map*, which correspond to the 50 top-ranked hot spots, $\langle T_h \rangle$, and 50 top-ranked cold spots,$\langle T_c \rangle$, (identified in the *tracer map*), separately;

(d) we also compute the mean temperature as $\langle T_m \rangle = (\langle T_h \rangle - \langle T_c \rangle)/2$;

We consider different cases, which differ from each other depending on the components we consider in the *temperature map*: (**Case 0**) the temperature map has only the ISW-SDSS component (same as tracer map). (**Case A**) the temperature map has only late-ISW, but integrated over the whole redshift range $0 < z < 10$. (**Case B**) the temperature map actually describes the CMB sky (comparable with GNS08, PLK13 and our analysis on CMB data). The results of our simulations are shown in the right panel of Fig. 1. For the interesting case (Case B), which best describes the observed sky, we find $\langle T_h \rangle = -\langle T_c \rangle = 2.30 \pm 3.1\mu K$ and $\langle T_m \rangle = 2.32 \pm 2.32\mu K$. We notice that this realistic case is in agreement with previous estimates by FHN13. However, we can argue that the filter (which suppresses the largest temperature fluctuations), removes possible issues related to the modes larger than the size of the N-body simulations. A different (and possibly more suitable) filter may not satisfy such a condition. In order to test the simulative pipeline against the data, we also reanalyze the CMB data using our *harmonic-space* filtering procedure and the GNS08 catalog. We find great consistency with the previous works by GNS08 and PLK13. Specifically, we report the following temperature values $\langle T_h \rangle = 6.89 \pm 3.1\mu K$, $\langle T_c \rangle = -9.33 \pm 3.1\mu K$ and $\langle T_m \rangle = 8.11 \pm 2.32\mu K$.

4. Conclusions and Future Prospectives

In this work, based on the upcoming paper (Aiola *et al.* 2014), we presented a detailed analysis of the stacked late-ISW signature, within flat ΛCDM cosmology. From the analysis of available CMB sky maps, we report a departure from the null signal at a level of 3.5σ level, which shows a 2.5σ tension with the expected maximum late-ISW signal. For the future, we plan to optimize and use this method as a cosmological probe for the equation of state of Dark Energy. Methods based on extreme-value statistics are intrinsically sensitive to different expansion histories, therefore useful to explore the parameter space.

References

Aiola, S., Kosowsky, A., & Wang, B. 2014, *in preparation*
Flender, S., Hotchkiss, S., & Nadathur, S. 2013, *JCAP*, 2, 013, 13
Granett, B. R., Neyrinck, M. C., & Szapudi, I. 2008, *ApJ* (Letters), 683, L99
Planck Collaboration 2013, arXiv:1303.5079
Sachs, R. K. & Wolfe, A. M. 1967, *ApJ*, 147, 73

Statistical Challenges in 21st Century Cosmology
Proceedings IAU Symposium No. 306, 2014
A. F. Heavens, J.-L. Starck & A. Krone-Martins, eds.

doi:10.1017/S1743921314013611

Transformed Auto-correlation

Jianfeng Zhou[1,2] **and Yang Gao**[1,2]

[1]Key Laboratory of Particle & Radiation Imaging (Tsinghua University)

[2]Department of Engineering Physics and Center for Astrophysics, Tsinghua University, Beijing 100084, China

email: zhoujf@tsinghua.edu.cn, gaoyang12@mails.tsinghua.edu.cn

Abstract. A transformed auto-correlation method is presented here, where a received signal is transformed based on a priori reflecting model, and then the transformed signal is cross-correlated to its original one. If the model is correct, after transformation, the reflected signal will be coherent to the transmitted signal, with zero delay. A map of transformed auto-correlation function with zero delay can be generated in a given parametric space. The significant peaks in the map may indicate the possible reflectors nearby the central transmitter. The true values of the parameters of reflectors can be estimated at the same time.

Keywords. methods: data analysis, techniques: radar astronomy

1. Introduction

Echo signals probably exist in the light curves of some sorts of astronomical objects, such as supernovae (Rest *et al.* 2005, Krause *et al.* 2008a, Krause *et al.* 2008b), x-ray binaries(Greiner 2001) and AGNs(Perterson 1993) etc. If the relative location between a central transmitter and a reflecting object is fixed, then auto-correlation (Edelson & Krolik 1988) can be used to detect the reflected signal and estimate the distance thereafter. However, in common situation, the nearby reflecting object is always moving (Perterson 2008), therefore, the delay between transmitted and reflected signal is variable, which means the coherence is broken and no echo signals could be detected by auto-correlation.

Here, we are going to introduce a method which is called transformed auto-correlation. Its aim is to rebuild the coherence and perform cross-correlation between the reflected and transmitted signals.

2. Basic Conception

Suppose there are only one transmitter and one reflector, and $S(t)$ is a transmitted signal, and $R(t)$ is the received signal. Since astronomical objects are usually far far away, the received light curve is the combination of transmitted and reflected signals, i.e.,

$$R(t) = S(t) + \kappa S(t - \tau(t, a, b, c, ...)) \tag{2.1}$$

where κ is reflectance, and $\tau(t, a, b, c, ...)$ is the variable delay which is model dependent, $a, b, c,$ are the parameters of the model.

Due to variable delay, a reflected signal will probably loss the coherence to the transmitted signal. In such situation, it is impossible to detect a reflected signal by auto-correlation. A transformation is needed to rebuild the coherence. Let's set:

$$\tilde{t} = t - \tau(t, a, b, c, ...) \tag{2.2}$$

$$t = \Gamma(\tilde{t}, a, b, c, ...) \tag{2.3}$$

where $\Gamma()$ is the inversion function of t in term of $\tilde{t}$.

Applying Equation. 2.2 and 2.3 into Equation 2.1, then we obtain a transformed received signal:

$$R^T(\tilde{t}) = S(\Gamma(\tilde{t}, a, b, c, ...)) + \kappa S(\tilde{t}) \tag{2.4}$$

The definition of transformed auto-correlation is the cross-correlation between received signal $R(t)$ and its transformation $R^T(t)$, i.e.,

$$C_{ta}(\tau, a, b, c, ...) = \langle R(t), R^T(t+\tau)\rangle = \int R(t)R^T(t+\tau)dt \tag{2.5}$$

If the parameters are correct, then after transformation, the reflected signal is coherent to transmitted signal now, with zero delay. So, only $C_{ta}(0, a, b, c, ...)$ is important. We can calculate all $C_{ta}(0, a, b, c, ...)$ in a given parametric space. The significant peaks of $C_{ta}(0, a, b, c, ...)$ may indicate possible reflectors existed nearby the central transmitter, and the true parameters of these reflectors cab be estimated at the same time.

3. Simulations

Here, an one-dimension simulation is used as an example, where a transmitter is fixed at $x = 0$, and several reflectors are moving or fixed in x axis. The transmitted signal, generated by convolving a Gaussian white noise with a Gaussian point spread function ($\sigma = 10$), is stochastic and band-limited. The sampling interval is $1ms$. and 2, 048, 000 sampled data are generated.

Four reflectors are set in the simulation. The model parameters including Initial Position, Velocity and Reflectance are listed in Table 1, columns 2-4.

Table 1. The parameters of the Reflectors. Columns 2-4 are values of a model. Columns 5-7 are estimated values by transformed auto-correlation.

Reflector	Position[1] (M) (lc[3])	Velocity(M) (c[4])	R^2(M)	**Position(E)** (lc)	**Velocity(E)** (c)	**R(E)**
1	0.8	-0.02	0.04	**0.80**	**-0.02**	**0.040**
2	0.9	0.0	0.08	**0.90**	**0.0**	**0.077**
3	1.0	0.01	0.08	**1.00**	**0.01**	**0.078**
4	1.2	0.03	0.06	**1.2**	**0.03**	**0.054**

Notes:
[1] Initial Position of the reflector. [2] Reflectance. [3] Light Second. [4] Speed of Light.

A part of received signal and its auto-correlation function are displayed in Figure 1. As shown in the auto-correlation function, although there are four reflectors, only one reflector, which is fixed at position of 0.9 light second, has a fringe. Other reflectors have variable delays, so the reflected signals are incoherent to the transmitted signals, and no fringes exist.

Now, it is able to search the possible reflectors and their parameters by transformed auto-correlation. Here, $C_{ta}(0)$ is normalized by $C_a(0) = \langle R(t), R(t)\rangle$. Thus the intensity of the peak of a reflector is equal to reflectance κ. The generated map in parametric space (initial position and velocity), which is called reflectance map, is shown in Figure 2. It is clearly seen that all of four reflectors have been detected. The estimated parameters (initial position, velocity and reflectance) are listed in Table 1, columns 5-7. The standard deviation of the background in reflectance map is about 0.0036.

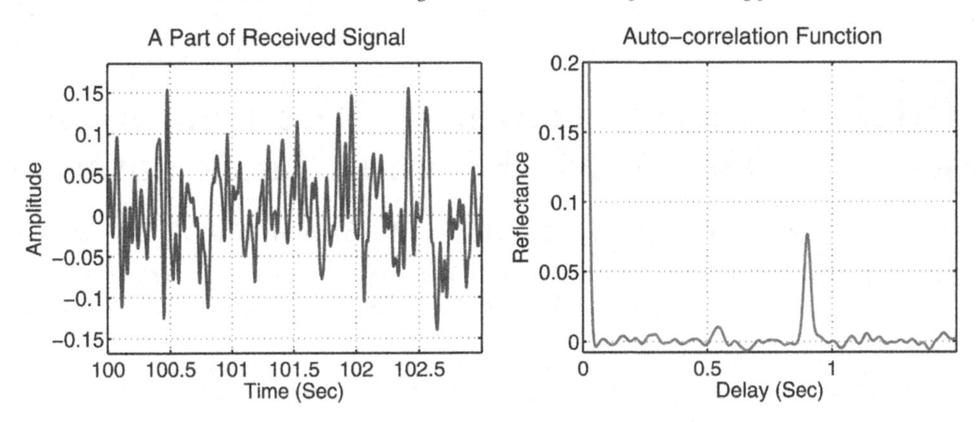

Figure 1. A part of received signal(left) and its auto-correlation function(right).

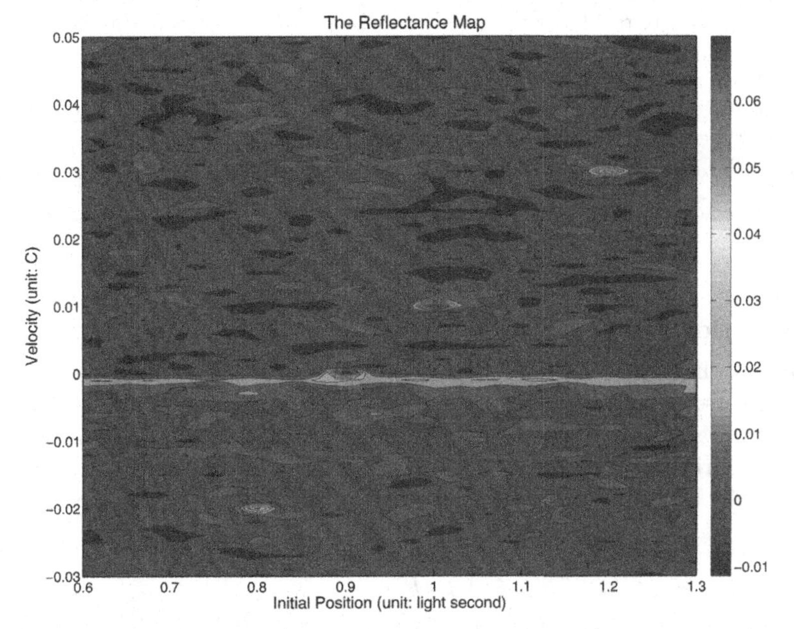

Figure 2. Four reflectors with their initial positions and velocities are found by transformed auto-correlation method.

4. Conclusions

Transformed auto-correlation is able to detect reflected signals when there are constant or variable delays. With a priori reflecting model, the parameters (such as position, velocity etc.) of reflectors can also be estimated.

References

Edelson, R. A. & Krolik, J. H. 1988, *ApJ*, 333, 646

Greiner, J. 2001, *arXiv preprint* , astro-ph/0111540

Krause, O., Birkmann, S. M., Usuda, T., *et al.* 2008, *Science*, 320, 1195

Krause, O., Tanaka, M., Usuda T., *et al.* 2008, *Nature*, 456, 617

Peterson, B. M. 1993, *PASP*, 247-268

Peterson, B. M. 2008, *New Astron. Revs*, 52, 240

Rest, A., Suntzeff, N. B., Olsen, K., *et al.* 2005, *Nature*, 438, 1132

Statistical Challenges in 21st Century Cosmology
Proceedings IAU Symposium No. 306, 2014
A. F. Heavens, J.-L. Starck & A. Krone-Martins, eds.

doi:10.1017/S1743921314010837

PRISM: Sparse recovery of the primordial spectrum from WMAP9 and Planck datasets

P. Paykari, F. Lanusse, J.-L. Starck, F. Sureau and J. Bobin

Service d'Astrophysique, CEA Saclay, F-91191 Gif sur Yvette cedex, France.
email: `paniez.paykari@cea.fr`

Abstract. The primordial power spectrum is an indirect probe of inflation or other structure-formation mechanisms. We introduce a new method, named **PRISM**, to estimate this spectrum from the empirical cosmic microwave background (CMB) power spectrum. This is a sparsity-based inversion method, which leverages a sparsity prior on features in the primordial spectrum in a wavelet dictionary to regularise the inverse problem. This non-parametric approach is able to reconstruct the global shape as well as localised features of the primordial spectrum accurately and proves to be robust for detecting deviations from the currently favoured scale-invariant spectrum. We investigate the strength of this method on a set of WMAP nine-year simulated data for three types of primordial spectra and then process the WMAP nine-year data as well as the Planck PR1 data. We find no significant departures from a near scale-invariant spectrum.

Keywords. Cosmology: Primordial Power Spectrum, Methods: Data Analysis, Statistical

---

1. Introduction

Inflation (Guth 1981; Linde 1982) is currently the most favoured model describing the early Universe, in which perturbations are produced by quantum fluctuations during the epoch of an accelerated expansion. The simplest models of inflation predict a near scale-invariant primordial power spectrum described in terms of a spectral index n_s, an amplitude of the perturbations A_s and a possible 'running' α_s of the spectral index

$$P(k) = A_s \left(\frac{k}{k_p}\right)^{n_s - 1 + \frac{1}{2}\alpha_s \ln(k/k_p)} , \tag{1.1}$$

where k_p is a pivot scale. The near scale-invariant spectrum with $n_s < 1$ fits the current observations very well (Ade *et al.* 2013a, Ade *et al.* 2013b).

The recent Planck data, combined with the WMAP large-scale polarisation, constrain the spectral index to $n_s = 0.9603 \pm 0.0073$ (Ade *et al.* 2013a), ruling out exact scale invariance at over 5σ. Also, Planck does not find a statistically significant running of the scalar spectral index, obtaining $\alpha_s = -0.0134 \pm 0.0090$. In addition, high-resolution CMB experiments, such as the South Pole Telescope (SPT)†, report a small running of the spectral index; $-0.046 < \alpha_s < -0.003$ at 95% confidence (Hou *et al.* 2012). However, in general, any such detections have been small and consistent with zero.

Determining the shape of the primordial spectrum generally consists of two approaches, one by parametrisation and the second by reconstruction. Non-parametric methods suffer from the non-invertibility of the transfer function that descries the *transfer* from $P(k)$ to the CMB spectrum: $C_\ell^{\rm th} = 4\pi \int_0^\infty d\ln k \Delta_\ell^2(k) P(k)$, where ℓ is the angular wavenumber and $\Delta_\ell(k)$ is the radiation transfer function. Because of the singularity of the transfer function and the limitations on the data due to effects such as projection, cosmic variance,

† `http://pole.uchicago.edu/spt/index.php`

instrumental noise, point sources, and etc., a sensitive algorithm is necessary for an accurate reconstruction of the primordial power spectrum from CMB data.

2. PRISM Algorithm

A CMB experiment measures the anisotropies in the CMB temperature $\Theta(\vec{p})$ in direction $\vec{p}$, which is described as $T(\vec{p}) = T_{\rm CMB}[1 + \Theta(\vec{p})]$. This field can be expanded in terms of spherical harmonic functions $Y_{\ell m}$ as $\Theta(\vec{p}) = \sum_\ell \sum_m a_{\ell m} Y_{\ell m}(\vec{p})$, where $a_{\ell m}$ are the spherical harmonic coefficients, that have a Gaussian distribution with zero mean, $\langle a_{\ell m} \rangle = 0$, and variance $\langle a_{\ell m} a^*_{\ell' m'} \rangle = \delta_{\ell\ell'}\delta_{mm'}C_\ell^{\rm th}$. In practice, we are restricted by cosmic variance, additive instrumental noise and also, due to the different Galactic foregrounds, we need to mask the observed CMB map, which induces correlations between different modes. Taking these effects into account and following the MASTER method from Hivon *et al.* (2002), the pseudo power spectrum $\widetilde{C}_\ell$ and the empirical power spectrum $\widehat{C}_\ell^{\rm th}$, which is defined as $\widehat{C}_\ell^{\rm th} = 1/(2\ell+1)\sum_m |a_{\ell m}|^2$, can be related through their ensemble averages $\langle \widetilde{C}_\ell \rangle = \sum_{\ell'} M_{\ell\ell'} \langle \widehat{C}_{\ell'}^{\rm th} \rangle + \langle \widetilde{N}_\ell \rangle$, where $M_{\ell\ell'}$ describes the mode-mode coupling due to the mask. We note that in this expression $\langle \widehat{C}_{\ell'}^{\rm th} \rangle = C_{\ell'}^{\rm th}$, and we set $C_\ell = \langle \widetilde{C}_\ell \rangle$ and $N_\ell = \langle \widetilde{N}_\ell \rangle$, where C_ℓ and N_ℓ refer to the CMB and the noise spectra of the masked maps, respectively. We assume the pseudo spectrum $\widetilde{C}_\ell$ follows a χ^2 distribution with $2\ell+1$ degrees of freedom and can be modelled as $\widetilde{C}_\ell = \left(\sum_{\ell'} M_{\ell\ell'} C_{\ell'}^{\rm th} + N_\ell\right) Z_\ell$, where Z_ℓ is a random variable representing a multiplicative noise.

The relation between the discretized primordial spectrum P_k and the pseudo spectrum $\widetilde{C}_\ell$, computed on a masked noisy map of the sky, can be condensed in the following form

$$\widetilde{C}_\ell = \left(\sum_{\ell' k} M_{\ell\ell'} T_{\ell' k} P_k + N_\ell \right) Z_\ell \, , \tag{2.1}$$

where $T_{\ell k} = 4\pi\Delta \ln k\, \Delta_{\ell k}^2$. Because of the non-invertibility of $\mathbf{T}$, recovering P_k constitutes an ill-posed inverse problem. This inverse problem can be regularised in a robust way by using the sparse nature of the reconstructed signal as a prior. Furthermore, sparse recovery has already been successfully used in the TOUSI algorithm (Paykari *et al.* 2012) to handle the multiplicative noise term, where after the stabilisation, the noise can be treated as an additive Gaussian noise with zero mean and unit variance. The sparse regularisation framework means that if P_k can be sparsely represented in an adapted dictionary $\mathbf{\Phi}$, then this problem, known as the basis pursuit denoising BPDN, can be recast as an optimisation problem, formulated as

$$\min_X \frac{1}{2} \parallel C_\ell - (\mathbf{MT}X + N_\ell) \parallel_2^2 + \lambda \parallel \mathbf{\Phi}^t X \parallel_0 \, , \tag{2.2}$$

where X is the reconstructed estimate for P_k. The first term imposes a ℓ_2 fidelity constraint to the data while the second term promotes the sparsity constraint of the solution in dictionary $\mathbf{\Phi}$, by tuning λ. The ℓ_0 optimisation problem in the second term cannot be solved directly, therefore, we estimate the solution by solving a sequence of relaxed problems using a re-weighted ℓ_1 minimisation technique (Candes 2007)

$$\min_X \frac{1}{2} \parallel \frac{1}{\sigma_\ell} \overline{R}_\ell(X) \parallel_2^2 + K \sum_i \lambda_i |[\mathbf{W\Phi}^t X]_i| \, , \tag{2.3}$$

where $\mathbf{W}$ is a diagonal matrix applying a different weight for each wavelet coefficient, $\overline{R}_\ell(X)$ is our estimate of the residual in the fidelity term in Eq. (2.2) and σ_ℓ is the noise on the residual. Our specific choice of the λ_i has allowed us to use a single regularisation

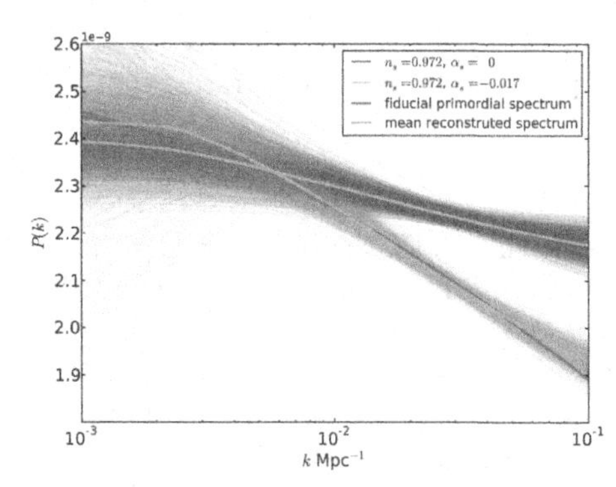

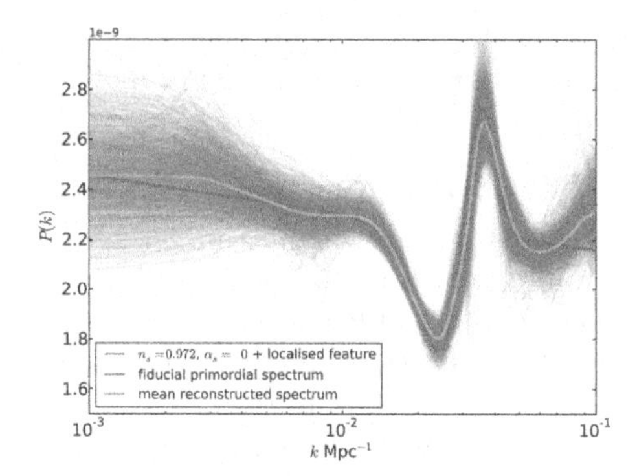

Figure 1. Reconstructions for three different types of primordial spectra: near scale-invariant spectrum, spectrum with a small running, and spectrum with a localised feature. Left: blue lines show the 2000 reconstructed spectra with $n_s = 0.972$ and $\alpha_s = 0$ and the cyan lines show the 2000 reconstructions for $n_s = 0.972$ and $\alpha_s = -0.017$. As can be seen, for $k > 0.015$ Mpc^{-1} PRISM can reconstruct the primordial spectra with such accuracy that the two are easily distinguishable. Right: the 2000 reconstructions of a spectrum with $n_s = 0.972$, $\alpha_s = 0.0$, and an additional feature around $k = 0.03$ Mpc^{-1} is shown in green. As can be seen, PRISM is able to recover both the position and the amplitude of the feature with great accuracy. In all cases the mean of the reconstructions is shown in orange and the fiducial input spectrum in red.

parameter K which translates into a significance level threshold for feature detection. The relaxed problem (2.3) can be solved by ISTA (iterative soft-thresholding algorithm)

$$\widetilde{X}^{n+1} = X^n + \mu \mathbf{T}^t \mathbf{M}^t \frac{1}{\sigma_\ell} \overline{R}_\ell(X^n) , \tag{2.4}$$

$$X^{n+1} = \text{prox}_{K\mu \|\lambda \odot W \Phi^t \cdot\|_1} \left(\widetilde{X}^{n+1} \right) , \tag{2.5}$$

where μ is an adapted step size and $\text{prox}_{K\mu\|\lambda\odot W\Phi^t\cdot\|_1}$ is the proximal operator corresponding to the sparsity constraint. Full details of PRISM is in Paykari *et al.* (2014).

3. Results and Conclusion

We apply PRISM to 2000 simulated pseudo spectra for three different types of P_k. The reconstructions are shown in Fig. 1 with the details being explained in the caption. We also apply PRISM to the WMAP nine-year and Planck PR1 LGMCA CMB pseudo spectra (Bobin *et al.* 2014) and the results are shown in Fig. 2. We have not detected any significant deviations, whether local or global, from a scale-invariant primordial spectrum.

The reconstruction of the primordial spectrum is limited by different effects on different scales. On very large scales, there are fundamental physical limitations due to the cosmic variance and the more severe geometrical projection of the modes, meaning the primordial spectrum cannot be fully recovered on these scales, even in a perfect CMB measurement. On small scales we are limited by the instrumental noise, point sources, beam uncertainties, and etc.. This leaves us with a window through which we can recover the primordial spectrum with a good accuracy. Nevertheless, as can be seen in the left plot of Fig. 1, for $k > 0.015$ Mpc^{-1}, PRISM can easily distinguish the two very similar spectra and performs very well in reconstructing the position and amplitude of the featured spectra, as shown on the right hand plot.

To conclude, the PRISM algorithm uses the sparsity of the primordial spectrum as well as an adapted modelling of the noise of the CMB spectrum to recover the primordial spectrum. This algorithm assumes no prior shape for the primordial spectrum and does

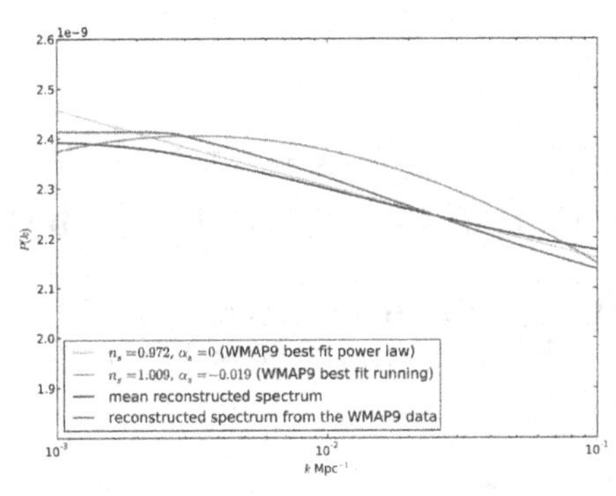

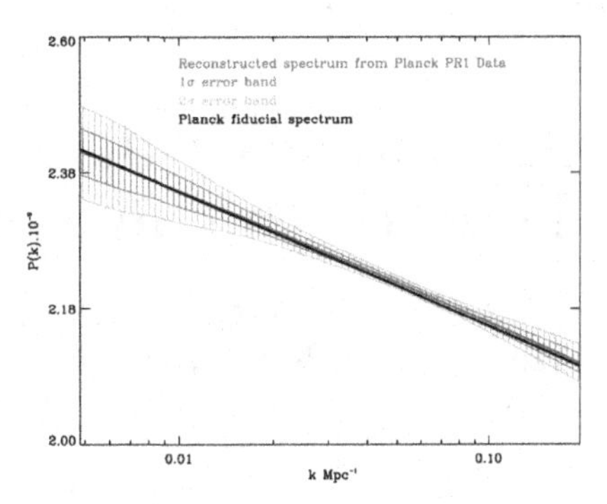

Figure 2. Left: Reconstruction of the primordial spectrum from the WMAP nine-year spectrum is shown in red. The mean of the reconstructions for $n_s = 0.972$ and $\alpha_s = 0$ is shown in solid dark blue line, with the 1σ interval around the mean shown as a shaded blue region. The WMAP nine-year fiducial primordial spectrum with $n_s = 0.972$ and $\alpha_s = 0$ is shown in yellow and in cyan we show the best-fit primordial spectrum with a running with $n_s = 1.009$ and $\alpha_s = -0.019$. Right: Reconstructed primordial spectrum from Planck PR1 data is shown in red. The 1σ and 2σ errors are also shown as green and yellow bands respectively. The fiducial primordial power spectrum with $n_s = 0.9626$ is shown in black solid line. We note that the error bands do not include the errors due to point sources and beam uncertainties. We have not detected any significant deviations, whether local or global, from a scale-invariant spectrum.

not require a coarse binning of the primordial spectrum, making it sensitive to both global smooth features (e.g. running of the spectral index) as well as local sharp features (e.g. a bump or an oscillatory feature). Another advantage of this method is that the regularisation parameter can be specified in terms of a signal-to-noise significance level for feature detection. These advantages make this technique very suitable for investigating different types of departures from scale invariance, whether it is the running of the spectral index or some localised sharp features as predicted by some of the inflationary models. We have applied PRISM to LGMCA WMAP nine-year and Planck PR1 spectra and have not detected any significant deviations from a scale-invariant power spectrum, whether local or global such as a running of the spectral index.

The developed C++ and IDL codes are released in iSAP (Interactive Sparse astronomical data Analysis Packages) via the web site `http://cosmostat.org/isap.html`.

References

Bobin, J., Sureau, F., Starck, J.-L., Rassat, A., & Paykari, P. 2014, *AA*, 563, A105
Candes, E. J., Wakin, M. B., & Boyd, S. P. 2008, *JFAA*, 14(5):877
Guth, A. H. 1981, *Phys. Rev. D*, 23, 347
Hivon, E., G?orski, K. M., Netterfield, C. B., *et al.* 2002, *ApJ*, 567, 2
Hou, Z., Reichardt, C. L., Story, K. T., *et al.* 2012, *ArXiv e-prints*
Linde, A. D. 1982, *Physics Letters B*, 108, 389
Paykari, P., Lanusse, F., Starck, J.-L., Sureau, F., & Bobin, J. 2014, *AA*, 566, A77
Paykari, P., Starck, J.-L., & Fadili, M. J. 2012, *AA*, 541, A74
Planck Collaboration, Ade, P. A. R., Aghanim, N., *et al.* 2013a, *ArXiv e-prints*
Planck Collaboration, Ade, P. A. R., Aghanim, N., *et al.* 2013b, *ArXiv e-prints*

Statistical Challenges in 21st Century Cosmology
Proceedings IAU Symposium No. 306, 2014
A. F. Heavens, J. -L. Starck & A. Krone-Martins, eds.

doi:10.1017/S1743921314011107

On spin scale-discretised wavelets on the sphere for the analysis of CMB polarisation

Jason D. McEwen[1], Martin Büttner[2], Boris Leistedt[2], Hiranya V. Peiris[2], Pierre Vandergheynst[3] and Yves Wiaux[4]

[1]Mullard Space Science Laboratory (MSSL), University College London (UCL), Surrey RH5 6NT, UK
email: jason.mcewen@ucl.ac.uk

[2]Department of Physics and Astronomy, University College London (UCL), London WC1E 6BT, UK
email: {martin.buettner.11, boris.leistedt.11, h.peiris}@ucl.ac.uk

[3]Institute of Electrical Engineering, Ecole Polytechnique Fédérale de Lausanne (EPFL), CH-1015 Lausanne, Switzerland
email: pierre.vandergheynst@epfl.ch

[4]Institute of Sensors, Signals & Systems, Heriot-Watt University, Edinburgh EH14 4AS, UK
email: y.wiaux@hw.ac.uk

Abstract. A new spin wavelet transform on the sphere is proposed to analyse the polarisation of the cosmic microwave background (CMB), a spin ± 2 signal observed on the celestial sphere. The scalar directional scale-discretised wavelet transform on the sphere is extended to analyse signals of arbitrary spin. The resulting spin scale-discretised wavelet transform probes the directional intensity of spin signals. A procedure is presented using this new spin wavelet transform to recover E- and B-mode signals from partial-sky observations of CMB polarisation.

Keywords. cosmology: cosmic microwave background, cosmology: observations, cosmology: early universe, methods: data analysis, techniques: image processing

1. Introduction

The polarisation of the cosmic microwave background (CMB) is a powerful probe of the physics of inflation (Spergel & Zaldarriaga 1997) and the reionisation history of the Universe (Zaldarriaga 1997). Numerous experiments have now measured CMB polarisation (some of the more recent include: Hanson *et al.* 2013; Naess *et al.* 2014; BICEP2 Collaboration 2014). Although the Planck satellite also measured CMB polarisation, polarisation data were not included in the Planck 2013 release (Planck Collaboration I 2013) but are anticipated later this year.

Since different physical processes often exhibit different symmetries, their signatures in observables like CMB polarisation may behave differently under a parity transform. CMB polarisation can be separated into parity even and parity odd components, so called E- and B-mode components, respectively (Zaldarriaga & Seljak 1997). Density perturbations in the early Universe provide no mechanism to generate B-mode polarisation in the CMB, whereas gravitational waves can induce both E- and B-mode components. The detection of primordial B-mode polarisation would thus provide evidence for gravitational waves and would provide a powerful probe of the physics of inflation.

In these proceedings we outline a new spin wavelet transform on the sphere to analyse observations of CMB polarisation, a spin ± 2 signal observed on the celestial sphere. In addition, we describe a simple technique based on this wavelet framework to separate

E- and B-mode CMB polarisation components from partial-sky observations. We present a preliminary discussion only; further details of these methods, fast implementations, and a rigorous evaluation of their performance will be given in a series of forthcoming articles.

2. Spin scale-discretised wavelets on the sphere

Scalar wavelets on the sphere (e.g. Antoine & Vandergheynst 1998, 1999; Baldi *et al.* 2009; McEwen *et al.* 2006a, 2007a, 2013; Marinucci *et al.* 2008; Narcowich *et al.* 2006; Starck *et al.* 2006; Wiaux *et al.* 2005, 2006, 2008; Leistedt *et al.* 2013) have proved an effective tool for analysing the temperature anisotropies of the CMB (e.g. Vielva *et al.* 2004; Vielva *et al.* 2006; McEwen *et al.* 2005, 2006b, 2008a, 2006c, 2007b, 2008b; Pietrobon *et al.* 2006; Faÿ *et al.* 2008; Feeney *et al.* 2011a,b; Bobin *et al.* 2013; Planck Collaboration XII 2013; Planck Collaboration XXIII 2013; Planck Collaboration XXIV 2014; Planck Collaboration XXV 2013). For a somewhat dated review see McEwen *et al.* (2007c). Spin wavelets to analyse the polarisation of the CMB have been constructed by Geller *et al.* (2008) and Starck *et al.* (2009). However, a spin wavelet transform on the sphere capable of probing the directional intensity of signals does not yet exist.† We propose such a transform here by extending the directional scale-discretised wavelet transform of Wiaux *et al.* (2008) to signals of arbitrary spin on the sphere.

Spin scale-discretised wavelets ${}_s\Psi^{(j)} \in \mathrm{L}^2(\mathbb{S}^2)$ can be constructed on the sphere $\mathbb{S}^2$ in an analogous manner to the scalar wavelet construction (Wiaux *et al.* 2008; Leistedt *et al.* 2013; McEwen *et al.* 2013), that is simply by defining the spin harmonic coefficients of the wavelets in the factorised form:

$$ {}_s\Psi^{(j)}_{\ell m} \equiv \sqrt{\frac{2\ell+1}{8\pi^2}}\, \kappa^{(j)}(\ell)\, \zeta_{\ell m} \,, \tag{2.1} $$

where ${}_s\Psi^{(j)}_{\ell m} = \langle {}_s\Psi^{(j)},\, {}_sY_{\ell m}\rangle$ are the spin $s \in \mathbb{Z}$ spherical harmonic coefficients of the wavelets, with ${}_sY_{\ell m}$ denoting the spherical harmonic functions and $\ell \in \mathbb{N}_0$, $m \in \mathbb{Z}$, such that $|s| \leqslant \ell$ and $|m| \leqslant \ell$. The *kernel* $\kappa^{(j)} \in \mathrm{L}^2(\mathbb{R}^+)$ controls the angular localisation of the wavelets, while their directional properties are controlled by the *directionality component* $\zeta \in \mathrm{L}^2(\mathbb{S}^2)$, with harmonic coefficients $\zeta_{\ell m} = \langle \zeta,\, Y_{\ell m}\rangle$. The wavelet scale $j \in \mathbb{N}_0$ encodes the angular localisation of $\Psi^{(j)}$. The kernel and directionality component are defined as in the scalar setting (Wiaux *et al.* 2008; McEwen *et al.* 2013).

The wavelet transform of a spin signal ${}_sf \in \mathrm{L}^2(\mathbb{S}^2)$ on the sphere is defined by the directional convolution of ${}_sf$ with the wavelet ${}_s\Psi^{(j)} \in \mathrm{L}^2(\mathbb{S}^2)$. The wavelet coefficients $W^{{}_s\Psi^{(j)}} \in \mathrm{L}^2(\mathrm{SO}(3))$ thus read

$$ W^{{}_s\Psi^{(j)}}(\rho) \equiv ({}_sf \star {}_s\Psi^{(j)})(\rho) \equiv \langle {}_sf,\, \mathcal{R}_\rho\, {}_s\Psi^{(j)}\rangle = \int_{\mathbb{S}^2} \mathrm{d}\Omega(\omega)\, {}_sf(\omega) (\mathcal{R}_\rho\, {}_s\Psi^{(j)})^*(\omega) \,, \tag{2.2} $$

where $\omega = (\theta, \varphi) \in \mathbb{S}^2$ denotes spherical coordinates with colatitude $\theta \in [0, \pi]$ and longitude $\varphi \in [0, 2\pi)$, $\mathrm{d}\Omega(\omega) = \sin\theta\, \mathrm{d}\theta\, \mathrm{d}\varphi$ is the usual rotation invariant measure on the sphere, and $\cdot^*$ denotes complex conjugation. The rotation operator is defined by

$$ \left(\mathcal{R}_\rho\, {}_s\Psi^{(j)}\right)(\omega) \equiv {}_s\Psi^{(j)}(\mathbf{R}_\rho^{-1} \cdot \omega) \,, \tag{2.3} $$

where $\mathbf{R}_\rho$ is the three-dimensional rotation matrix corresponding to $\mathcal{R}_\rho$. Rotations are

† Spin curvelets (Starck *et al.* 2009) could be used for a directional analysis however these are constructed on the base pixels of Healpix (Górski *et al.* 2005) and so do not live naturally on the sphere.

specified by elements of the rotation group SO(3), parameterised by the Euler angles $\rho = (\alpha, \beta, \gamma) \in \mathrm{SO}(3)$, with $\alpha \in [0, 2\pi)$, $\beta \in [0, \pi]$ and $\gamma \in [0, 2\pi)$. Note that the wavelet coefficients are a scalar signal defined on the rotation group SO(3). The wavelet transform of Eqn. (2.2) thus probes the directional intensity of the signal of interest ${}_sf$.

Provided the wavelets satisfy an admissibility property analogous to the scalar setting, the original signal can be synthesised exactly from its wavelet coefficients by

$$ {}_sf(\omega) = \sum_{j=J_0}^{J} \int_{\mathrm{SO}(3)} \mathrm{d}\varrho(\rho) W^{{}_s\Psi^j}(\rho) (\mathcal{R}_\rho \, {}_s\Psi^j)(\omega) \, , \tag{2.4} $$

where $\mathrm{d}\varrho(\rho) = \sin\beta \, \mathrm{d}\alpha \, \mathrm{d}\beta \, \mathrm{d}\gamma$ is the usual invariant measure on SO(3) and J_0 and J are the minimum and maximum wavelet scales considered, respectively, i.e. $J_0 \leqslant j \leqslant J$. Throughout this description we have neglected to include a scaling function, which must be introduced to capture the low-frequency content of the analysed signal ${}_sf$.

3. E- and B-mode separation

CMB experiments measure the scalar Stoke parameters $I, Q, U \in \mathrm{L}^2(\mathbb{S}^2)$, where I encodes the intensity and Q and U the linear polarisation of the incident CMB radiation (the circular polarisation component of the four Stokes parameters $V \in \mathrm{L}^2(\mathbb{S}^2)$ is zero). The linear polarisation signal that is observed depends on the choice of local coordinate frame. The component $Q \pm \mathrm{i}U$ transforms under a rotation of the local coordinate frame by $\chi \in [0, 2\pi)$ as $(Q \pm \mathrm{i}U)'(\omega) = \exp(\mp \mathrm{i}2\chi)(Q \pm \mathrm{i}U)(\omega)$ and is thus a spin ± 2 signal on the sphere (Zaldarriaga & Seljak 1997). The quantity $Q \pm \mathrm{i}U$ can be decomposed into parity even and odd components by $\tilde{E}(\omega) = -\frac{1}{2}\left[\bar{\eth}^2(Q+\mathrm{i}U)(\omega) + \eth^2(Q-\mathrm{i}U)(\omega)\right]$ and $\tilde{B}(\omega) = \frac{\mathrm{i}}{2}\left[\bar{\eth}^2(Q+\mathrm{i}U)(\omega) - \eth^2(Q-\mathrm{i}U)(\omega)\right]$ respectively, where $\tilde{E}, \tilde{B} \in \mathrm{L}^2(\mathbb{S}^2)$ and $\eth$ and $\bar{\eth}$ are spin raising and lowering operators, respectively (Zaldarriaga & Seljak 1997). Recovering E- and B-modes from full-sky observations is relatively straightforward, however in practice we observe the CMB over only part of the sky, since microwave emissions from our Galaxy obscure our view. A number of techniques have been developed to recover E- and B-modes from Q and U maps observed on the partial-sky (e.g. Lewis *et al.* 2002; Bunn *et al.* 2003; Kim 2011; Bowyer *et al.* 2011). Here we propose a simple alternative approach using the spin scale-discretised wavelet transform described above (a similar approach using needlets has been proposed by Geller *et al.* (2008), however there are some minor differences since spin needlets yield spin and not scalar wavelet coefficients).

First, consider the wavelet coefficients of the observable $Q + \mathrm{i}U$ signal computed by a *spin* wavelet transform: $W^{{}_2\Psi^{(j)}}_{Q+\mathrm{i}U}(\rho) \equiv \langle Q + \mathrm{i}U, \, \mathcal{R}_\rho \, {}_2\Psi^{(j)} \rangle$. Second, consider the wavelet coefficients of the unobservable $\tilde{E}$ and $\tilde{B}$ signals computed by a *scalar* wavelet transform: $W^{{}_0\tilde{\Psi}^j}_{\tilde{E}}(\rho) \equiv \langle \tilde{E}, \, \mathcal{R}_\rho \, {}_0\tilde{\Psi}^j \rangle$ and $W^{{}_0\tilde{\Psi}^j}_{\tilde{B}}(\rho) \equiv \langle \tilde{B}, \, \mathcal{R}_\rho \, {}_0\tilde{\Psi}^j \rangle$. If the wavelet used in the scalar wavelet transform is a spin lowered version of the wavelet used in the spin wavelet transform, i.e. ${}_0\tilde{\Psi}^j = \bar{\eth}^2 \, {}_2\Psi^j$, then the wavelet coefficients of $\tilde{E}$ and $\tilde{B}$ are simply related to the wavelet coefficients of $Q + \mathrm{i}U$ by $W^{{}_0\tilde{\Psi}^j}_{\tilde{E}}(\rho) = -\mathrm{Re}\left[W^{{}_2\Psi^j}_{Q+\mathrm{i}U}(\rho)\right]$ and $W^{{}_0\tilde{\Psi}^j}_{\tilde{B}}(\rho) = -\mathrm{Im}\left[W^{{}_2\Psi^j}_{Q+\mathrm{i}U}(\rho)\right]$, respectively.

This leads to an elegant procedure to recover E- and B-modes from Q and U maps observed over the partial-sky. Firstly, compute the spin wavelet transform of $Q + \mathrm{i}U$. Secondly, mitigate the impact of the partial sky coverage in wavelet space, where signal content (and thus the influence of the mask) is localised in scale and position simultaneously. Thirdly, reconstruct $\tilde{E}$ and $\tilde{B}$ maps by inverse scalar wavelet transforms of the real and imaginary components, respectively, of the processed spin wavelet coefficients.

References

Antoine & Vandergheynst 1998, *J. Math. Phys.*, 39, 8, 3987

Antoine & Vandergheynst 1999, *ACHA*, 7, 1

Baldi, Kerkyacharian, Marinucci, & Picard 2009, *Ann. Stat.*, 37 No.3, 1150, arXiv:math/0606599

BICEP2 Collaboration 2014, submitted, arXiv:1403.3985

Bobin, Starck, Sureau, & Basak 2013, *A&A*, 550, A73,arXiv:1206.1773

Bowyer, Jaffe, & Novikov 2011, arXiv, 1009, arXiv:1101.0520

Bunn, Zaldarriaga, Tegmark, & de Oliveira-Costa 2003, *PRD*, 67, 2, 023501, astro-ph/0207338

Faÿ, Guilloux, Betoule, Cardoso, Delabrouille, & Le Jeune 2008, *PRD*, 78, 8, 083013, arXiv:0807.1113

Feeney, Johnson, Mortlock, & Peiris 2011a, *Phys. Rev. Lett.*, 107, 071301, 1012.1995

Feeney, Johnson, Mortlock, & Peiris 2011b, *PRD*, D84, 043507, 1012.3667

Geller, Hansen, Marinucci, Kerkyacharian, & Picard 2008, *PRD*, 78, 12, 123533, arXiv:0811.2881

Górski, Hivon, Banday, Wandelt, Hansen, Reinecke, & Bartelmann 2005, *ApJ*, 622, 759, astro-ph/0409513

Hanson, *et al.* 2013, *PRL*, 111, 14, 141301, arXiv:1307.5830

Kim 2011, *A&A*, 531, A32, arXiv:1010.2636

Leistedt, McEwen, Vandergheynst, & Wiaux 2013, *A&A*, 558, A128, 1, arXiv:1211.1680

Lewis & Challinor, Turok 2002, *PRD*, 65, 2, 023505, astro-ph/0106536

Marinucci, *et al.* 2008, *MNRAS*, 383, 539, arXiv:0707.0844

McEwen, Hobson, & Lasenby 2006a, ArXiv, astro-ph/0609159

McEwen, Hobson, Lasenby, Mortlock, 2005, *MNRAS*, 359, 1583, astro-ph/0406604

McEwen, Hobson, Lasenby, & Mortlock 2006b, *MNRAS*, 371, L50, astro-ph/0604305

McEwen, Hobson, Lasenby, & Mortlock 2006c, *MNRAS*, 369 , 1858, astro-ph/0510349

McEwen, Hobson, Lasenby, & Mortlock 2008a, *MNRAS*, 388, 2, 659, arXiv:0803.2157

McEwen, Hobson, Mortlock, & Lasenby 2007a, *IEEE TSP*, 55, 2, 520, astro-ph/0506308

McEwen, Vandergheynst, & Wiaux 2013, in *SPIE Wavelets and Sparsity XV*, arXiv:1308.5706

McEwen, Vielva, Hobson, Martínez-González, & Lasenby 2007b, *MNRAS*, 373, 1211, astro-ph/0602398

McEwen, Wiaux, Hobson, Vandergheynst, & Lasenby 2008b, *MNRAS*, 384, 4, 1289, arXiv:0704.0626

McEwen, *et al.* 2007c, *JFAA*, 13, 4, 495, arXiv:0704.3158

Naess, *et al.* 2014, ArXiv e-prints, arXiv:1405.5524

Narcowich, Petrushev, & Ward 2006, *SIAM J. Math. Anal.*, 38, 2, 574

Pietrobon, Balbi, & Marinucci 2006, *PRD*, 74, 4, 043524, astro-ph/0606475

Planck Collaboration I 2013, *A&A*, in press, arXiv:1303.5062

Planck Collaboration XII 2013, *A&A*, in press, arXiv:1303.5072

Planck Collaboration XXIII 2013, *A&A*, in press, arXiv:1303.5083

Planck Collaboration XXIV 2014, *A&A*, in press, arXiv:1303.5084

Planck Collaboration XXV 2013, *A&A*, in press, arXiv:1303.5085

Spergel & Zaldarriaga 1997, *PRL*, 79, 2180, astro-ph/9705182

Starck, Moudden, & Bobin 2009, *A&A*, 497, 931, arXiv:0902.0574

Starck, Moudden, Abrial, & Nguyen 2006, *A&A*, 446, 1191, astro-ph/0509883

Vielva, Martínez-González, Barreiro, Sanz & Cayón 2004, *ApJ*, 609, 22, astro-ph/0310273

Vielva, Wiaux, Martínez-González, & Vandergheynst 2006, *New Astronomy Review*, 50, 880, astro-ph/0609147

Wiaux, Jacques & Vandergheynst 2005, *ApJ*, 632, 15, astro-ph/0502486

Wiaux, Jacques, Vielva, & Vandergheynst 2006, *ApJ*, 652, 820, astro-ph/0508516

Wiaux, McEwen, Vandergheynst, & Blanc 2008, *MNRAS*, 388, 2, 770, arXiv:0712.3519

Zaldarriaga 1997, *PRD*, 55, 1822, astro-ph/9608050

Zaldarriaga & Seljak 1997, *PRD*, 55, 4, 1830, astro-ph/9609170

Statistical Challenges in 21st Century Cosmology
Proceedings IAU Symposium No. 306, 2014
A. F. Heavens, J.-L. Starck & A. Krone-Martins, eds.

doi:10.1017/S1743921314013568

Estimating the distribution of Galaxy Morphologies on a continuous space

Giuseppe Vinci[1], Peter Freeman[1], Jeffrey Newman[2], Larry Wasserman[1] and Christopher Genovese[1]

[1] Dept. of Statistics, Baker Hall, Carnegie Mellon University,
5000 Forbes Avenue, Pittsburgh, PA 15213, USA
email: gvinci@andrew.cmu.edu

[2] Dept. of Physics & Astronomy, University of Pittsburgh,
310 Allen Hall 3941 O'Hara St., Pittsburgh, PA 15260, USA

Abstract. The incredible variety of galaxy shapes cannot be summarized by human defined discrete classes of shapes without causing a possibly large loss of information. Dictionary learning and sparse coding allow us to reduce the high dimensional space of shapes into a manageable low dimensional continuous vector space. Statistical inference can be done in the reduced space via probability distribution estimation and manifold estimation.

Keywords. dictionary learning, manifold estimation, Radon transform, redshift, sparse coding, galaxies: statistics

1. Introduction

The evolution of the Universe has led to the formation of complex objects apparently without any regular shape, which our mind would just classify as *irregular*. Thus, the incredible variety of galaxy shapes cannot be summarized by human defined discrete classes of shapes (e.g. "Hubble sequence") without causing a possibly large loss of information. Our human concept of shape could limit the complete understanding of the complex structure of the galaxies. Estimating the distribution of galaxy morphologies is one means to test theories of the formation and the evolution of the Universe. We estimate the distribution of morphologies on a *continuous* Euclidean space, such that a particular shape will be viewed as a point in a continuous space. This task must be performed in an *unsupervised* way, i.e. free from any human judgement. Galaxy images are intrinsically high-dimensional data, and we use *dictionary learning* and *sparse coding* [Mairal *et al.* (2010)] to reduce the high dimensional space of shapes into a manageable low dimensional one. Essentially, galaxy images will be approximated by sparse linear combinations of basis pictures, which are *learned* from the data. Statistical inference on the reduced space can be performed via probability distribution estimation. We propose a testing procedure and analyse a dataset of galaxy images† to show some examples.

2. Dictionary Learning and Sparse Coding - Radon Transform

The general idea of dictionary learning and sparse coding is to approximate images by *sparse* linear combinations of a fixed number of *basis images*, which are not predefined,

† GOODS-South Early Release Science Field dataset observed in the near-infrared regime by the Wide Field Camera 3 on-board the Hubble Space Telescope [see Windhorst *et al.* (2011), Freeman *et al.* (2013)].

but are *learned* from the data. Let $x_i \in \mathbb{R}^{a \times b}$ be an image, which has $a \times b$ dimensions. For $m << a \times b$, we want to approximate x_i as:

$$x_i \approx \sum_{j=1}^{m} \alpha_{ij} B_j \tag{2.1}$$

where $\alpha_i = (\alpha_{i1}, ..., \alpha_{im}) \in \mathbb{R}^m$ is a sparse vector of coefficients, and $\{B_j\}_{j=1}^m$ is a collection of basis images $B_j \in \mathbb{R}^{a \times b}$. Notice that the basis images will not be imposed to be orthogonal such that the dictionary can easily adapt to the structure of the data [Mairal *et al.* (2010)]. Moreover, learning the bases from the data was shown to perform better in signal reconstruction with respect to using predefined bases [Elad *et al.* (2006)].

2.1. *Optimization problem*

From a dataset of galaxy images $\{x_i\}_{i=1}^n$, we can estimate the dictionary $D = \{B_j\}_{j=1}^m$ and the vectors of coefficients $A = \{\alpha_i\}_{i=1}^n$ by solving the following optimization problem:

$$\begin{cases} \min\limits_{\{\alpha_i\}_{i=1}^n, \{B_j\}_{j=1}^m} \sum\limits_{i=1}^{n} \left[\frac{1}{2} \left\| x_i - \sum\limits_{j=1}^{m} \alpha_{ij} B_j \right\|_2^2 + \underbrace{\lambda \|\alpha_i\|_1}_{\text{SPARSITY}} \right] \\ \textbf{s.t.} \quad \|B_j\|_2^2 \leqslant 1, \forall j = 1, ..., m \end{cases} \tag{2.2}$$

where $\lambda \geqslant 0$ is a sparsity parameter and $\| * \|_2^2$ is the Frobenius norm [Mairal *et al.* (2010); R package "spams"]. We suggest to choose m and λ via *cross validation.* See Mairal *et al.* (2010) for other configurations of problem (2.2).

2.2. *Standardization of the images. Radon transform*

Before solving problem (2.2), images must be standardized to eliminate any spurious dimensionality and improve the quality of the approximations (2.1). We are talking about: *centring*, *resizing* and *rotation orientation.* While the first one can be easy to perform, the two others are not. Images can be rotated and resized by using Radon Transform (RT) and Inverse RT (IRT). The RT of a function f is $\mathcal{R}_f(t,\theta) = \int_{-\infty}^{\infty} f(t\cos\theta - u\sin\theta, t\sin\theta + u\cos\theta)du$, where $(t,\theta) \in \mathbb{R}^2$. An image can be viewed as the discrete evaluation of a function. The orientation of the texture of an image can be estimated by $\theta^* = \arg\min_\theta \frac{\partial^2 \sigma_\theta^2}{\partial \theta^2}$, where σ_θ^2 is the variance of $\mathcal{R}_f(t,\theta)$ at angle θ [Jafari-Khouzani *et al.* (2005), Arodź (2012); R package "PET"]. Rotating images by angle $-\theta^*$ essentially makes all the pictures *horizontally oriented.* To rotate an image we need to: 1) evaluate its RT on a discrete grid, say $\hat{R}_{M \times (\omega 180+1)} = \{\mathcal{R}_f(t,\theta)\}$ with $t \in \{t_1, ..., t_M\}$, $\theta \in \{\frac{j}{\omega 180}\pi\}_{j=0}^{\omega 180}$, and $\omega \in \mathbb{N}^+$; 2) find θ^* and move the first $k^* = \theta^* \frac{\omega 180}{\pi}$ columns of $\hat{R}$ as described in Figure 1 to get $\tilde{R}$ ("rotation" in the Radon domain); 3) computing the IRT of $\tilde{R}$ on a grid of desired resolution ("resizing"). In Figure 2 we show some effects of images standardization.

3. Statistical inference on the reduced space

In this section we propose a method to estimate the distribution of galaxy morphologies on a low-dimensional space, and we use the GOODS-S dataset to perform a simulation.

3.1. *Probability distribution of galaxy morphologies.*

For a dataset of n images $\{x_i\}_{i=1}^n$, where x_i is a matrix of nonnegative light intensity:

(a) Standardize all the images as described in paragraph 2.2;

(b) Obtain the dictionary D and the vectors $A = \{\alpha_i\}_{i=1}^n$ according to paragraph 2.1;

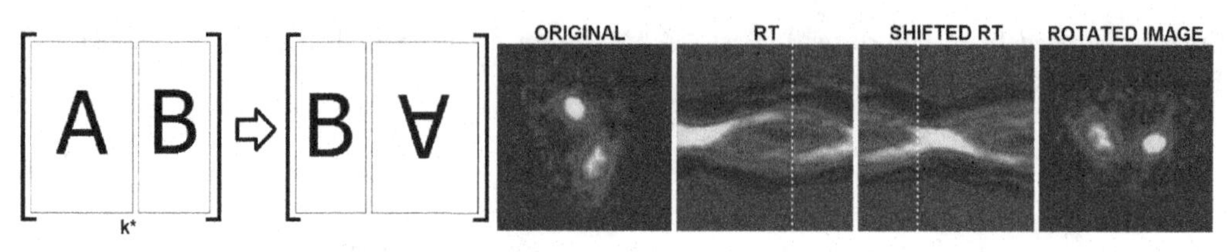

Figure 1. Left: vectors A are moved after vectors B with values moved up and down. Right: starting from an original image, we compute its Radon transform on a discrete grid, then by shifting the vectors of this matrix according to the orientation θ^*, we can obtain a standardized rotated version of the image as the IRT of the shifted RT.

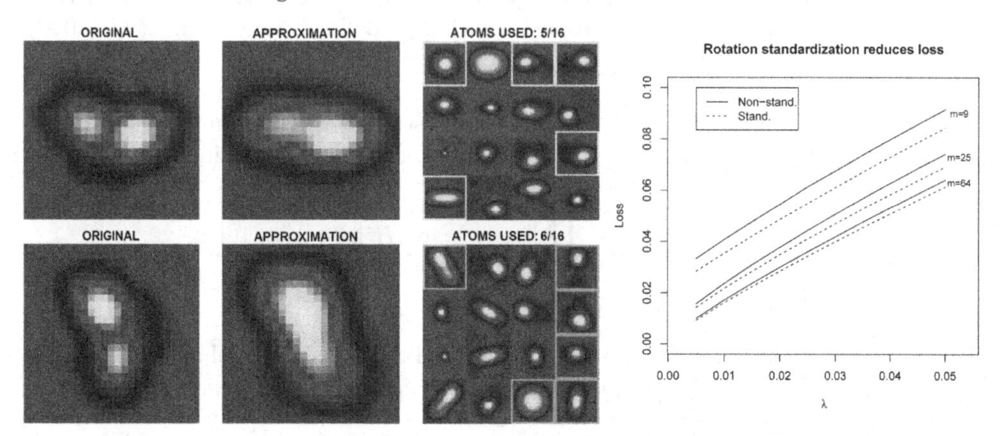

Figure 2. Rotation standardization improves the fit. Left: an image approximated using a dictionary learned with rotation standardization (top) and not (bottom). Spurious dimensionality negatively affects the dictionary at the bottom, while rotation standardization may lead to more refined approximations. Right: for different numbers of atoms ($m = 9, 25, 64$), the minimum loss (2.2) is smaller when using standardized images. Images are from the GOODS-S dataset, H-band.

(c) Estimate the joint distribution of vector $\alpha_i \in \mathbb{R}^m$. Call it $\hat{P}_\alpha$.

Given the fitted dictionary D, estimate $\hat{P}_\alpha$ can be viewed as an approximation of the distribution of galaxy morphologies.

3.2. *Comparing populations of shapes*

In this section we propose a method to compare the distributions of two collections of images. Let X, Y be two collections of images. Suppose we want to test hypothesis $X \overset{\mathcal{D}}{=} Y$, i.e. a distribution test. We propose the following method:

(a) Pool X and Y into a unique dataset $Z = [X, Y]$

(b) From Z, fit dictionary D and vectors of coefficients $\{\alpha_{Z,k}\} = [\{\alpha_{X,i}\}, \{\alpha_{Y,j}\}]$.

(c) Implement a distribution test $\alpha_X \overset{\mathcal{D}}{=} \alpha_Y$.

For step (c), we suggest to use the nonparametric test based on the Maximum Mean Discrepancy (MMD) statistic (Gretton *et al.* (2012); R package "kernlab"). We can call this testing procedure "DSM test" (Dictionary Learning - Sparse Coding - MMD).

3.2.1. *Simulation*

We selected two subsets of images of the GOODS-S dataset in the H-band (see Figure 3): X_1 with 25 images of non-mergers, and X_2 with 25 images of mergers. To generate n images of non-mergers and n images of non-mergers we: 1) randomly sample with replacement n images from X_1 and n images from X_2, respectively; 2) randomly rotate them by angles $\theta \sim$Unif$(0, 2\pi)$, i.i.d.; 3) add heteroscedastic noise: $\epsilon_{jk} \overset{\text{indep}}{\sim} N(0, \beta^2 \times I_{jk})$,

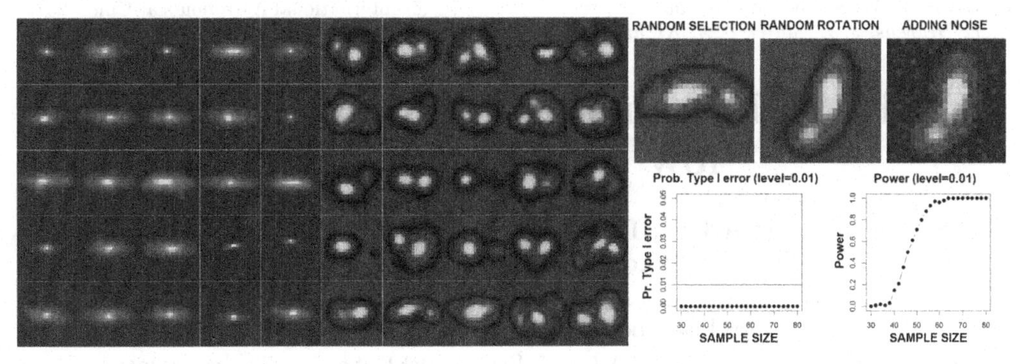

Figure 3. Left: selected non-mergers (X_1) and mergers (X_2) from the GOODS-S dataset, H-band. Top right: procedure to simulate an image from X_i. An image is randomly selected from the subset, randomly rotated and heteroscedastic Gaussian noise is added to each pixel. Bottom right: the DSM test helps to distinguish different shapes. The probability of Type I error of the DSM test is always smaller than the level of the test; the power of the test is increasing in the sample size. The shape of the power function depends on the original sets X_1, X_2.

where $I_{jk} \geqslant 0$ is the light intensity at position jk in a matrix. We repeat comparisons (via DSM test) of samples of the same kind (Mer Vs Mer, NMer Vs NMer) and different one (Mer Vs NMer) to estimate the probability of Type I error and the power of the test as functions of the sample size (see Figure 3). We chose $m = 4$ and $\lambda = 0.05$ via 10-CV.

4. Conclusions and future work

An unsupervised analysis based on dictionary learning and sparse coding allows us to approximate the distribution of galaxy morphologies by a multivariate distribution defined on a subset of $\mathbb{R}^m$, where dimension m is much smaller than the dimension of a galaxy image. Hypothesis testing on the reduced space can help to distinguish the distributions of two sets of images. Current and future work is: using dictionary learning and sparse coding to put constraints on the parameters of cosmological models; comparing the distribution of galaxy shapes at different redshift ranges; manifold estimation: some clusters may correspond to some human defined shapes (e.g. spiral, elliptical) and filaments [see Chen *et al.* (2013)] may describe the transition from a shape to another one; analysing images of other astronomical objects and 3D images.

References

Arodź, T. 2012, *Computing and Informatics*, 24 no. 2 (2012): 183-199.

Chen, Y.-C., Genovese, C. R., & Wasserman, L. 2013, *arXiv preprint* arXiv:1312.2098 (2013).

Elad, M. & Aharon, M. 2006, *Image Processing, IEEE Transactions on* 15, no. 12 (2006): 3736-3745.

Freeman, P. E., R. Izbicki, A. B. Lee, J. A. Newman *et al.* 2013, *MNRAS* (2013): stt1016.

Gretton, A., Sejdinovic, D., Strathmann, H., Balakrishnan, S., Pontil, M., Fukumizu, K., & Sriperumbudur, B. K. 2012, *Advances in neural information processing systems*, pp. 1205-1213. 2012.

Jafari-Khouzani, K. & Soltanian-Zadeh , H. 2005, *Pattern Analysis and Machine Intelligence*, IEEE Transactions on 27, no. 6 (2005): 1004-1008.

Mairal, J., Bach, F., Ponce, J., & Sapiro, G. 2010, *The Journal of Machine Learning Research* 11 (2010): 19-60.

Windhorst, R. A., Cohen, S. H., Hathi, N. P., McCarthy, P. J. *et al.* 2011, *ApJS* 193, no. 2 (2011): 27.

Statistical Challenges in 21st Century Cosmology
Proceedings IAU Symposium No. 306, 2014
A. F. Heavens, J.-L. Starck & A. Krone-Martins, eds.

doi:10.1017/S1743921314013581

Darth Fader: Analysing galaxy spectra at low signal-to-noise

Adrienne Leonard*[1,2], Daniel P. Machado[2], Filipe B. Abdalla[1] and Jean-Luc Starck[2]

[1]Department of Physics and Astronomy, University College London
Gower Place, London WC1E 6BT, United Kingdom

[2]Laboratoire AIM, UMR CEA-CNRS-Paris 7, IRFU, Service d'Astrophysique
CEA Saclay, F-91191 Gif-sur-Yvette CEDEX, France

*email: adrienne.leonard@ucl.ac.uk

Abstract. Spectroscopic redshift surveys are an incredibly valuable tool in cosmology, allowing us to trace the distribution of galaxies as a function of distance and, thus, trace the evolution of structure formation in the Universe. However, estimating the redshifts from spectra with low signal-to-noise is difficult, and such data are often either discarded or require human classification of spectral lines to obtain the galaxy redshift. Darth Fader offers an automated method for estimating the redshifts of galaxies in the low signal-to-noise regime. Using a sophisticated, wavelet-based technique, galaxy spectra can be separated into continuum, line and noise components, and the lines can then be cross-correlated with template spectra in order to estimate the redshifts. Cross-matching of the identified lines then allows for a cleaning of the resulting catalogue, effectively removing the vast majority of erroneous redshift estimates and resulting in a highly pure, highly accurate redshift catalogue. Darth Fader allows us to effectively use low signal-to-noise galaxy spectra, and dramatically reduces the number of human hours required to do this, allowing spectroscopic surveys to probe deeper into the formation history of the Universe.

Keywords. galaxies: distances and redshifts, methods: data analysis, techniques: spectroscopic surveys

---

1. Introduction

Automated spectroscopic redshift estimation is typically carried out using either modelling, template matching, or cross-correlation techniques. In Darth Fader (Machado *et al.* 2013), we employ a cross-correlation method (see, e.g., Glazebrook *et al.* 1998). We assume that any galaxy spectrum can be represented as a linear combination of template spectra:

$$S_\lambda = \sum_i a_i T_{i\lambda} \tag{1.1}$$

where the index λ runs over the wavelengths sampled by the spectrograph, S_λ is the true spectrum of the galaxy and $T_{i\lambda}$ is a representative set of template spectra.

If the templates and galaxy spectra are binned on a logarithmic wavelength axis, a shift along this axis is directly proportional to $\log(1+z)$. The redshift can then be obtained by considering the cross-correlation between the galaxy spectra and the templates. However, cross-correlation methods require the templates and galaxy spectra to be continuum free. Templates can be obtained either from simulations or high signal-to-noise data. We use a principal component analysis to reduce the dimensionality of the problem, and to extract the important features of the template spectra.

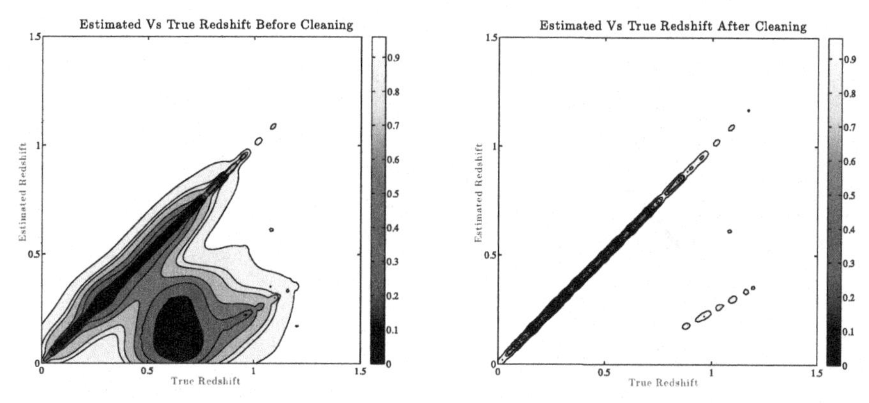

Figure 1. The distribution of estimated vs. true redshift for simulated galaxy spectra at a signal-to-noise of 2 in the r-band. The left panel shows the distribution before cleaning, where we obtain a success rate of 65.5%, while the right panel shows the distribution after cleaning, where we have retained 76.2% of the galaxy spectra and reached a success rate of 94.9%.

Darth Fader works by exploiting the sparsity properties of spectra in order to separate the line signal, continuum and noise from each spectrum in a blind, nonparametric way. Spectra are decomposed in a wavelet basis; the continuum is represented by the largest wavelet scale, and the noise is removed using an iterative procedure that identifies the most significant wavelet coefficients (those containing the sparse signal) using a False Discovery Rate method (Benjamini & Hochberg, 1995).

Thus Darth Fader is able to separate a spectrum into three components: lines, continuum and noise. The line spectra are then used to estimate the redshifts of the galaxies by cross-correlation. At high signal-to-noise, cross-correlation will yield a highly pure sample of redshift estimates. However, this method begins to fail in certain cases when the signal-to-noise of the galaxy spectrum is low. In this case, Darth Fader employs a cleaning step, whereby galaxies whose line spectra show very few features, or whose blue-shifted features do not match any expected or prevalent lines, are excluded from further analysis.

2. Darth Fader Performance

To test the Darth Fader algorithm, we first generated a sample of simulated galaxy spectra (Jouvel *et al.* 2009) at a signal to noise of 2 in the r-band†. In Figure 1 we show the distribution of estimated vs. true redshift after cross-correlation with the templates. We obtained a correct redshift estimate for around 65.5% of the galaxies in the sample. The figure also shows the distribution of redshift estimates after cleaning. To clean the catalogue, we selected galaxies with 6 or more line features. Thus, we retained 76.2% of the sample and attained a success rate of 94.9% for this subsample.

We also applied Darth Fader to spectroscopic data from the WiggleZ survey (Drinkwater *et al.* 2010, 2014), which contains 225,415 low signal-to-noise spectra. The redshifts for all the galaxies in this survey were obtained by eye. Figure 2 shows the results of application of Darth Fader to a subsample of 1000 randomly-selected WiggleZ galaxies. Before cleaning, using cross-correlation we correctly matched redshifts for only $\sim$ 50% of the galaxies. To clean the catalogue, we selected galaxies whose rest-frame spectra showed features consistent with OII or Hα. We then matched 88% of the redshift estimates, retaining around $\sim$ 60% of the galaxies.This demonstrates the effectiveness of Darth Fader

† See Machado *et al.* (2013) for additional experiments with simulated data.

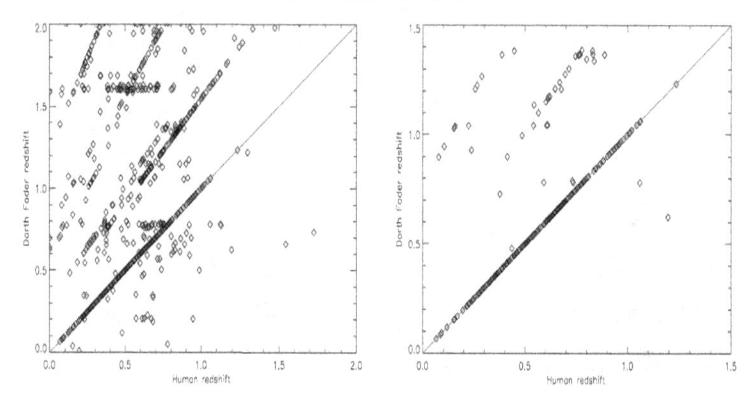

Figure 2. Darth Fader redshift (y-axis) vs Human redshift (x-axis) for a subsample of 1000 WiggleZ galaxy spectra. The left panel shows the distribution before cleaning, for which we obtain matches for around 50% of the galaxies, while the right panel shows the distribution after cleaning, where we retain around 60% of the galaxies with a redshift match rate of 88%.

to obtain accurate redshift estimates in the low-signal-to-noise regime with minimal data loss, and could represent a dramatic reduction in the number of human hours required to analyse such data, and a valuable cross-check on human-identified redshift estimates.

3. Summary

Darth Fader is a powerful tool for the improvement of redshift estimation without any a priori knowledge of galaxy composition, type or morphology. Our algorithm allows us to successfully estimate the continuum without the need for detailed modelling of the galaxy spectra, and to confidently make use of data at signal-to-noise levels that were previously beyond the reach of other techniques. This is demonstrated through extensive experiments using simulated data in Machado *et al.* (2013).

We have successfully applied this technique to real data from the WiggleZ survey (Leonard *et al.* 2014). Though this research is still in a preliminary stage, we are already able to correctly estimate the redshifts for around half of the galaxies in the survey in an automated and therefore fast way. Moreover, we can effectively separate the galaxies for which we expect to obtain a reliable redshift estimate, generating a high-purity subsample of galaxy redshift estimates. This is useful both for verification of human redshift estimates and to reduce the number of human hours required to analyse a survey such as WiggleZ. The Darth Fader software is publicly available at `http://cosmostat.org/darth_fader.html`

References

Benjamini, Y. & Hochberg, Y. 1995, *Journal of the Royal Statistical Society B*, 57, 289
Drinkwater, M. J. *et al.* 2010, *MNRAS*, 401, 1429
Drinkwater, M. J. *et al.* 2014, *in prep.*
Glazebrook, K., Offer, A. R., & Deeley, K. 1998, *ApJ*, 492, 98
Jouvel, S. *et al.* 2009, *A&A*, 504, 359
Leonard, A. *et al.* 2014, *in prep.*
Machado, D. P., Leonard, A., Starck, J.-L., Abdalla, F. B., & Jouvel, S. 2014, *A&A*, 560, 83

Statistical Challenges in 21st Century Cosmology
Proceedings IAU Symposium No. 306, 2014
A. F. Heavens, J.-L. Starck & A. Krone-Martins, eds.

doi:10.1017/S1743921314010801

Radial 3D-Needlets on the Unit Ball

Claudio Durastanti[1], Yabebal T. Fantaye[2], Frode K. Hansen[3], Domenico Marinucci[4] and Isaac Z. Pesenson[5]

[1]Department of Mathematics, University of Tor Vergata, Rome,
email: durastan@mat.uniroma2.it

[2]Department of Mathematics, University of Tor Vergata, Rome,
email: fantaye@mat.uniroma2.it

[3]Institutt for teoretisk astrofysikk, University of Oslo,
email: f.k.hansen@astro.uio.no

[4]Department of Mathematics, University of Tor Vergata, Rome,
email: marinucc@mat.uniroma2.it

[5]Department of Mathematics, Temple University, Philadelphia,
email: isaak.pesenson@temple.edu

Abstract. We present a simple construction of spherical wavelets for the unit ball, which we label Radial 3D Needlets. We envisage an experimental framework where data are collected on concentric spheres with the same pixelization at different radial distances from the origin. The unit ball is hence viewed as a tensor product of the unit interval with the unit sphere: a set of eigenfunctions is therefore defined on the corresponding Laplacian operator. Wavelets are then constructed by a smooth convolution of the projectors defined by these eigenfunctions. Localization properties may be rigorously shown to hold in the real and harmonic domain, and an exact reconstruction formula holds; the system allows a very convenient computational implementation.

Keywords. methods: data analysis, methods: statistical, cosmology: observations

---

1. Motivations and background

It is well-known that Cosmology has recently experienced a golden era where datasets of unprecedented accuracy have become available, for instance on Cosmic Microwave Background radiation (see for instance Bobin *et al.* (2013), Planck XXIII (2013), and the references therein). These datasets are typically collected over the full-sky, usually covering thousands of square degrees, and hence data analysis methods based on flat sky approximations have become unsatisfactory; procedures which take into account the spherical nature of these observations have become mandatory. These methods are usually based on spherical Fourier analysis, and thus they are described in the frequency domain in terms of the spherical harmonics; this framework, however, can often turn out to be inadequate, due to the lack of localization properties in the real domain. Indeed, real data are typically characterized by huge regions of masked data and/or other features for which localization in the real domain is highly desirable; for this reason, several procedures involving spherical wavelets have become rather popular in astrophysical data analysis, see for instance McEwen *et al.* (2007), Starck *et al.* (2006), Donzelli *et al.* (2012) , Faÿ *et al.* (200), Marinucci *et al.* (2008), Pietrobon *et al.* (2008) and Starck *et al.* (2010) for a review.

The next decade will probably be characterized by an equally amazing improvement on the quality of observational data: in particular three-dimensional investigation of weak gravitational lensing and large scale structure are expected to be implemented by experiments such as Euclid, see for instance Laureijs *et al.* (2010). These developments clearly motivate the implementation of three-dimensional wavelet systems; to this issue our paper is devoted.

2. The radial 3D-needlet construction

Some important efforts have already been spent for the construction of three-dimensional wavelets in an astrophysical context, especially in the last few years, see for instance Lanusse *et al.* (2012) and Leistedt *et al.* (2012). In the former reference, wavelets are implemented using a frequency filter on the Fourier-Bessel transform of the three-dimensional field, while in the latter the authors focus on the discretization of the Fourier-Bessel transforms in terms of damped Laguerre polynomials, which allow for an exact quadrature rule.

Our construction is somewhat related to these proposals, and it is based on the astrophysical applications we have in mind Consider for instance an observer located at the centre of a unit ball: we assume that this observer is collecting observations on a set of concentric spheres centred at the origin. At a given resolution level, we also assume that the pixelization on each of these spheres is the same. It is then natural to exploit this implicit radial symmetry, and to view the unit ball as a manifold $M = [0,1] \times \mathbb{S}^2$,: the standard spherical Laplacian has then to be modified, implying that the distance between two points on the same spherical shell depends only on the angular part and not on the radius of the shell. The corresponding eigenfunctions have very simple expressions in terms of sines and spherical harmonics: in particular the set of elements given by

$$u_{\ell,m,n}(r,\vartheta,\phi) = \sin(nr)\, Y_{\ell,m}(\vartheta,\phi)$$

provides an orthonormal basis on the space of square-integrable functions over M. The system of 3D radial needlets is then built using the same procedures as for needlets on the sphere (see Narcowich *et al.* (2006a), Narcowich *et al.* (2006b)): we compute the convolution of a projection operator by means of a smooth window function $b(.)$, and discretize the system by means of an explicitly provided set of cubature points. More precisely, for any resolution level j , let the indexes k and q denote respectively the cardinality over the spherical shell and over the radius. Fix a scale parameter $B > 1$, and define the radial 3D-needlet as follows:

$$\Phi_{j,q,k}(r,\vartheta,\varphi) = \sqrt{\lambda_{j,q,k}} \sum_{[\ell,n]_j} \sum_{m=-\ell}^{\ell} b\left(\frac{\sqrt{-e_{\ell,n}}}{B^j}\right) \overline{u}_{\ell,m,n}(\xi_{j,q,k})\, u_{\ell,m,n}(x),$$

where $\lambda_{j,q,k}$ and $\xi_{j,q,k}$ are the *pixel volume* and the *pixel center* and $e_{\ell,n}$ is the eigenvalue of the Laplacian operator associated to $u_{\ell,m,n}$; finally, observe that $[\ell,n]_j$ denotes the *pairs* of ℓ and n s.t. $B^{2(j-1)} \leqslant e_{\ell,n} \leqslant B^{2(j+1)}$. The window function $b(u)$, $u \in \mathbb{R}$ is a positive kernel satisfying three standard properties, namely $b(\cdot)$ has compact support in $[1/B, B]$, it is infinitely differentiable in $(0,\infty)$ and the following partition of unity property holds: for all $x > B$,

$$\sum_{j=-\infty}^{\infty} b^2\left(\frac{x}{B^j}\right) = 1.$$

Cubature points and weights are obtained by the tensor products of cubature points on the sphere (as provided by HealPix in Gorski *et al.* (2005), for instance), and a uniform discretization on the radial part, sufficient for exact integration of trigonometric polynomials. Some features of this construction can be illustrated as follows:

(*a*) Extremely good localization properties hold in both frequency and real domains; these results can be established rigorously by exploiting related arguments on the construction of wavelets for general compact manifolds in Geller *et al.* (2009), see also Pesenson (2014);

(*b*) An exact reconstruction formula holds

(*c*) Computational implementation is simple and effective exploiting existing packages

(*d*) A direct correspondence holds with standard experimental designs in an astrophysical environment.

References

Bobin, J., Sureau, F., Paykari P., Rassat, A., Basak, S., & Starck, J. -L. 2013, *A&A*, 553, L4

Donzelli, S., Hansen, F. K., Liguori, M., Marinucci, D., & Matarrese, S. 2012, *ApJ*, 755,19

Faÿ, G., Guilloux, F., Betoule, M., Cardoso, J.-F., Delabrouille, J., & Le Jeune, M. 2008, *Phys. Rev. D*, D78:083013

Geller, D. & Mayeli, A. 2009, *Math. Z.*1, 263

Gorski, K. M. , Hivon, E., Banday, A. J., Wandelt, B. D., Hansen F. K., Reinecke, M., & Bartelman M. 2005, *ApJ*, 622

Lanusse, F., Rassat, A., & Starck, J. L. 2012, *A& A* 540, A9

Laureijs, R., Duvet, L., Escudero Sanz, I., Gondoin, P., Lumb, D. H., Oosterbroek T., & Saavedra Criado G. 2010, *The Euclid Mission* SPIE Proceedings 7731

Leistedt, B. & McEwen, J. D. 2012, IEEE Trans. on Sign. Proc. 60, 12

Marinucci, D., Pietrobon, D., Balbi, A., Baldi, P., Cabella, P., Kerkyacharian, G., Natoli, P., Picard, D. & Vittorio N. (2008). 2008, *MNRAS*, 383, 2

McEwen, J. D., Vielva, P., Wiaux, Y., Barreiro, R. B., Cayón, I., Hobson, M. P., Lasenby, A. N., Martí nez-González, E., & Sanz, J. L. 2007, *J. Fourier Anal. Appl.* , 13, 4

Narcowich, F. J., Petrushev, P., & Ward, J. D. 2006, *SIAM Journal of Mathematical Analysis*, 38

Narcowich, F. J., Petrushev, P., & Ward, J. D. 2006, *Journal of Functional Analysis*, 238, 2

Pesenson, I. Z. 2014, submitted

Pietrobon, D., Amblard, A., Balbi, A., Cabella, P., Cooray, A., & Marinucci, D. 2008, *Phys. Rev. D*, D78:103504

Planck Collaboration, 2013, to appear on *A&A*

Starck, J.-L., Moudden, Y., Abrial P., & Nguyen, M. 2006,*A&A*, 446

Starck, J.-L., Murtagh, F., & Fadili, J. *Sparse Image and Signal Processing: Wavelets, Curvelets, Morphological Diversity* (Cambridge University Press)

Statistical Challenges in 21st Century Cosmology
Proceedings IAU Symposium No. 306, 2014
A. F. Heavens, J.-L. Starck & A. Krone-Martins, eds.

doi:10.1017/S1743921314013878

Statistical challenges in weak lensing cosmology

Masahiro Takada

Kavli Institute for the Physics and Mathematics of the Universe (WPI), Todai Institutes for Advanced Study The University of Tokyo, Chiba 277-8583, Japan
email:masahiro.takada@ipmu.jp

Abstract. Cosmological weak lensing is the powerful probe of cosmology. Here we address one of the most fundamental, statistical questions inherent in weak lensing cosmology: whether or not we can *recover* the initial Gaussian information content of large-scale structure by combining the weak lensing observables, here focused on the weak lensing power spectrum and bispectrum. To address this question we fully take into account correlations between the power spectra of different multipoles and the bispectra of different triangle configurations, measured from a finite area survey. In particular we show that super-survey modes whose length scale is larger than or comparable with the survey size cause significant sample variance in the weak lensing correlations via the mode-coupling with sub-survey modes due to nonlinear gravitational clustering – the so-called *super-sample variance*. In this paper we discuss the origin of the super-sample variance and then study the information content inherent in the weak lensing correlation functions up to three-point level.

Keywords. methods: statistical, gravitational lensing, cosmology: theory, (cosmology:) large-scale structure of universe

---

1. Introduction

Cosmological weak lensing is one of the most powerful cosmological probes, as it allows us to directly map out the distribution of matter in the universe without assumptions about galaxy biases (see Heymans *et al.* 2013; More *et al.* 2014 for the recent results). Upcoming galaxy surveys such as the Subaru Hyper Suprime-Cam Survey (Takada 2010) aim to use the high-precision weak lensing measurements to tackle questions on fundamental physics including the origin of cosmic acceleration and neutrino masses.

Most of the useful weak lensing signals are in the nonlinear clustering regime, over a range of multipoles around $\ell \sim$ a few thousands. Due to the mode-coupling nature of nonlinear structure formation, the weak lensing field at the scales of interest display prominent non-Gaussian features. Thus the two-point correlation function or the Fourier-transformed counterpart, the power spectrum, no longer fully describes the statistical properties of the weak lensing field. Which statistical quantities or their combination can be optimal to extract a maximum information of the weak lensing field is still an open question and needs to be carefully explored in order to attain the full potential of the weak lensing surveys. Although weak lensing cosmology involves various statistical issues such as an accurate measurement of galaxy shapes, astronomical data reduction, and parameter estimation, in this paper we focus on the above statistical question.

2. Weak lensing cosmology

The weak lensing convergence field is expressed as a weighted projection of the three-dimensional mass density fluctuation field along the line of sight. For a source galaxy at the radial distance χ_s and in the angular direction $\boldsymbol{\theta}$, the convergence field is given by

$$\kappa(\boldsymbol{\theta}) = \frac{3}{2}\Omega_{\rm m0}H_0^2 \int_0^{\chi_s} d\chi\, a^{-1}\chi \left(1 - \frac{\chi}{\chi_s}\right) \delta_m(\chi, \boldsymbol{\theta}), \tag{2.1}$$

where δ_m is the mass density fluctuation field along the line of sight and we assumed a flat geometry universe. Although the weak lensing is observationaly estimated from the ellipticities of source galaxy shapes, the so-called weak lensing shear field, the shear field is equivalent to the convergence field in the weak lensing regime, so we throughout this paper work on the convergence field. As obvious from the above equation, the statistical properties of the weak lensing field reflect those of the underlying mass density. If the mass density field is a Gaussian random field, which is the case in the linear regime, the weak lensing field is also Gaussian. If the mass field is non-Gaussian, which is the case in the nonlinear regime, the weak lensing should inevitably display non-Gaussian features.

The weak lensing field is measurable only in a statistical way. The most conventional method used in the literature is the two-point correlation function. Using the Limber's approximation and the flat-sky approximation, the Fourier-transformed counterpart, the power spectrum is given as

$$P_\kappa(\ell) = \int_0^{\chi_s} d\chi\, W_{\rm GL}(\chi)^2 \chi^{-2} P_\delta\left(k = \frac{l}{\chi}; \chi\right), \tag{2.2}$$

where we defined the lensing efficiency function $W_{\rm GL}(\chi) \equiv (3/2)\Omega_{\rm m0}H_0^2 a^{-1}\chi(1 - \chi/\chi_s)$, and $P_\delta(k;a)$ is the mass power spectrum at redshift $a = a(\chi)$. Similarly, the lensing bispectrum, which is the lowest-order correlation function to measure the non-Gaussianity, is defined as

$$B_\kappa(\mathbf{l}_1, \mathbf{l}_2, \mathbf{l}_3) = \int_0^{\chi_s} d\chi\, W_{\rm GL}(\chi)^3 \chi^{-4} B_\delta(\mathbf{k}_1, \mathbf{k}_2, \mathbf{k}_3; \chi), \tag{2.3}$$

where $\mathbf{k}_i = \mathbf{l}_i/\chi$ and $B_\delta(\mathbf{k}_i)$ is the mass bispectrum, and a set of the three wavevectors satisfies the triangle condition in Fourier space: e.g., $\mathbf{l}_1 + \mathbf{l}_2 + \mathbf{l}_3 = \mathbf{0}$. While the power spectrum is a one-dimensional function of the wavelength l, the bispectrum is given as a function of triangle configurations. Likewise the n-point correlation function of the weak lensing field arises from the n-point function of the mass density field.

The weak lensing correlation functions are sensitive to both the geometry of the Universe and the growth of matter clustering via the lensing efficiency function and the mass correlation functions. With these dependences weak lensing cosmology is expected to be one of the most powerful probes for constraining cosmological parameters as well as testing theory of gravity on cosmological scales (Takada & Jain 2004; Oguri & Takada 2011).

3. Statistical power of weak lensing correlation functions

3.1. *Super-sample covariance*

In order to realize the constraining power of the weak lensing correlation functions for a given survey, we need to quantify statistical uncertainties in the measured correlation functions. The important source of the statistical uncertainties is the *sample variance* arising due to a finite number of Fourier modes sampled from a finite survey volume. Recently we developed a simple, unified approach to describing the impact of super-sample covariance, which arises from modes that are larger than or comparable with the survey size, on the correlation functions. The method is written in a general form and can be applied to any large-scale structure probe. In this section we briefly review the theory following Takada & Hu (2013) (also see Hamilton *et al.* 2006 for the pioneer work).

Since the statistical properties of the weak lensing field reflect those of the mass density field as we discussed above, we here consider the power spectrum of the three-dimensional mass density field. For a finite volume survey, the observed field is generally expressed as

$$\delta_{m,W}(\mathbf{x}) = \delta_m(\mathbf{x})W(\mathbf{x}), \tag{3.1}$$

where $W(\mathbf{x})$ is a survey window function; $W(\mathbf{x}) = 1$ if $\mathbf{x}$ is in the survey region, otherwise $W(\mathbf{x}) = 0$. For a finite-area weak lensing survey, one can think of the windowed mass density field as the mass density field in the finite volume around a certain lens redshift and confined with the survey area. The Fourier-transformed density field is given as $\tilde{\delta}_{m,W}(\mathbf{k}) = \int d^3\mathbf{q}/(2\pi^3)\tilde{W}(\mathbf{k}-\mathbf{q})\tilde{\delta}_m(\mathbf{q})$. Through the window function that has support for $q \lesssim 1/L$ where $L = V^{1/3}$ is the typical size of the survey, we can properly take into account the effects of super-survey modes that are comparable with or larger than the survey size.

Then we can define the power spectrum estimator as

$$\hat{P}_\delta(k_i) \equiv \frac{1}{V_W}\int_{|\mathbf{k}|\in k_i}\frac{d^3\mathbf{k}}{V_{k_i}}\tilde{\delta}_{m,W}(\mathbf{k})\tilde{\delta}_{m,W}(-\mathbf{k}), \tag{3.2}$$

where the integral is over a shell in k-space of width Δk, volume $V_{k_i} \simeq 4\pi k_i^2\Delta k$ for $\Delta k/k_i \ll 1$, and the effective survey volume is defined as $V_W = \int d^3\mathbf{x}W(\mathbf{x})$. The ensemble average of its estimator is a convolution of the underlying power spectrum with the window

$$\left\langle \hat{P}_\delta(k_i)\right\rangle = \int_{|\mathbf{k}|\in k_i}\frac{d^3\mathbf{k}}{V_{k_i}}\int\frac{d^3\mathbf{q}}{(2\pi)^3}|\tilde{W}(\mathbf{q})|^2P_\delta(\mathbf{k}-\mathbf{q}), \tag{3.3}$$

Thus for $k \sim 1/L$ this estimator is biased low compared to the true power spectrum due to transfer of power into the fluctuation in the spatially-averaged density of the survey volume. For $k \gg 1/L$ this bias becomes progressively smaller since the underlying power spectrum is expected to be smooth across $\Delta k \sim 1/L$.

The covariance matrix describes statistical uncertainties in the power spectrum estimation, defined as

$$\begin{aligned} C^P_{ij} \equiv \mathrm{Cov}[P_\delta(k_i), P_\delta(k_j)] &= \left\langle \hat{P}_\delta(k_i)\hat{P}_\delta(k_j)\right\rangle - \left\langle \hat{P}_\delta(k_i)\right\rangle\left\langle \hat{P}_\delta(k_j)\right\rangle \\ &\simeq C^G_{ij} + \frac{1}{V_W}\bar{T}_W(k_i, j_j). \end{aligned} \tag{3.4}$$

Here the Gaussian piece is

$$C^G_{ij} = \frac{1}{V_W}\frac{(2\pi)^3}{V_{k_i}}2P_\delta(k_i)^2\delta^K_{ij}, \tag{3.5}$$

with $\delta^K_{ij} = 1$ if $k_i = k_j$ to within the bin width, otherwise $\delta^K_{ij} = 0$. Here $V_{k_i}/[(2\pi)^3/V_W]$ is the number of Fourier modes used in the power spectrum estimation at the bin k_i. The prefactor "2" in C^G_{ij} arises from the fact that the density field is real, yielding $\tilde{\delta}_m(\mathbf{k}) = \tilde{\delta}^*(-\mathbf{k})$, and therefore the modes of $\mathbf{k}$ and $-\mathbf{k}$ are not independent. The Gaussian piece has only the diagonal components, ensuring that the power spectra of different bins are uncorrelated with each other. The second term, proportional to $\bar{T}_W(k_i, k_j)$, is the non-Gaussian contribution arising from the connected 4-point function or trispectrum, convolved with the survey window function:

$$\bar{T}_{\delta,W}(k_i,k_j) = \frac{1}{V_W}\int_{|\mathbf{k}|\in k_i}\frac{d^3\mathbf{k}}{V_{k_i}}\int_{|\mathbf{k}|\in k_j}\frac{d^3\mathbf{k}'}{V_{k_j}}\int\left[\prod_{a=1}^{4}\frac{d^3\mathbf{q}_a}{(2\pi)^3}\tilde{W}(\mathbf{q}_a)\right] \times(2\pi)^3\delta_D^3(\mathbf{q}_{1234})T_\delta(\mathbf{k}+\mathbf{q}_1,-\mathbf{k}+\mathbf{q}_2,\mathbf{k}'+\mathbf{q}_3,-\mathbf{k}'+\mathbf{q}_4), \quad (3.6)$$

where T_δ is the mass trispectrum and the notation $\mathbf{q}_{1\ldots n}=\mathbf{q}_1+\cdots+\mathbf{q}_n$. The convolution with the window function means that different 4-point configurations separated by less than the Fourier width of the window function contribute to the covariance. We call this aspect of the covariance the super sample covariance (SSC) effect.

The trispectrum consistency introduced in Takada & Hu (2013) asserts that the SSC term in the trispectrum must be consistent with the response of the power spectrum to change in the background density:

$$\bar{T}_\delta(\mathbf{k},-\mathbf{k}+\mathbf{q}_{12},\mathbf{k}',-\mathbf{k}'-\mathbf{q}_{12}) \simeq T_\delta(\mathbf{k},-\mathbf{k},\mathbf{k}',-\mathbf{k}') + \frac{\partial P_\delta(k)}{\partial\delta_b}\frac{\partial P_\delta(k')}{\partial\delta_b}P_\delta^L(q_{12}), \quad (3.7)$$

where the mode of $\mathbf{q}_{12}$ is a super-survey mode satisfying $k,k'\gg q_{12}$, and the overbar refers to an angle average over the direction of $\mathbf{q}_{12}$. The background density δ_b is the average density fluctuation in the survey region. Here $P_\delta^L(q)$ is is the linear power spectrum and is designated that for this relation to be applicable δ_b must be a mode in the linear regime, i.e. the survey scale must be much larger than the nonlinear scale. With this consistency prescription, the power spectrum covariance is simplified as

$$C_{ij}^P = C_{ij}^G + C_{ij}^{T0} + \sigma_b^2\frac{\partial P_\delta(k)}{\partial\delta_b}\frac{\partial P_\delta(k')}{\partial\delta_b}, \quad (3.8)$$

where σ_b^2 is the variance of the background density in the survey window, defined as

$$\sigma_b^2 \equiv \frac{1}{V_W}\int\frac{d^3\mathbf{q}}{(2\pi)^3}|\tilde{W}(\mathbf{q})|^2P_\delta^L(q). \quad (3.9)$$

Here C_{ij}^{T0} is the standard non-Gaussian term arising from the mass trispectrum of sub-survey modes: $C_{ij}^{T0}=(1/V_W)\int_{|\mathbf{k}|\in k_i}(d^3\mathbf{k}/V_{k_i})\int_{|\mathbf{k}'|\in k_j}(d^3\mathbf{k}'/V_{k_j})T_\delta(\mathbf{k},-\mathbf{k},\mathbf{k}',-\mathbf{k}')$. The linear variance σ_b can be easily computed for any survey geometry, either by evaluating the above equation directly, or using Gaussian realizations of the linear density field. The SSC term scales with the survey volume only through σ_b^2 whereas the other terms scale like white noise $1/V_W$. Thus, even if the initial density field is Gaussian, the nonlinear structure formation induces non-Gaussian contributions to the sample variance. In other words, the non-Gaussian sample variance depends on the nature of large-scale structure formation that governs how the different Fourier modes are correlated with each other via nonlinear gravity.

To compute the power spectrum response for a given cosmological model, we can use the *separate universe approach* developed in Li *et al.* (2014). In this approach the impact of the super-box mode δ_b is absorbed into changes in the background cosmological parameters in a finite-volume simulation with periodic boundary condition:

$$\frac{\delta a}{a} = -\frac{1}{3}\delta_b, \quad \frac{\delta h}{h} = -\frac{5\Omega_m}{6}\frac{\delta_b}{D}, \quad \frac{\delta\Omega_m}{\Omega_m} = \frac{\delta\Omega_\Lambda}{\Omega_\Lambda} = \frac{\delta\Omega_K}{1-\Omega_K} = -\frac{\delta h}{h}. \quad (3.10)$$

Here D is the linear growth rate and we have introduced the notations such as $\delta h/h=(H_{0W}-H_0)/H_0$, where H_{0W} denotes the parameter in a separate universe (a finite volume region at the fixed δ_b). Our convention is to set the scale factor of the separate universe a_W to agree with the global one at high redshift: $\lim_{a\to 0}a_W(a,\delta_b)=a$. Since

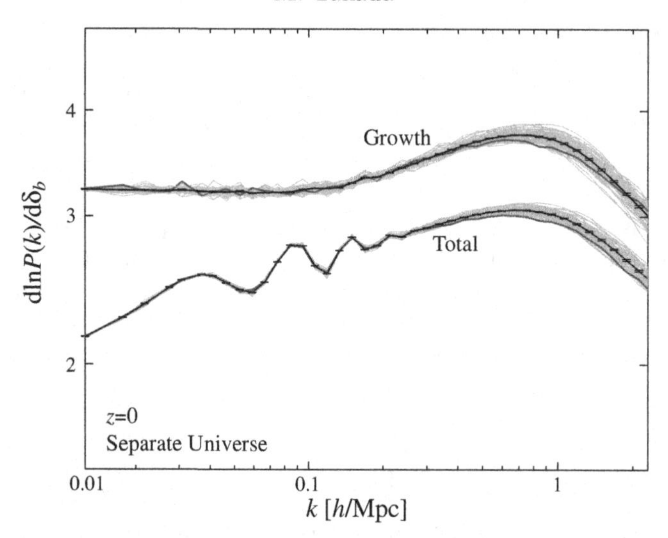

Figure 1. The response of the mass power spectrum to the super-box mode δ_b, computed using the separate universe approach. In addition to the fiducial run with $\delta_b = 0$, we ran 64 separate universe pairs with $\delta_b = \pm 0.01$, where we used the same initial seeds to reduce the stochasticity (each simulation has a box size of 500 h^{-1}Mpc and 256^3 particles). There are two distinct effects of the super-box mode (here treated as a DC mode): the growth effect and the dilation effect (see text for details). The curve labeled by "total" is a sum of these two effects, leaving characteristic scale dependence in the response. The bold curve is the average of the 64 pairs, and the thin curve is the result for each pair. The red curve is one particular realization. This plot is taken from Li *et al.* (2014).

the linear background density δ_b evolves with D, so $\delta_b/D = \delta_{b0}/D_0$; the relations about cosmological parameters hold independently of the redshift at which δ_b and D are defined. Thus, even if the global universe has a flat geometry, $\Omega_K = 0$, the separate universe with non-zero δ_b is realized as a non-zero curvature universe, $\Omega_{KW} \neq 0$. Because this is the curvature effect, time evolution of all the sub-box Fourier modes is affected by the super-box mode, due to the modified expansion history.

Fig. 1 shows the power spectrum response computed using the separate universe approach. There are two distinct contributions to the power spectrum response. First, the presence of super-survey mode modifies the growth of sub-volume modes via the mode coupling in nonlinear structure formation. If the survey region is embedded in a coherently overdense region, i.e. $\delta_b > 0$, the growth of sub-volume modes is enhanced. We call this effect "growth". Second, the super-survey mode causes remapping of physical and comoving length scales. An overdense region expands less quickly than in the global universe. We call this effect "dilation" as it changes the comoving scale corresponding to features in the power spectrum. The figure shows that, in the total, these two effects partially cancel, leaving a characteristic scale-dependence in the response. We also note that, if we use the halo model to estimate the power spectrum response by directly computing the windowed trispectrum, $\bar{T}_W(k_i, k_j)$, the analytical prediction gives about 10%-level agreement with the separate universe result over the range of k shown in Fig. 1.

Upcoming wide-area galaxy surveys require an accurate estimation of the power spectrum covariance or more generally the covariance of any large-scale structure probes. This is indeed computationally challenging. With the unified theory of the covariance, Eq. (3.8), we can propose a way of calibrating the power spectrum covariance at computationally reasonable expense. To compute the standard part, the Gaussian piece and the trispectrum piece of sub-volume modes, we can use mock catalogs of a galaxy survey,

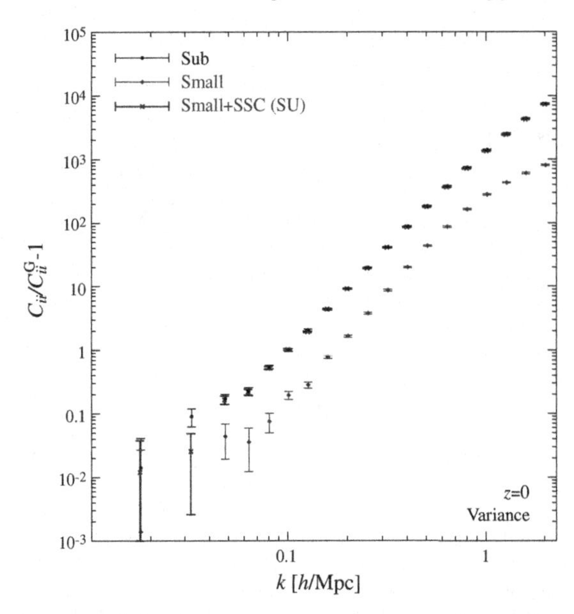

Figure 2. Diagonal elements of the mass power spectrum covariance, C_{ij}, relative to the Gaussian term C_{ii}^G at $z = 0$. The result denoted as "Sub" is the covariance estimated from subvolumes of 7 large-volume simulations; each of the simulations has a 4 Gpc/h box size and is divided into $8^3 = 512$ subvolumes of size 500 Mpc/h each (3584 subvolumes in total). Thus each subvolume includes the super-box mode effects. The result "Small" is the covariance estimated from small-box simulations of 500 Mpc/h each, with periodic boundary conditions. The result "Small+SSC" shows the covariance computed by adding the SSC effect, calibrated based on the separate universe approach in the previous figure, to the small-bx variance. The "Small+SSC" result is in nice agreement with the "Sub" result to within the bootstrap errors. Note that bootstrap errors between bins are highly correlated. This plot is taken from Li *et al.* (2014).

based on N-body simulations of small boxes. To compute the SSC effect, we can use the separate universe simulations for the fiducial cosmology. In doing this, we can properly take into account the survey window to compute the linear variance, σ_b^2. This method does not require huge-volume simulations whose size is designed to be well larger than the size of survey volume in order to include the super-survey effects.

In Fig. 2, we indeed show that the above method combining the small-box simulations and the separate universe simulations well reproduces the covariance matrix from the large-volume simulations. The figure also shows that the SSC effect boosts the covariance amplitude by up to an order of magnitude over the range of wavenumbers we consider. Hence the SSC effect is the dominant source of the sample variance. This results hold for any size of survey volumes relevant for upcoming galaxy surveys (see Fig. 1 in Takada & Hu 2013).

If the power spectrum needs to be estimated with respect to the local average density within the finite-volume survey region, which is the case for the galaxy power spectrum (Takada *et al.* 2014), the power spectrum response is modified as

$$\frac{\partial \ln P^W(k)}{\partial \delta_b} = \frac{\partial \ln P(k)}{\partial \delta_b} - 2, \tag{3.11}$$

where $P^W(k) = P(k)/(1 + \delta_b)^2$ is the power spectrum with respect to the local average density. The SSC effect is reduced in the covariance of the local power spectrum, but still gives a dominant contribution in the nonlinear regime (Li *et al.* 2014).

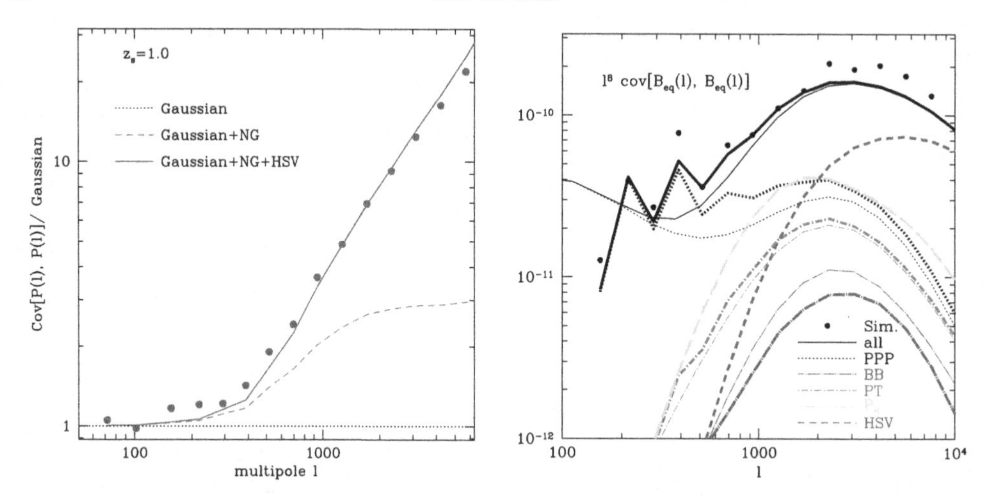

Figure 3. *Left panel*: Diagonal elements of the weak lensing power spectrum covariance, relative to the Gaussian covariance. Note that we here ignored shape noise contribution. The circle points show the results measured from 1000 realizations of the ray-tracing simulations for the ΛCDM model, each of which is for source redshift $z_s = 1$ and has an area of 25 sq. degrees. Note that we used the logarithimcally-spacing bins of $\Delta \ln \ell = 0.3$. The solid or dashed curves are the analytical predictions with or without the HSV contribution, which is the small-scale, non-linear version of the SSC effect (see text for details). *Right panel*: Diagonal elements of the weak lensing bispectrum for equilateral triangle configurations against the triangle side length. The data points are the results measured from the 1000 ray-tracing simulations. The bold solid curve is the total power including the HSV effect, and fairly well reproduces the simulation results. The other curves are each contribution that arises from the HSV effect or the 2-, 3-, 4- and 6-point correlation functions as indicated.

The small-scale, nonlinear version of the SSC effect can also be realized within the framework of the halo model approach – the halo sample variance (HSV) (Hu & Kravtsov 03; Takada & Bridle 2007; Takada & Jain 2009; Sato *et al.* 2009; Kayo *et al.* 2013; Takada & Hu 2013; Takada & Spergel 2014; Schaan *et al.* 2014). In the halo model formulation, the 1-halo term of the mass power spectrum, which describes correlations between dark matter in the same halo, is expressed as

$$P_\delta^{1h}(k) = \int dM \frac{dn}{dM} \left(\frac{M}{\bar{\rho}_m}\right)^2 |\tilde{u}_M(k)|^2, \tag{3.12}$$

where dn/dM is the halo mass function in the mass range $[M, M+dM]$, $\bar{\rho}_m$ is the cosmic mean mass density, and $\tilde{u}_M(k)$ is the Fourier transform of the mass density profile of halos of mass M. However, the above equation is correct only in an ensemble average sense. For a finite-volume survey, the coherent density fluctuation across the survey window, δ_b, would change the abundance of halos in the survey region along the peak-background splitting theory:

$$\left.\frac{dn}{dM}\right|_{\delta_b} = \frac{dn}{dM}\left[1 + b(M)\delta_b + \cdots\right], \tag{3.13}$$

where the notation $|_{\delta_b}$ denotes the average over the realizations of different sub-survey modes at fixed δ_b, and $b(M)$ is the halo bias. Thus, e.g., if the survey region is in a coherent over-density region, it enhances the number of halos on average.

By inserting Eqs. (3.12) and (3.13) into the covariance formula (Eq. 3.8) via the trispectrum consistency relation, we find that the change in the halo mass function via the super-survey modes causes co-variant scatters in the power spectrum estimation:

$$
\begin{aligned}
C_{ij}^{\rm HSV} &= \sigma_b^2 \frac{\partial P^{1h}(k_i)}{\partial \delta_b}\frac{\partial P^{1h}(k_j)}{\partial \delta_b} \\
&= \sigma_b^2 \left[\int dM \frac{dn}{dM} b(M)|\tilde{u}_M(k_i)|^2\right]\left[\int dM' \frac{dn}{dM'} b(M')|\tilde{u}_{M'}(k_j)|^2\right], \qquad (3.14)
\end{aligned}
$$

where we have assumed that the super-survey modes do not affect the halo mass profile. We found that the HSV effect gives a dominant contribution of the SSC effect in the power spectrum covariance at $k \gtrsim$ a few 0.1 h/Mpc, fairly well reproducing the separate universe simulation results at the scales (see Fig. 2 Takada & Hu 2013 or Fig. 1 in Li *et al.* 2014).

3.2. *Information content of lensing power spectrum and bispectrum*

Similarly to the mass power spectrum covariance, we can compute the covariance matrices for the weak lensing power spectrum and bispectrum (Takada & Bridle 2007; Takada & Jain 2009; Sato *et al.* 2009; Kayo *et al.* 2013; Takada & Spergel 2014; Schaan *et al.* 2014). The left panel of Fig. 3 shows diagonal elements of the weak lensing power spectrum covariance relative to the Gaussian covariance, measured from the ray-tracing simulations that are built using a suite of N-body simulations for the WMAP ΛCDM model (Sato *et al.* 2009). We used 1000 realizations for source redshift $z_s = 1$ each of which has an area of 25 sq. degrees corresponding to the fundamental Fourier mode $l_f \simeq 72$. The ray-tracing simulations were done in a light cone volume with an observer's position being its cone vertex, and therefore include contributions from N-body Fourier modes with length scales greater than the light-cone volume at each lens redshift (see Fig. 1 in Sato *et al.* 2009). Thus the simulations are suitable to study the SSC effect. As can be found from the figure, the weak lensing power spectrum covariance shows significant non-Gaussian errors at $\ell \gtrsim$ a few hundreds. The solid curve denotes the analytical prediction including the HSV effect, showing remarkably nice agreement with the ray-tracing simulation result. We note that the HSV effect causes highly correlated scatters between different multipoles. If we ignore the HSV effect, i.e. include the standard non-Gaussian error alone arising from the lensing trispectrum of sub-survey modes, the model prediction significantly underestimates the total power.

Similarly the right panel of Fig. 3 shows the results for the bispectrum covariance matrix. Here we consider the equilateral triangle configurations. The HSV effect gives a dominant contribution to the total power of the covariance matrix at $\ell \gtrsim 1000$, compared to other terms up to the 6-point correlation function. If we include the HSV contribution, the analytical model gives a 10-20% level agreement with the simulation results.

Once the covariance matrices for the weak lensing power spectrum and bispectrum are computed, we can address the information content carried by the weak lensing correlation functions. For a given range of multipoles, the cumulative signal-to-noise ratio or the information content for the power spectrum measurement is defined as

$$\left(\frac{S}{N}\right)_P^2 \equiv \sum_{l_{\rm min} \leqslant l, l' \leqslant l_{\rm max}} P_\kappa(l)[\mathbf{C}^{-1}]_{ll'} P_\kappa(l'), \qquad (3.15)$$

where the summation runs over all multipole bins in the range $l_{\rm min} \leqslant l \leqslant l_{\rm max}$ and $\mathbf{C}^{-1}$ is the inverse of the covariance matrix. The inverse of S/N is equivalent to a precision of measuring the logarithmic amplitude of the power spectrum up to a given maximum multipole $l_{\rm max}$, assuming that the shape of the power spectrum is perfectly known. Similarly we can define the S/N values for the bispectrum measurement and for a joint

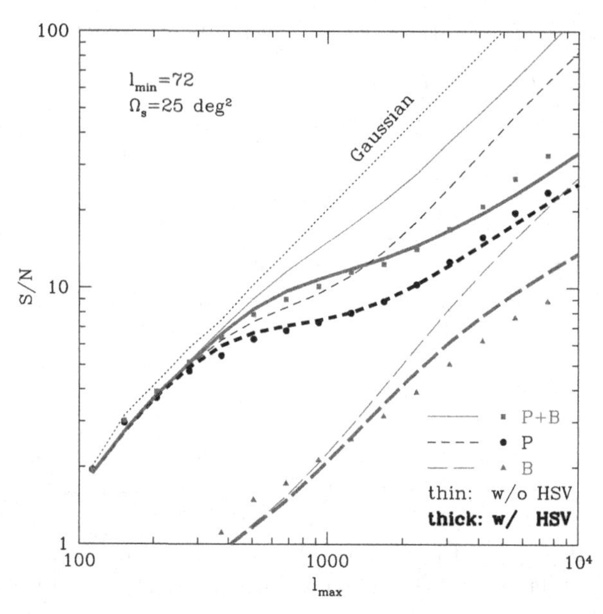

Figure 4. Cumulative signal-to-noise ratios (S/N) for the power spectrum (P), the bispectrum (B) and the joint measurement ($P+B$) for a survey area of 25 deg^2 and source redshift $z_s = 1$. The circle, triangle and square symbols are the simulation results for P, B and $P+B$ measurements, respectively, computed from the 1000 realizations. To account for the full bispectrum information, we included the bispectra of all-available triangle configurations from the multipole range, up to 204 triangles for $l_{\rm max} = 8745$. The thick short-dashed, long-dashed and solid curves are the corresponding halo model predictions. The corresponding thin curves are the results without the HSV contributions. For comparison, the dotted curve shows the S/N for the power spectrum for the Gaussian field, which the primordial density field should have contained. The Gaussian information follows a simple scaling as $S/N|_{\rm Gaussian} \propto l_{\rm max}\Omega_s^{1/2}$, where Ω_s is the survey area. This plot is from Kayo *et al.* (2013).

measurement of the power spectrum and the bispectrum. For the latter case, we need to properly take into account their cross-covariance.

Fig. 4 shows the expected S/N for measurements of the weak lensing power spectrum and bispectrum for a survey area of 25 square degrees (i.e. the area of the ray-tracing simulation), as a function of the maximum multipole $l_{\rm max}$ up to which the power spectrum and/or bispectrum information are included. The minimum multipole is fixed to $l_{\rm min} = 72$. Roughly speaking the S/N value scales with survey area as $S/N \propto \Omega_s^{1/2}$ (exactly speaking the scaling does not hold due to the different dependence of the SSC effect on survey area). The circle, triangle and square symbols are the simulation results for the power spectrum, the bispectrum and the joint measurement, respectively. For the bispectrum measurement we included the bispectra of all triangle configurations available from the multipole range, and considered up to 204 triangles for $l_{\rm max} = 8745$. The thick/thin short-dashed, long-dashed and solid curves are the analytical predictions with/without the HSV effects. First, the figure clearly shows that the lensing bispectra add new information content to the power spectrum measurement. To be more quantitative, adding the bispectrum measurement increases the S/N by about 50 per cent for $l_{\rm max} \simeq 10^3$ compared to the power spectrum measurement alone. This improvement is equivalent to about 2.3 larger survey area for the power spectrum measurement alone; that is, the same data sets can be used to obtain the additional information, if the bispectrum measurement is combined with the power spectrum measurement. Secondly, the analytical

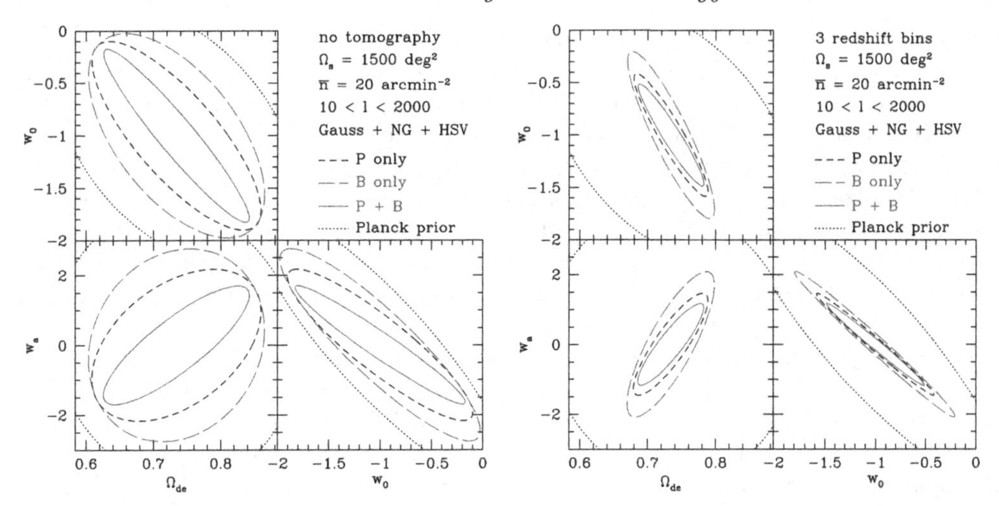

Figure 5. Expected accuracies of dark energy parameters ($\Omega_{\rm de}, w_0, w_a$) for a galaxy survey that resemble the Subaru Hyper Suprime-Cam survey ($\Omega_s = 1500$ sq. degrees, $\bar{n}_g = 20$ arcmin^{-2} and $\sigma_\epsilon = 0.22$). For this plot we included the shape noise contribution to the covariance. The error ellipses include marginalization over other parameters, and we included the CMB information expected for the Planck experiment. We here show the parameter forecasts for the power spectrum information alone (P), the bispectrum alone (B) and the joint information ($P + B$) as in the previous plot. The left and right panels show the results for no tomography (single redshift bin) and three-redshift tomography case. As in the previous figure, we took into account the full covariance between the observables including the HSV effects. For the tomography case, we included 6525 different bispectra in the multipole range $10 \leqslant l \leqslant 2000$ that are constructed from all combinations of different redshift bins and triangle configurations. This plot is from Kayo & Takada (2014).

predictions are in nice agreement with the simulation results. Note that the total S/N for the joint measurement ($P + B$) is close to the linear sum of the S/N values, not the sum of their squared values $(S/N)^2$, due to the significant cross-covariance. If ignoring the cross-covariance, adding the bispectrum measurement does not much improve the S/N (only by 5 per cent or so).

The top, dotted lines shows the information content for a Gaussian field, which the initial density field of large-scale structure should have contained – therefore can be considered as a maximum information content we could ultimately extract. The Gaussian information content depends only on the number of Fourier modes available from the range of multipoles up to $l_{\rm max}$: it has a simple scaling given by $S/N|_{\rm Gaussian} \propto \Omega_s^{1/2} l_{\rm max}$. The figure shows that the joint measurement can recover only about 50% or less of the Gaussian information at $l_{\rm max} \gtrsim 1000$. The information loss is mainly due to the HSV effect, as can be found from the thin curves. If we ignore the HSV effect, the joint measurement recovers about 70% of the Gaussian information at $l_{\rm max} \simeq 1000$. These results imply that further higher-order functions such as the 4-point function may be important and add the information (see Seo *et al.* 2011). Or some of the initial Gaussian information is lost or cannot be recovered due to the nonlinear clustering. This is not yet known, and still an open question.

3.3. *Weak lensing tomography*

Adding redshift information to the weak lensing correlation functions greatly improves the cosmological sensitivity – the so-called weak lensing tomography (Takada & Jain 2004). However, to realize the genuine power of the weak lensing tomography, we need to consider all the spectra available from all possible combinations of different redshift bins and multipole bins. For the bispectrum case, adding the lensing tomography easily leads to more than 1000 bispectra, and this is even worse for the higher-order correlations. Hence, an accurate calibration of the covariance matrix for the lensing tomography would require a huge number of independent ray-tracing simulations, e.g., a factor 10 more realizations than the number of bispectra to achieve about 10% accuracy. This is computationally very challenging, and is even impossible if we need to compute the covariance as a function of cosmological models. So again a hybrid method combining small-box simulations, separate-universe simulations, and the analytical model would be useful and tractable in practice.

In Kayo & Takada (2013), we used the halo model approach, which gives a fairly good agreement with the simulation results for no tomography case as shown in Fig. 4, in order to estimate the cosmological power of the weak lensing bispectrum tomography. Fig. 5 shows expected accuracies of dark energy parameters assuming survey parameters that resemble the Subaru Hyper Suprime-Cam, characterized by $\Omega_{\rm s} = 1500$ sq. degrees, $n_g = 20$ arcmin^{-2} and $\sigma_\epsilon = 0.22$ for the survey area, the mean number density of source galaxies and the rms intrinsic ellipticities per component, respectively. The bispectrum further adds the information to improve the parameter constraints compared to the power spectrum alone. To be more precise, for the three redshift bin case (the right panel), the joint measurement leads to about 60% improvement in the dark energy figure-of-merit (FoM) that is defined by FoM $= 1/[\sigma(w_0)\sigma(w_a)]$. Again this is equivalent to about 60% larger survey area for the power spectrum tomography alone. In this case we considered 6525 triangle configurations, and we take into account the non-Gaussian correlations between the different spectra including the HSV effects.

4. Discussion

Can we recover the initial Gaussian field from observables of the present-day large-scale structure? This is an unresolved, open question. In this paper we discussed the example of the weak lensing field that is the line-of-sight projection of the three-dimensional mass density field in large-scale structure. We showed that the information content inherent in the power spectrum, which is the two-point correlation function in Fourier space, is smaller than the Gaussian information by more than a factor of 2 at $l \gtrsim 1000$. We showed that the bispectrum, which is the three-point correlation function, does add the information to the power spectrum, but the combined information does not fully recover the Gaussian information – still only 50% of the Gaussian information at $l_{\rm max} \sim 1000$. In order to derive this conclusion, we included all the two- and three-point level information in a sense that we included the bispectra of all available triangle configurations for a given range of multipoles as well as properly took into account the auto- and cross-covariances between the two- and three-point correlation functions. This implies that the higher-order correlation functions are further needed to recover the Gaussian information. Or our result implies a limitation of the information recovery; some of the initial memory is lost due to the strong mode coupling in the deeply nonlinear regime. Alternatively, the nonlinear mapping of the cosmological field, such as the log-normal mapping (Seo *et al.* 2012), might be a more practically useful way.

However, this conclusion is a bit misleading. Most of the information loss is caused mainly by the super-sample covariance. As we showed, the super-survey effects are parameterized mainly by the average density fluctuation in the survey volume, δ_b, on each realization basis. Hence, by treating δ_b as an additional parameter together with cosmological parameters, we may be able to calibrate most of the super-sample variance effects in the correlation measurements (e.g., see Takada & Spergel 2013; Schaan *et al.* 2014; Li *et al.* in preparation). We can even treat the super-survey mode as "signal" and then estimate its value for a given survey volume by fitting the measurements with the model. This is an interesting possibility, and needs to be further explored. A physical understanding of the super-survey effects is also important to explore an optimal survey design that allows an efficient operation of massive cosmological surveys (Takahashi *et al.* 2014).

Acknowledgments. - MT was supported by World Premier International Research Center Initiative (WPI Initiative), MEXT, Japan, by the FIRST program "Subaru Measurements of Images and Redshifts (SuMIRe)", CSTP, Japan, and by Grant-in-Aid for Scientific Research from the JSPS Promotion of Science (Nos. 23340061 and 26610058).

References

Hamilton, A. J. S., Rimes, C. D., & Scoccimarro, R. 2006, *MNRAS*, 371, 1188
Heymans, C., *et al.* 2013, *MNRAS*, 432, 2433
Hu, W. & Kravtsov, A. 2003, *ApJ*, 584, 702
Kayo, I. & Takada, M. 2013, arXiv:1306.4684
Kayo, I., Takada, M., & Jain, B. 2013, *MNRAS*, 429, 344
Li, Y., Hu, W., & Takada, M. 2014, *Phys. Rev. D*, 89, 083519
Li, Y., Hu, W., & Takada, M. 2014b, in preparation
More, S., *et al.* 2014, arXiv:1407.1856
Oguri, M. & Takada, M. 2011, *Phys. Rev. D*, 83, 023008
Sato, M., Hamana, T., Takahashi, R., Takada, M., *et al.* 2009, *ApJ*, 701, 945
Seo, H.-J., Sato, M., Dodelson, S., Jain, B., & Takada, M. 2011, *ApJ Letters*, 729, 11
Schaan, E., Takada, M., & Spergel, D. N. 2014, arXiv:1406.3330
Takada, M. 2010, *AIP Conference Proceedings*, 1279, 120
Takada, M. & Bridle, S. 2007, *New Journal of Physcis*, 9, 446
Takada, M. & Hu, W. 2013, *Phys. Rev. D*, 87, 123504
Takada, M. & Jain, B. 2004, *MNRAS*, 348, 897
Takada, M. & Jain, B. 2009, *MNRAS*, 395, 2065
Takada, M. & Spergel, D. N. 2014, *MNRAS*, 441, 2456
Takada, M., Ellis, R. S., Chiba, M., Greene, J. E., *et al.* 2014, *Publ. Astron. Soc. Japan*, 66, R1
Takahashi, R., Soma, S., Takada, M., & Kayo, I. 2014, arXiv:1405.2666

Statistical Challenges in 21st Century Cosmology
Proceedings IAU Symposium No. 306, 2014
A. F. Heavens, J.-L. Starck & A. Krone-Martins, eds.

doi:10.1017/S1743921314011016

The potential of likelihood-free inference of cosmological parameters with weak lensing data

Michael Vespe

Department of Statistics, Carnegie Mellon University,
Baker Hall 132, Pittsburgh, PA 15232
email: mvespe@andrew.cmu.edu

Abstract. In the statistical framework of likelihood-free inference, the posterior distribution of model parameters is explored via simulation rather than direct evaluation of the likelihood function, permitting inference in situations where this function is analytically intractable. We consider the problem of estimating cosmological parameters using measurements of the weak gravitational lensing of galaxies; specifically, we propose the use a likelihood-free approach to investigate the posterior distribution of some parameters in the ΛCDM model upon observing a large number of sheared galaxies. The choice of summary statistic used when comparing observed data and simulated data in the likelihood-free inference framework is critical, so we work toward a principled method of choosing the summary statistic, aiming for dimension reduction while seeking a statistic that is as close as possible to being sufficient for the parameters of interest.

1. Introduction

Weak gravitational lensing, also known as cosmic shear, is the distortionary effect on images of distant galaxies by matter in between the galaxy and the observer. The ensemble behavior of this distortionary effect, which would render a circular object slightly elliptical, can yield insight into the distribution of dark matter and permit constraint of the parameters in a cosmological model. However, galaxies are not intrinsically circular; in fact, the signal from cosmic shear is very faint compared to the intrinsic variability in the ellipticity of galaxies. Thus, a large number of galaxies must be observed in order to isolate the shear signal.

Once the galaxies are observed and catalogued, weak lensing analyses traditionally proceed by summarizing these galaxies via some summary statistic, often referred to as an *observable* or a *data vector*. Common examples of observables include, among others, estimates of the two-point correlation functions $\xi_\pm$ or power spectrum modes C_ℓ (as in, e.g., Lin *et al.* 2012), among others. Then, the summary statistic is assumed to have a multivariate Gaussian distribution, so that the likelihood of a set of values $\tilde{\theta}$ for cosmological parameters is given by

$$\mathcal{L}(\tilde{\theta}; \hat{d}) = \frac{1}{\sqrt{(2\pi)^p |\mathbf{C}|}} \exp\left((\hat{d} - d(\tilde{\theta}))^T \mathbf{C}^{-1} (\hat{d} - d(\tilde{\theta})) \right) \tag{1.1}$$

where $\hat{d}$ is the observable estimated from data, $d(\tilde{\theta})$ is the theoretical value of the observable given parameters $\tilde{\theta}$, and $\mathbf{C}$ is the covariance matrix of the observable, often estimated via some simulation approach. This likelihood can be evaluated either as part of a frequentist maximum likelihood analysis or, in a Bayesian framework, as one step toward deriving a posterior distribution for cosmological parameter values.

The quality of any inferences resulting from this procedure hinges on several factors: the ability of the chosen summary statistic to capture the information in the raw data relevant to the parameters; the accuracy of the estimated covariance matrix; and the validity of the assumption that the observable has a multivariate Gaussian distribution. This last assumption is not equivalent to assuming that the *parameters* have a Gaussian distribution, but it does impose some indirect constraints on their distribution.

These concerns motivate our desire to explore likelihood-free inference methods, which were introduced in contexts where methods exist for simulating data given parameter values, but evaluation of a likelihood is analytically or computationally intractable.

2. Methodology

Approximate Bayesian computation. We focus on approximate Bayesian computation (ABC), a particular likelihood-free inference method, introduced by Pritchard *et al.* (1999) in the biology literature. For full details, as well as generalizations and improvements, we refer to that paper as well as Blum *et al.* (2012) and Beaumont *et al.* (2009). Thus far, ABC has found limited use in astronomy settings, as in Cameron and Pettitt (2012) and Weyant *et al.* (2013).

The algorithm aims to generate a sample from a desired posterior distribution $\pi(\theta|\mathbf{x})$ given observed data $\mathbf{x}$; in the weak lensing setting, we aim to sample from the posterior distribution of the cosmological parameters of interest given a catalogue of galaxies with position and shear.

In its simplest form, it proceeds via the repeated execution of three steps. Step one is to generate candidate parameter values $\tilde{\theta}$ from a prior distribution. Step two is to simulate a realization $\tilde{\mathbf{x}}$ of the data set using $\tilde{\theta}$ as input parameters. Step three is to compare the simulated data $\tilde{\mathbf{x}}$ to $\mathbf{x}$ and retain $\tilde{\theta}$ if and only if $\tilde{\mathbf{x}}$ and $\mathbf{x}$ match. Formally, we retain $\tilde{\theta}$ if and only if $\rho(S(\tilde{\mathbf{x}}), S(\mathbf{x})) \leqslant \epsilon$, for some distance metric ρ, summary statistic S, and tolerance threshold ϵ.

The resulting retained parameter values constitute a sample from an approximation $\pi_\epsilon(\theta|\mathbf{x})$ to the posterior $\pi(\theta|\mathbf{x})$. It can be shown that if S is a sufficient statistic for θ, then $\pi_\epsilon(\theta|\mathbf{x}) \to \pi(\theta|\mathbf{x})$ as $\epsilon \to 0$. Any ABC analysis will be sensitive to the choice of ϵ, ρ, and S. As in the traditional analysis framework, it is desirable to choose the summary statistic S so that it captures the information from $\mathbf{x}$ relevant for inference on θ while discarding any useless information. Due to the manner in which ABC algorithms rely on accepting/rejecting simulated parameter values, there is a particular need to reduce the dimensionality of the summary statistic to the greatest extent possible, lest the simulations be burdened by the inefficiency of repeatedly rejecting parameter candidates.

Exponential family approximation. Our proposed approach to approximating sufficient statistics of low-dimension is built upon the following idea. Suppose summary statistics $\mathbf{s}$ are available that are sufficient for θ but may be high-dimensional and thus contain some redundant information. For j = 1, 2, ..., J, we seek a mapping of θ, denoted $\eta_j(\theta)$, along with a mapping of $\mathbf{s}$, denoted $T_j(\mathbf{s})$, such that $\mathbb{E}(\eta_j(\theta)|\mathbf{s}) = T_j(\mathbf{s})$. These can be thought of as compressed versions of the summary statistics $\mathbb{E}(\theta|\mathbf{s})$ that have been shown (Fearnhead and Prangle 2012) to be optimal under a reasonable choice for the loss function.

We refer to this approach as an *exponential family approximation.* The standard exponential family form for the distribution of $\mathbf{s}$ given parameter θ is

$$f(\mathbf{s}|\theta) = h(\mathbf{s}) \exp\left(\sum_{j=1}^{J} \eta_j(\theta) T_j(\mathbf{s}) - A(\theta)\right)$$

If θ is modeled as a random variable drawn from prior distribution π, the joint distribution of $(\mathbf{s}, \theta)$ is given by

$$f(\mathbf{s}, \theta) = h(\mathbf{s}) \exp\left(\sum_{j=1}^{J} \eta_j(\theta) T_j(\mathbf{s}) - A(\theta)\right) \pi(\theta)$$

or, equivalently,

$$f(\mathbf{s}, \theta) = \exp\left(\sum_{j=1}^{J} \eta_j(\theta) T_j(\mathbf{s}) - A^*(\theta) - h^*(\mathbf{s})\right)$$

with $h^*(\mathbf{s}) = -ln(h(\mathbf{s}))$ and $A^*(\theta) = A(\theta) - ln(\pi(\theta))$. It is assumed that one can simulate pairs $(\mathbf{s}_i, \theta_i)$ for i = 1, 2, ..., n. This will be accomplished by drawing θ from a prior distribution π, simulating $\mathbf{x}$ conditional on θ, and then computing the available $\mathbf{s}(\mathbf{x})$. Then, the joint log-likelihood for this synthetic "data set" is given by

$$f(\mathbf{s}, \theta) = \sum_{i=1}^{n} \left[\sum_{j=1}^{J} \eta_j(\theta) T_j(\mathbf{s}) - A^*(\theta) - h^*(\mathbf{s})\right] \tag{2.1}$$

Heuristically, we will seek to maximize this joint log-likelihood over the space of mappings η_j and T_j. The resulting mapping $T_j(\mathbf{s})$ for j = 1, ..., J would be an approximately sufficient statistic of dimension J, where J is chosen intentionally to be closer to the dimension of the intrinsic parameter space Θ than to that of $\mathcal{S}$, the domain of $\mathbf{s}$.

3. Example application

We present an application of this method in a simple, stylized cosmic shear analysis, simulating data from known inputs. Specifically, we generate a random realization of a shear field using input cosmological parameters $\Omega_M = 0.25, \sigma_8 = 0.8$, and all other inputs (including survey redshift distribution) chosen to replicate those in the simulation exercises of Lin *et al.* (2012). In this case, we model cosmic shear as a Gaussian random field (GRF) on a grid of pixels, although future analyses would incorporate more realistic models (see Kiessling *et al.* 2011.) We add i.i.d. shape noise ($\sigma_{\text{int}} = 0.37$) to represent the effect of intrinsic galaxy ellipiticity.

For our ABC analysis, we sample candidate parameter values $(\tilde{\Omega}_M, \tilde{\sigma}_8)$ from a uniform prior distribution on the rectangle $[0.1, 8] \times [0.5, 1]$. In Fig. 1 we compare samples from the approximate posterior distributions using two summary statistics: at left, estimates $\hat{\xi}_\pm(\theta)$ of the two-point correlation functions – evaluated in eight logarithmically spaced bins – and, at right, the first two coordinates of $\hat{T}$ learned via the exponential family method. In each case, we take ρ to be standard Euclidean distance and we choose ϵ so that 5% (blue) and 10% (green) of the candidate samples are retained for each case.

As Ω_M and σ_8 are known to be degenerate (these data can only distinguish the value of $\Omega_M^{0.7}\sigma_8$), we display the degeneracy curve corresponding to the input parameters in orange. Simple inspection suggests that using $\hat{T}$ as learned via the exponential family approximation as the summary statistic is preferable to simply using $\hat{\xi}_\pm(\theta)$, because the samples from the former assemble more tightly around the true degeneracy curve than those from the latter.

4. Discussion and future directions

The exponential family approximation-derived statistic seems to improve upon the canonical $\hat{\xi}_\pm(\theta)$ statistic, even in the simple case where shear is assumed to follow a GRF

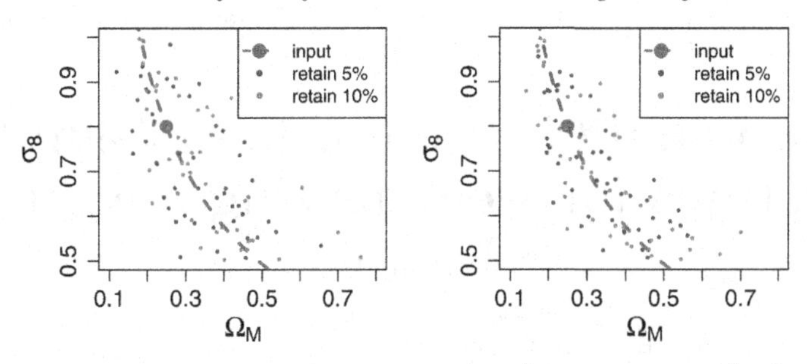

Figure 1. ABC-derived posterior samples using $\hat{\xi}_{\pm}(\theta)$ (left) and $\hat{T}_1, \hat{T}_2$ (right).

model. This is noteworthy because the true correlation function $\xi_{\pm}(\theta)$ is known to be a sufficient statistic for the GRF, so the improvement likely stems from the act of reducing the dimension of the data vector to mitigate the effect of noisy discretized estimates. Put another way, when the observable consists of noisy, somewhat redundant estimators, it is preferable to have fewer of those provided that no information is being discarded.

Future work will proceed in both theoretical and applied directions. In the former direction, we aim to better understand the theoretical properties of the summary statistic resulting from the exponential family approximation method. In addition, we hope to assess the method's feasibility under particular circumstances.

Regarding applications, we intend to apply the method in more sophisticated, realistic settings. One such setting would be that of tomographic weak lensing analysis, as in Heymans *et al.* (2013), wherein the data vector consists of auto- and cross-correlation functions $\hat{\xi}_{\pm}^{i,j}$ within or between various bins in redshift. In this setting, where the dimension of the data vector is inherently much larger, the motivation for principled dimension reduction is apparent. We also hope to incorporate more complex simulation models, such as the SUNGLASS pipeline of Kiessling *et al.* (2011) to better account for non-Gaussianity of the true shear field.

5. Acknowledgements

This work is done is collaboration with Peter Freeman, Rachel Mandelbaum, and Chad Schafer. Support comes from DOE Grant #ER26170.

References

Beaumont, M. A., Cornuet, J., Marin, J., & Robert, C. 2009, *Biometrika*, 96, 983
Blum, M. G. B., Nunes, M. A., Prangle, D., & Sisson, S. A. 2012, `arXiv:1202.3819`
Cameron, E. & Pettitt, A. N. 2012, *MNRAS*, 425, 44
Fearnhead, P. & Prangle, D. 2012, *J. Roy. Stat. Soc. B Met.*, 74, 419
Heymans, C., Grocutt, E., Heavens, A., Kilbinger, M., *et al.* 2013, *MNRAS*, 432, 2433
Kiessling, A., Heavens, A. F., & Taylor, A. N. 2011, *MNRAS*, 414, 2235
Lin, H., Dodelson, S., Seo, H., Soares-Santos, M., Annis, J., *et al.* 2012, *ApJ*, 761, 15
Pritchard, J. K., Seielstad, M. T., *et al.* 2012, *Mol. Biol. Evol.*, 16, 1791
Weyant, A., Schafer, C., & Wood-Vasey, W. M. 2013, *ApJ*, 764, 116

Statistical Challenges in 21st Century Cosmology
Proceedings IAU Symposium No. 306, 2014
A. F. Heavens, J.-L. Starck & A. Krone-Martins, eds.

doi:10.1017/S174392131401360X

The probability distribution of ellipticity: implications for weak lensing measurement

Massimo Viola

Leiden Observatory, Leiden University,
Niels Bohrweg 2, 2333 CA Leiden, The Netherlands
email: viola@strw.leidenuniv.nl

Abstract. The weak lensing effect generates spin-2 distortions, referred to as shear, on the observable shape of distant galaxies, induced by intervening gravitational tidal fields. Traditionally, the spin-2 distortion in the light distribution of distant galaxies is measured in terms of a galaxy ellipticity. This is a very good unbiased estimator of the shear field in the limit that a galaxy is measured at infinite signal-to-noise. However, the ellipticity is always defined as a ratio between two quantities (for example, between the polarisation and measurement of the galaxy size, or between the semi-major and semi-minor axis of the galaxy) and therefore requires some non-linear combination of the image pixels. This means, in any realistic case, this would lead to biases in the measurement of the shear (and hence in the cosmological parameters) whenever noise is present in the image. This type of bias can be understood from the particular shape of the 2D probability distribution of the ellipticity of an object measured from data. Moreover this probability distribution can be used to explore strategies for calibration of noise biases in present and future weak lensing surveys (e.g. KiDS, DES, HSC,Euclid, LSST...)

Keywords. methods: data analysis, cosmology, weak-lensing

---

1. Introduction

Weak gravitational lensing is a very powerful tool to study properties of dark matter halos (e.g Hoekstra *et al.* 2013) as well to investigate the growth-rate of structures in the Universe (e.g. Schrabback *et al.* 2010, Kilbinger *et al.* 2013). However it is quite challenging from a practical point of view since it relies on measurements of tiny choerent distortions (shear) in the shapes of background galaxies.

One of the reasons why measuring galaxy shapes (ellipticities) is notoriously challenging is because the measurements have to be done from noisy pixels.

Since 2007 it has become common practice in the weak-lensing community to test and validate different algorithms used to infer the gravitational shear from measurements of the shape of galaxies, on some sets of common image simulations (Heymans *et al.* (2007), Massey *et al.* (2007), Bridle *et al.* (2010), Kitching *et al.* (2012)).

Some important lessons were learned: the bias in shear measurements is a strong function of the galaxy signal-to-noise and of the galaxy size with respect to the size of the point spread function (PSF): galaxies with sizes closer to the size of the PSF have more bias then larger objects at equal signal-to-noise. This behaviour is common to all algorithms tested on those simulations and usable on real data.

The bias in the shear is generally parameterised in terms of a multiplicative term m and an additive term c: $\mathrm{g^{obs}} = (1 + \mathrm{m})\mathrm{g^{true}} + \mathrm{c}$.

The additive bias c can be estimated from the data itself, making use of the fact that over the full survey area, the average of each component of the ellipticity must vanish.

For galaxies with low signal-to-noise (15 or below) the multiplicative bias in the shear is typically very large, of the order of 20% or larger. If not accounted for, this bias propagates directly into potentially large biases in the cosmological parameters or in derived properties of dark matter halos.

Ideally, any bias in the shear measurements should be smaller than the measurement statistical error, $\sigma_\gamma \simeq \sigma_\epsilon/\sqrt{\mathrm{N}}$, where $\sigma_\epsilon \simeq 0.3$ is the intrinsic ellipticity dispertion and N the number of galaxies used to infer the shear.

Even for existing surveys, like CFHTLenS, the amplitude of the bias is too large at low signal noise for not being corrected, and some calibration needs to be applied (Miller *et al.* 2013). Ongoing larger surveys, like Dark Energy Survey (DES),Kilo Degree Survey (KiDS),Hyper Suprime-Cam (HSC), have even more stringent requirements on the amplitude of the shear bias, which poses greater challenges about calibration of existing shape measurements methods.

2. The Marsaglia-Tin distribution

At a very fundamental level, the bias in shear measurements is a consequence of the fact that it is not possible to measure an unbiased ellipticity in presence of noise (if the effect of the noise is not properly accounted for). The reason is that an ellipticity measurement involves some non-linear transformation on the noisy pixels of an astronomical image.

In order to further investigate this problem, we derive the probability distribution of the observed ellipicity in presence of noise given a true ellipticity.

Since the shear is derived as an average over an ensemble of measured ellipticities in a region of the sky, this probability lies behind any weak lensing analysis.

This particular probability distribution was first derived by Marsaglia (1965) and Tin (1965) and re-derived in a weak lensing context by Melchior & Viola (2012) and Viola, Kitching, Joachimi (2014).

Here we summarise the derivation of the Marsaglia-Tin distribution.

2.1. *Some definitions*

We start by defining the $i+j$ order moments of the object surface brightness $I(x,y)$:

$$\{Q\}_{i,j} = \int I(x,y) x^i y^j \,\mathrm{d}x\mathrm{d}y \tag{2.1}$$

Note that in reality what are measured are weighted moments of the convolved surface brightness. A weighting function has to be employed in order to suppress the pixel noise at large distances from the galaxy centre, and its effect has to be accounted for (this involves measuring higher order moments of the surface brightness). Moreover, the contribution of the PSF has to be removed.

The second-order moments can be used to characterise the object's normalised polarisation χ and the ϵ-ellipticity of the object:

$$\chi := \frac{\{Q\}_{20} - \{Q\}_{02} + 2\mathrm{i}\{Q\}_{11}}{\{Q\}_{20} + \{Q\}_{02}} \quad \text{and} \quad \epsilon := \frac{\{Q\}_{20} - \{Q\}_{02} + 2\mathrm{i}\{Q\}_{11}}{\{Q\}_{20} + \{Q\}_{02} + 2\sqrt{\{Q\}_{20}\{Q\}_{02} - \{Q\}_{11}^2}}. \tag{2.2}$$

The two definitions are related through:

$$\chi = \frac{2\epsilon}{1 + |\epsilon|^2}. \tag{2.3}$$

The shear is then derived by averaging many galaxies' ellipticities under the assumption that the the intrinsic orientation of galaxies in the universe is random.

Both definitions are used in literature. However only an unbiased measurement of ϵ is indeed an unbiased estimate of the shear (Seitz & Schneider (1997)).

We start focusing on what is the probability distribution for χ, which can also be written as a ratio of the Stokes parameters $u = \{Q\}_{20} - \{Q\}_{02}$,$v = 2\{Q\}_{11}$ and s=$\{Q\}_{20} + \{Q\}_{02}$.

The probability distribution function of the Stokes parameters in presence of homoscedastic noise (i.e. uncorrelated and gaussian) can be described in terms of a trivariate Gaussian with correlation coefficients ρ_{ij} between each of the variables.

The 2-dimensional probability distribution for the normalised polarisation χ defined as $(u/s, v/s)$ can be derived starting from the three-dimensional probability distribution for the Stokes parameters.

First of all we transform the distribution $p_{u,v,s}(u,v,s)$ into $p(\chi_1, \chi_2, s)$ by a change of variable and then we marginalise over s

$$p_\chi(\chi_1, \chi_2) = \int_{-\infty}^{\infty} \mathrm{d}s s^2 p_{u,v,s}(\chi_1 s, \chi_2 s, s) \; ; \tag{2.4}$$

this is the form of a two dimensional quotient distribution. The result of this integration is the so-called Marsaglia-Tin distribution. It has an analytical (even though not simple) expression, that interested readers can find in Section 3.1 of Viola, Kitching, Joachimi (2014).

The probability of measuring a χ polarisation can be transformed into the probability of measuring an ϵ-ellipticity using Equation 2.3.

We note here that in the case of uncorrelated variables which are gaussian distributed with zero-mean the Marsaglia-Tin distribution reduces to the Cauchy distribution.

2.2. *Properties of the Marsaglia-Tin distribution*

We summarise here the main properties of the Marsaglia-Tin distribution:

• Only the amplitude of the ellipticity is generally biased, while the angle $(1/2)\tan^{-1}(\chi_2/\chi_2)$ is always unbiased Wardle & Kronberg (1974);

• In the case an 'optimal' weighting function (i.e. matching exactly the radial profile, ellipticity and size of the object) and χ is used as a definition for the ellipticity, both the mean and the maximum of the Marsaglia-Tin distribution are biased, while in the case that ϵ is used only the maximum is biased while the mean is unbiased *independent* of the signal-to-noise level;

• Truncation of the ϵ-ellipticity distribution, for example by removing very elliptical objects, introduces a bias in the measurements of the mean ϵ even in the ideal case of a weighting function that perfectly matches the galaxy profile;

• In the case where a circular weighting is employed in the moment measurements, the correlation between the Stokes parameters deviates from the true one. The larger this deviation, the larger is the bias;

• For a fixed value of signal-to-noise, the amplitude of the bias is determined by two factors: the correlation between the Stokes parameters, and the signal-to-noise on the quadrupole moments (or the ellipticity of the object);

• If the size of the object becomes comparable to the size of the PSF then the probability distribution of the convolved ellipticity is the convolution of the Marsaglia-Tin distribution with the probability distribution of the ratio of the size of the galaxy and the size of the PSF (the so-called resolution): the lower the resolution, the larger the bias.

The probability distribution of the absolute value of the ϵ-ellipticity is shown in left panel of Figure 1 for different choices of the weighting function.

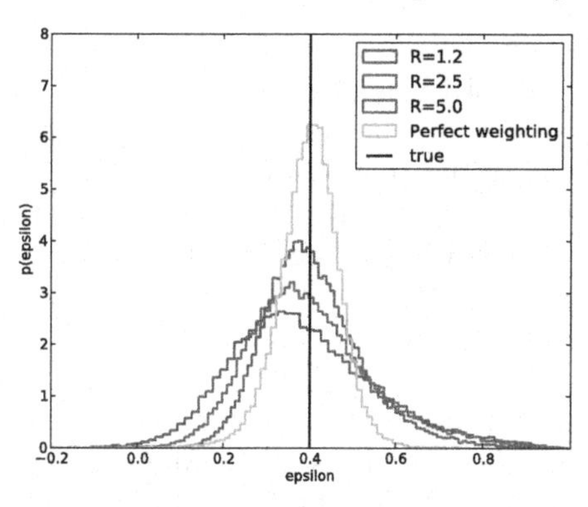

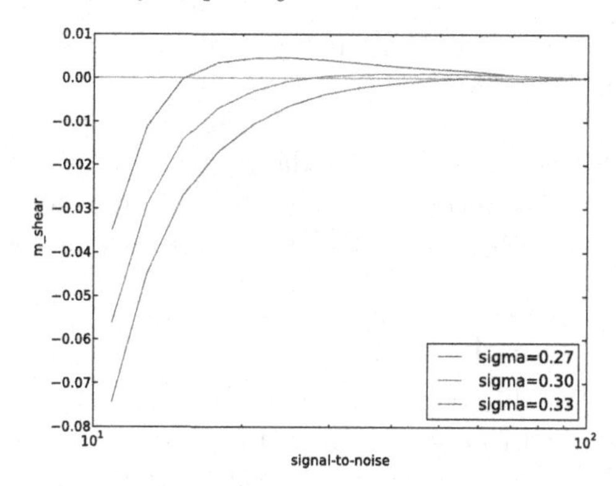

Figure 1. *Left panel:* Probability distribution of the absolute value of the ϵ-ellipticity given a true galaxy ellipticity of $\epsilon = (0.4, 0.0)$ and zero PSF ellipticity. The cyan line corresponds to the case of using a weighting function which is matched perfectly to the galaxy profile and no PSF convolution, the red line to the case of using a circular weighting function with size 1.2 times the galaxy semi-major axis and a resolution of $R = 5.0$, the green line to a resolution of 2.5 and the blue line to a resolution of 1.2. Note that in all cases we removed objects having an unphysical combination of second-order moments ($Q_{20}Q_{02} - Q_{11}^2 < 0$). *Right panel:* Shear multiplicative Marsaglia bias as a function of signal-to-noise. The three curves represent the case of galaxies with intrinsic ellipticities following a Rayleigh distribution with $\sigma_\epsilon = 0.27$ (blue), $\sigma_\epsilon = 0.3$ (green) and $\sigma_\epsilon = 0.33$ (red). The width of the weighting function has been chosen to be 1.2 times the object semi-major axis. This plot highlights the importance of knowing the ellipticity distribution in order to calibrate the shear bias. These figures are adapted from Viola, Kitching & Joachimi 2014.

3. Implication for current and future surveys

If we assume that the galaxy profile is known (i.e. we neglect the so-called model bias), the amplitude of the multiplicative bias in the shear measurements depends essentially on the galaxy resolution, the intrinsic ellipticity distribution (since the bias in the ellipticity measurements is a function of the ellipticity) and on the object signal-to-noise and it can be numerically computed starting from the Marsaglia-Tin distribution (for details we refer to Section 3 of Viola, Kitching & Joachimi.)

Therefore any attempt to characterise and calibrate the noise-bias by means of image simulations requires knowledge of these three quantities.

How well those three quantities need to be known depends on the statistical power of a survey (given by its area and its depth) which sets requirements on the knowledge of the shear multiplicative bias $\sigma_{\rm m}$.

For a surveys like CFHTLenS this number is of order $\sigma_m \simeq 10^{-2}$, for current surveys like KiDS, DES, HSC, $\sigma_m \simeq 3 \times 10^{-3}$, and for a Euclid-like survey $\sigma_m \simeq 5 \times 10^{-4}$.

Hence the requirements on the knowledge of a quantity $\vec{x} = (\sigma_\epsilon, \nu, R, ..)$ can be computed as:

$$\sigma_{\vec{x}_{i0}} = \sigma_m \left[\left| \frac{\partial \mathrm{m}}{\partial \vec{x}_i} \right|_{\vec{x}_{i0}} \right]^{-1} \tag{3.1}$$

from which it is clear that the requirements on the knowlege of the intrinsic ellipticity distribution, noise level and resolution are driven by the steepness of the multiplicative-bias as a function of this quantity and not by its amplitude.

In other words among methods with similar amplitude of the multiplicative bias, it is preferred, in the sense that it is more calibratable, the one with the shallower derivative of m as a function of $\vec{\mathrm{x}}$

The effect of the intrnsic ellipticity distribution on the shear multiplicative bias as a function of signal to noise is shown in the right panel of Figure 1.

In Viola, Kitching & Joachimi we investigated the requirements on the knowledge of the intrinsic ellipticity distribution and we found that it has to be known with a precision of $\sim 5\%$ in order to properly calibrate shear estimates for current surveys, for upcoming surveys with a precision of $\sim 1\%$ and for future surveys with a precision of $\sim 0.3\%$.

4. Conclusions

We showed in this work how the bias in shear measurements, affecting all methods applied to data so far, can be understood studying the properties of the Marsaglia-Tin distribution (which the probability distribution of measuring an ellipticity in presence of noise).

In particular we show how the amplitude of the multiplicative shear bias strongly depends on the intrinsic ellipticity distribution, the resolution and the signal-to-noise of the objects.

Hence these properties of galaxies need to be known with great precision and accuracy in order to calibrate the bias using image simulation.

One way to avoid the noise-bias in shear measurements would be using avoiding taking ratios, for example using the un-normalised stokes parameters (i.e. not normalised by the galaxy flux). However, it has been shown (Viola, Kitching & Joachimi), that the price paid for this is an increased variance in the shear estimate. This can be understood by the fact that no information about the object flux is used.

References

Bridle, S., Balan, S. T., Bethge, M., Gentile, M., Harmeling, S., Heymans, C., Hirsch, M., Hosseini, R., Jarvis, M., & Kirk, D. 2009, ArXiv e-prints 0908.0945

Heymans, C., Van Waerbeke, L., Bacon, D., Berge, J., Bernstein, G., Bertin, E., Bridle, S., Brown, M. L., Clowe, D., *et al.* 2006, *MNRAS*, 368, 1323

Hoekstra, H., Bartelmann, M., Dahle, H., Israel, H., Limousin, M., & Meneghetti, M. 2013, *SSR*, 177, 75

Kilbinger, M., Fu, L., Heymans, C., Simpson, F., Benjamin, J., Erben, T., Harnois-Déraps, J., Hoekstra, H., *et al.* 2013, *MNRAS*, p. 735

Kitching, T. D., Balan, S. T., Bridle, S., Cantale, N., Courbin, F., Eifler, T., Gentile, M., Gill, M. S. S., Harmeling, S., Heymans, C., *et al.* 2012, *MNRAS*, 423, 3163

Marsaglia, G., 1965, j-J-AM-STAT-ASSOC, 60, 193

Massey, R., Heymans, C., Bergé, J., Bernstein, G., Bridle, S., Clowe, D., Dahle, H., Ellis, R., Erben, T., *et al.* 2007, *MNRAS*, 376, 13

Melchior, P. & Viola, M. 2012, *MNRAS*, 424, 2757

Miller, L., Heymans, C., Kitching, T. D., van Waerbeke, L., Erben, T., & Hildebrandt, H. 2013, *MNRAS*, 429, 2858

Schneider, P. & Seitz, C. 1995, *A&A*, 294, 411

Schrabback, T., Hartlap, J., Joachimi, B., Kilbinger, M., Simon, P., Benabed, K., Bradač, M., Eifler, T., Erben, T., *et al.* 2010, *A&A*, 516, A63+

Seitz, C. & Schneider, P. 1997, *A&A*, 318, 687

Tin, M., 1965, j-J-AM-STAT-ASSOC, 60, 294

Wardle, J. F. C. & Kronberg, P. P. 1974, *ApJ*, 194, 249

Viola, M., Kitching, T. D., & Joachimi, B. 2014, *MNRAS*, 439, 1909

Statistical Challenges in 21st Century Cosmology
Proceedings IAU Symposium No. 306, 2014
A. F. Heavens, J.-L. Starck & A. Krone-Martins, eds.

doi:10.1017/S1743921314013428

Errors on errors – Estimating cosmological parameter covariance

Benjamin Joachimi[1] and Andy Taylor[2]

[1]Department of Physics & Astronomy, University College London,
Gower Place, London WC1E 6BT, United Kingdom
email: b.joachimi@ucl.ac.uk

[2]Institute for Astronomy, University of Edinburgh,
Royal Observatory, Blackford Hill, Edinburgh EH9 3HJ, United Kingdom
email: ant@roe.ac.uk

Abstract. Current and forthcoming cosmological data analyses share the challenge of huge datasets alongside increasingly tight requirements on the precision and accuracy of extracted cosmological parameters. The community is becoming increasingly aware that these requirements not only apply to the central values of parameters but, equally important, also to the error bars. Due to non-linear effects in the astrophysics, the instrument, and the analysis pipeline, data covariance matrices are usually not well known a priori and need to be estimated from the data itself, or from suites of large simulations. In either case, the finite number of realisations available to determine data covariances introduces significant biases and additional variance in the errors on cosmological parameters in a standard likelihood analysis. Here, we review recent work on quantifying these biases and additional variances and discuss approaches to remedy these effects.

Keywords. Methods: statistical, methods: data analysis, cosmological parameters

1. What noise does to your covariance matrix

The variance-covariance of cosmological data is often complicated and not known analytically, receiving contributions from sample variance coupled with complex survey masks, instrumental effects, as well as measurement and shot noise. Therefore it is customary to estimate data covariance matrices from the data itself (via resampling techniques) or from suites of realistic mock datasets. The latter are computationally expensive to generate, so that there is a strong drive to keep the number of simulated realisations of the data, N_S, to a minimum. Conversely, the ever increasing size of surveys and number of mature cosmological probes to be extracted leads to data vectors which can easily exceed dimensions of $N_D > 1000$ in the near future.

Despite the pressure to keep N_S small while creating data vectors with large N_D, cosmologists have until recently almost exclusively employed the standard sample covariance estimator and proceeded to perform a likelihood analysis without further consideration of the statistical uncertainty and potential biases of their covariance estimate. Only recently, beginning with the work of Hartlap *et al.* (2007), has there been an increased awareness of long-established results on this topic in the statistics literature.

If the elements of a data vector $\boldsymbol{D}$ are Gaussian distributed, the data sample covariance estimate $\mathbf{M} = \langle \Delta\boldsymbol{D}\, \Delta\boldsymbol{D}^{\tau} \rangle$, where $\Delta\boldsymbol{D} = \boldsymbol{D} - \langle \boldsymbol{D} \rangle$, follows a Wishart distribution (the multivariate generalisation of a χ^2 distribution) with $N_S - 1$ degrees of freedom (Wishart 1928). The inverse data covariance, $\boldsymbol{\Psi} \equiv \mathbf{M}^{-1}$, which is required in least squares and

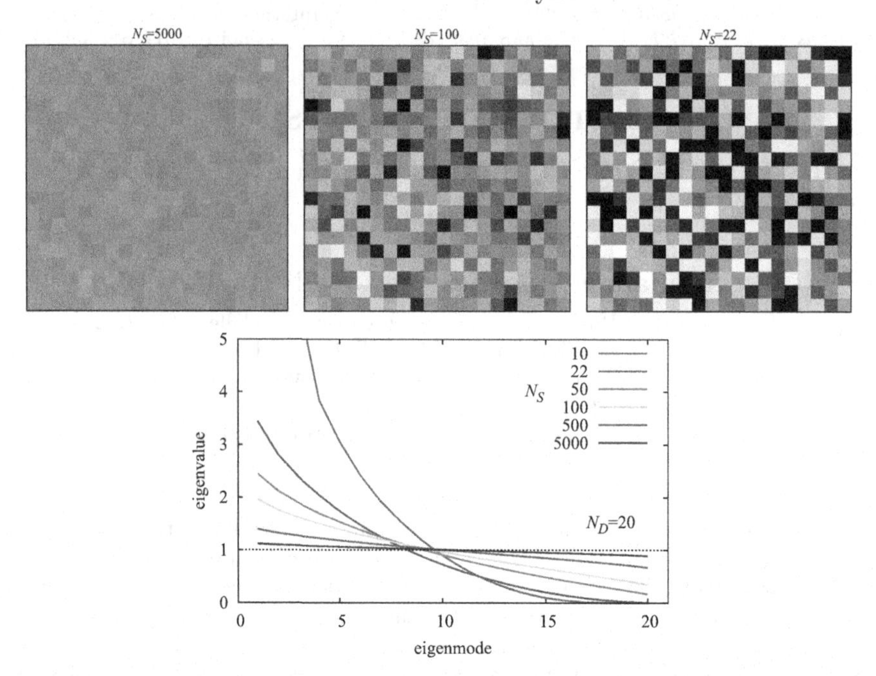

Figure 1. Illustration of the impact of noise on a covariance matrix, for the toy case of a 20-dimensional identity matrix. *Top*: Realisations of Wishart-distributed random matrices with expectation subtracted, generated for $N_S = 5000, 100, 22$ (from left to right) to illustrate the increasing levels of noise. *Bottom*: Ordered eigenvalues, averaged over 1000 realisations of covariance matrices. The lines show eigenvalues for different values of the number of realisation used for computing the covariance, N_S, as indicated in the legend. The dashed horizontal line indicates the eigenvalues in the noise-free case.

likelihood expressions, then follows an Inverse-Wishart distribution with $N_S - N_D - 2$ degrees of freedom. The moments of this distribution were first derived by Kaufman (1967) who found for the mean

$$\left\langle \hat{\mathbf{\Psi}} \right\rangle = \frac{N_S - 1}{N_S - N_D - 2} \mathbf{\Psi} \,, \tag{1.1}$$

demonstrating that $\hat{\mathbf{\Psi}}$ is increasingly biased high for decreasing N_S and diverges at $N_S = N_D + 2$. Fig. 1 provides an illustration for this non-intuitive behaviour. For decreasing N_S the largest (smallest) eigenvalues of a noisy covariance matrix are biased increasingly high (low), and the condition number dramatically increases. The smallest eigenvalue drops to zero at $N_S = N_D + 2$, rendering the covariance singular. Even after correcting for the bias, the variance in the covariance estimate diverges at a very similar rate (see again Kaufman 1967).

One may wonder about the impact on early cosmological analyses which generally boasted only a small number of simulations and seem to have ignored the biases in the inverse data covariance. Since the estimated covariances would have been very noisy, one can hypothesise that researchers would have calculated a singular value decomposition and set the smallest eigenvalues to zero before proceeding with a Moore-Penrose pseudo-inverse. Fig. 2 illustrates two possible scenarios, alongside with the bias expected from Eq. (1.1). Depending on the exact value of N_S and N_D, and the details of the noise-suppression measures, biases may have been large and both positive and negative. The

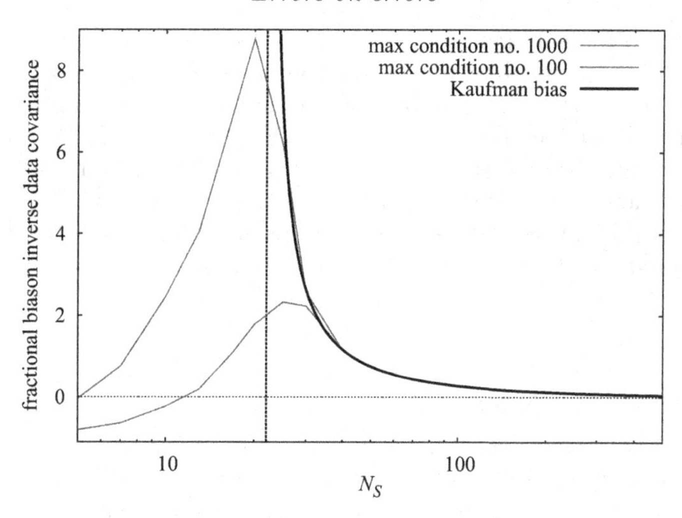

Figure 2. The effect of noise stabilising measures (via singular value decomposition) on the bias of the inverse covariance. Shown is the average fractional bias on the diagonal elements of the inverse covariance matrix (for $N_D = 24$; indicated by the vertical line), as a function of the number of realisation used for computing the covariance, N_S. The black solid line corresponds to the bias of the inverse of the sample covariance matrix, as calculated by Kaufman (1967). The red (blue) line shows the bias for the case that the smallest eigenvalues of the covariance have been set to zero before calculating a pseudo-inverse, such that the condition number does not exceed 1000 (100).

increase in the largest eigenvalues would have remained untreated, and consequently the inverse covariance would have tended to zero (or the fractional bias to -1) for $N_S \to 0$.

Precise and accurate cosmological analysis requires a more careful treatment of these noise effects, either quantifying their effect on cosmological parameter errors or mitigating them by employing alternatives to the standard sample covariance estimator.

2. Impact on the errors of cosmological parameters

Taylor & Joachimi (2014, TJ14 hereafter) calculated the impact of biases and variances in the data sample covariance on the cosmological parameter covariance, finding that the latter is generally biased high and takes up additional variance (see Dodelson & Schneider 2013 and Percival *et al.* 2014 for earlier, approximate results). For the case that the parameter covariance is estimated from the curvature of the likelihood at its peak (similar to estimates from MCMC samples), they derived its full distribution, which is again a Wishart distribution with $N_S - N_D + N_P - 1$ degrees of freedom, where N_P is the number of parameters, i.e. the dimension of the parameter covariance matrix. The mean reads

$$\left\langle \hat{\mathbf{C}}^W_{\mu\nu} \right\rangle = \frac{N_S - N_D + N_P - 1}{N_S - N_D - 2}\, \mathbf{C}_{\mu\nu} \,, \tag{2.1}$$

while exact expression for the variance can also be derived. Additionally, TJ14 found an exact expression for the mean of the parameter covariance estimated from the scatter in likelihood peaks,

$$\left\langle \hat{\mathbf{C}}^P_{\mu\nu} \right\rangle = \frac{N_S - 2}{N_S - N_D + N_P - 2}\, \mathbf{C}_{\mu\nu} \,, \tag{2.2}$$

which they determined from simulations based on random Wishart matrices. De-biasing the parameter covariances based on these expressions, TJ14 derived constraints on the minimum number of simulations required to reach a maximum contribution, ν, to the error on cosmological parameters originating from noise in the data covariance (see their Fig. 5). In the case of the parameter covariance derived from the likelihood curvature, $\mathbf{C}^W$, the minimum number is given by $N_S \approx N_D + N_P/\nu + 2\nu^{-2}$. For forthcoming cosmological surveys, where easily $N_D \sim 1000$, this implies that, if one tolerates a 10% increase in parameter errors due to noise in the data covariance, N_S has to be only slightly larger than N_D, whereas if one restricts this increase to 1%, a challenging number of $N_S > 10^4$ would result.

3. Beyond the sample covariance

There are various alternatives to using the standard sample covariance estimator in a likelihood analysis, which Taylor *et al.* (2013, TJK13 hereafter) presented in schematic form in their Fig. 8. Below we summarise the most relevant points.

Resampling methods: We have concentrated here on the generation of realisations of the data from simulations, i.e. externally. Resampling methods like bootstrap or jackknife allow for the estimation of covariances internally from the data. To preserve the correlations in the data, one has to define blocks of survey volume, or patches on the sky, which are quasi-independent. This requirement limits the number of blocks for a given survey volume/area. Moreover, long-range correlations will invalidate this assumption to some degree and cause biases. In addition, bootstrap and jackknife estimates are generally biased and, although consistent, can converge slowly. For more details on resampling methods see the contribution by P. Arnalte-Mur.

Data compression: Data compression alleviates the problem of noise effects originating from the data covariance as biases and variances scale approximately with $N_S - N_D$. Maximal data compression, i.e. $N_D \to N_P$, even eliminates all adverse noise effects, as can be seen from Eqs. (2.1) and (2.2). Note that in the latter case the factor from the Kaufman bias in Eq. (1.1) remains, which can be removed as the parameter covariance is now a linear transformation of the data covariance, so no inversion is necessary. However, data compression techniques require some information about the data covariance. TJK13 show for maximal Karhunen-Loeve compression that, using the sample covariance in the compression operation, exactly reproduces the original noise effects – nothing is gained. Employing a noise-free model covariance instead renders the compression suboptimal, thus increasing errors on cosmological parameters to a yet unknown degree.

Shrinkage: We are not completely ignorant about the form of covariance matrices for cosmological data. Their elements vary smoothly across angular and redshift scales, and sometimes it is fair to assume that the diagonal dominates the errors, e.g. if shot noise is prominent. In some regimes, such as on large, nearly Gaussian scales, we may even have good, if not exact, analytic models. This is valuable prior information that can be used to improve covariance estimates. One such approach is shrinkage estimation, building a linear combination of the sample covariance and a model covariance (which can contain free parameters). The weighting of the linear combination can be estimated analytically from the data. TJK13 test several shrinkage estimators of covariance for a toy cosmological case (see also Pope & Szapudi 2008). Their worst-performing estimator was derived by Stein *et al.* (1972) who proved that nonetheless their estimator outperforms the sample estimator with respect to a 'natural' loss function based on the mean square error of the mean vector of the data, which implies the sample covariance estimator is inadmissible (see also D. van Dyk's contribution).

References

Dodelson, S. & Schneider, M. D. 2013, *Phys Rev D*, 88, 063537
Hartlap, J., Simon, P., & Schneider, P. 2007, *A&A*, 464, 399
Kaufman, G. M. 1967, *Report No. 6710*, Center for Operations Research and Econometrics, Catholic University of Louvain, Heverlee, Belgium
Percival, W., Ross, A. J., Sanchez, A. G., *et al.* 2014, *MNRAS*, 439, 2531
Pope, A. C. & Szapudi, I. 2008, *Tech. Report No. 37*, Depart. Statist., Stanford University
Stein, C., Efron, B., & Morris, C. 1972, 389, 766
Taylor, A. N. & Joachimi, B. 2014, *MNRAS*, accepted (TJ14)
Taylor, A. N., Joachimi, B., & Kitching, T. 2013, *MNRAS*, 432, 1928 (TJK13)
Wishart, J. 1928, *Biometrika*, 20A, 32

Statistical Challenges in 21st Century Cosmology
Proceedings IAU Symposium No. 306, 2014
A. F. Heavens, J. -L. Starck & A. Krone-Martins, eds.

doi:10.1017/S1743921314010965

Density reconstruction from 3D lensing: Application to galaxy clusters

François Lanusse[1], **Adrienne Leonard**[2], **and Jean-Luc Starck**[1]

[1]Laboratoire AIM, UMR CEA-CNRS-Paris 7, Irfu, Service d'Astrophysique, CEA Saclay, F-91191 Gif-Sur-Yvette CEDEX, France
email: `francois.lanusse@cea.fr`

[2]Department of Physics and Astronomy, University College London, Gower Place, London WC1E 6BT, U.K.
email: `adrienne.leonard@ucl.ac.uk`

Abstract. Using the 3D information provided by photometric or spectroscopic weak lensing surveys, it has become possible in the last few years to address the problem of mapping the matter density contrast in three dimensions from gravitational lensing. We recently proposed a new non linear sparsity based reconstruction method allowing for high resolution reconstruction of the over-density. This new technique represents a significant improvement over previous linear methods and opens the way to new applications of 3D weak lensing density reconstruction. In particular, we demonstrate that for the first time reconstructed over-density maps can be used to detect and characterise galaxy clusters in mass and redshift.

Keywords. gravitational lensing, methods: statistical, galaxies: clusters: general

---

1. Introduction

While techniques exist to generate high-fidelity two-dimensional projected weak-lensing maps, it was not until recently that several linear methods were developed to reconstruct a full three-dimensional density map from photometric weak lensing measurements (Simon *et al.* 2009; VanderPlas *et al.* 2011). However, even the most successful linear methods for 3D lensing density mapping suffer important limitations. In particular they present a broad smearing of structures along the line of sight, the amplitude of the density contrast can be severely underestimated and they remain very noisy. To address these issues we recently proposed GLIMPSE (Leonard *et al.* 2014), a non linear, sparsity based method, which drastically improves the quality of reconstructed 3D maps. Contrary to the linear methods, structures are no longer smeared along the line of sight and we see no systematic bias in the reconstructed redshifts. More importantly, the results of previous linear methods could only be exploited as signal to noise maps to place significance on the detection of structures while GLIMPSE is the first method being able to reconstruct the value of the matter density contrast and with sufficient quality to allow direct mass measurement of the reconstructed structures.

In particular, we demonstrate here that the reconstructed over-density maps can be used to detect and measure the masses and redshifts of galaxy clusters. We test the accuracy of these measurements by simulating the lensing signal generated by a range of clusters of different masses and redshifts in the context of a wide weak lensing survey with photometric redshift information. Potentially, this technique could be used to complement X-ray or optical cluster studies as it does not rely on the assumption of scaling relations or hydrodynamical properties and could help to constrain the cluster mass function.

2. The GLIMPSE algorithm

The 3D weak lensing reconstruction problem can be condensed in the following form, relating the measured shear $\gamma(\boldsymbol{\theta}, z)$ to the unknown matter over-density field $\delta(\boldsymbol{\theta}, z)$:

$$\gamma(\boldsymbol{\theta}, z) = \mathbf{P}_{\gamma\kappa}\mathbf{Q}\delta(\boldsymbol{\theta}, z) + n_\gamma(\boldsymbol{\theta}, z), \tag{2.1}$$

where n_γ is a Gaussian noise on the shear measurements, $\mathbf{P}_{\gamma\kappa}$ is an angular convolution while $\mathbf{Q}$ is a line of sight integration of the density contrast δ with a lensing kernel. The aim of the 3D lensing reconstruction is to invert this linear relation in order to reconstruct δ. However, the lensing operator $\mathbf{Q}$, which spreads out the redshift information of a localized halo at redshift z over all measurements $\gamma(z')$ along the line of sight for $z' > z$, leads to an ill-posed inverse problem i.e. the problem does not accept a unique, stable solution. This very general class of problems can be addressed through the framework of sparse regularisation. This approach aims to recover a robust solution of an inverse problem by promoting solutions that are sparse in an adapted dictionary (e.g. wavelets, discrete cosines, ...). In this framework, the reconstruction problem stated in Equation (2.1) can be recast as the following optimisation problem:

$$\arg\min_{\alpha} \frac{1}{2} \parallel \gamma - \mathbf{P}_{\gamma\kappa}\mathbf{Q}\mathbf{\Phi}\alpha \parallel^2_{\mathbf{\Sigma}^{-1}} + \lambda \parallel \alpha \parallel_1, \tag{2.2}$$

where $\mathbf{\Sigma}$ is the diagonal covariance matrix of the noise, λ is a parameter tuning the sparsity constraint, and $\mathbf{\Phi}$ is a wavelet based dictionary allowing for sparse representation of the dark matter density $\delta = \mathbf{\Phi}^*\alpha$, α being the coefficients of the reconstructed density field δ in $\mathbf{\Phi}$. We build an appropriate dictionary $\mathbf{\Phi}$ to represent the 3D density contrast by combining isotropic undecimated wavelets in the 2D angular domain and Dirac δ-functions along the radial dimension. Finally, the optimisation is performed using the Fast Iterative Soft Thresholding Algorithm (FISTA) (Beck & Teboulle 2009) which is widely used in sparse linear inversion and relies on the following simple iteration:

$$\alpha_{n+1} = ST_\lambda\left(\alpha_n + \mu\mathbf{\Phi}^t\mathbf{Q}^t\mathbf{P}^t_{\gamma\kappa}\mathbf{\Sigma}^{-1}\left[\gamma - \mathbf{P}_{\gamma\kappa}\mathbf{Q}\mathbf{\Phi}\alpha_n\right]\right), \tag{2.3}$$

where $ST_\lambda(x) = \text{sgn}(x)\max(|x| - \lambda, 0)$ is the soft thresholding operator and μ is a gradient descent step ensuring convergence of the algorithm. Full description of the algorithm is provided in Leonard *et al.* (2014).

3. Reconstruction of simulated galaxy clusters

To statistically assess the performance of GLIMPSE for the reconstruction of galaxy clusters, we generate a large number of simulated weak lensing fields for typical NFW halos. Each field contains the lensing signal for a single cluster of virial mass ranging from $3 \times 10^{13} h^{-1}$ $\mathrm{M}_\odot$ to $1 \times 10^{15} h^{-1}$ $\mathrm{M}_\odot$ and redshifts between $z = 0.05$ and $z = 0.75$, taking into account photometric redshift errors. Assuming a typical level of noise for space based weak lensing surveys ($n_{gal} = 30$ arcmin^{-2} and intrinsic galaxy ellipticity of $\sigma_\epsilon = 0.25$) we generate for each field 1000 independent noise realisations and perform a 3D reconstruction for each of them.

Fig. 1(a) shows the fraction of detected clusters over the 1000 noise realisations for each cluster in our sample. We detect all clusters of mass above $10^{15} h^{-1}\mathrm{M}_\odot$ up to redshift $z = 0.55$. However, the mass limit decreases with the redshift and bellow $z = 0.25$, all clusters of mass above $4 \times 10^{14} h^{-1}\mathrm{M}_\odot$ are detected. Fig. 1(b) shows the estimated versus true redshift for a cluster of virial mass $7 \times 10^{14} h^{-1}\mathrm{M}_\odot$. Shown on the figure are the peak of an adaptive kernel density estimate (AKDE) of the reconstructed redshift distribution and its median, with $1 - \sigma$ error bars for both statistics. Both estimates

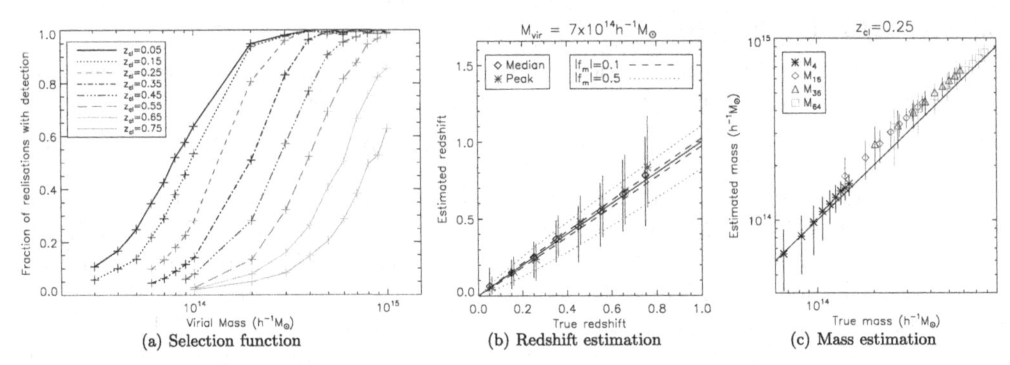

Figure 1. Left: Fraction of detected cluster over 1000 noise realisations as a function of virial mass and redshift. Center: Redshift of reconstructed halos as a function of true redshift for a cluster of $7 \times 10^{14} h^{-1} \mathrm{M}_{\odot}$. Red points indicate the peak of the AKDE distribution while the black points indicate the median of the reconstruction. The dashed and doted lines correspond to the error in redshift that would result in respectively 10 % and 50% error in mass estimate. Right: Estimated mass against true mass for several mass proxies obtained by integrating the density over 4, 16, 32 and 64 pixels for a cluster at $z = 0.25$

clearly yield an accurate estimate of the redshift and we see no systematic bias. We find that the standard deviation of the redshift distribution increases with redshift and for lower masses. Fig. 1(c) shows the estimated mass for a cluster at $z = 0.25$, assuming that the reconstructed redshift is accurate, using as a mass proxy the integrated density within the central 4, 16, 32 and 64 pixels of the density field. As can be seen, this very simple mass proxy gives a very good handle on the true mass of the clusters. However, if the reconstructed redshift is inaccurate, the mass estimate will be biased. Over-plotted in dashed and dotted lines on Fig. 1(b) are the redshift ranges within which the resulting fractional error on the mass estimates remains below 10% (dashed) and 50 % (dotted).

4. Conclusion

The improvement in three-dimensional map quality allowed by the GLIMPSE algorithm opens the way for new applications of 3D weak lensing map making. We have demonstrated that in the context of a wide weak lensing survey, which implies a limited number of galaxies per square degrees and photometric redshift errors, we are able to detect a large proportion, over 50 %, of all galaxy clusters with virial masses above $5 \times 10^{14} h^{-1} \mathrm{M}_{\odot}$ up to redshifts of $z = 0.6 - 0.7$. We find that statistically the distribution of reconstructed redshifts is largely unbiased with a standard deviation for massive clusters (virial mass of $10^{15} h^{-1} \mathrm{M}_{\odot}$) of about $\sigma_z \sim 0.1$ at $z = 0.25$. More importantly, we show that we can use the reconstructed density to build a good proxy for the true halo mass. From these results, we believe that 3D density mapping from weak gravitational lensing has now entered the regime where it can be used as a useful cosmological probe, in particular as a complement of other cluster studies.

This work is supported by the European Research Council grant SparseAstro (ERC-228261).

References

Beck, A. & Teboulle, M. 2009, *SIAM J. Img. Sci.*, 2:1, 183
Leonard, A., Lanusse, F., & Starck, J.-L. 2014 *MNRAS*, 440, 1281
Simon, P., Connolly, A. J., Jain, B., & Jarvis, M. 2009, *MNRAS*, 399, 48
VanderPlas, J. T., Taylor, A. N., & Hartlap, J. 2011, *ApJ*, 118, 727

Statistical Challenges in 21st Century Cosmology
Proceedings IAU Symposium No. 306, 2014
A. F. Heavens, J.-L. Starck & A. Krone-Martins, eds.

doi:10.1017/S1743921314013477

A New Model to Predict Weak Lensing Peak Counts

Chieh-An Lin and Martin Kilbinger
Service d'Astrophysique, CEA Saclay,
Orme des Merisiers, Bât 709, 91191 Gif-sur-Yvette, France
email: chieh-an.lin@cea.fr

Abstract. Peak statistics from weak gravitational lensing have been shown to be a promising tool for cosmology. Here we propose a new approach to predict weak lensing peak counts. For an arbitrary cosmology, we draw dark matter halos from the halo mass function, and calculate the number of peaks from the projected halo mass distribution. This procedure is much faster than time-consuming N-body simulations. By comparing these "fast simulations" to N-body runs, we find that the peak abundance is in very good agreement. Furthermore, our model is able to discriminate cosmologies with different sets of parameters, using high signal-to-noise peaks ($\gtrsim 4$). This encourages us to examine the optimal combinations of parameters to this approach in the future.

Keywords. peak counting, mass function, weak lensing

1. Introduction

Weak gravitational lensing (WL) uses the deflection of light from background galaxies to probe the Universe. It contains the information about the non-linear regime which has been encoded in cosmological structures. Peak statistics from WL is a measure of this non-Gaussian information. Peaks are local maxima of the projected mass distribution, therefore they probe the mass function, which is sensitive to cosmology.

To predict WL peak counts, Fan *et al.* (2010, hereafter FSL10) gave an analytical model using Gaussian random field theory. Their model computes the peak number density function by taking into account Gaussian shape noise. Meanwhile, Maturi *et al.* (2010) provided another analytical model by defining peaks as a contiguous area, and considering aperture-mass peaks with different filter functions. However, analytical models rely on linear filters to suppress noise, which may not be optimal. Furthermore, such models are strongly limited in more realistic scenarios, e.g. in the presence of galaxy intrinsic alignment. This motivated us to adopt a new approach, which is probabilistic.

The idea is to create "fast simulations" using a halo sampling technique. This avoids time-consuming N-body simulations and keeps open the possibility for using non-linear filters. The only requirement is a cosmology with a known mass function.

2. The model

The fast simulations are generated by sampling halos from the mass function of Jenkins *et al.* (2001), in 10 redshift bins from 0 to 1. These halos are randomly placed across the field of view, thus their angular positions are not correlated. The NFW density profile is chosen.

We validate this approach by comparing our fast simulations to N-body runs and two intermediate cases as follows: full N-body runs (case 1); replacing N-body halos with

NFW profiles with the same masses (case 2); randomizing angular positions of halos from the case 2 (case 3); fast simulations, corresponding to our model (case 4). In the case 2 and the case 3, the halo population and their redshift are identical to N-body runs. This allow us to study impacts from two hypothesis on which our model is based: (1) diffuse, unbound matter, for example filaments, does not significatly contribute to peak counts, and (2) the spatial correlation of halos has a minor influence on peak counts.

In this study, we follow Hamana *et al.* (2004) and add the shape noise as Gaussian noise with variance $\sigma_{\rm pix}^2 = \sigma_\epsilon^2/(2n_{\rm g}A_{\rm pix})$, where $\sigma_\epsilon = 0.4$ is the intrinsic ellipticity dispersion of our choice, $n_{\rm g}$ is the galaxy number density, and $A_{\rm pix}$ is the surface of a pixel. To supress the noise, we applied a Gaussian filter function $W_{\rm G}(\boldsymbol{\theta}) = \exp(-\theta^2/\theta_{\rm G}^2)/\pi\theta_{\rm G}^2$, with $\theta_{\rm G} = 1$ arcmin.

We measure peaks as local maxima on a gridded convergence (κ) map. A peak is defined to have a signal-to-noise ratio (SNR) higher than its 8 neighbors. The SNR is given by $\nu = \kappa/\sigma_{\rm noise}$ where $\sigma_{\rm noise}^2 = \sigma_\epsilon^2/(4\pi n_{\rm g}\theta_{\rm G}^2)$. Finally, we compress the information from the WL maps into abundance histograms of peak counts.

3. Simulations

The N-body simulations "Aardvark" used in this study are provided by A. Evrard and generated by GADGET-2 (Springel 2005). The Aardvark parameters had been chosen to be a WMAP-like ΛCDM cosmology, with $\Omega_{\rm m} = 0.23$, $\Omega_\Lambda = 0.77$, $\Omega_{\rm b} = 0.047$, $\sigma_8 = 0.83$, $h = 0.73$, $n_{\rm s} = 1.0$, and $w_0 = -1.0$.

Halos were identified using ROCKSTAR (Behroozi *et al.* 2013). The field of view is 859 deg^2. Ray-tracing for the case 1 was performed with CALCLENS (Becker 2013), available only on a subset of 53.7 deg^2 (16 times smaller). For the other cases, ray-tracing is performed on the entire field. Source galaxies are regularly placed at redshift $z_{\rm s} = 1.0$, and their number density $n_{\rm g}$ is 21.69 arcmin^{-2}.

4. Results and conclusion

Fig. 1 shows the SNR histograms from the 4 studied cases, from which we can make the following observations. First, all cases agree with each other in low ν bins, since peaks with $\nu \lesssim 3{\sim}4$ are dominated by noise (see FSL10). Second, the randomization of halo angular positions is a good approximation (compare the green circles to the red squares). Third, our model is in a good agreement with N-body simulations, except for very high SNR.

The left panel of Fig. 2 gives a comparison between FSL10 and this study. Overall, our model performs better than FSL10. The right panel is a sensitivity test with different inputs of $\Omega_{\rm m}$ and σ_8. One can observe that different cosmologies are discernible for $\nu \gtrsim 4$ and that our model is accurate enough to distinguish between cases.

As a result, we provide a new model which can accurately recover lensing peak abundance. Our model opens the possibility to model non-linearly reconstructed convergence fields. It allows us to study optimal combinations of parameters sensitive to this approach. In this proceeding, the analysis is done for a flat universe without galaxy intrinsic alignment. However, further studies for more complex scenarios are in the progress.

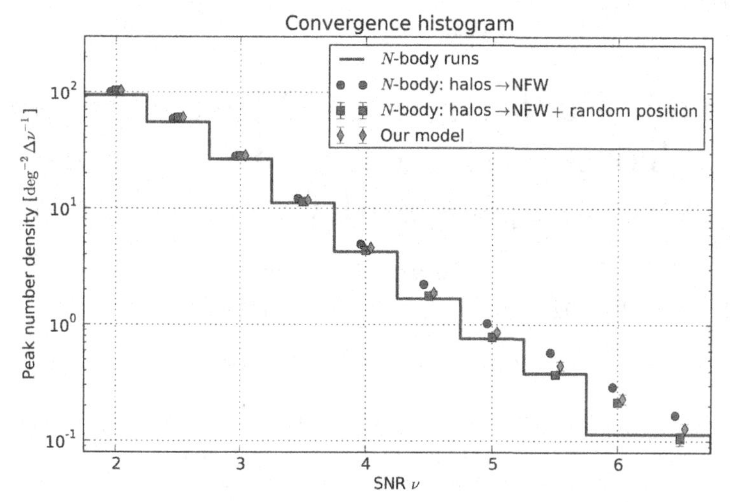

Figure 1. SNR histogram for 4 studied cases. Blue solid line: N-body runs; green circles: replacement of halos by NFW profiles; red squares: replacement of halos by NFW profiles and randomization of halo angular positions; cyan diamonds: our model. The error bars are standard deviations calculated from 8 different realizations. The N-body runs corresponds to 53.7 deg^2, while for the other cases the field of view is 859 deg^2.

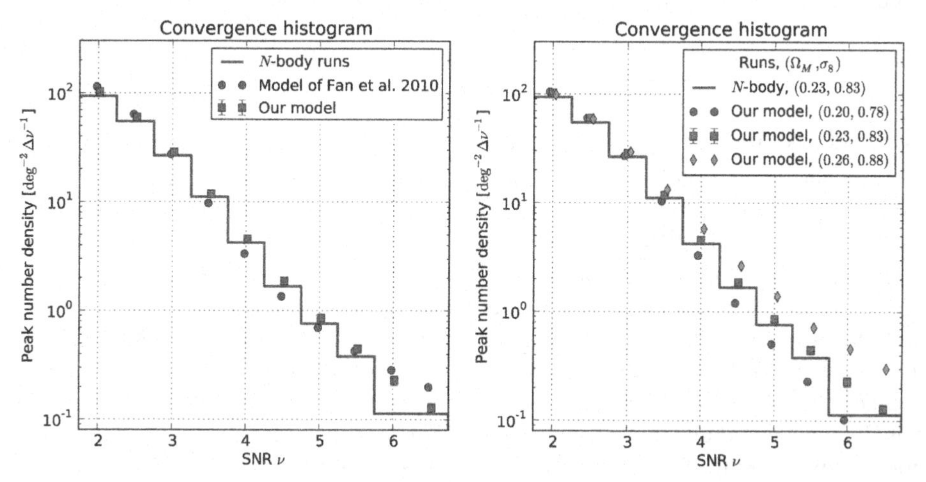

Figure 2. Left: comparison between the model from Fan *et al.* 2010 and this study. Right: the sensitivity test on $(\Omega_{\rm m}, \sigma_8)$. In both plots, the blue solid line corresponds to the N-body runs, and the red squares are our model with the same parameters as the N-body simulations.

References

Becker, Matthew R. 2013, *MNRAS*, 435, 115

Behroozi, P. S., Wechsler, R. H., & Wu, H.-Y. 2013, *ApJ*, 762, 109

Fan, Z., Shan, H., & Liu, J. 2010, *ApJ*, 719, 1408

Hamana, T., Takada, M., & Yoshida, N. 2004, *MNRAS*, 350, 893

Jenkins, A., Frenk, C. S., White, S. D. M., Colberg, J. M., Cole, S., Evrard, A. E., Couchman, H. M. P., & Yoshida, N. 2001, *MNRAS*, 321, 372

Maturi, M., Angrick, C., Pace, F., & Bartelmann, M. 2010, *A&A*, 519, A23

Navarro, J. F., Frenk, C. S., & White, S. D. M. 1996, *ApJ*, 462, 563

Springel, V. 2005, *MNRAS*, 364, 1105

Statistical Challenges in 21st Century Cosmology
Proceedings IAU Symposium No. 306, 2014
A. F. Heavens, J.-L. Starck & A. Krone-Martins, eds.

doi:10.1017/S1743921314013544

Detecting Particle Dark Matter Signatures via Cross-Correlation of Gamma-Ray Anisotropies and Cosmic Shear

Stefano Camera
CENTRA, Instituto Superior Técnico, Universidade de Lisboa,
Avenida Rovisco Pais 1, 1049-001 Lisboa, Portugal
email: stefano.camera@tecnico.ulisboa.pt

Abstract. Similarly to gravitational lensing effects like cosmic shear, cosmological γ-ray emission too is to some extent a tracer of the distribution of dark matter (DM) in the Universe. Intervening DM structures source gravitational lensing distortions of distant galaxy images, and those same galaxies can emit γ rays, either because they host astrophysical sources, or directly by particle DM annihilations or decays occurring in the galactic halo. If such γ rays exhibit correlation with the cosmic shear signal, this will provide novel information on the composition of the extragalactic γ-ray background.

Keywords. Dark matter, gravitational lensing, gamma rays: observations, methods: statistical

1. Introduction

Our current understanding of the Universe is encoded in the concordance cosmological model, which requires a large abundance of cold dark matter (DM) in the present cosmos (Planck Collaboration, 2013). However, very little is known about its nature. One of the most promising interpretations of DM sees it as a Weakly Interacting Massive Particle (WIMP). Many investigation strategies are currently being pursued to the aim of improving our knowledge on the subject. One of the theoretically more promising approaches is the search for WIMP DM signatures in the extragalactic γ-ray background (see e.g. Fornengo, Pieri & Scopel, 2004). The underlying hypothesis is that the DM structures in the Universe can emit light at various wavelengths, including the γ-ray range. Indeed, besides γ rays produced by astrophysical sources hosted inside DM haloes, WIMP DM itself may be a source of γ rays through its self-annihilation (or decay, depending on the properties of the DM particle).

2. Auto-Correlation of Gamma-Ray Anisotropies

The most recent measurement of the extragalactic γ-ray background (EGB) was performed by Abdo *et al.* (2010) for the Fermi Large Area Telescope (LAT), covering a range between 200 MeV and 100 GeV. The emission is obtained by subtracting from the whole Fermi-LAT data the contribution of resolved sources (both point-like and extended) and the Galactic foreground (due to cosmic-ray interaction with the interstellar medium). Unresolved astrophysical sources like blazars and radio galaxies still contribute to the EGB, but the exact amount of their contribution is unknown. Similarly, γ rays produced by DM annihilation or decay in principle also contribute to the EGB.

However, the fact that the measured EGB energy spectrum is well compatible with a power law suggests that DM cannot play a leading rôle in the whole energy range (Abdo *et al.*, 2010). In the angular anisotropies of the EGB emission, DM is even more subdominant. Indeed, a detection of a significant auto-correlation angular power spectrum has been recently reported for multipoles $\ell > 100$, but the absence of distinctive spectral features in the signal point towards an interpretation in terms of point-like sources like blazars (Ackermann *et al.*, 2012; Harding & Abazajian, 2012).

3. Cross-Correlation of Gamma-Ray Anisotropies and Cosmic Shear

Since the study of γ-ray anisotropies does not seem, in itself, capable of detecting WIMP DM signatures, a possible solution is to cross-correlate the γ-ray signal with other tracers of the underlying DM structure. Camera *et al.* (2013) proposed for the first time to employ weak gravitational lensing to this purpose. Weak lensing refers to the small distortions of images of distant galaxies, produced by the distribution of matter located between galaxies and the observer (Bartelmann & Schneider, 2000; Bartelmann, 2011). A distorted image can be parameterised in terms of convergence κ (controlling modifications in the size of the image) and shear γ (accounting for shape distortions). Whilst the former is a direct estimator of matter density fluctuations integrated along the line of sight, the latter is easier to measure, through correlations in the pattern of observed source ellipticities. In the flat-sky approximation, the two generate identical angular power spectra and we thus focus on the shear as an estimator of the convergence. Present and planned surveys like the Dark Energy Survey (DES) and Euclid will reconstruct two-dimensional shear maps at various redshifts, from which we can extract the auto-correlation cosmic shear power spectrum (Abbott *et al.*, 2005; Laureijs *et al.*, 2011; Amendola *et al.*, 2013).

Camera *et al.* (2013) demonstrated that WIMP DM distinctive signatures can be detected in the cross-correlation between the cosmic shear and the EGB. The strength of this method—which can also provide novel information on the composition of the EGB—resides in the fact that the shear signal is stronger for larger halo masses and most of the γ-ray emission from annihilating/decaying DM is produced in large-mass haloes as well. Thus, their cross-correlation is more significant than for the case of astrophysical sources, associated with galaxy-mass haloes. (For more details see Camera *et al.*, 2013, and references therein.)

Since then, Shirasaki *et al.* (2014) measured the 2-point cross-correlation function of Fermi-LAT data and cosmic shear, as detected by the Canada-France-Hawaii Telescope Lensing Survey (Heymans *et al.*, 2012). Their measurement is consistent with no signal and the null detection was thus used to derive constraints on the WIMP annihilation cross section. The upper limits excluded values smaller than the thermal cross section, 3×10^{-26} cm^3/s. This is an additional proof of the potential of such a technique for indirect detection of DM.

3.1. *Tomographic Cross-Correlation of Gamma-Ray Anisotropies and Cosmic Shear*

An even more effective technique is proposed by Camera *et al.* (2014). It involves a tomographic approach to the study of the cross-correlation between cosmic shear and γ-ray emission. To better model the bulk of unresolved astrophysical sources, they also include the contribution of unresolved misaligned Active Galactic Nuclei (MAGNs). This component, together with that of unresolved blazars and star-forming galaxies (SFGs), is associated to a quite large cross-correlation power spectrum. However, the abundance of those classes of sources as a function of redshift or observed γ-ray energy is quite different from that of WIMP DM. Tomography takes advantage of this behaviour by computing the cross-correlation power spectrum in different redshift and energy bins. The study

Table 1. Forecast 1σ marginal errors on decaying DM model parameters for the fiducial values $\{m_\chi, \langle\sigma_a v\rangle, \mathcal{A}_B, \mathcal{A}_{\rm SFG}, \mathcal{A}_{\rm MAGN}\} = \{100\,{\rm GeV}, 3\times10^{-26}\,{\rm cm}^3/{\rm s}, 1, 1, 1\}$.

Binning	m_χ [GeV]	$\langle\sigma_a v\rangle$ [10^{-26} cm^3/s]	$\mathcal{A}_B$ [−]	$\mathcal{A}_{\rm SFG}$ [−]	$\mathcal{A}_{\rm MAGN}$ [−]
—	2.1×10^5	8.0×10^3	4.7×10^7	7.8×10^4	8.2×10^3
E_γ-z	11	0.44	1.3×10^4	1.4×10^2	3.2

of how the power spectrum changes as a function of z or E_γ significantly boosts the capability of the cross-correlation power spectrum to determine the composition of the EGB and thus to distinguish DM from astrophysical components.

This is particularly useful because, on the one hand, the redshift dependence of γ rays generated by astrophysical objects is peculiarly different from what expected for DM-sourced γ rays, and, on the other hand, the energy spectrum of DM-produced γ rays peaks at an energy proportional to the WIMP mass, whereas unresolved astrophysical sources contribute to the EGB with well defined power laws. Preliminary results have been obtained by performing a Fisher matrix analysis for the cross-correlation of γ-ray anisotropies and cosmic shear to forecast the capabilities of a weak lensing survey like DES and a γ-ray experiment as Fermi-LAT in constraining the WIMP DM mass, m_χ, and its annihilation cross-section, $\langle\sigma_a v\rangle$, or its decay rate, Γ_d (Camera *et al.*, 2014). To account for the underlying ignorance on the relative abundances, the amplitudes of the one-halo power spectra of astrophysical sources are included as nuisance parameters. The cosmic shear signal is divided into 3 redshift bins of width $\Delta z = 0.4$ from 0.3 to 1.5, whilst γ-ray measurements into 6 energy bins between 1 and 300 GeV. As an example, the forecast marginal errors on annihilating DM model parameters are summarised in Table 1, without and with the energy-redshift binning to better illustrate impact of tomography. We assume the halo substructure scenario of Gao *et al.* (2012).

Acknowledgments

SC is funded by FCT-Portugal under Post-Doctoral No. Grant SFRH/BPD/80274/2011. SC warmly thanks Mattia Fornasa, Nicolao Fornengo and Marco Regis for allowing him to show results from a common project.

References

Abbott, T., *et al.* 2005, *arXiv:astro-ph/0510346*
Abdo, A. A., *et al.* 2010, *Phys. Rev. Lett.* 104,101101
Ackermann, M., *et al.* 2012, *Phys. Rev.* D85, 083007
Amendola, L., *et al.* 2013, *Living Rel. Rev.* 16, 6
Bartelmann, M., 2010, *Class. Quant. Grav.* 27, 233001
Bartelmann, M. & Schneider, P. 2001, *Phys. Rept.*340, 291
Camera, S., Fornasa, M., Fornengo, N. & Regis, M. 2013, *Astrphys. J.* 771, L5
Camera, S., Fornasa, M., Fornengo, N. & Regis, M. 2014, *arXiv:astro-ph:1411.4651*
Fornengo, N., Pieri, L., & Scopel, S. 2004, *Phys. Rev.* D70, 103529
Gao, L., *et al.* 2012, *Mon. Not. R. Astron. Soc.* 419, 1721
Harding, J. P. & Abazajian, K. N. 2012, *J. Cosmol. Astrop. Phys.* 1211, 026
Heymans, C., *et al.* 2012, *Mon. Not. R. Astron. Soc.* 427, 46
Laureijs, R., *et al.* 2011, *ESA-SRE* 12
Planck Collaboration 2013, *arXiv:1303.5076*
Shirasaki, M., Horiuchi, S., & Yoshida, N. 2014, *arXiv:1404.5503*

Statistical Challenges in 21st Century Cosmology
Proceedings IAU Symposium No. 306, 2014
A. F. Heavens, J.-L. Starck & A. Krone-Martins, eds.

doi:10.1017/S1743921314010916

Cluster strong lensing: a new strategy for testing cosmology with simulations

M. Killedar[1,2,3], S. Borgani[2,3,4], D. Fabjan[6,7], K. Dolag[1,5], G. Granato[2,3] M. Meneghetti[8,9,10], S. Planelles[2,3] and C. Ragone-Figueroa[3,11]

[1]Universitäts-Sternwarte München, Scheinerstrasse 1, D-81679, München, Germany
e-mail: killedar@usm.lmu.de

[2]Astronomy Unit, Department of Physics, University of Trieste, Via Tiepolo 11, I-34131 Trieste, Italy

[3]INAF - Osservatorio Astronomico di Trieste, Via G.B. Tiepolo 11, I-34131 Trieste, Italy

[4]INFN - National Institute for Nuclear Physics, Via Valerio 2, I-34127 Trieste, Italy

[5]Max-Planck-Institut für Astrophysik, Garching, Germany

[6]SPACE-SI, Slovenian Centre of Excellence for Space Sciences and Technologies, Aškerčeva 12, 1000 Ljubljana, Slovenia

[7]Faculty of Mathematics and Physics, University of Ljubljana, Jadranska 19, 1000 Ljubljana, Slovenia

[8]INAF - Osservatorio Astronomico di Bologna, Via Ranzani 1, I-40127 Bologna, Italy

[9]INFN - Sezione di Bologna, Viale Berti Pichat 6/2, I-40127 Bologna, Italy

[10]Jet Propulsion Laboratory, 4800 Oak Grove Dr. Pasadena, CA 91109, USA

[11]Instituto de Astronomía Teórica y Experimental (IATE), Consejo Nacional de Investigaciones Científicas y Técnicas de la República Argentina (CONICET), Observatorio Astronómico, Universidad Nacional de Córdoba, Laprida 854, X5000BGR, Córdoba, Argentina

Abstract. Comparisons between observed and predicted strong lensing properties of galaxy clusters have been used to claim either tension or consistency with ΛCDM cosmology. However, standard approaches to such tests are unable to quantify the preference for one cosmology over another. We advocate a Bayesian approach whereby the parameters defining the scaling relation between Einstein radii and cluster mass are treated as the observables. We demonstrate a method of estimating the likelihood for observing these parameters under the ΛCDM framework, using the X-ray selected $z > 0.5$ MACS clusters as a case in point and employing both N-body and hydrodynamic simulations of clusters. We account for cluster lens triaxiality within the modelling of the likelihood function. Cluster selection criteria is found to play as important a role as the uncertainty related to the description of star formation and feedback.

Keywords. cosmology: miscellaneous, gravitational lensing, galaxies: clusters: general, methods: n-body simulations, methods: statistical, X-rays: galaxies: clusters

---

1. Introduction

Galaxy clusters gravitationally lens and distort the images of background galaxies; their lensing efficiency is a powerful probe of cosmology with the ability to constrain structure formation parameters. The earliest comparisons between simulated clusters and the observed frequency of arc-like lensed galaxy images in a cluster sample revealed an order of magnitude difference between the observations and ΛCDM predictions (Bartelmann *et al.* (1998), Li *et al.* (2005)). This discrepancy was dubbed the 'arc-statistics problem', and now is part of a broader study of cluster concentrations and strong lensing efficiencies.

In the present work, we take the well-studied $z > 0.5$ MACS clusters as our case in point (Horesh *et al.* (2010), Meneghetti *et al.* (2011), Zitrin *et al.* (2011), Waizmann *et al.* (2014)). We propose a Bayesian approach to the strong lensing cosmological test, employing clusters modelled within ΛCDM hydrodynamic simulations which include the effects of stellar and AGN feedback. Massive clusters have been modelled in four different flavours of smooth-particle hydrodynamic (SPH) simulations: with only dark matter (*DM*); including non-radiative hydrodynamics (*NR*); including cooling, star formation and supernova feedback (*CSF*); and further including AGN feedback (*AGN*). For more details on the simulations, see Planelles *et al.* (2014).

2. ΛCDM strong lensing likelihood

Strong lensing efficiencies, as characterised by the Einstein radii, scale well with the mass of clusters at large overdensities (Killedar *et al.* (2012)). If the $z > 0.5$ MACS sample are, in fact, stronger lenses than predicted by the ΛCDM model, they will have larger Einstein radii for a given total mass at low overdensities (or a proxy thereof).

A Bayesian approach to cosmological parameter estimation is advocated, in which one determines the relative preference of two hypothetical choices of cosmological parameters, C_1 and C_2, in light of the data D, by calculating the likelihood ratios: $\mathcal{L}(D|C_1)/\mathcal{L}(D|C_2)$, and subsequently multiplying by their relative priors. The initial aim is to calculate the likelihood of observing the Einstein radii of the high-z MACS sample under one chosen hypothesis: ΛCDM$\{\Omega_{\Lambda,0} = 0.76; \Omega_{\mathrm{M},0} = 0.24; \Omega_{\mathrm{b},0} = 0.04; h = 0.72; \sigma_8 = 0.8; P(k) \propto k^n$ with $n = 0.96\}$, with the aid of mock samples from numerical simulations. The likelihood function related to the observables (θ_E and M_{500}) is intractable because there are a finite number of objects from the simulations. Instead, we assume **a power-law relation between the strong lensing and mass**:

$$\log\left[\frac{M_{500}}{9\times 10^{14} M_\odot}\right] = \alpha \log\left[\frac{\theta_E}{20''}\sqrt{\frac{D_d}{D_{ds}}}\right] + \beta\,, \tag{2.1}$$

and use the MCMC sampler `emcee` (Foreman-Mackey *et al.* (2013)) to infer the joint posterior, $P(\alpha,\beta)$ while marginalising out the intrinsic scatter (Hogg *et al.* 2010). We average over the posteriors of many mock samples and re-interpret this as a 'likelihood function': the probability that one would observe the scaling relation $\{\alpha,\beta\}$ under the hypothesis that ΛCDM is the true description of cosmology. From the MACS clusters, we obtain the posterior $P(\alpha,\beta)$ which is interpreted as a single 'data point'. We calculate the likelihood, $\mathcal{L}$, of observing $\{\alpha,\beta\}$ by convolving the 'data-point' and 'likelihood function'.

Note that one cannot comment on whether the likelihood is *large* or *small*. One cannot use this value to claim 'consistency' or 'tension' with ΛCDM. However, if the same process is repeated for simulations under a different cosmological model then the Bayes factor can be calculated and, after accounting for priors, it may (or may not) reveal a preference for one of the cosmologies, in light of this data.

3. Cluster Selection

In the left panel of Fig. 1 we show the relation between the Einstein radii and the cluster mass M_{500}. Measuring the strong lensing likelihood using $z = 0.5$ clusters from the *AGN* simulations that exceed the X-ray flux threshold, we find: $\mathcal{L} = 0.27$.

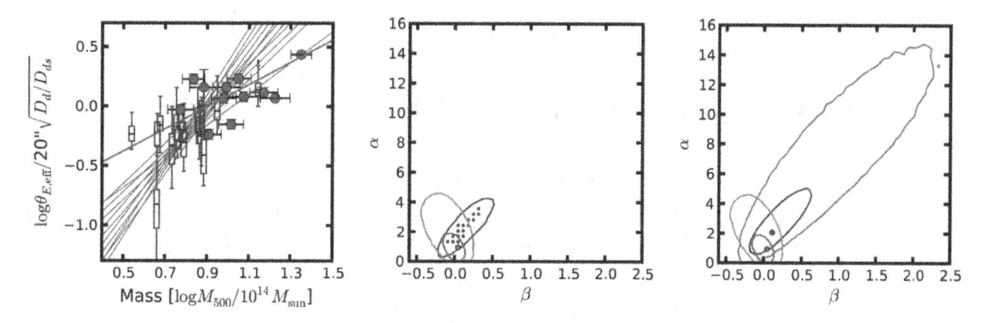

Figure 1. *Left*: Einstein radii, $\theta_{\rm E,eff}$, plotted as a function of M_{500}. The range of Einstein radii for each $z = 0.5$ cluster from the *AGN* simulations is shown by a box-plot. The circles represent the MACS $z > 0.5$ clusters (Mantz *et al.* (2010)). The thick line marks the maximum a-posteriori fit to observational data, while the thin lines mark the fit to 20 randomly chosen mock samples from simulations. *Middle*: 1-σ and 2-σ constraints on parameters of the strong lensing - mass relation given the MACS $z > 0.5$ cluster data (contours), with a maximum a posteriori fit marked by a filled circle. Overplotted in dots are the best fits to 80 mock observations of the simulated cluster sample. A typical 1-σ error is shown as an ellipse. *Right*: Constraints from the MACS $z > 0.5$ cluster data are the same as in the middle panel, but the dark filled circle and curves mark respectively the maximum and the 1 and 2-σ contours of the likelihood function found by combining all 80 mocks.

We then consider how the measured likelihood for ΛCDM may depend on other details, such as cluster selection criteria and the numerical implementation of hydrodynamics.

Selection by dynamical state: Consider now the effect of applying the incorrect selection criteria when modelling lenses. Excluding the most disturbed clusters, the likelihood function derived from simulations is more sharply peaked than that which is derived without this additional selection. If this relaxed simulated sample is used to analyse the full observational sample, one would incorrectly derive a likelihood of $\mathcal{L} = 1.23$.

Selection by mass: Since the clusters in the MACS survey are selected by flux rather than luminosity, there is no corresponding mass threshold, yet it is common practise to select simulated clusters by mass for similar studies. We find that when selecting simulated clusters by mass (instead of flux), the likelihood increases to $\mathcal{L} = 0.54$.

Hydrodynamics: We determine the sensitivity of our conclusions to the inclusion of baryonic processes and the resulting gas distribution. Using cluster counterparts from the *CSF* simulations leads to $\mathcal{L} = 1.28$, four times that derived from the *AGN* simulations, while the *DM* and *NR* simulations result in $\mathcal{L} = 0.16$ and 0.11 respectively.

References

Bartelmann, M., Huss, A., Colberg, J. M., Jenkins, A. , & Pearce, F. R. 1998, *A&A*, 330, 1

Foreman-Mackey D., Hogg D. W., Lang D., & Goodman J. 2013, *PASP*, 125, 306

Hogg, D. W., Bovy, J., & Lang, D. (2010), ArXiv: 1008.4686

Horesh, A., Maoz, D., Ebeling, H., Seidel, G., & Bartelmann, M. 2010, *MNRAS*, 406, 1318

Killedar, M., Borgani, S., Meneghetti, M., Dolag, K., Fabjan, D., & Tornatore, L. 2012, *MNRAS*, 427, 533

Li, G.-L., Mao, S., Jing, Y. P., Bartelmann, M., Kang, X., & Meneghetti, M. 2005, *ApJ*, 635, 795

Mantz, A., Allen, S. W., Ebeling, H., Rapetti, D., & Drlica-Wagner, A. 2010, *MNRAS*, 406, 1773

Meneghetti, M., Fedeli, C., Zitrin, A., *et al.* 2011, *A&A*, 530, A17

Planelles S., Borgani S., Fabjan D. *et al.* 2014, *MNRAS*, 438, 195

Waizmann, J.-C., Redlich, M., Meneghetti, M., & Bartelmann, M. 2014, *A&A*, 565, A28

Zitrin, A., Broadhurst, T., Barkana, R., Rephaeli, Y., & Benítez, N. 2011, *MNRAS*, 410, 1939

Statistical Challenges in 21st Century Cosmology
Proceedings IAU Symposium No. 306, 2014
A. F. Heavens, J.-L. Starck & A. Krone-Martins, eds.

doi:10.1017/S1743921314011119

Statistics of cosmological fields

Sabino Matarrese[1,2]

[1]Dipartimento di Fisica e Astronomia "G. Galilei", Università degli Studi di Padova and INFN Sezione di Padova, via Marzolo 8, 35131 Padova, Italy

[2]Gran Sasso Science Institute, INFN, viale F. Crispi 7, 67100 L'Aquila, Italy
email: sabino.matarrese@pd.infn.it

Abstract. The general problem of the statistics of the primordial curvature perturbation field in cosmology is reviewed. The search for non-Gaussian signatures in cosmological perturbations, originated from inflation in the early Universe is discussed both from the theoretical point of view and in connection with constraints coming from recent observations and future prospects for observing/constraining them.

Keywords. Cosmology: early Universe: Inflation; Cosmic Microwave Background Radiation; Large Scale Structure of Universe; non-Gaussianity

1. Introduction

A very relevant fraction of the datasets used to obtain cosmological parameters and to understand the overall cosmological scenario relies on cosmological random fields, like the matter density fluctuation field, the gravitational potential field, the galaxy number density fluctuation field and their peculiar velocity field. These random fields are considered to be mutually connected either because of some common origin or owing to some dynamical mechanism.

In most theoretical treatments these fields are considered to be well described by Gaussian random fields prior to the action of gravity, that tends to create mode-coupling, owing to its intrinsically non-linear character. Of course, we see phase coherence (i.e. deviation from Gaussianity) in the sky, even on very large scales. A simple – and by many respects attractive – idea is that all the non-Gaussianity we observe is of gravitational origin and that perturbations (e.g. in the matter density field or in the so-called peculiar gravitational potential) were primordially Gaussian.

Historically, the first determination of the galaxy 3-point correlation function, i.e. the first evidence for a non-Gaussian signal in the galaxy distribution, was obtained in the late seventies, when Groth and Peebles (Groth & Peebles 1977) analysed the high-resolution Shane-Wirtanen catalog of galaxies and fitted the spatial three-point function ζ (indirectly obtained from angular correlations, thanks to the Limber equation) by the so-called hierarchical model: $\zeta(1,2,3) = Q\,(\xi(1,2) + \xi(1,3) + \xi(2.3))$, where $\xi(i,j)$ indicates the spatial two-point function at separation $|\mathbf{r}_i - \mathbf{r}_j|$ and Q is a phenomenological constant that they determined to be $Q = 1.29 \pm 0.21$.

The expectation was (and still is) that such a hierarchical formula arises due to the non-linear gravitational action. At that time, however, there also existed some radically alternative ideas on the origin of the large-scale structure of the Universe, which made of use of *strongly* non-Gaussian initial conditions for clustering: the so-called explosion scenario and various scenarios based upon the idea that cosmic defects (strings, textures) could have triggered structure formation in the Universe. Numerical N-body simulations of Large-Scale Structure (LSS), assuming some simple non-Gaussian models for the initial conditions were also performed (Moscardini *et al.* 1991; Weinberg & Cole 1992). Simple ideas were also proposed to study non-Gaussianity in Cosmic Microwave Background (CMB) temperature anisotropies (Coles & Barrow 1987).

Independently on the specific reasons which led to abandon these alternative ideas, we can say that the incredible improvement on the amount and quality of data on galaxy clustering and CMB temperature anisotropies (which had not been yet observed at that time) implied that the "rule of the game" on non-Gaussianity has completely changed! Nowadays we restrict ourselves to consider only slight variations on the Gaussian paradigm. Primordial fluctuations are assumed to be either exactly Gaussian or very mildly non-Gaussian, as described in the next section.

The plan of the paper is as follows. In Section 2 Ithe early-Universe model that is used to describe the mildly non-Gaussian random field considered in the cosmological framework will be introduced. Section 3 deals with CMB constraints on primordial non-Gaussianity. Section 4 briefly describes how LSS constrains primordial non-Gaussianity. Section 5 provides some general conclusions.

2. Origin of perturbations and primordial non-Gaussianity

It has now become standard practice to parametrise primordial non-Gaussianity by means of a Taylor expansion in powers of a Gaussian zero-mean field φ. One writes (Gangui *et al.* 1994; Wang & Kamionkowski 2000; Komatsu & Spergel 2001)

$$\Phi(\mathbf{x}) = \varphi(\mathbf{x}) + f_{\rm NL} \star \left(\varphi(\mathbf{x})^2 - \langle\varphi^2\rangle\right) + g_{\rm NL} \star \left(\varphi(\mathbf{x})^3 - \langle\varphi^2\rangle\varphi(\mathbf{x})\right) + \dots \quad (2.1)$$

Here the potential Φ is defined in terms of the "comoving curvature perturbation" ζ on super-horizon scales by $\Phi \equiv (3/5)\zeta$. In matter domination, on super-horizon scales, Φ is equivalent to Bardeen's gauge-invariant gravitational potential (Bardeen 1980), and I adopt this notation for historical consistency. The non-linearity parameters $f_{\rm NL}$ and $g_{\rm NL}$ set the amplitude of quadratic and cubic non-Gaussianity are constants in the simplest case (dubbed *local* non-Gaussianity) or may depend on space themselves for more general shapes. In all generality the symbol $\star$ denotes a convolution which reduces to a product in the local case (see e.g. Bartolo *et al.* 2004). Notice that, owing to the smallness of the gravitational potential itself (and hence of the Gaussian field φ) whose r.m.s. is smaller than 10^{-5}, our Taylor expansion makes sense even for relatively large values of $f_{\rm NL}$ and even larger values of $g_{\rm NL}$. In other words the percent of quadratic non-Gaussianity in the model is $f_{\rm NL}\Phi_{\rm rms}$ (and $g_{\rm NL}\Phi^2_{\rm rms}$ for cubic non-Gaussianity).

Besides having the great advantage of simplicity, the model above is actually well-motivated in the frame of inflationary models for the early Universe (see e.g. Bartolo *et al.* 2004, Chen 2010 for a review), which indeed predict non-Gaussianity of this type. The shapes of non-Gaussianity are fully described by their impact on the bispectrum (we restrict ourselves to quadratic non-Gaussianity, for simplicity; the case of cubic NG follows very similar lines, making use of the trispectrum). which is defined as

$$\langle\Phi(\vec{k}_1)\Phi(\vec{k}_2)\Phi(\vec{k}_3)\rangle = (2\pi)^3\delta^{(3)}(\vec{k}_1 + \vec{k}_2 + \vec{k}_3)B_\Phi(k_1, k_2, k_3), \quad (2.2)$$

The bispectrum $B_\Phi(k_1, k_2, k_3)$ measures the correlation among 3 perturbation modes. Assuming translational and rotational invariance, it depends only on the magnitudes of the three wave-vectors. In general, the bispectrum can be written as

$$B_\Phi(k_1, k_2, k_3) = f_{\rm NL}F(k_1, k_2, k_3)\,. \quad (2.3)$$

The bispectrum is measured by sampling triangles in Fourier space. The dependence of the function $F(k_1, k_2, k_3)$ on the type of triangle (i.e., the configuration) formed by the three wave-vectors describes the *shape* (and the scale-dependence) of the bispectrum, which encodes a lot of physical information. Different non-Gaussianity shapes are linked to distinctive physical mechanisms that can generate such non-Gaussian fingerprints in the early Universe. Notice that, according to the latter notation, we treat $f_{\rm NL}$ as a constant and ascribe the shape dependence to the function $F(k_1, k_2, k_3)$. Let us provide here the most important shapes. We have

$$\begin{aligned} B_\Phi^{\rm local}(k_1,k_2,k_3) &= 2f_{\rm NL}^{\rm local}\Big[P_\Phi(k_1)P_\Phi(k_2)+P_\Phi(k_1)P_\Phi(k_3) \\ &+ P_\Phi(k_2)P_\Phi(k_3)\Big] \\ &= 2A^2 f_{\rm NL}^{\rm local}\left[\frac{1}{k_1^{4-n_s}k_2^{4-n_s}}+{\rm cycl.}\right], \end{aligned} \tag{2.4}$$

for the local case, havine parametrised the Φ power-spectrum as $P_\Phi(k) = Ak^{n_s-4}$, with A is a constant amplitude and n_s is the spectral index of scalar perturbations.

"Equilateral" non-Gaussianity is described via

$$\begin{aligned} &B_\Phi^{\rm equil}(k_1,k_2,k_3) = 6A^2 f_{\rm NL}^{\rm equil} \\ &\times \left\{-\frac{1}{k_1^{4-n_s}k_2^{4-n_s}}-\frac{1}{k_2^{4-n_s}k_3^{4-n_s}}-\frac{1}{k_3^{4-n_s}k_1^{4-n_s}}-\frac{2}{(k_1k_2k_3)^{2(4-n_s)/3}}\right. \\ &\left.+\left[\frac{1}{k_1^{(4-n_s)/3}k_2^{2(4-n_s)/3}k_3^{4-n_s}}+(5\text{ permutations})\right]\right\}, \end{aligned} \tag{2.5}$$

"Orthogonal" non-Gaussianity can be described by the template

$$\begin{aligned} &B_\Phi^{\rm ortho}(k_1,k_2,k_3) = 6A^2 f_{\rm NL}^{\rm ortho} \\ &\times \left\{-\frac{3}{k_1^{4-n_s}k_2^{4-n_s}}-\frac{3}{k_2^{4-n_s}k_3^{4-n_s}}-\frac{3}{k_3^{4-n_s}k_1^{4-n_s}}-\frac{8}{(k_1k_2k_3)^{2(4-n_s)/3}}\right. \\ &\left.+\left[\frac{3}{k_1^{(4-n_s)/3}k_2^{2(4-n_s)/3}k_3^{4-n_s}}+(5\text{ perm.})\right]\right\}. \end{aligned} \tag{2.6}$$

Other bispectrum shapes are of course allowed. We refer to ... for a general discussion of this important problem, as well as to the general discussion of trispeectrum shapes.

3. Cosmic Microwave Background constraints on primordial non-Gaussianity

As described above, quadratic inflationary non-Gaussianity in Bardeen's gravitational potential can be characterised by the dimensionless non-linearity parameter $f_{\rm NL}$, for any given non-Gaussianity shape. The *Planck* collaboration (see Planck Collaboration 2013a, for general overview) estimated $f_{\rm NL}$ for various non-Gaussianity shapes – including the three fundamental ones, local, equilateral, and orthogonal – predicted by different classes of inflationary models, using nominal mission CMB temperature maps.

Results for these three fundamental shapes are $f_{\rm NL}^{\rm local} = 2.7 \pm 5.8$, $f_{\rm NL}^{\rm equil} = -42 \pm 75$, $f_{\rm NL}^{\rm ortho} = -25 \pm 39$ (Planck Collaboration 2013b). These results were obtained using a suite of optimal bispectrum estimators. The reported values are obtained after marginalising over the bispectrum contribution of diffuse point-sources – assumed to be Poissonian – and subtracting the bias due to the secondary bispectrum arising from the coupling of the Integrated Sachs-Wolfe (ISW) effect and the weak gravitational lensing of CMB photons.

The *Planck* collaboration (Planck Collaboration 2013b) also obtained constraints on key, primordial, non-Gaussian models and provided a survey of scale-dependent features and resonance models of inflation.

It is worth mentioning here that the *Planck* collaboration will soon release results based on full mission CMB temperature data plus CMB polarisation. The analysis of such an extended dataset is expected to improve accuracy in constraining the non-linearity parameter $f_{\rm NL}$ by a factor $\sim 30\%$.

Let me now very synthetically describe the way the analysis of CMB data – to the goal of searching for primordial non-Gaussiani signals described by Eq. (2.1) – is performed. The CMB temperature field can be characterised using the multipoles of a spherical harmonic decomposition of the CMB temperature map

$$\frac{\Delta T}{T}(\mathbf{x}_O, \hat{n}) = \sum_{\ell m} a^T_{\ell m}(\mathbf{x}_O) Y^m_\ell(\hat{n}) \,, \tag{3.1}$$

where Y^m_ℓ are spherical harmonics and $\mathbf{x}_O$ is the observer's position. At linear order, the relation between the primordial perturbation field and the CMB multipoles reads

$$a_{\ell m}(\mathbf{x}_O) = 4\pi(-i)^\ell \int \frac{d^3k}{(2\pi)^3} e^{\mathbf{k}\cdot\mathbf{x}_O} \, \Phi(\mathbf{k}) Y^m_\ell(\hat{\mathbf{k}}) \Delta_\ell(k) \,, \tag{3.2}$$

where Φ is our primordial (Bardeen's) gravitational potential and Δ_ℓ the linear CMB radiation temperature transfer function and, with our any loss of generality, we can set the observer's position in the origin $\mathbf{x}_O = \mathbf{0}$.

The CMB angular bispectrum is defined as

$$B^{m_1 m_2 m_3}_{\ell_1 \ell_2 \ell_3} \equiv \langle a_{\ell_1 m_1} a_{\ell_2 m_2} a_{\ell_3 m_3} \rangle \,. \tag{3.3}$$

If the CMB sky is rotationally invariant the angular bispectrum can be factorised as

$$\langle a_{\ell_1 m_1} a_{\ell_2 m_2} a_{\ell_3 m_3} \rangle = G^{\ell_1 \ell_2 \ell_3}_{m_1 m_2 m_3} \, b_{\ell_1 \ell_2 \ell_3} \,, \tag{3.4}$$

where $b_{\ell_1 \ell_2 \ell_3}$ is the so-called *reduced bispectrum*, and $G^{\ell_1 \ell_2 \ell_3}_{m_1 m_2 m_3}$ is the Gaunt integral, defined as the integral over the solid angle of the product of three spherical harmonics.

$$G^{\ell_1 \ell_2 \ell_3}_{m_1 m_2 m_3} \equiv \int Y_{\ell_1 m_1}(\hat{n}) \, Y_{\ell_2 m_2}(\hat{n}) \, Y_{\ell_3 m_3}(\hat{n}) \, d^2\hat{n} \,. \tag{3.5}$$

The Gaunt integral, which can also be written in terms of Wigner 3j-symbols, enforces rotational symmetry, and allows us to restrict attention to a tetrahedral domain of multipole triplets $\{\ell_1, \ell_2, \ell_3\}$, satisfying both a triangle condition and a limit given by some maximum resolution $\ell_{\rm max}$ of the given experiment.

To the goal of extracting the non-linearity parameter $f_{\rm NL}$ from the data, for different primordial shapes, one fits a theoretical CMB bispectrum $b_{\ell_1 \ell_2 \ell_3}$ to the observed 3-point function (see e.g. Liguori *et al.* 2010, for a general introduction). Theoretical predictions for CMB angular bispectra arising from inflation models can be obtained by applying the relation between $a_{\ell m}$ and Φ to the primordial bispectra of the previous section.

A very general expression for the optimal estimator of non-linearity parameter has been recently derived by (Verde *et al.* 2013). Such an expression is based on a second-order Edgeworth expansion for the multivariate PDF of the harmonic coefficients $a_{\ell m}$; it reads

$$\mathcal{P}(a|f_{\rm NL}) = \frac{(\det C^{-1})^{1/2}}{(2\pi)^{n/2}} \exp\left[-\frac{1}{2} \sum_{\ell\ell' m m'} a^{*m}_\ell (C^{-1})_{\ell m \ell' m'} a^{m'}_{\ell'})\right] \times$$

$$
\Big\{1+\frac{1}{6}\sum_{\mathrm{all}\,\ell_i m_j}\langle a_{\ell_1}^{m_1}a_{\ell_2}^{m_2}a_{\ell_3}^{m_3}\rangle\Big[(C^{-1}a)_{\ell_1}^{m_1}(C^{-1}a)_{\ell_2}^{m_2}(C^{-1}a)_{\ell_3}^{m_3}-3(C^{-1})_{l_1,l_2}^{m_1m_2}(C^{-1}a)_{l_3}^{m_3}\Big]+
$$

$$
\frac{1}{24}\sum_{\mathrm{all}\,\ell m}\langle a_{\ell_1}^{m_1}a_{\ell_2}^{m_2}a_{\ell_3}^{m_3}a_{\ell_4}^{m_4}\rangle\big[3(C^{-1})_{\ell_1\ell_2}^{m_1m_2}(C^{-1})_{\ell_3,\ell_4}^{m_3m_4}
$$

$$
-6(C^{-1})_{\ell_1,\ell_2}^{m_1m_2}(C^{-1}a)_{\ell_3}^{m_3}(C^{-1}a)_{\ell_4}^{m_4}+(C^{-1}a)_{\ell_1}^{m_1}(C^{-1}a)_{\ell_2}^{m_2}(C^{-1}a)_{\ell_3}^{m_3}(C^{-1}a)_{\ell_4}^{m_4}\big]+
$$

$$
\frac{1}{72}\sum_{l_1,\dots,l_6}\langle a_{\ell_1}^{m_1}a_{\ell_2}^{m_2}a_{\ell_3}^{m_3}\rangle\langle a_{\ell_4}^{m_4}a_{\ell_5}^{m_5}a_{\ell_6}^{m_6}\rangle\times
$$

$$
\big[(C^{-1}a)_{\ell_1}^{m_1}(C^{-1}a)_{\ell_2}^{m_2}(C^{-1}a)_{\ell_3}^{m_3}(C^{-1}a)_{\ell_4}^{m_4}(C^{-1}a)_{\ell_5}^{m_5}(C^{-1}a)_{\ell_6}^{m_6}
$$

$$
-15(C^{-1})_{\ell_1\ell_2}^{m_1m_2}\left((C^{-1}a)_{\ell_3}^{m_3}(C^{-1}a)_{\ell_4}^{m_4}(C^{-1}a)_{\ell_5}^{m_5}(C^{-1}a)_{\ell_6}^{m_6}+(C^{-1})_{\ell_3\ell_4}^{m_3m_4}(C^{-1})_{\ell_5\ell_6}^{m_5m_6}\right)
$$

$$
+45(C^{-1})_{\ell_1\ell_2}^{m_1m_2}(C^{-1})_{\ell_3\ell_4}^{m_3m_4}(C^{-1}a)_{\ell_5}^{m_5}(C^{-1}a)_{\ell_6}^{m_6}\big]\Big\}, \tag{3.6}
$$

where C^{-1} is the inverse of the covariance matrix $C_{\ell_1 m_1,\ell_2 m_2}\equiv\langle a_{\ell_1 m_1}a_{\ell_2 m_2}\rangle$,.

In the second line of the latter equation one can recognise the standard formulation for approximating the PDF (Babich 2005) which is the starting point to derive the standard $f_{\rm NL}$ estimator (see e.g. Komatsu 2010). applied by the *Planck* collaboration (see Planck Collaboration 2013b, for details), which can be written as (Babich 2005; Creminelli *et al.* 2006; Yadav *et al.* 2008; Senatore, Smith & Zaldarriaga 2010)

$$
\hat{f}_{\rm NL}=\frac{1}{N}\sum_{\ell_i,m_i}G^{\ell_1\,\ell_2\,\ell_3}_{m_1m_2m_3}b^{\rm th}_{\ell_1\ell_2\ell_3}\times \tag{3.7}
$$

$$
\left[C^{-1}_{\ell_1m_1,\ell_1'm_1'}a_{\ell_1'm_1'}C^{-1}_{\ell_2m_2,\ell_2'm_2'}a_{\ell_2'm_2'}C^{-1}_{\ell_3m_3,\ell_3'm_3'}a_{\ell_3'm_3'}-3\,C^{-1}_{\ell_1m_1,\ell_2m_2}C^{-1}_{\ell_3m_3,\ell_3'm_3'}a_{\ell_3'm_3'}\right],
$$

where N is a suitable normalisation chosen to produce unit response to $b^{\rm th}_{\ell_1\ell_2\ell_3}$.

It should be mentioned here that such a standard estimator can be expressed in different ways, depending on how one expands the reduced bispectrum: KSW (Komatsu, Spergel & Wandelt 2005), modal (Fergusson, Liguori & Shellard 2010, 2012), binned (Bucher, Van Tent & Carvalho 2010), skew-C_ℓ (Munshi & Heavens 2010), needlet (Donzelli *et al.* 2012) and wavelet (Curto, Mafrtinez-Gonzalez & Barreiro 2009) bispectrum algorithms represent different representations of the same underlying optimal bispectrum estimator.

In the following lines of Eq. (3.6) one can recognise Eq. (32), but more specifically Eq. (158), of (Regan *et al.* 2010) as well as "new" terms that arise from expanding the exponential to second order in $f_{\rm NL}$, thus involving a term proportional to the bispectrum squared. Interpreting this as a likelihood for $f_{\rm NL}$ enables one to combine optimally bispectrum and trispectrum measurements and obtain both best-fit value and confidence intervals for the non-Gaussianity parameter.

This expression is valid for general non-Gaussianity shapes, as long as deviations from Gaussianity are small. This expression is strictly speaking second order in $f_{\rm NL}$. Within a given non-Gaussian model where the bispectrum and trispectrum amplitudes are specified by a single parameter, this PDF can be used to constrain such a parameter. Moreover, in a model-independent approach one could use the above PDF to find joint constraints on the amplitude of the bispectrum and trispectrum for comparison with theory.

4. Large-Scale Structure constraints on primordial non-Gaussianity

The search for signatures of primordial non-Gaussianity in LSS data is made more complex by the very fact that non-linear gravity adds up its own non-Gaussian imprints on LSS observables during the evolution of perturbations. A second complication arises from the unavoidably "biased" which connects the galaxy to the matter distributions. Such a relation (called "galaxy bias" by cosmologists) may well involve non-linear, hence non-Gaussian terms (e.g. Verde *et al.* 2000). On the other hand, the LSS datasets has the obvious statistical advantage of being intrinsically 3-dimensional, contrary to the CMB pattern, which is restricted to a sphere around the observer; this fact can be in principle exploited to improve our determination of e.g. $f_{\rm NL}$ (e.g. Sefusatti & Komatsu 2007).

This field, however, experienced a remarkable boost (see e.g. Komatsu *et al.* 2009; Liguori *et al.* 2010; Verde 2010, for recent reviews on the field) when theoretical cosmologists started to work-out the consequences of a well-known fact: cosmic objects preferentially form on matter density fluctuation *peaks.* See also Licia Verde's contribution to these proceedings.

This idea can be traced back to the Press & Schechter model (Press & Schechter 1974), as well as to the Kaiser model for the formation of galaxy clusters (Kaiser 1984), later extended to the galaxies themselves (Bardeen *et al.* 1986).

Needless to say, peaks represent *rare events* in the underlying dark matter density field: such peaks, while having the apparent disadvantage of being rare they have the obvious advantage of probing the tails of the underlying PDF, hence being more sensitive to deviations from Gaussianity of the underlying matter distribution. If we make the rough approximation that all existing galaxies reside In suitably high peaks of the dark matter density field, there is no loss of information arising from considering these "rare" events (unless one can somehow observe the underlying gravitational potential field itself, as is the case of gravitational lensing).

Moreover, on suitably large scales, one should expect that matter density peaks are less affected by non-linear gravitational evolution than the underlying (smoothed) matter density field, hence preserving a better memory of their initial conditions, including their primordial statistical distribution.

It is this very fact which makes the analysis of galaxy clustering a potential gold mine from the point of view of the search for primordial non-Gaussian signatures. The effects of primordial non-Gaussianity on the clustering of peaks was studied in the eighties (Grinstein & Wise 1986; Matarrese, Lucchin & Bonometto 1986; Lucchin & Matarrese 1987; Lucchin, Matarrese & Vittorio 1988), when very general relations were obtained. The implementation of the physically motivated non-Gaussian model of Eq.(2.1) to analyse the clustering of peaks led to some very interesting results for observables such as the mass-function of cosmic objects (Matarrese, Verde & Jimenez 2000; Verde *et al.* 2001) the linear bias of dark mater halos (Dalal *et al.* 2008; Matarrese & Verde 2008), as appearing in the galaxy power-spectrum, as well as some promising results on higher-order correlations (Giannantonio & Porciani 2010; Baldauf, Seljak & Senatore 2011).

A very promising technique to constrain primordial non-Gaussianity is that of studying the large-scale limit of galaxy biasing, which, in the presence of primordial non-Gaussianity as described by the model of Eq. (2.1) gets an extra contribution, linearly proportional to the parameter $f_{\rm NL}$. In the local case, one indeed gets a scale-dependent term which, for small wave numbers k goes like $\Delta b_{\rm NG} \propto f_{\rm NL} k^{-2}$ (Dalal *et al.* 2008; Matarrese & Verde 2008).

Analyses of available LSS datasets have so far led to interesting and promising results (see, e.g. Slosar *et al.* 2008; Xia *et al.* 2011, Giannantonio *et al.* 2014; Karagiannis, Shanks

& Ross 2014; Leistedt, Peiris & Roth 2014). Owing to uncertainties on systematics affecting galaxy surveys, the present limits on $f_{\rm NL}$ are still uncertain. For instance, Leistedt *et al.* (Leistedt, Peiris & Roth 2014), analysing the clustering of 800,000 photometric quasars from the Sloan Digital Sky Survey in the redshift range $0.5 < z < 3.5$ obtain $-49 < f_{\rm NL} < 31$ (95% CL) for the local case.

One should mention here that future prospects in this field are extremely exciting. Future galaxy surveys are indeed expected to provide constraints to values of local $f_{\rm NL}$ around unity (e.g. Carbone, Verde & Matarrese 2008), hence opening the window to the possibility of probing signatures of General Relativity on LSS (Verde & Matarrese 2009; Bruni, Hidalgo & Wands 2014).

5. Conclusions

The analysis of primordial non-Gaussianity in cosmology proved to be an extremely relevant source of information on the physics of the early Universe. Indeed, contrary to earlier naive expectations of the eighties, some level of non-Gaussianity is generically present in all inflation models. The level of non-Gaussianity predicted in the simplest (single-field, slow-roll, canonical kinetic term, "Bunch-Davies" initial state, General Relativity as the correct theory of gravitation up to the energy scale at which inflation took place) inflation is below the minimum value detectable by *Planck*. However, even the expected amount of non-Gaussianity of the simplest inflation models is at reach of future galaxy surveys, if one accounts for general relativistic effects which held a contribution to $f_{\rm NL}$ of order unity.

Constraining/detecting non-Gaussianity is a powerful tool to discriminate among competing scenarios for perturbation generation some of which imply large non-Gaussianity. Thanks to the analysis of *Planck* data, non-Gaussianity has become the smoking-gun for non-standard inflation models and a powerful tool to probe fundamental physics and the highest energy scales.

Primordial non-Gaussianity appears in a surprisingly large variety of cosmic phenomena, hence opening the possibility to constraining it by several complementary techniques.

Acknowledgments

I would like to thank all my collaborators in the work mentioned here and in particular Nicola Bartolo, Alan Heavens, Raul Jimenez, Michele Liguori and Licia Verde. I acknowledge partial financial support by the ASI/INAF Agreement 2014-024-R.0 for the Planck LFI Activity of Phase E2.

References

Babich, D. 2005, *Phys. Rev. D* **72**, 043003
Baldauf, T., Seljak U. & Senatore, L. 2011, *JCAP* **1104**, 006
Bardeen, J. M. 1980, *Phys. Rev. D* **22**, 1882
Bardeen, J. M., Bond, J. R., Kaiser, N., & Szalay, A. S. 1986, *Astrophys. J.* **304**, 15
Bartolo, N., Komatsu, E., Matarrese, S., & Riotto, A. 2004, *Phys. Rept.* **402**, 103
Bruni, M., Hidalgo, J. C., & Wands, D. 2014, arXiv:1405.7006 [astro-ph.CO].
Bucher, M., Van Tent, B., & Carvalho, C. S. 2010, *Mon. Not. Roy. Astron. Soc.* **407**, 2193
Carbone, C., Verde, L., & Matarrese, S. 2008, *Astrophys. J.* **684**, L1
Chen, X. 2010, Adv. Astron. **2010**, 638979
Coles, P. & Barrow, J. D. 1987, *Mon. Not. Roy. Astron. Soc.* **228**, 407
Creminelli, P., Nicolis, A., Senatore, L., Tegmark, M., & Zaldarriaga, M. 2006, *JCAP* **0605**, 004

Curto, A., Martinez-Gonzalez, E., & Barreiro, B., 2009, *Astrophys. J.* **706**, 399
Dalal, N., Dore, O., .Huterer D., & Shirokov, A. 2008, *Phys. Rev. D* **77**, 123514
Donzelli, S., Hansen, F. K., Liguori, M., Marinucci, D., & Matarrese, S. 2012 *Astrophys. J.* **755**, 19 [arXiv:1202.1478 [astro-ph.CO]].
Fergusson, J. R., Liguori, M., & Shellard, E. P. S.. 2010, *Phys. Rev. D* **82**, 023502
Fergusson, J. R., Liguori, M., & Shellard, E. P. S.. 2012, *JCAP* **1212**, 032
Gangui, A., Lucchin, F., Matarrese, S., & Mollerach, S. 1994, *Astrophys. J.* **430**, 447
Giannantonio, T. & Porciani, C. 2010, *Phys. Rev. D* **81**, 063530
Giannantonio, T., Ross, A. J., Percival, W. J., Crittenden, R., Bacher, D., Kilbinger, M., Nichol R., & Weller, J. 2014, *Phys. Rev. D* **89**, 023511
Grinstein, B. & Wise, M. B. 1986, *Astrophys. J.* **310**, 19
Groth, E. J. & Peebles, P. J. E.. 1977, *Astrophys. J.* **217**, 385
Kaiser, N. 1984, *Astrophys. J.* **284**, L9
Karagiannis, D., Shanks, T., & Ross, N. P. 2014, *Mon. Not. Roy. Astron. Soc.* **441**, 486
Komatsu, E. *et al.* 2009, arXiv:0902.4759 [astro-ph.CO].
Komatsu, E. 2010, *Class. Quant. Grav.* **27**, 124010
Komatsu, E. & Spergel, D. N. 2001, *Phys. Rev. D* **63**, 063002
Komatsu, E., Spergel, D. N., & Wandelt, B. D. 2005, *Astrophys. J.* **634**, 14
Leistedt, B., Peiris, H. V., & Roth, N. 2014, arXiv:1405.4315 [astro-ph.CO].
Liguori, M., Sefusatti, E., Fergusson, J. R., & Shellard, E. P. S.. 2010, *Adv. Astron.* **2010**, 980523
Lucchin, F. & Matarrese, S. 1987, *Astrophys. J.* **330**, 535 .
Lucchin, F., Matarrese, S., & Vittorio, N. 1988 *Astrophys. J.* **330**, L21
Matarrese, S., Lucchin, F., & Bonometto, S. A. 1986, *Astrophys. J.* **310**, L21
Matarrese, S. & Verde, L. 2008, *Astrophys. J.* **677**, L77
Moscardini, L., Matarrese, S., Lucchin, F., & Messina, A. 1991, *Mon. Not. Roy. Astron. Soc.* **248**, 424
Munshi, D. & Heavens, A., 2010, *Mon. Not. Roy. Astron. Soc.* **401**, 2406
Planck collaboration 2013a arXiv:1303.5062 [astro-ph.CO].
Planck collaboration 2013b, arXiv:1303.5084 [astro-ph.CO].
Press, W. H. & Schechter, P., 1974, *Astrophys. J.* **187**, 425
Regan, D. M., Shellard, E. P. S.., & Fergusson, J. R. 2010, *Phys. Rev. D* **82**, 023520
Sefusatti, E., & Komatsu, E., *Phys. Rev. D* **76**, 083004
Senatore, L., Smith K. M. & Zaldarriaga, M. 2010, *JCAP* **1001**, 028
Slosar, A., Hirata, C., Seljak, U., Ho, S., & Padmanabhan, N., *JCAP* **0808**, 031
Verde, L., Wang, L.M., Heavens, A., & Kamionkowski, M. 2000 *Mon. Not. Roy. Astron. Soc.* **313**, L141
Verde L., Jimenez, R., Kamionkowski, M., & Matarrese, S. 2001 *Mon. Not. Roy. Astron. Soc.* **325**, 412
Verde, L. 2010, *Adv. Astron.* **2010**, 768675
Verde, L., Jimenez, R., Alvarez-Gaume, L., Heavens, A. F., & Matarrese, S. 2013, *JCAP* **1306**, 023
Verde, L., & Matarrese, S. 2009 *Astrophys. J.* **706**, L91
Wang, L.M., & Kamionkowski, M. 2000 *Phys. Rev. D* **61**, 063504
Weinberg, D. H. & Cole, S. 1992, *Mon. Not. Roy. Astron. Soc.* **259**, 652
Xia, J.-Q., Baccigalupi, C., Matarrese, S., Verde, L., & Viel, M. 2011, *JCAP* **1108**, 033
Yadav, A. P. S., Komatsu, E., Wandelt, B. D., Liguori, M., Hansen, F. K., & Matarrese, S. 2008 *Astrophys. J.* **678**, 578

Statistical Challenges in 21st Century Cosmology
Proceedings IAU Symposium No. 306, 2014
A. F. Heavens, J. -L. Starck & A. Krone-Martins, eds.

doi:10.1017/S1743921314011132

Considerations in the Interpretation of Cosmological Anomalies

Hiranya V. Peiris

Department of Physics and Astronomy, University College London,
Gower Street, London WC1E 6BT, United Kingdom
email: h.peiris@ucl.ac.uk

Abstract. Anomalies drive scientific discovery – they are associated with the cutting edge of the research frontier, and thus typically exploit data in the low signal-to-noise regime. In astronomy, the prevalence of systematics — both "known unknowns" and "unknown unknowns" — combined with increasingly large datasets, the widespread use of *ad hoc* estimators for anomaly detection, and the "look-elsewhere" effect, can lead to spurious false detections. In this informal note, I argue that anomaly detection leading to discoveries of new physics requires a combination of physical understanding, careful experimental design to avoid confirmation bias, and self-consistent statistical methods. These points are illustrated with several concrete examples from cosmology.

Keywords. cosmic microwave background, cosmological parameters, early universe, cosmology: miscellaneous, methods: statistical, methods: data analysis

---

1. Introduction

In the next decade, Big Data will form a pivotal part of the experimental landscape in cosmology, with a multitude of large surveys, from cosmic microwave background (CMB) experiments, photometric and spectroscopic galaxy surveys, 21 cm arrays, and direct-detection gravitational wave observatories, expected to yield a deluge of information about the origin and evolution of the Universe. The associated high data rates and huge data volumes, bringing with them the curse of dimensionality, pose challenging problems for data analysis and scientific discovery. Some of the challenges include: (1) data compression (*i.e.*, the formulation of almost-sufficient statistics), filtering, sampling, associated with very large datasets; (2) making robust and accurate inferences or conclusions from such datasets; (3) the small signal-to-noise regime in which discoveries at the research frontier are often made; (4) a very large model space; (5) cosmic variance (*i.e.*, the fact that we have access to a single realisation of an inherently stochastic cosmological model).

We can imagine two qualitatively different kinds of modelling that would be required to extract the maximum information from this cornucopia of cosmological data. **Mechanistic** (physical) modelling is what has driven cosmological discovery thus far – in this case, forward modelling based on physics is feasible, leading directly to standard parameter estimation and model comparison analyses that form the bread and butter of modern cosmology. However, as cosmological data probe deeper into the non-linear regime of the evolution of structure, against complex foregrounds, **empirical** (data-driven) modelling — characterising relationships discovered in the data — will gain increasing prominence. Such models may be purely data-driven, or qualitatively based on physics but be required in cases where forward modelling is infeasible; they may be used to postulate new theories, or generate statistical predictions for new observables.

In this context, the treatment of data anomalies acquires great importance. Anomalies — unusual data configurations — can consist of statistical outliers, unusual concentrations of data points or sudden behaviour changes. They may arise from chance configurations due to random fluctuations, systematics (unmodelled astrophysics; instrument/detector artefacts; data processing distortions), or they can indicate genuinely new discoveries. Determining into which of these categories a given anomaly falls is fraught with difficulty, especially as humans have evolved in such a way that we can often spuriously identify patterns in data where none exist, a phenomenon known as *pareidolia*. When considering whether a cosmological anomaly indicates new physics, one must consider that many such anomalies are often identified using *a posteriori* estimators, which leads to spurious enhancements of detection significance. In the absence of an alternative model for comparison, one often cannot account for the "look-elsewhere effect" arising from multiple testing, or formulate model priors to compare with the standard model. In the absence of an alternative theory, how can we judge if a given anomaly represents new physics? In the rest of this article, I will summarise several case studies demonstrating different approaches to answering this question.

2. Assessing anomalies

There are two distinct steps associated with anomalies in the cosmological context. The first involves finding data anomalies in the first place — this provides an area of rich algorithmic development related to assessing measures of irregularity, unexpectedness, unusualness etc., especially in the coming "Big Data" era. The second involves drawing inferences from the results of the search for anomalies — *i.e.*, assessing whether the anomaly is due to random chance, or whether it represents an unknown mechanism which may point to new physics (or, more prosaically, systematics). In making these inferences, the particle physicists' "look-elsewhere effect", or multiple testing, comes into play. One must correct for any *a posteriori* choices that were made in the process of detecting an anomaly, and properly account for the possible ways in which an anomaly could have shown up (but did not).

In cases where an alternative model motivated the search for an anomaly or the formulation of the statistic in which an anomaly was discovered, the probability that the anomaly represents a new physical mechanism (versus random chance) can be assessed using Bayesian model comparison: the model and parameter priors account for the uncertainties associated with multiple-testing. Even in this case, however, encapsulating all the relevant uncertainties can be quite difficult. For example, Cruz *et al.* (2007) carried out a Bayesian model comparison analysis of the well-known cosmic microwave background (CMB) anomaly termed the *Cold Spot* and concluded that it was likely a texture (a type of spatially-localised cosmic defect). This was a highly sophisticated analysis — one of the first principled Bayesian model comparison analyses in the context of CMB anomalies. Nevertheless, it represents an incomplete attempt to account for *a posteriori* selection effects associated with basing the analysis on a single feature at a particular location which is known to be a statistical outlier.

The texture model predicts a statistically isotropic distribution of spatially-localised features in the CMB sky (with an expectation value for the sky fraction covered by textures), and textures can lead to both hot and cold spots in the CMB. Interpreted as a correction for the "look-elsewhere" effect of having analysed a patch of sky containing the Cold Spot, the Cruz *et al.* (2007) analysis accounts for the expected sky fraction covered by textures in a patch, but does not account for the fact that textures could be placed anywhere on the sky. The coordinates of the central position of the texture template should be marginalised over in the context of statistical isotropy, whereas the texture template was centred on the Cold Spot in the analysis. There is also a factor of two associated with the fact that the feature could have been a hot spot, rather than a cold spot. In Feeney *et al.* (2012, 2013), where the texture hypothesis is formulated in terms of a hierarchical Bayesian model, the impact of these *a posteriori* selection effects becomes clear and, when fully accounting for these uncertainties in computing the marginal likelihood, there is no preference for augmenting the standard cosmological model with textures.

This example highlights how the "look-elsewhere" effect can spuriously enhance the significance of the anomaly when all the relevant uncertainties are not incorporated into the formulation of an alternative model representing new physics. Nevertheless, if one possesses alternative model(s) with well-motivated priors, the standard Bayesian model comparison framework can be brought to bear on the problem, providing the most straightforward scenario for assessing anomalies.

3. "Just-so" models

If the anomaly is detected as an unusual property of a large dataset and is not motivated by an alternative model, accounting for the "look-elsewhere" effect can be highly non-trivial. In this context, designer theories that stand in for best-possible explanations — "just-so" models — can prove useful in terms of gaining an intuition for whether the anomaly provides evidence for new physics.

In this case one would proceed as follows: (a) find a *designer theory* or "just-so" model which maximises the likelihood of the anomaly; (b) thus determine the maximum available likelihood gain for this particular anomaly with respect to the standard model, or null hypothesis; (c) judge whether this is compelling given the baroqueness of the designer theory.

Such an example can be found in the context of another well-known CMB anomaly often termed the $C(\theta)$ anomaly (Copi *et al.* 2007). It is a long-standing observation that the statistic $S_{1/2} = \int_{60^\circ}^{180^\circ} [C(\theta)]^2 \cos(\theta) d\theta$, where $C(\theta)$ is the angular correlation function of the CMB, is anomalous when evaluated outside typical Galactic sky cuts. Here, the lower integration limit is an *a posteriori* choice (Bennett *et al.* 2011). Under the standard ΛCDM model, the probability of obtaining a cut-sky $S_{1/2}$ statistic of the observed value or less is $\sim 0.03\%$; however, it is notable that the *full sky* value of $S_{1/2}$ (evaluated on reconstructed full sky maps or Galactic foreground-cleaned maps) is not anomalous.

It is possible that such an observation can arise in a cosmological model which breaks statistical isotropy, in contrast to the standard ΛCDM model; however, a specific well-motivated broken-isotropy model predicting the observed characteristics of the large-angle CMB sky is not currently available. Pontzen & Peiris (2010) used a convex optimisation algorithm to maximise the likelihood of the cut sky $S_{1/2}$ statistic subject to fixed full sky angular power spectrum multipoles C_ℓ over all anisotropic Gaussian models with zero mean†. Using this technique, they constructed a designer anisotropic model (Fig. 1) which gives the maximum likelihood improvement over ΛCDM: this model, which had ~ 6900 degrees of freedom (compared to the isotropic case with eight) gave an improvement in likelihood of $\ln \mathcal{L} \sim 5$. This allowed for a finite limit to be placed on the Bayesian statistical gain available under a wide class of alternative straw-man models, providing a plausible way to probe the significance of this *a posteriori* anomaly.

† The covariance matrix of spherical harmonic coefficients $a_{\ell m}$ can be can be arbitrarily correlated, as long as it is positive-definite.

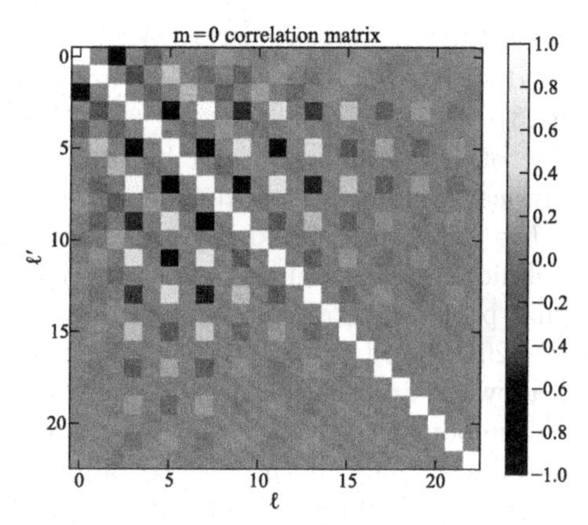

Figure 1. The $m = 0$ component of the covariance matrix defining the designer anisotropic model derived for the $S_{1/2}$ anomaly (Pontzen & Peiris 2010).

It is interesting to note that the decision-making process after computing the likelihood gain supplied by the just-so model involves a subjective judgement. For example, Pontzen & Peiris (2010) concluded that the level of fine-tuning involved in this designer model in order to provide a relatively small likelihood gain did not make for a compelling pointer towards new physics. However, a more speculative scientist may make a different decision to continue the search for an alternative theory, assuming that the number of degrees of freedom can be dramatically reduced by unknown symmetries. In this case, the designer model may provide physical intuition about the correlation properties necessary in such a theory in order to reproduce the observations.

A further example of the utility of just-so models in cosmology where physically-motivated model priors are not available is found in the context of the number of relativistic degrees of freedom $N_{\rm eff}$, which may differ from the standard value of 3.046 due to, *e.g.*, extra neutrino species. The fundamental debate here focuses on whether tensions observed in cosmological datasets require the standard cosmological model to be augmented with a non-standard $N_{\rm eff}$. While this may appear to be a straightforward Bayesian model comparison problem, the model space leading to $N_{\rm eff}$ differing from the standard value is vast, and simple priors on $N_{\rm eff}$ may not capture the physical prior uncertainties. In this context, Verde *et al.* (2013) show that the *profile likelihood ratio* (Wilks 1938) — the ratio of the maximum likelihood conditioned on a particular value of the parameter of interest and the overall, unconditional maximum likelihood — provides the parameter value at which a designer model would maximise the Bayesian evidence. While it is unreasonable to assess a model by tuning it to the same data being used to test it, this upper limit on the marginal likelihood provides an heuristic assessment of the possible need for an extra parameter in the absence of well-motivated parameter priors, as well as a pointer towards its most likely value.

4. Data-driven models

As mentioned above, data-driven models are often constructed *a posteriori* to fit various statistical anomalies observed in large data sets. In order to be testable, such tuned models must, of necessity, make predictions for new data beyond those they were designed to fit. In the context of CMB anomalies, several studies demonstrate how models designed to mimic apparent isotropy-breaking features in the CMB temperature field and features in the CMB temperature power spectrum can be tested with new data not used in constructing the model, such CMB polarisation (Dvorkin *et al.* 2008; Mortonson *et al.* 2009), large-scale structure data (Pullen & Hirata 2010), and non-Gaussianity (Adshead *et al.* 2011; Peiris *et al.* 2013).

The statistical technique of cross-validation can be used as a powerful method to test the consistency of physical inferences from datasets by constructing a data-driven model describing part of the data, and checking how well it predicts the rest of the data. For example, this technique has been used to characterize the deviation from scale invariance in the primordial power spectrum in a minimally-parametric way (Verde & Peiris 2008; Peiris & Verde 2010; Bird *et al.* 2011). The idea is as follows: (1) Choose a functional form which allows a great deal of freedom in the form of the deviation from scale invariance (*e.g.*, smoothing splines)†. (2) Now perform cross-validation: set aside some of the data (validation set), fit the rest (training set), and see how well it predicts the validation set. A very good fit to the training set, which poorly predicts the validation set, indicates over-fitting of noisy data. (3) The final ingredient is a parameter that penalises fine-scale structure of the functional form. By performing cross-validation as a function of this parameter, one can judge when the structure in the smoothing spline is what the data requires without fitting the noise. The technique is also very powerful in detecting systematic issues in data analysis: Verde & Peiris (2008) identified a "kink" in the reconstructed power spectrum from WMAP at a particular scale which subsequently turned out to be a problem in point source subtraction (Huffenberger *et al.* 2008; Peiris & Verde 2010).

5. Blind analysis

The data analysis challenges of next-generation cosmological surveys require careful experimental design to minimize false detections due to experimenters' (subconscious) bias. This has been an integral part of the particle physics and medical research culture for decades, but has yet to find broad adoption within cosmology. As cosmological constraints make the transition from *precise* to *accurate*, and the search for new physics leads to the hunt for small signals embedded within the ΛCDM cosmology, the systematic bias due to non-blind experimental design can no longer be ignored if we hope to make convincing claims for paradigm-shifting new physics from cosmological data.

Blind analysis is based on the simple idea that the *value* of a measurement does not contain any information about its *correctness*. Knowing the value of a measurement is therefore of no use in performing the analysis itself. Blind analysis is necessary because data collection, analysis and inference necessarily involves a human stage, which can lead to unquantifiable subjective inferences. Some examples of the origin of such bias include: looking for bugs when a result does not conform to expectation (and not looking for them when it does); looking for additional sources of systematic uncertainty when a result does not conform to expectation; deciding whether to publish, or wait for more data; choosing cuts while looking at the data and knowing whether it fits expectations; preferentially keeping / dropping outlier data. Examples of good experimental design which are used in cosmology, such as "double unblind" (doing two independent analyses in parallel), "mock data" analysis, and "semi-blind" (using a fraction of the data to calibrate the analysis and freezing the pipeline before doing the final analysis), while extremely helpful in their own right, nevertheless do not constitute blind analysis.

† Naively fitting this to the data will lead one to fit the fluctuations due to noise, with arbitrary improvement in the goodness-of-fit due to over-fitting.

Harrison (2002) and Roodman (2003) give examples of systematic trends in measurements (periods of surprisingly small variation, followed by jumps of several standard deviations) in the presence of experimenters' bias using examples from particle physics, and provide some excellent strategies for implementing blind analysis, including:

- **Encrypting the science result.** *e.g.*, adding a non-changing random number (not revealed to the analyst) to a numerical result or transforming a variable, in order to thwart preconceptions due to "expected" results. It is important to note that it is not necessary to blind how the result *changes* due to changes in the analysis pipeline, or to blind calibration data.
- **Hiding the "signal region".** This is useful when the observable is a peak or localised feature in datasets.
- **Blind injection of signals into the data.** This is useful to test biases in rare event searches, and has been adopted by the gravitational wave direct detection community.
- **Mixing in an unknown fraction of simulated data** during calibration, etc.

The first of these, the so-called *hidden offset* method, is a straightforward technique to implement at the level of the parameter estimation step for essentially all cosmological data types; it works as follows. The parameter estimation code adds a fixed, unknown random number to the fitted value of the measured parameter, $x^* = x + \mathcal{R}$; x^* is returned with the true error and the likelihood value instead of x. $\mathcal{R}$ can be set *e.g.*, by sampling from a Gaussian of mean zero and standard deviation $\sim$ few times the experimental standard deviation. Relative changes in the result as the analysis changes can be hidden using a second offset. Informative visual aspects in plots may need to be hidden as well.

While blind analysis within observational cosmology may be conceptually more difficult than in an experimental field such as particle physics, nevertheless numerous considerations motivate us to adopt at least some practices that lead to the mitigation of experimenters' bias. Such considerations include: (i) the accuracy and reliability of scientific inferences from next-generation surveys; (ii) best return on the enormous investment of public funds in cosmological surveys, including satellite experiments and large data analysis efforts requiring huge teams, which may be very difficult — if not impossible — to replicate due to cost considerations; (iii) the substantial wasted scientific effort in going down blind alleys due to premature announcements of false detections, a side effect of which is the potential damage to the public perception of science; and (iv) missing discoveries of new physics due to overly-cautious treatment of data that do not agree with expectations.

Even simply thinking about how to blind an analysis pipeline can lead to a greater understanding of potential pitfalls. Adopting blind analysis practices requires a shift in community standards, and leadership by large data collaborations. A very welcome shift in this direction has been made by the weak lensing community, *e.g.*, in the recent CFHTLenS analyses (Heymans *et al.* 2012; Fu *et al.* 2014). The POLARBEAR B-mode polarisation blind analysis (Ade *et al.* 2014) represents a milestone in the CMB context, and such techniques have been in use for some time in supernova cosmology (Conley *et al.* 2006). In general, setting all the free parameters in analysis pipelines using only null-tests is a good blind analysis practice. An example of blind mitigation of systematics in constraining primordial non-Gaussianity using quasar surveys (Leistedt & Peiris 2014; Leistedt *et al.* 2014) is presented elsewhere in these proceedings.

6. Summary

I have presented an informal summary, from a practitioner's viewpoint, of techniques for evaluating cosmological anomalies. The discussion has emphasised the pitfalls associated with multiple-testing, or the "look-elsewhere" effect. I have given practical examples of techniques that can be used in the absence of alternative models in order to gain intuition on whether or not a given cosmological anomaly represents new physics. Finally, I have discussed blind analysis as a strategy to guard against experimenters' subconscious bias, which introduces "unknown unknowns" into physical inferences. These techniques do not represent a cure-all for data analysis problems, but rather cultivate a mindset that attempts to rise to Feynman's challenge: *"The first principle is that you must not fool yourself — and you are the easiest person to fool."*

Acknowledgements— I thank Stephen Feeney, David Hand, Boris Leistedt, Daniel Mortlock, Andrew Pontzen, Aaron Roodman, and Licia Verde for influencing my evolving take on this topic. I am grateful to Andrew Pontzen for generating Fig. 1, which was previously unpublished.

References

Ade, P. *et al.* 2014, *ApJ*, 794, 171
Adshead, P., Hu, W., Dvorkin, C., & Peiris, H. V. 2011, *Phys. Rev.*, D84, 043519
Bennett, C. L., Hill, R. S., Hinshaw, G., *et al.* 2011, *ApJS*, 192, 17
Bird, S., Peiris, H. V., Viel, M., & Verde, L. 2011, *MNRAS*, 413, 1717
Conley, A. J. *et al.* 2006, *ApJ*, 644, 1
Copi, C., Huterer, D., Schwarz, D., & Starkman, G. 2007, *Phys. Rev.*, D75, 023507
Cruz, M., Turok, N., Vielva, P., Martinez-Gonzalez, E., & Hobson, M. 2007, *Science*, 318, 1612
Dvorkin, C., Peiris, H. V., & Hu, W. 2008, *Phys. Rev.*, D77, 063008
Feeney, S. M., Johnson, M. C., McEwen, J. D., Mortlock, D. J., & Peiris, H. V. 2013, *Phys. Rev.*, D88, 043012
Feeney, S. M., Johnson, M. C., Mortlock, D. J., & Peiris, H. V. 2012, *Phys. Rev. Lett.*, 108, 241301
Fu, L., Kilbinger, M., Erben, T., *et al.* 2014
Harrison, P. F. 2002, *Journal of Physics G: Nuclear and Particle Physics*, 28, 2679
Heymans, C., Van Waerbeke, L., Miller, L., *et al.* 2012, *MNRAS*, 427, 146
Huffenberger, K. M., Eriksen, H. K., Hansen, F. K., Banday, A. J., & Górski, K. M. 2008, *ApJ*, 688, 1
Leistedt, B. & Peiris, H. V. 2014, *MNRAS*, 444, 2
Leistedt, B., Peiris, H. V., & Roth, N. 2014, ArXiv e-prints
Mortonson, M. J., Dvorkin, C., Peiris, H. V., & Hu, W. 2009, *Phys. Rev.*, D79, 103519
Peiris, H., Easther, R., & Flauger, R. 2013, *JCAP*, 1309, 018
Peiris, H. V. & Verde, L. 2010, *Phys. Rev.*, D81, 021302
Pontzen, A. & Peiris, H. V. 2010, *Phys. Rev.*, D81, 103008
Pullen, A. R. & Hirata, C. M. 2010, *JCAP*, 5, 27
Roodman, A. 2003, in Statistical Problems in Particle Physics, Astrophysics, and Cosmology, ed. L. Lyons, R. Mount, & R. Reitmeyer, 166
Verde, L., Feeney, S. M., Mortlock, D. J., & Peiris, H. V. 2013, *JCAP*, 9, 13
Verde, L. & Peiris, H. V. 2008, *JCAP*, 0807, 009
Wilks, S. S. 1938, *The Annals of Mathematical Statistics*, 9, 60

Statistical Challenges in 21st Century Cosmology
Proceedings IAU Symposium No. 306, 2014
A. F. Heavens, J.-L. Starck & A. Krone-Martins, eds.

doi:10.1017/S1743921314013465

From data to science: Planck data and the CMB non-Gaussianity

Anna Mangilli[1] on behalf of the Planck collaboration

[1]Institut d'Astrophysique de Paris, CNRS and UPMC Sorbonne Universites Paris VI, Bd. Arago , 75014, Paris, France
email: mangilli@iap.fr

Abstract. Studying the non-Gaussianity (NG) of the Cosmic Microwave Background (CMB) is an extremely powerful tool to investigate the properties of the very early Universe. The Planck nominal mission CMB maps yielded unprecedented constraints on primordial non-Gaussianity providing with the highest precision test of the standard model of inflation. Planck's high sensitivity also allowed to find evidence for the first time of the late-time non-Gaussianity arising from the Lensing-Integrated Sachs Wolfe (ISW) cross correlation. In this talk I will give details on the Planck data analysis and I will discuss the theoretical implications of the results.

Keywords. (cosmology:) cosmic microwave background, methods: data analysis, cosmology: observations, (cosmology:) early universe

1. Overview

One of the main reasons of the popularity of the standard inflationary model relies on the fact that *inflation* provides a mechanism that can explain the production of the first density perturbations in the early Universe which are the seeds of the large scale structure in the distribution of galaxies and of the Cosmic Microwave Background (CMB) temperature anisotropies that we observe today. Studying the CMB non-Gaussianity, i.e. the statistical properties of the CMB anisotropies beyond the linear regime, emerged as one of the highest precision test for the standard inflationary framework as a key prediction of the simplest models of inflation is a very small level of CMB non-Gaussianity (Acquaviva *et al.* (2003); Maldacena (2003)). In general the CMB non-Gaussianity is related to non-linear mechanisms and it can be roughly divided into two main categories:

- **Primordial type non-Gaussianity:** refers to the CMB non-Gaussianity due to the non-linear processes happening between the end of inflation (or equivalent mechanism) and the last scattering surface. A detection of the primordial type non-Gaussianity in the CMB would rule out the standard single field slow roll inflation and would be the smoking gun of a different scenario for the generation of the primordial perturbations. Constraining primordial non-Gaussianity (NG) is fully complementary e.g. to looking for the imprint of the primordial gravity waves in the B-modes CMB polarization. *An example: the* **local type** *non-Gaussainity.* One of the most common example of primordial NG is given by the so called *local* type, where the non-Gaussianity arises because of the non-linear corrections to the gravitational potential's perturbations $\Phi(\mathbf{x})$, parametrized by f_{NL}, the amplitude of the quadratic non-linear contribution (Salopek & Bond (1990); Gangui *et al.*(1994); Verde *et al.* (2000); Komatsu & Spergel (2001)):

$$\Phi(\mathbf{x}) = \Phi_L(\mathbf{x}) + f_{NL}(\Phi_L^2(\mathbf{x}) - \langle \Phi_L^2(\mathbf{x}) \rangle) \quad (1.1)$$

This type of non-Gaussianity is very small in the case of the standard slow roll inflation but can be big for alternatives models as for example multi field inflationary models.

- **Late-time type non-Gaussianity:** refers to the CMB non-Gaussianity that arises because of the non-linear processes in the late time stage of the Universe evolution when the structures form. *The lensing-ISW late-time NG:* an extremely interesting late time non-Gaussian signal arises in the CMB because of the correlation of two effects: the lensing and the Integrated Sachs Wolfe (ISW) effect. Weak lensing is responsable for the deflection of the CMB photons trajectories when they pass by the gravitational potential wells of the structures, while the ISW effect is related to the fact that in the late time evolution, when dark energy starts dominating ($z < 2$), the gravitational potentials are "stretched" by the accelerated expansion causing a net gain of energy for the CMB photons that climb out these less and less deep overdensities. Both lensing and the ISW effect are related to the same gravitational potentials distribution, lensing by itself is a non-linear and non-Gaussian process, therefore the combination of these two effects gives rise to the lensing-ISW non-Gaussian signal in the CMB. This signal is extremely interesting because it provides with a new direct probe of the action of dark energy on the evolution of the structures. Also, it has been shown that the lensing-ISW non-Gaussianity, if not properly accounted for, can be a serious contaminant for the estimation of the primordial *local* signal, so it is extremely important to correctly model it.

1.1. *The CMB bispectrum*

In the case of gaussian perturbations the CMB power spectrum encloses all the statistical information. On the other hand, even a small deviation from gaussianity would imply that the CMB sky encodes more informations and in order to access them it is necessary to study the higher order statistics of the CMB anisotropies beyond the power spectrum. The CMB three-points function, or its spherical harmonic projection called the CMB angular *bispectrum*, is the best candidate preposed for this task being the lowest order non-gaussian statistic and in principle the most easily detectable. The CMB bispectrum is proportional to the non-linear amplitude parameter f_{NL} and it is characterized by a shape function. There are many different (triangle) shapes that can be formed and, more importantly, *different shapes are related to different phenomena.* There are three main fundamental CMB bispectrum shapes motivated by different classes of models:

Squeezed-local bispectrum shape: this shape is very important because it is the shape of the local type non-Gaussianity mentioned in Sec. 1 and it is motivated e.g. by multi-fields inflationary models or alternatives like the ekpirotic/cyclic models. Also, the squeezed shape characterizes the late-time lensing-ISW bispectrum. This shape degeneracy explains why the lensing-ISW bispectrum can contaminate the estimation of the primordial local bispectrum amplitude with a bias on f_{NL}^{local} that has been estimated to be $\Delta f_{NL} \simeq 10$ (Mangilli & Verde (2009); Hanson *et al.* (2009)). Considering that the errors on f_{NL}^{local} from Planck and future CMB experiments are $\sigma_{f_{NL}^{local}} < 10$, the lensing-ISW signal must be correctly included in the analysis in order to get unbiased estimates of the local type primordial non-Gaussianity. *Equilateral bispectrum shape:* this shape is motivated by classes of inflationary models as for example the DBI and ghost inflation or in general models with e.g. non-canonical kinetic terms. *Orthogonal bispectrum shape:* also this shape is motivated by inflationary models with a more complicated inflationary dynamics with higher order derivatives and/or non-canonical kinetic terms in the inflaton's action.

1.1.1. *Optimal estimators for f_{NL}*

One of the main goal of the CMB non-Gaussian analysis is to constrain with high accuracy the amplitude parameter f_{NL} for different bispectrum shapes. In order to do so, an optimal bispectrum estimator for f_{NL} can be constructed. This is based on a

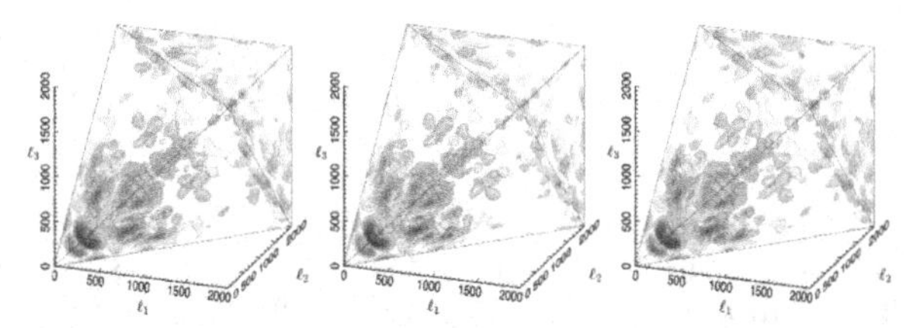

Figure 1. The *Planck* bispectrum reconstruction for three different component separation methods: SMICA, SEVEM, NILC (The Planck collaboration XXIV (2013)).

least-squares fit comparison of the observed CMB bispectrum with a given theoretical bispectrum template $b^{th}_{\ell_1\ell_2\ell_3}$ and takes the form:

$$\begin{aligned}\hat{f}_{NL} &\equiv \frac{1}{N}\sum_{all\ lm} \mathcal{G}^{m_1m_2m_3}_{l_1l_2l_3} b^{th}_{\ell_1\ell_2\ell_3}\Big[(C^{-1}a)_{\ell_1m_1}(C^{-1}a)_{\ell_2m_2}(C^{-1}a)_{\ell_3m_3} \\ &- 3(C^{-1})_{\ell_1m_1,\ell_2m_2}(C^{-1}a)_{\ell_3m_3}\Big]. \end{aligned} \tag{1.2}$$

The linear term $\hat{f}^{linear}_{NL} = -3(C^{-1})_{\ell_1m_1,\ell_2m_2}(C^{-1}a)_{\ell_3m_3}$ is needed in the case of a realistic experimental setting in order to correct for spurious terms due to e.g. the anisotropic noise and/or the incomplete sky cuts when applying a mask. Different types of optimal estimators can be defined depending on the properties of the theoretical bispectrum templates.

2. *Planck* data analysis and results

The results presented here are a summary of the *Planck* 2013 results detailed in: The Planck collaboration XXIV (2013); The Planck collaboration XIX (2013); The Planck collaboration XVII (2013); The Planck collaboration XXIII (2013).

Planck† is the ESA telescope that observed the sky in 9 frequency bands from 30 to 857 GHz with an unprecedented sensitivity, opening for the first time the possibility to constrain the CMB non-Gaussianity with very high precision. In order to constrain the CMB bispectrum the cleanest as possible CMB map is needed so that the *Planck* data have been processed through the component separation pipelines in order to correctly remove the foregrounds contamination. Fig. 1 presents the *Planck* bispectrum reconstruction in the case of three different methods of foregrounds removal (SMICA, SEVEM, NILC), showing that the bispectrum reconstruction is extremely robust with respect to foreground cleaning. In order to constrain the bispectrum amplitude f_{NL}, different type of estimators have been used for the *Planck* data analysis. In particular for the analysis of the primordial bispectrum shapes the main optimal estimators used are: the separable KSW estimator (Komatsu, Spergel & Wandelt (2005)), the skew-C_ℓ (Munshi & Heavens (2009)), the modal (Fergusson *et al.* (2010)) and binned (Bucher *et al.* (2010)) estimators. Also, in order to correctly account for lensing-ISW contribution, the following optimal estimators specifically built for the lensing-ISW signal have been used: Mangilli *et al.*(2013) for the KSW and Lewis *et al.* (2011) for the lensing reconstruction estimator. All the estimators have been massively tested on realistic *Planck* FFP6 lensed simulations processed the same as the data and they showed a remarkable map-by-map agreement and consistency. The main results of the *Planck* data analysis on the constraints of f_{NL} for the three fundamental primordial shapes and for the lensing-ISW bispectrum are reported in Table 1. *Planck* finds evidence of the squeezed local type non-Gaussianity: as expected

† http://sci.esa.int/planck/

Table 1. Summary of the *Planck* constraints on f_{NL} for the three primordial fundamental shapes and for the lensing-ISW bispectrum (The Planck collaboration XXIV (2013); The Planck collaboration XIX (2013))

	KSW	Binned	Modal
Local biased lensing-ISW	9.8 ± 5.8	9.2 ± 5.9	8.3 ± 5.9
Local unbiased	2.7 ± 5.8	2.2 ± 5.9	1.6 ± 6
Lensing-ISW bias	7.7 ± 1.5	7.7 ± 1.6	10 ± 3
Lensing-ISW	0.85 ± 0.32	1.03 ± 0.37	0.93 ± 0.37
Equilateral	-37 ± 75	-20 ± 73	-20 ± 77
Orthogonal	-46 ± 39	-39 ± 41	-36 ± 41

this is due to the lensing-ISW bias that is found to be compatible with the theoretical expectations (Mangilli & Verde(2009), Hanson *et al.*(2009)). Only after the subtraction of the lensing-ISW contribution the *Planck* analysis shows that there is no evidence of primordial local type non-Gaussianity. Also, *Planck* does not find evidence of primordial equilateral and orthogonal non-Gaussianity. This means that the *Planck* analysis favors the simplest models for inflation that in this way has passed the most stringent test to date. Also, *Planck* finds evidence for the first time of the late-time lensing-ISW bispectrum, compatible with the signal expected in the ΛCDM scenario. In general the *Planck* data analysis showed a very good internal consistency, stability across frequencies, stability for changes of the masks and proved to be robust with respect to the contamination from noise and foregrounds. The *Planck* constraints on CMB non-Gaussianity provided with the highest precision test to date of the physical mechanism of the origin of cosmic structure.

References

The Planck collaboration, *Planck* 2013 Results XXIV *A&A* 2013/21554
The Planck collaboration, *Planck* 2013 Results XIX *A&A* 2013 arxiv:1303.5079
The Planck collaboration, *Planck* 2013 Results XVII *A&A* 2013 arXiv:1303.5077
The Planck collaboration, *Planck* 2013 Results XXIII *A&A* 2013 arXiv:1303.5083
Acquaviva V., N. Bartolo, S. Matarrese, & A. Riotto, Nuclear Physics B **667**, 119 (2003)
Bucher, M., van Tent, B., & Carvalho, C. S. 2010, *MNRAS*, 407, 2193
Fergusson, J. R., Liguori, M., & Shellard, E. P. S. 2010a, *Phys. Rev. D*, 82, 023502
Gangui, A., F. Lucchin, S. Matarrese, & S. Mollerach, *Astrophys. J.* **430**, 447 (1994)
Hanson, D., K. M. Smith, A. Challinor, & M. Liguori, *Physical Review D* **80** :083004 (2009)
Komatsu, E., D. Spergel, & B. Wandelt, *Astrophys. J.* **634**, 14-19 (2005)
Komatsu, E. & Spergel, D. N. 2001, *Phys. Rev. D*, 63, 063002
Lewis, A., *et al.* JCAP, 03(2011)018
Maldacena, J., *Journal of High Energy Physics* **5**, 13 (2003), arXiv:astro-ph/0210603.
Mangilli, A., B. Wandelt, F. Elsner, & M. Liguori *A&A* 555, A82 (2013)
Mangilli, A., L. Verde, *Physical Review D* **80**, 123007 (10/2009)
Munshi, D. & Heavens, A. 2010, *MNRAS*, 401, 2406, arXiv:0904.4478
Salopek, D. S. & J. R. Bond, *Phys. Rev. D* **42**, 3936 (1990).
Verde, L., Wang, L., Heavens, A. F., & Kamionkowski, M. 2000, *MNRAS*, 313, 141

Statistical Challenges in 21st Century Cosmology
Proceedings IAU Symposium No. 306, 2014
A. F. Heavens, J.-L. Starck & A. Krone-Martins, eds.

doi:10.1017/S1743921314010795

Cosmological Applications of the Gaussian Kinematic Formula

Yabebal T. Fantaye[1] and Domenico Marinucci[1]†

[1]Dipartimento di Matematica, Universit di Roma"Tor Vergata,
Via della Ricerca Scientifica 1, I-00133 Roma, Italy
email: fantaye@mat.uniroma2.it, marinucc@axp.mat.uniroma2.it

Abstract. The Gaussian Kinematic Formula (GKF, see Adler and Taylor (2007,2011)) is an extremely powerful tool allowing for explicit analytic predictions of expected values of Minkowski functionals under realistic experimental conditions for cosmological data collections. In this paper, we implement Minkowski functionals on multipoles and needlet components of CMB fields, thus allowing a better control of cosmic variance and extraction of information on both harmonic and real domains; we then exploit the GKF to provide their expected values on spherical maps, in the presence of arbitrary sky masks, and under nonGaussian circumstances.

Keywords. methods: analytical,methods: data analysis,methods: numerical ,methods: statistical, techniques: image processing, cosmology: cosmic microwave background

1. Introduction

The expected values of the Minkowski functionals in the planar case and under Gaussianity is analytically known to the literature since the work of Adler in the early 80's Adler 1981, see also Tomita 1986. A major advancement of this field in the last dacade, however, was brought by the discovery of the Gaussian Kinematic Formula (GKF) by Adler and Taylor (2003, 2007,2009). GKF allows a simple computation of the expected values for Lipschitz-Killing curvatures (LKCs) (which are equivalent to Minkowski functionals up to a constant factor) under an impressive variety of extremely different circumstances, covering arbitrary manifolds with and without masked regions and a broad class of nonGaussian models.

One of our purposes in this paper is to exploit these recent results to develop a number of analytic predictions on functionals tailored to test nonGaussianities and asymmetries on CMB data. Due to page constraints, here we will show only results for Gaussian fields convolved with a realistic mask in needlet domain. For detail analysis and explicit analytical expressions of the expected values of LKCs for Gaussian and nonGaussian cases on multipole and needlet domains and including an arbitrary sky cuts, please refer to our paper Fantaye et. al. 2014.

2. Gaussian Kinematic Formula (GKF)

The GKF is about expected values of Lipschitz-Killing curvatures for excursion regions. The great power of the GKF is that it allows for a full decoupling of the expected values of LKCs of an excursion set $\mathbb{E}\mathcal{L}_i(A_u(g(T), M))$ into components which are completely independent: the LKCs of the original manifold $\mathcal{L}_{i+l}(M)$, the Gaussian Minkowski Functionals which depends only threshold value $\mathcal{M}(u)$, and the covariance structure λ of the field. The independence of these components to each other means an enormous

† Research supported by ERC Grant 277742 Pascal.

computational advantages to derive analytical results under a variety of circumstances, including masked regions as we shall see below. Before giving the expression of the GKF, let us define the important components in turn.

Exursion sets: $A_u(f(x), M))$

On the sphere, the excursion sets $A_u(f)$ of a given (possibly random) function f are defined as

$$A_u(f) := \left\{x \in S^2 : f(x) \geqslant u\right\} .$$

Of course, in the limit where we take $u = -\infty$, we have that $A_u(f) = S^2$.

Lipschitz-Killing Curvatures (LKCs): $\mathbb{E}\mathcal{L}_i(A_u(f(x), M))$

The LKCs for the region A with dimension $\dim(A) = n$, are defined as the coefficients of a Taylor expansion of a *Tube* of radius r around A.

$$Vol\left[Tube(A, r)\right] = \sum_{k=0}^{n} \mathcal{L}_{n-k}(A)\omega_k r^k.$$

where ω_k are the volume of a unit ball in $\mathcal{R}^k$. LKCs depend on the Riemannian metric, and are a measure of the k-dimensional size of the Riemannian manifold M. For instance, let A be the unit square on the plane; by elementary geometry, the volume of the Tube is then given by

$$\mathcal{L}_2(A) + 2\mathcal{L}_1(A)r + \mathcal{L}_0(A)\pi r^2 = 1 + 2 \cdot 2 \cdot r + \pi r^2,$$

the LKCs $\mathcal{L}_0, \mathcal{L}_1, \mathcal{L}_2$ correspond to Euler-Poincaré characteristic, half the boundary length and area, respectively.

Gaussian Minkowski Functionals (GMFs): $\mathcal{M}_k(U)$

The Gaussian Minkowski Functionals (GMFs) $\mathcal{M}_k(U)$ are defined as the Taylor coefficients in the expansion of the Tube probabilities $\Pr\{Z \in Tube(U, r)\}$, the probability that a zero-mean standard Gaussian variable belongs to $Tube(U, r)$; for instance, for $U = [u, \infty)$. The GMFs dependence only on the excursion threshold u and can be easily computed using the following expression

$$\mathcal{M}_j^{\gamma_k}([u, \infty)) = (2\pi)^{-1/2} H_{j-1}(u) e^{-u^2/2}.$$

where H_j is the Hermite polynomials: $H_0(u) = 1$, $H_1(u) = 2u$, $H_2(u) = 4u^2 - 1$, $H_3(u) = 8u^3 - 12u$.

Metric scaling coefficients: $\lambda^{1/2}$

The metric scaling coefficient λ represents the covariance structure of the manifold we are working with, and are simply given by the second derivative of the covariance function at the origin. On a sphere the scaling λ required to go to a needlet domain is given by:

$$\lambda_j = \sqrt{\frac{\sum_\ell b^2(\frac{\ell}{2^j}) C_\ell \frac{2\ell+1}{4\pi} \frac{\ell(\ell+1)}{2}}{\sum_\ell b^2(\frac{\ell}{2^j}) C_\ell \frac{2\ell+1}{4\pi}}},$$

where $b(.)$ is the needlet weight function, j is the needlet frequency parameter, and $B > 1$ is some fixed bandwidth.

Under these circumstances, the Gaussian kinematic formula takes the form

$$\lambda^{i/2}\mathbb{E}\mathcal{L}_i(A_u(g(T),M)) = \sum_{k=0}^{\dim(M)-i} \lambda^{(i+k)/2}\mathcal{L}_{i+k}(M)\mathcal{M}_k(g^{-1}[u,\infty)) \; ; \qquad (2.1)$$

3. LKCs in the presence of sky cuts

Assume we observe only $M := S^2\backslash G$. Recall that $\mathcal{L}_0(M), \mathcal{L}_1(M)\mathcal{L}_2(M)$ are the LKCs of the unmasked region. Then the expected values of LKCs in the presence of mask G are:

the Euler characteristic is given by

$$\mathbb{E}\mathcal{L}_0(A_u(f(x),M)) = \{1-\Phi(u)\}\,\mathcal{L}_0(M) + \frac{1}{4}\lambda_s e^{-u^2/2}\mathcal{L}_1(M) + \lambda_s^2\frac{ue^{-u^2/2}}{\sqrt{(2\pi)^3}}\mathcal{L}_2(M) \; ; \; (3.1)$$

half the boundary length

$$\mathbb{E}\mathcal{L}_1(A_u(f(x),M)) = 2\,\{1-\Phi(u)\}\,\mathcal{L}_1(M) + \frac{1}{4}\lambda_s e^{-u^2/2}\mathcal{L}_2(M) \; ; \qquad (3.2)$$

the area

$$\mathbb{E}\mathcal{L}_2(A_u(f(x),S^2)) = (1-\Phi(u))\mathcal{L}_2(S^2). \qquad (3.3)$$

For an arbitrary mask G what is unknown in the above equations are the input LKCs $\mathcal{L}_i(S^2\backslash G)$. A very simple solution to derive these parameters can be provided by exploiting one more time Gaussian Kinematic Formula, following an idea discussed in Adler & Taylor 2011, chapter 5.4. The strategy is as follows:

(*a*) Fix a simple power spectrum C_ℓ, for instance with $L_{\max} = 10$, and generate Gaussian maps out of it

(*b*) Fix a limited number of threshold values u and perform a Monte Carlo evaluation of the LKCs evaluated on the excursion set of the fields generated according to (a) and with the mask G applied

(*c*) Use least square regression to estimate $\mathcal{L}_i(S^2\backslash G)$, $i = 0, 1, 2$

(*d*) Use the estimates obtained in point 3 as an input for equation (2.1) for any arbitrary power spectrum (for instance, multipole or needlet components on realizations of a ΛCDM model, under Gaussian and nonGaussian circumstances).

4. Numerical results

To compare analytical results of LKCs with Monte Carlo simulations, we generated 100 realizations of an input power spectrum using the HEALpix Górski et. al. 2005 package. A root mean square normalized Gaussian needlet maps, which we used for all our analysis, are obtained by first generating spherical harmonic coefficients from the input power spectrum through the HEALpix *create_alm* routine up to a maximum multipole of $\ell_{\max} = 2000$; and then by convolving these coefficients with the needlet filter as

$$\beta_j(x) = \frac{1}{\sigma_j}\sum_{\ell=B^{j-1}}^{B^{j+1}}\sum_m b^2\left(\frac{\ell}{B^j}\right)a_{\ell m}Y_{\ell m}(x) \; , \qquad (4.1)$$

where $\sigma_j^2 = \frac{1}{4\pi}\sum_\ell b^4(\frac{\ell}{B^j})(2\ell+1)C_\ell$ is the variance. We then applied a realistic mask on these maps. The mask used is shown in the left panel of Fig. (1).

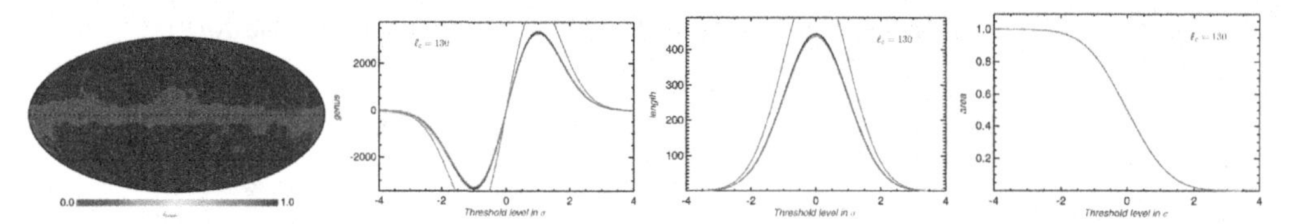

Figure 1. Masked LKCs on needlet space for a Gaussian field. Right panel shows the mask used in the analysis. The other three panels are for the Euler-Poincaré characteristic (genus), boundary length and area functionals. Analytical results are shown in red (assuming full sky) and blue (for mask corrected). Simulations are shown in black (mean of the simulations) and grey shades, which are 68, 95 and 99% percentiles estimated from 100 simulations. The needlet parameters used are $B = 1.5$ and $j = 12$, which corresponds to multipole range $\ell = [87, 195]$.

From these normalized and masked needlet maps we computed the three Minkowski Functionals, which as argued earlier are equivalent to the LKCs up to constant factors. This implementation is achieved exploiting the algorithms described in Eriksen et. al. 2004. The results we will show here are for a needlet frequency parameter $j = 12$ and the band width $B = 1.5$; this particular needlet map has a compact support for multipole ranges $\ell = [87, 195]$. More details can be found in Fantaye et. al. 2014.

To obtain the analytical results corresponding to our simulations, we first computed the mask dependent input LKCs, $\mathcal{L}_{\rangle}(S^2/G)$, using a single multipole sumulation at multipole $\ell = 15$ and following the procedures given in Section 3. We then substituted back these values to Eqns. (3.1, 3.2, 3.3). The comparison of the analytical and simulation results for Euler-Poincaré characteristic, boundary length, and area are shown in Fig. (1).

5. Summary and Conclusion

The Gaussian Kinematic Formula allows to evaluate exact expected values for Lipschitz-Killing curvatures (Minkowski functionals) in a number of circumstances of applied interest, covering in particular full-sky experiments (accounting for the geometry of the sphere), nonlinear statistics and masked data.

We found an excellent agreement in all the cases that we investigated; more precisely, the analytical estimates are always well within the 68% Confidence Interval (CL) estimated from simulations, and as shown in the figures they are for practical purposes indistinguishable from the theoretical predictions even with a relatively low number of Monte Carlo simulations.

References

Planck Collaboration *ArXiv e-prints*: 1303.5083

Yabebal Fantaye, Frode Hansen, Davide Maino, Domenico Marinucci *ArXiv e-prints*: 1406.5420

R. J. Adler, *John Wiley & Sons Inc, ISBN: 0471278440* The Geometry of Random Fields 1981

J. E. Taylor & R. J. Adler, *The Annals of Probability* 31, 533, 2003

Adler, Ewing, & Taylor, *Statistical Science* 24, 1, 1009

Adler & Taylor, *Springer Berlin Heidelberg, ISBN: 978-3-642-19579-2* Topological Complexity of Smooth Random Functions 2007

Adler & Taylor *Springer New York, ISBN: 9780387481128* Random Fields and Geometry 2007

H. Tomita, *Progress of Theoretical Physics* 76, 952, 1986

Eriksen, Hansen, Banday, Gorski, and Lilje, *ApJ* 605,14, 2004

Statistical Challenges in 21st Century Cosmology
Proceedings IAU Symposium No. 306, 2014
A. F. Heavens, J.-L. Starck & A. Krone-Martins, eds.

doi:10.1017/S1743921314010989

Detectability of Torus Topology

Ophélia Fabre[1,2], Simon Prunet[1,3] and Jean-Philippe Uzan[1,3]

[1]Institut d'Astrophysique de Paris,
98 bis boulevard Arago, 75014, Paris, France

[2]Observatoire de Lyon, Université Claude Bernard, Lyon 1, CNRS-UMR 5574,
Centre de Recherche Astrophysique de Lyon
9 avenue Charles André, Saint-Genis Laval, F-69230, France

[3]UPMC Université Paris 06, UMR7095,
98 bis Boulevard Arago, F-75014, Paris, France
email: {fabre, prunet, uzan}@iap.fr

Abstract. The global shape, or topology, of the universe is not constrained by the equations of General Relativity, which only describe the local universe. As a consequence, the boundaries of space are not fixed and topologies different from the trivial infinite Euclidean space are possible. The cosmic microwave background (CMB) is the most efficient tool to study topology and test alternative models. Multi-connected topologies, such as the 3-torus, are of great interest because they are anisotropic and allow us to test a possible violation of isotropy in CMB data. We show that the correlation function of the coefficients of the expansion of the temperature and polarization anisotropies in spherical harmonics encodes a topological signature. This signature can be used to distinguish an infinite space from a multi-connected space on sizes larger than the diameter of the last scattering surface (D_{LSS}). With the help of the Kullback-Leibler divergence, we set the size of the edge of the biggest distinguishable torus with CMB temperature fluctuations and E-modes of polarization to 1.15 D_{LSS}. CMB temperature fluctuations allow us to detect universes bigger than the observable universe, whereas E-modes are efficient to detect universes smaller than the observable universe.

Keywords. cosmic microwave background, methods: statistical, topology, Kullback-Leibler divergence, detectability

---

1. Introduction

In the standard model of cosmology, the universe is described by a 6-parameter Λ-CDM model. The universe is assumed to be isotropic, homogeneous and Gaussian at very large scales. These properties are expected to be found in the CMB. The CMB is emitted from a sphere all around us, the last scattering surface (of diameter D_{LSS}). The last scattering surface is the limit of the observable universe. By studying the CMB, *via* its temperature and polarization anisotropies, we can extract information about the properties of our universe. This work used the `Healpix` package and `CAMB` software.

<u>*CMB*</u>. The most convenient way to work with the CMB is to project its temperature and polarization fluctuations on the spherical harmonic basis to get the $T_{\ell,m}$, $E_{\ell,m}$ and $B_{\ell,m}$ coefficients (see Hu & White 1997). These coefficients arise from initial conditions in the primordial universe to which a transfer function, describing the local perturbation (Sachs-Wolfe, ISW, Doppler), is applied and finally the eigenmodes of the Laplacian of the space. The E-modes are generated by scalar, vector and tensor perturbations whereas B-modes are generated by vector and tensor perturbations. In *Planck* results XXII (2013), the limit of the tensor-to-scalar ration r evaluated with the *Planck* temperature data is found to be $r < 0.11$ with a 95% C.L. More recently, the first detection of B-modes in the CMB by the BICEP2 collaboration set $r = 0.20^{+0.07}_{-0.05}$ (BICEP2 2014). As a consequence,

even if B-modes are supposed to take part in the CMB fluctuations, their contribution is sub-dominant compared to E-modes. That is why in the remaining part of the study, we will only consider E-modes as in Riazuelo *et al.* (2006). Then, we use the covariance matrix $(C_{\ell m,\ell' m'})$ divided in 4 blocks $C^{XY}_{\ell m,\ell' m'} =< X_{\ell m} Y^*_{\ell' m'} >$, with $X, Y \in \{T, E\}$. For an isotropic space, each covariance matrix reduces to a pure diagonal matrix, $C^{XY}_{\ell m,\ell' m'} = \delta_{\ell\ell'}\delta_{mm'}C^{XY}_\ell$, where C^{XY}_ℓ is the power spectrum.

Topology. The topology is the *global* shape of the universe and it is not constrained by General Relativity, which is a set of differential equations that describes space *locally, via* the metric, but not the boundary conditions. That is why for the same metric, different topologies are possible. The standard model of cosmology assumes an isotropic infinite universe, but some anomalies, hints of the violation of the global isotropy, have been discovered in the WMAP data (Bielewicz *et al.* 2004). That is why multi-connected flat spaces have been considered (Riazuelo *et al.* 2004a, Riazuelo *et al.* 2004b) because they are anisotropic models of flat universes. A complete review of their characteristics can be found in Lachieze-Rey & Luminet (1995), Levin (2002).

Multi-connected spaces are anisotropic and the covariance matrix of an isotropic space is purely block-diagonal. That is why it is very important to take into account the full covariance for any topological study. The restriction to the power spectrum is not enough because topological information will consequently be lost. As the CMB is emitted from the last scattering surface, one could imagine that any search of topology would not be able to detect a topology bigger than the observable universe. However the correlations between the modes inside the observable universe and outside should help us extracting information above the last scattering surface limit. That is why we looked how far we can see a topology beyond the observable universe, and we illustrate here this analysis with the example of the cubic 3-torus (see Fabre *et al.* 2013).

2. The Kullback-Leibler divergence $D_{\rm KL}$

Kullback-Leibler divergence. We would like to compare two theories that predict that the coefficients of the expansion of the temperature anisotropies in spherical harmonics, $a_{\ell m}$, are Gaussian and satisfy $\langle a_{\ell m} a^*_{\ell' m'}\rangle_1 = C^{(1)}_{\ell\ell' mm'} = C^{(1)}_\ell \delta_{\ell\ell'}\delta_{mm'}$ for the isotropic model and $\langle a_{\ell m} a^*_{\ell' m'}\rangle_2 = C^{(2)}_{\ell\ell' mm'}$ for the non-trivial topology, where the ensemble average are taken for each theory respectively. Such a comparison can be performed in terms of the Kullback-Leibler divergence (Kullback) for two probability distribution functions p_1 and p_2 defined by

$$D_{\rm KL}(p_1||p_2) = \int p_1(x) \ln\left[\frac{p_1(x)}{p_2(x)}\right] dx. \tag{2.1}$$

This divergence is the expectation value of $\ln(p_1/p_2)$ with the ensemble average related to p_1. Due to the Gibbs inequality, $D_{\rm KL}$ is always positive and equal to zero if and only if $p_1 = p_2$. In terms of information theory, $D_{\rm KL}(p_1||p_2)$ quantifies the amount of information lost when the data (p_1) is represented by the model (p_2). Comparing any multi-connected space (2) with the Euclidean trivial space (1) is interesting because the latter has a rotationally invariant covariance matrix. Consequently the Kullback-Leibler divergence does not depend on the relative orientation of the two spaces and thus quantifies how much information "separates" model 2 from model 1. Furthermore the flat Euclidean space is the most probable topology given previous studies. It is important to see if a deviation from it easily can be detected. When working with a 3-torus whose edge bigger than D_{LSS}, $D_{\rm KL}(1||2)$ is obviously close to zero. The main interest of this approach is that, unlike the circles-in-the-sky method (see Cornish *et al.* 1998), one can measure a

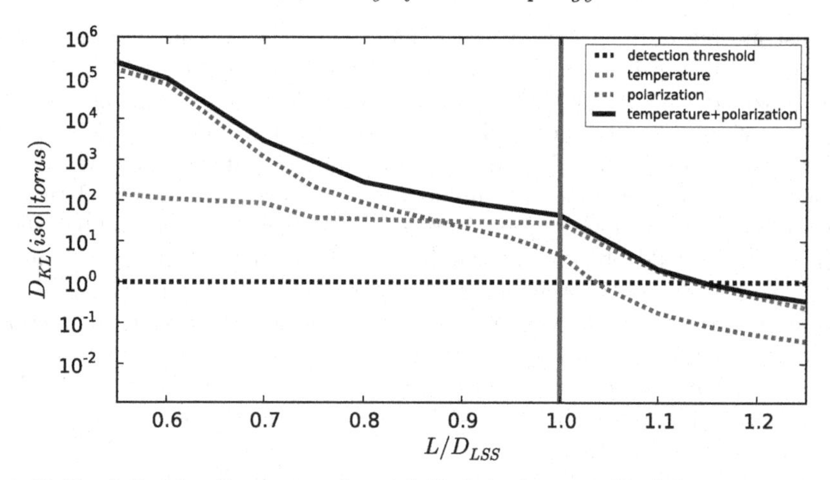

Figure 1. Kullback-Leibler divergence of a trivial infinite isotropic Euclidean space compared to a cubic 3-torus in function of the size L of the edge of the torus in units of D_{LSS}. The black line is the threshold of detection above which the divergence should be in order to detect a 3-torus from a Euclidean space. The green dotted line is the contribution of the temperature fluctuations only, the blue dotted line is the contribution of the E-modes polarization fluctuations only and the black line is taken into account both temperature and polarization fluctuations.

distance even for spaces with a size larger than $D_{\rm LSS}$. Furthermore, the Kullback-Leibler divergence is not affected by instrumental noise or galactic masking and the small scales only improve the detection of universes smaller than the observable universe, which is very interesting computationally speaking (see Fabre *et al.* 2013 for more details).

Detection threshold. Let us introduce the Bayes factor $B_{12} = \frac{P_1(d|M_1)}{P_2(d|M_2)}$. If $B_{12} > 1$ (resp. $B_{12} < 1$) it represents the increase (resp. decrease) of the credence in favour of model 1 (M_1) versus model 2 (M_2) given the observed data (see Trotta 2008). It gives the factor by which the relatives odds between the two models have changed after taking into account the data. The data are the $a_{\ell m}$ in this experiment. If we take into account formula (2.1), we have $D_{\rm KL}(1||2) = \langle \ln(B_{12}) \rangle_1$. There is thus a direct link between the Kullback divergence and the Bayes factor. The Jeffrey scale, usually used to interpret the Bayes factor, is not modified if we consider $\langle \ln(B_{12}) \rangle_1$ instead of $\ln(B_{12})$. As a consequence we obtain the same levels of significance, with a threshold of detectability (represented in black dotted line in Fig.1) for $D_{\rm KL} = 1$. This threshold of detectability quantifies the level at which we can distinguish a torus topology from the isotropic model. If $D_{\rm KL} < 1$, the result is inconclusive and the torus topology cannot be distinguished from a Euclidean space.

3. Results and discussion

Results. The three D_{KL} curves decrease as the size of the 3-torus increase which is logical since we are getting close to an infinite isotropic universe. There is also a change of slope at $L = D_{LSS}$ (red vertical line in Fig. 1), especially for the temperature curve. It is because we are crossing the last scattering surface, *i.e.* the limit of the observable universe, and losing an important amount of information. It appears that the biggest distinguishable 3-torus has an edge of size 1.15 D_{LSS} given by the temperature fluctuations contribution only. Similar results can be found in Ben-David *et al.* (2012) with other topologies. With the E-modes polarization contribution only, the biggest distinguishable torus is only of size 1.04 D_{LSS}. The CMB temperature fluctuations are the best to constrain big universes and CMB E-modes polarization fluctuations are very efficient to constrain universes smaller than the last scattering surface. There are similar results in Riazuelo *et al.* (2006), Bielewicz *et al.* (2011) where it was noticed that the circle-in-the-sky method (only adapted to universes smaller than D_{LSS}) is more efficient with polarization data. Nonetheless, it was only about the efficiency of the circle-in-the-sky method itself and not about the influence of the size of the space. Using of both temperature and polarization data will improve the detectability around 0.9 D_{LSS}.

Discussion. One could object that the B-modes were not taken into account in our study but in addition to our explanation in Section 1, we can cite Kunz *et al.* 2008: gravitational waves only add a noise-like like contribution and lower the detectability of topologies. Multi-connected topologies have not been detected yet in CMB data. On the one side, in the *Planck* paper on cosmology (*Planck* results XXVI 2013), with CMB temperature data only, no back-to-back pairs of circles of correlation was found. Nevertheless it was possible to constrain the lower spatial dimension to $L_{min} = 0.94\ D_{LSS}$ at the 99% C.L. On the other side, the likelihood maximization analysis performed was unable to make any decisive detection: there is only a faint hint of a 3-torus universe bigger than the observable universe, but without any statistical significance. That is why we can consider that the *Planck* E-modes data will only improve constraints on small universes, pushing the actual *Planck* constraint on the lower boundary of the size of the universe L_{min} =0.94 D_{LSS} closer to D_{LSS}. Furthermore, although the isotropy anomalies arising from the WMAP data are also found in the *Planck* data, they seem to be fainter.

4. Conclusion

We present a test of cosmology beyond the standard model, with an exploration of a possible violation of the isotropy of the CMB with the help of a multi-connected topology, the cubic 3-torus. The size L_* of the edge of the biggest distinguishable 3-torus was found to be equal to 1.15 D_{LSS}. The CMB temperature fluctuations are the best tool to study topologies bigger that the observable universe whereas E-modes are more efficient to study smaller universes. As a consequence the joined contribution of temperature and E-modes fluctuations will not be helpful to detect a multi-connected universe of size bigger than D_{LSS}, but only to better constrain small universes.

References

Ben-David, A., Rathaus B. & Itzhaki, N. 2012, *JCAP*, 020, arXiv:1207.6218
BICEP2 collaboration 2014, arXiv:1403.3985
Bielewicz, P. & Banday, A. J. arXiv:1012.3549v1
Bielewicz, P., Górski, K. M., & Banday, A. J. 2004, *MNRAS*, 355, 1283
Bielewicz, P., Banday, A. J., & Górski, K. M. 2011, arXiv:1111.6046
CAMB `http://camb.info`
Cornish, N. J., Spergel, D. N., & Starkman, G. D. 1998, *Class. Quant. Grav.*, 15, 2657
Cornish, N., Spergel, D., Starkman, G., & Komatsu, E. 2004, *Phys. Rev. Lett.*, 92, 201302
Fabre, O., Prunet, P., & Uzan, J.-P. 2013, arXiv:1311.3509
Gorski, K. M. *et al.* 2005, *ApJ*, 622, 759 `http://healpix.jpl.nasa.gov/`
Hu, W. & White, M. 1997, *Phys. Rev. D*, 56, 596
Key, J. S., Cornish, N., Spergel, D., & Starkman, G. 2007, *Phys. Rev. D*, 75, 084034
Kullback, S. 1959, *Information theory and statistics*, Wiley Publications in Statistics
Kunz, M., Aghanim, N., Riazuelo, A., & Forni, O. 2008, *Phys. Rev. D*, 77, 023525
Lachieze-Rey, M. & Luminet, J.-P. 1995, *Phys. Rep.*, 254, 135, arXiv:gr-qc/9605010
Levin, J. J. 2002, *Phys. Rep.*, 365, 251, arXiv:gr-qc/0108043

Planck results XXII 2013, arXiv:1303.5082
Planck results XXVI 2013, arXiv:1303.5086
Riazuelo, A., Uzan, J.-P., Lehoucq, R., & Weeks, J. 2004, *Phys. Rev. D*, 69, 103514
Riazuelo, A. *et al.* 2004, *Phys. Rev. D*, 69, 103518
Riazuelo, A., Caillerie, S., Lachieze-Rey, M., Lehoucq, R., & Luminet, J.-P. 2006, arXiv:astro-ph/0601433
Trotta, R. 2008, *Contemp. Phys.*, 49, 71

Statistical Challenges in 21st Century Cosmology
Proceedings IAU Symposium No. 306, 2014
A. F. Heavens, J. -L. Starck & A. Krone-Martins, eds.

doi:10.1017/S1743921314011065

Cosmic infrared background measurements and star formation history from *Planck*

Paolo Serra, on behalf of the Planck Collaboration

Institut d'Astrophysique Spatiale (IAS),
Bâtiment 121, F- 91405, Orsay (France)
email: pserra@ias.u-psud.fr

Abstract. We present new measurements of Cosmic Infrared Background (CIB) anisotropies using *Planck*. Combining HFI data with *IRAS*, the angular auto- and cross-frequency power spectrum is measured from 143 to 3000 GHz. After careful removal of the contaminants (cosmic microwave background anisotropies, Galactic dust and Sunyaev-Zeldovich emission), and a complete study of systematics, the CIB power spectrum is measured with unprecedented signal to noise ratio from angular multipoles $\ell \sim 150$ to 2500. The interpretation based on the halo model is able to associate star-forming galaxies with dark matter halos and their subhalos, using a parametrized relation between the dust-processed infrared luminosity and (sub-)halo mass, and it allows to simultaneously fit all auto- and cross- power spectra very well. We find that the star formation history is well constrained up to redshifts around 2, and agrees with recent estimates of the obscured star-formation density using *Spitzer* and *Herschel*. However, at higher redshift, the accuracy of the star formation history measurement is strongly degraded by the uncertainty in the spectral energy distribution of CIB galaxies. We also find that the mean halo mass which is most efficient at hosting star formation is $\log(M_{\rm eff}/{\rm M}_{\odot}) = 12.6$ and that CIB galaxies have warmer temperatures as redshift increases.

Keywords. galaxies: high-redshift, (cosmology:) dark matter, (cosmology:) diffuse radiation

1. Introduction and data used

The relic emission from galaxies formed throughout cosmic history appears as a diffuse, cosmic background. The Cosmic Infrared Background (CIB) is the far-infrared part of this emission and it contains about half of its total energy (Dole *et al.* 2006). The anisotropies detected in this background light trace the large-scale distribution of star-forming galaxies and, to some extent, the underlying distribution of the dark matter halos in which galaxies reside. The CIB is thus a direct probe of the interplay between baryons and dark matter throughout cosmic time and it allows an accurate determination of the star formation history of the Universe, which has long been a fundamental goal in Astronomy. Combining *Planck* HFI data with IRAS, the CIB angular auto- and cross-frequency power spectra are measured from 143 to 3000 GHz and they are interpreted with a Halo Model that associates star-forming galaxies with dark matter halos and their sub-halos, using a parametrized relation between the dust-processed infrared luminosity and (sub-)halo mass. The star formation history is well constrained up to redshifts around z$\sim$2, and agrees with recent estimates of the obscured star formation density using Spitzer and Herschel. However, at higher redshift, the accuracy of the star formation history measurement is strongly degraded by the uncertainty in the spectral energy distribution of CIB galaxies.

We use *Planck* channel maps in Healpix (Gorski *et al.* 2005) format from the six HFI frequencies: 100, 143, 217, 353, 545, and 857 GHz, together with far-infrared data at 3000 GHz (100 m) from IRAS (IRIS, Miville-Deschenes & Lagache 2005). Although *Planck*

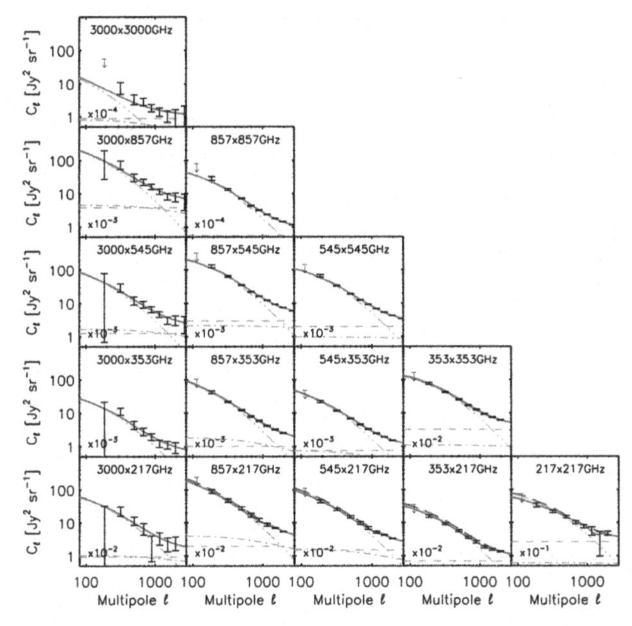

Figure 1. (Cross-) power spectra of the CIB anisotropies measured by *Planck* and *IRAS*, compared with the best-fit extended halo model. Data points are shown in black. The red line is the sum of the linear, 1-halo and shot-noise components, which is fitted to the data. The orange dashed, green dot-dashed, and cyan three-dots-dashed lines are the best-fit shot-noise level, the 1-halo and the 2-halo terms, respectively.

is an all-sky survey, we restricted our CIB anisotropy measurements to a few fields at high Galactic latitude, where foregrounds can be more easily controlled. The total area used to compute the CIB power spectrum is about 2240 deg^2, 16 times larger than in Planck Collaboration XVIII (2011). Measuring the CIB anisotropies is not easy, and it requires a very accurate component separation. Galactic dust, Cosmic Microwave Background (CMB) anisotropies, emission from galaxy clusters through the thermal Sunyaev-Zeldovich (tSZ) effect, and point sources all have a part to play. To remove the CMB we use a simple subtraction technique, using the HFI lowest frequency channel (100 GHz) as a CMB template and taking into account for its contamination due to CIB and tSZ residuals. For the Galactic dust we use an independent, external tracer of diffuse dust emission, the HI gas. From 100 μm to 1 mm at high Galactic latitude and outside molecular clouds, a tight correlation is observed between far-infrared emission from dust and the 21-cm emission from gas: HI can thus be used as a tracer of cirrus emission and the dust can be removed accordingly.

2. Results: constraints on Star Formation History

With a best-fit χ^2 of 100.7 and 98 degrees of freedom, we obtain a very good fit to the data (see Fig. 1) using an extended version of the standard halo model. The mean value of the most efficient halo mass for generating the CIB, $\log(\mathrm{M_{eff}}/\mathrm{M_\odot}) = 12.6 \pm 0.1$, is in good agreement with results obtained from a similar analysis using *Herschel* data. The star formation rate density (SFRD) predicted (see Fig. 2) is in agreement with other measurements up to redshift z$\sim$2, while at higher redshifts our model predicts higher values for the SFRD respect to other measurements (see e.g. Gruppioni *et al.* 2013) and also a different analysis of *Planck* data using only linear scales, i.e. multipoles $l < 500$).

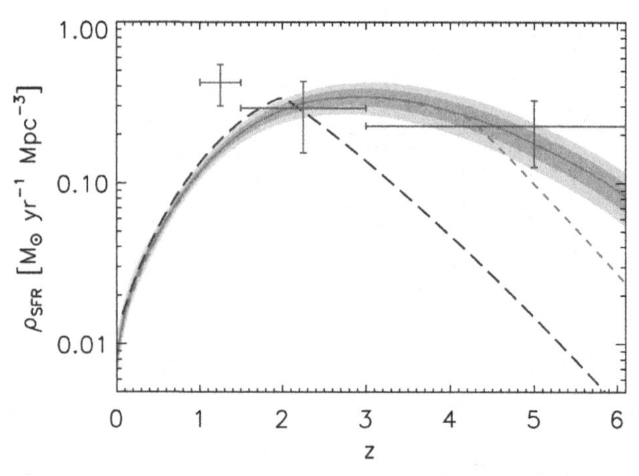

Figure 2. Marginalized constraints on the star formation rate density, as derived from our extended halo model (red continuous line with ± 1 and $\pm 2\,\sigma$ orange dashed areas). It is compared with mean values computed imposing different conditions on the redshift evolution of galaxy luminosities (black long-dashed line and blue dashed line). The violet points with error bars are the SFR density determined from the modeling of the CIB-CMB Lensing cross correlation by Planck 2013 Results XVIII.

However, the SFRD predicted is in very good agreement with the estimate obtained from the gravitational lensing-infrared background correlation with *Planck* data (see Planck Collaboration XVIII). Knowing that the SEDs of the galaxies responsible for the bulk of the CIB is the principal limitation in our modeling framework, the accurate measurement of the SEDs with future surveys will be crucial to properly estimate the obscured star formation rate density at high redshift from the CIB anisotropies and thus determine whether or not the bulk of the star formation is obscured at high redshift, and whether the UV and Lyman-break galaxy populations are a complete tracer of the star formation in the early Universe.

Acknowledgments

The development of *Planck* has been supported by: ESA; CNES and CNRS/INSU-IN2P3-INP (France); ASI, CNR, and INAF (Italy); NASA and DoE (USA); STFC and UKSA (UK); CSIC, MICINN and JA (Spain); Tekes, AoF and CSC (Finland); DLR and MPG (Germany); CSA (Canada); DTU Space (Denmark); SER/SSO (Switzerland); RCN (Norway); SFI (Ireland); FCT/MCTES (Portugal); and PRACE (EU). A description of the Planck Collaboration and a list of its members with the technical or scientific activities they have been involved into, can be found at ESA's RSSD website. The Parkes radio telescope is part of the Australia Telescope National Facility which is funded by the Commonwealth of Australia for operation as a National Facility managed by CSIRO. Some data used in this paper are based on observations with the 100-m telescope of the MPIfR (Max-Planck-Institut fur Radioastronomie) at Effelsberg.

References

Dole, H., Lagache, G., Puget, J. L., *et al.* 2006, *A&A*, 451, 417
Gorski, K. M., Hivon, E., Banday, A. J., *et al.* 2005, *ApJ*, 622, 759
Gruppioni, C., Pozzi, F., Rodighiero, G. *et al.*. 2013, *MNRAS*, Vol. 432, 23-52
Miville-Deschenes, M. -A. & Lagache, G. 2005, *ApJS*, 157, 302
Planck Collaboration XVIII 2013, accepted for publication in *A&A*

Statistical Challenges in 21st Century Cosmology
Proceedings IAU Symposium No. 306, 2014
A. F. Heavens, J.-L. Starck & A. Krone-Martins, eds.

doi:10.1017/S1743921314013829

Searching for non-Gaussianity in the Planck data

Marcelo J. Rebouças[1] and Armando Bernui[2]

[1]Centro Brasileiro de Pesquisas Físicas,
Rua Dr. Xavier Sigaud 150, 22290-180 Rio de Janeiro – RJ, Brazil
email: reboucas@cbpf.br

[2]Observatório Nacional
Rua General José Cristino 77, 20921-400 Rio de Janeiro – RJ, Brazil
email: bernui@on.br

Abstract. The statistical properties of the temperature anisotropies and polarization of the of cosmic microwave background (CMB) radiation offer a powerful probe of the physics of the early universe. In recent works a statistical procedure based upon the calculation of the kurtosis and skewness of the data in patches of CMB sky-sphere has been proposed and used to investigate the large-angle deviation from Gaussianity in WMAP maps. Here we briefly address the question as to how this analysis of Gaussianity is modified if the foreground-cleaned Planck maps are considered. We show that although the foreground-cleaned Planck maps present significant deviation from Gaussianity of different degrees when a less severe mask is used, they become consistent with Gaussianity, as detected by our indicators, when masked with the union mask U73.

Keywords. Non-Gaussianity, Cosmic Microwave Background Radiation, CMB Planck maps

1. Introduction

The statistical properties of the temperature fluctuations and polarization of cosmic microwave background (CMB) radiation offer a powerful probe of the physics of the early universe (Komatsu 2010). In this way, a detection of a significant level of primordial non-Gaussianity (NG) of local type ($f_{\rm NL}^{\rm local} \gg 1$) would rule out, for example, the entire class of single scalar field models (see, e.g., Creminelli & Zaldarriaga 2004 and Komatsu 2010).

It is conceivable, however, that no single statistical estimator can be sensitive and suitable to capture all forms of non-Gaussianity that may be present in the observed CMB data. Thus, it is important to test CMB data for non-Gaussianity by using different statistical indicators.

In a recent paper (Bernui & Rebouças 2009) statistical procedure based upon the calculation of the skewness and kurtosis by taking the values of the CMB temperatures fluctuations assigned to the pixels inside patches of CMB sky-sphere has been proposed and used to study deviation from Gaussianity in foreground-reduced WMAP maps (Bernui & Rebouças 2010) as well as in simulated maps (Bernui & Rebouças 2012). A pertinent question is how the analysis of Gaussianity made by using WMAP data is modified if the foreground-cleaned maps released by the Planck are considered. We have addressed this question and here we report partially the results of our analyses performed with the skewness estimator. For a comprehensive statistical analysis we refer the readers to our recent paper (Bernui & Rebouças 2014).

2. Statistical procedure and main results

Perhaps the simplest test for Gaussianity of a CMB map can be made by computing the skewness, S, and kurtosis, K from the whole set of CMB temperature fluctuations values of a given CMB map. However, one can go a step further and, instead of calculate two numbers, one can compute n values of the skewness as well as n values of the kurtosis, and obtain with directional information on NG, by dividing the CMB sphere $\mathbb{S}^2$ into a number n of uniformly distributed spherical patches of equal area that cover $\mathbb{S}^2$, and by calculating the skewness and the kurtosis

$$S_j = \frac{1}{N_{\rm p}\,\sigma_j^3} \sum_{i=1}^{N_{\rm p}} \left(T_i - \overline{T_j}\right)^3 , \tag{2.1}$$

$$K_j = \frac{1}{N_{\rm p}\,\sigma_j^4} \sum_{i=1}^{N_{\rm p}} \left(T_i - \overline{T_j}\right)^4 - 3 , \tag{2.2}$$

for each patch $j = 1, \ldots, n$. Here $N_{\rm p}$ is the number of pixels in the $j^{\,\rm th}$ patch, T_i is the temperature at the $i^{\,\rm th}$ pixel, $\overline{T_j}$ is the CMB mean temperature in the $j^{\,\rm th}$ patch, and σ is the standard deviation. In this work, we have chosen these patches to be spherical caps (calottes) with aperture $\gamma = 90°$.

The two set of n values (each) $\{S_j\}$ and $\{K_j\}$ along with the spherical coordinates of the center of the patches, θ_j, ϕ_j can then be employed to define two discrete functions on $\mathbb{S}^2$, namely $S(\theta_i, \phi_i)$ and $K(\theta_i, \phi_i)$ in such way that $S(\theta_j, \phi_j) = S_j$ and $K(\theta_j, \phi_j) = K_j$ for every $j = 1, \ldots, n$. These functions give local measurements of NG as functions of angular coordinates. The Mollweide projections of $S(\theta_j, \phi_j)$ and $K(\theta_j, \phi_j)$ are skewness and kurtosis maps, whose power spectra S_ℓ and K_ℓ can be used to study large-angle deviation from Gaussianity by determinig the goodness of fit of these power spectra obtained from the Planck maps as compared to the mean power spectra calculated from 1 000 simulated Gaussian maps ($\overline{S_\ell^G}$ and $\overline{K_\ell^G}$) through a χ^2 analysis. In this way, for S_ℓ obtained from a given Planck map one has

$$\chi^2_{S_\ell} = \frac{1}{N-1} \sum_{\ell=1}^{N} \frac{\left(S_\ell - \overline{S_\ell^G}\right)^2}{(\sigma_\ell^G)^2} , \tag{2.3}$$

where $\overline{S_\ell^G}$ are the mean multipole values for each ℓ mode, $(\sigma_\ell^G)^2$ is the variance computed from 1 000 Gaussian maps, and N is the highest multipole taken in the analysis of NG.

Clearly a similar expression and reasoning can be used for K_ℓ. In what follows, however, for the sake of brevity we will only briefly report the results of our analysis related to the skewness. For a comprehensive statistical analysis see our recent paper (Bernui & Rebouças 2014).

Fig. 1 shows the power spectra S_ℓ calculated from SMICA and NILC maps masked with INPMASK (left panel). The right panel of this figure shows the power spectra S_ℓ computed from SMICA, NILC and SEVEM maps with the U73 mask. This figure also contains the points of the averaged power spectra $\overline{S_\ell^G}$ calculated from 1 000 Gaussian simulated CMB maps and the 1σ error bars. To the extent that some of power spectra values S_ℓ fall off the 1σ error bars centered at $\overline{S_\ell^G}$ value, the left panel of this figure indicates departure from Gaussianity in both SMICA and NILC maps when masked with INPMASK. However, the right panel indicates that departure disappear when the more severe U73 mask is used.

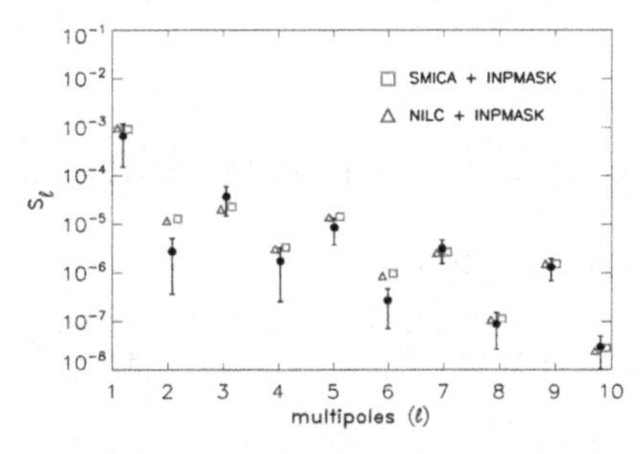

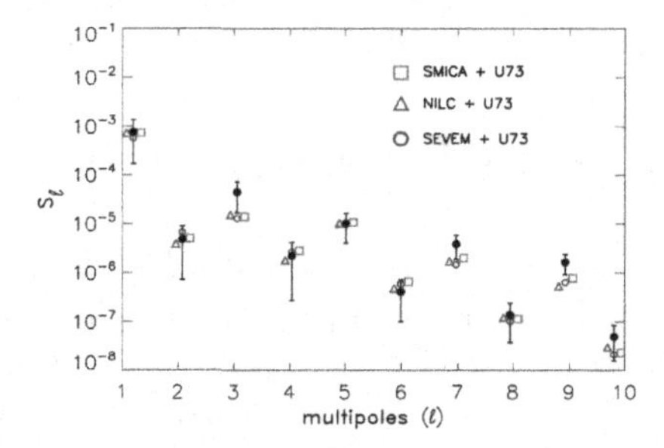

Figure 1. Low ℓ power spectra S_ℓ calculated from SMICA and NILC Planck maps equipped with INPMASK (left panel) and with U73 mask. We note that since there is no available INPMASK for the SEVEM map we have not included this map in the analysis with the INPMASK. Tiny horizontal shifts were used to avoid overlaps of symbols.

The above comparison of the power spectra by using Fig. 1 is useful as a qualitative indication of NG of Planck maps with different masks. However, to have a quantitative overall assessment of large-angle deviation from Gaussianity we have used the power spectra S_ℓ (calculated from the Planck maps) to carry out the above-mentioned χ^2 analysis to determine the goodness of fit of S_ℓ computed from the Planck maps as compared to the mean power spectra $\overline{S_\ell^G}$. Table 1 makes clear that although with different χ^2–probabilities the SMICA, NILC and SEVEM masked with INPMASK exhibit small level of NG, but when the union mask U73 is used these maps are consistent with Gaussianity as detected by our indicator S, in agreement with the results found by the Planck team (Ade *et al.* 2013).

Map & Mask	$\chi^2_{S_\ell}$– probability
SMICA–INPMASK	1.00×10^{-4}
NILC–INPMASK	1.80×10^{-3}
SMICA–U73	8.43×10^{-1}
NILC–U73	8.25×10^{-1}
SEVEM–U73	7.29×10^{-1}

Table 1. Results of the χ^2– probability test to determine the goodness of fit for S_ℓ multipole values, calculated from the SMICA, NILC and SEVEM with INPMASK and U73 masks, as compared to the mean power spectra $\overline{S_\ell^G}$ obtained from 1 000 simulated Gaussian maps.

Acknowledgements

M.J. Rebouças acknowledges the support of FAPERJ under a CNE E-26/102.328/2013 grant. M.J.R. and A.B. thank the CNPq for the grants under which this work was carried out. Some of the results were derived using the HEALPix package (Gorski *et al.* 2005).

References

Ade, P. A. R, *et al.* (Planck Collaboration) *arXiv*: 1303.5084
Bernui, A. & Rebouças, M. J. 2009, *Phys. Rev. D*, 79, 063528
Bernui, A. & Rebouças, M. J. 2010, *Phys. Rev. D*, 81, 063533
Bernui, A. & Rebouças, M. J. 2012, *Phys. Rev. D*, 85, 023522
Bernui, A. & Rebouças, M. J. 2014, *arXiv*: 1405.1128
Creminelli, P. & Zaldarriaga, M. 2004, *J. Cosmol. Astrop. Phys.* 2004, 10, 006
Gorski, K. M., *et al.* 2005, *ApJ*, 622, 759
Komatsu, E. 2010, *Class. Quant. Grav.* 27, 124010

Statistical Challenges in 21st Century Cosmology
Proceedings IAU Symposium No. 306, 2014
A. F. Heavens, J.-L. Starck & A. Krone-Martins, eds.

doi:10.1017/S1743921314013490

A Close Examination of CMB Mirror-Parity

Assaf Ben-David[1] and Ely D. Kovetz[2]

[1]Niels Bohr International Academy and Discovery Center, The Niels Bohr Institute, Blegdamsvej 17, DK-2100 Copenhagen Ø, Denmark; email: bendavid@nbi.dk

[2]Theory Group, Department of Physics and Texas Cosmology Center, The University of Texas at Austin, TX 78712, USA; email: elykovetz@gmail.com

Abstract. We revisit recent claims of significant mirror-parity in CMB data from *WMAP* and *Planck* with a careful analysis using statistical estimators in both harmonic and pixel spaces. While the data indeed show significant signs of odd mirror-parity under some circumstances, the broad study shows that the results are not in significant tension with ΛCDM.

Keywords. cosmic microwave background, methods: data analysis, methods: statistical

1. Method and Data

Pixel-Based Statistic: Under reflection through the plane normal to $\hat{\mathbf{n}}$, a direction $\hat{\mathbf{r}}$ in the sky transforms as $\hat{\mathbf{r}} \to \hat{\mathbf{r}}_{\hat{\mathbf{n}}} = \hat{\mathbf{r}} - 2(\hat{\mathbf{r}} \cdot \hat{\mathbf{n}})\,\hat{\mathbf{n}}$. The score for mirror-parity in pixel-space is the temperature difference between hemispheres, averaged over all unmasked spatial directions (de Oliveira-Costa et al. 2004; Finelli et al. 2012; Ade et al. 2013b; Ben-David & Kovetz 2014)

$$S_{\mathrm{p}}^{\pm}(\hat{\mathbf{n}}) = \overline{\left[\frac{T(\hat{\mathbf{r}}) \pm T(\hat{\mathbf{r}}_{\hat{\mathbf{n}}})}{2}\right]^2}. \tag{1.1}$$

With this statistic, a large degree of even (odd) mirror-parity is given by a low S_{p}^{-} (S_{p}^{+}). To apply a Galactic mask, the masked pixels are simply ignored in the summation.

Harmonic Statistic: Under reflection through the z-axis, $Y_{\ell m}(\hat{\mathbf{r}}) \to (-1)^{\ell+m} Y_{\ell m}(\hat{\mathbf{r}})$. The harmonic parity score compares for each symmetry-plane direction the relative power of even and odd multipoles (Ben-David et al. 2012; Ben-David & Kovetz 2014)

$$S_{\mathrm{h}}(\hat{\mathbf{n}}) = \sum_{\ell=2}^{\ell_{\max}} \sum_{m=-\ell}^{\ell} (-1)^{\ell+m} \frac{|a_{\ell m}(\hat{\mathbf{n}})|^2}{\widehat{C}_\ell} - (\ell_{\max} - 1), \tag{1.2}$$

where $\hat{\mathbf{n}}$ is the z-axis used in the harmonic expansion and the score is normalized by the actual *observed* power $\widehat{C}_\ell = (2\ell+1)^{-1} \sum_m |a_{\ell m}|^2$. The parameter $\ell_{\max}$ allows to easily test the scale dependence of the results. With this statistic, a large degree of even (odd) mirror-parity is given by a high (low) value for S_{h}. On a masked map a method for maximum likelihood reconstruction of the harmonic coefficients is used.

Significance Estimation: We use two complementing methods: (i) Compare the "best" raw score R, i.e. the maximum or minimum of the score map, to random ΛCDM realizations (Ade et al. 2013b). (ii) Normalize the "best" raw score as $\bar{R} = |R - \mu|/\sigma$, where μ and σ are the mean and standard deviation of the score-map over all directions (Ben-David et al. 2012), and compare $\bar{R}$ to random ΛCDM realizations. While the first method is straightforward, the second also tests whether the anomalous direction is unique. This is suited for a scenario in which the anomaly is of a cosmological origin, and a *single* anomalous direction is expected. We use a set of 10^4 random ΛCDM realizations using the best-fit power-spectrum. We mask randoms and the data with the same mask.

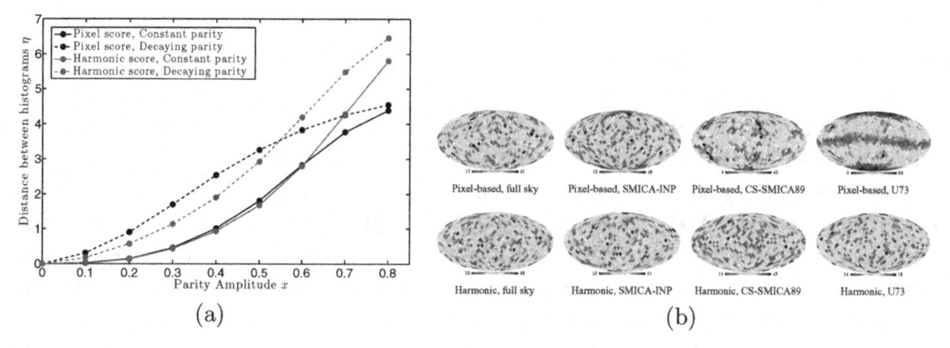

(a) (b)

Figure 1. (a) Estimate of the signal-to-noise ratio for the detection of mirror-parity as a function of the parity modulation amplitude. (b) Histograms of the direction of maximal parity for random realization with the various masks.

Data: We use the *Planck* SMICA and NILC maps (Ade et al. 2013a). We also use the LGMCA map released by Bobin et al. (2014). This map is a combination of data from *WMAP* and *Planck* that is claimed to be a rather clean full sky map. We also use three Galactic masks in our analysis (Ade et al. 2013a): the large unified U73 mask and the confidence and inpainting masks of the SMICA map, CS-SMICA89 and SMICA-INP. These masks have $f_{\rm sky} = 0.73, 0.89$ and 0.97, respectively. We smooth all maps and masks with a Gaussian beam with FWHM of $\sim 11^\circ$ and degrade them to $N_{\rm side} = 16$.

2. Tests on Random Simulations

We test our parity statistics on random simulations to assess their effectiveness.

Detectability of Mirror-Parity: Fig. 1(a) shows a rough estimate of the S/N for the detection of mirror-parity using the two statistics. We used random ΛCDM realizations modulated to contain traces of even mirror-parity and calculated the distances between the histograms for the modulated randoms and the unmodulated ones. The modulation amplitude is either constant or exponentially decaying with scale. Both statistics are effective in detecting mirror-parity, and perform similarly in this test.

Sensitivity to the Total Power: The theoretical power spectrum on large scales is not well determined. The power observed on these scales is somewhat low compared with expectations. How sensitive are the parity statistics to this uncertainty? The harmonic statistic is manifestly normalized by the total observed power, whereas the pixel-based statistic is not. We tested parity-modulated randoms and calculated the significance of the results once versus normal ΛCDM realizations and then versus realizations constrained to have low power, consistent with the levels measured by *Planck*. We find that when comparing to the "wrong" ensemble, the pixel-based significance changes by $\sim 0.5\sigma$. The harmonic significance level is, however, much more stable and only changes by $\sim 0.02\sigma$.

Bias on Masked Sky: We find the direction exhibiting maximal even parity for each of the isotropic ΛCDM realizations. We repeat the test using each of the masks, as well as using the full sky. In Fig. 1(b) we plot histograms of the results. Since the randoms are isotropic, a non-uniform distribution can only be attributed to the non-isotropic mask. While both statistics are isotropic on the full sky, as the masked area is increased the pixel-based statistic is increasingly biased, while the harmonic-statistic is unaffected.

3. Results

For all map and mask combinations and for both statistics, we identify a single direction of maximal even mirror-parity and a corresponding one for odd mirror-parity.

Even Mirror-Parity Direction $(l, b) \sim (260°, 48°)$**:** Regardless of which map and mask are considered, which statistic and which significance estimator, the significance level for the even parity direction never reaches 3σ and is therefore not anomalous.

Odd Mirror-Parity Direction $(l, b) \sim (264°, 17°)$**:** The significance levels of the odd parity direction using both the R and $\bar{R}$ estimators range from $1.5 - 3\sigma$. When masked with the large U73 mask and measured using the R estimator, the results are marginally significant, crossing the 3σ level (matching Ade et al. 2013b). However, with a smaller mask or no mask, the results are insignificant. When measured using the $\bar{R}$ estimator, the results are not significant for any of the masks. The significance levels for the harmonic statistic are not high either. No combination of map and mask shows a significant ($> 3\sigma$) level of odd parity regardless of the significance estimator used, for any $5 \leqslant \ell_{\max} \leqslant 9$.

4. Conclusion

We have reproduced the results of Ade et al. (2013b) showing a marginally significant ($\sim 3\sigma$) odd parity direction in the *Planck* data using a pixel-based statistic and the large U73 mask. However, we have shown that these results are biased due to the large mask, and are also sensitive to assumptions regarding the total power on large scales, which is weakly constrained by the data. Indeed, with a smaller mask, the significance level is much lower.

Using a harmonic statistic, which we have shown to be far more robust, the odd-parity direction is not significant, no matter the map and mask used, for any of the tested large scales ($5 \leqslant \ell_{\max} \leqslant 9$).

In light of these findings, we conclude that while there is some tendency for odd parity in the CMB data, when embracing a broader perspective and examining the complete set of data maps and Galactic masks and the properties of the statistical estimators, it appears that the evidence for anomalous mirror-parity is rather weak. It poses no real challenge to the concordance model, and should therefore not be considered a ΛCDM anomaly.

Acknowledgements

ABD was supported by the Danish National Research Foundation. EDK was supported by the National Science Foundation under Grant Number PHY-1316033.

References

Ade P. A. R., *et al.* (The Planck Collaboration) 2013a, preprint (arXiv:1303.5072)
Ade P. A. R., *et al.* (The Planck Collaboration) 2013b, preprint (arXiv:1303.5083)
Ben-David A., Kovetz E. D., & Itzhaki N. 2012, *ApJ*, 748, 39
Ben-David A. & Kovetz E. D. 2014, preprint (arXiv:1403.2104)
Bobin J., Sureau F., Starck J.-L., Rassat A., & Paykari P. 2014, *A&A*, 563, A105
de Oliveira-Costa A., Tegmark M., Zaldarriaga M., & Hamilton A. 2004, *Phys. Rev. D*, 69, 063516
Finelli F., Gruppuso A., Paci F., & Starobinsky A. A. 2012, *J. Cosmol. Astropart. Phys.*, 7, 49

Statistical Challenges in 21st Century Cosmology
Proceedings IAU Symposium No. 306, 2014
A. F. Heavens, J. -L. Starck & A. Krone-Martins, eds.

doi:10.1017/S1743921314013714

A Supervoid Explanation of the Cosmic Microwave Background Cold Spot

F. Finelli[1,2], J. García-Bellido[3], A. Kovács[4], F. Paci[5], I. Szapudi[6]

[1]INAF-IASF Bologna, Istituto di Astrofisica Spaziale e Fisica Cosmica di Bologna
Istituto Nazionale di Astrofisica, via Gobetti 101, I-40129 Bologna, Italy
[2]INFN, Sezione di Bologna, Via Irnerio 46, I-40126 Bologna, Italy
[3]Instituto de Física Teórica IFT-UAM/CSIC, Universidad Autónoma de Madrid, Cantoblanco 28049 Madrid, Spain
[4]Institute of Physics, ELTE, 1117 Pázmány Péter sétány 1/A Budapest, Hungary
[5]SISSA, Astrophysics Sector, Via Bonomea 265, 34136, Trieste, Italy
[6]Institute for Astronomy, University of Hawaii, 2680 Woodlawn Drive, Honolulu, HI, 96822

Abstract. The Cold Spot is an anomalously cold region in the Cosmic Microwave Background (Vielva *et al.* 2004), either caused by a structure in the line of sight or could be of primordial origin. We search for a supervoid aligned with the Cold Spot region, filling the gap in redshift at $z < 0.3$ which has never been explored in details. We find a large projected under density in the recently constructed WISE-2MASS catalogue, whose median redshift is $z \simeq 0.14$, with an angular size of 30 degrees. We show that a spherically symmetric Lemaitre-Tolman-Bondi (LTB) void model can simultaneously fit the $\delta_{gal}/b = \delta_{2D} \simeq -0.12$ underdensity in the WISE-2MASS catalogue, and the Cold Spot as observed by both the WMAP and Planck satellites. Such an LTB supervoid gives a plausible explanation of the Cold Spot anomaly, and is preferred over the null hypothesis or a texture model.

Keywords. cosmic microwave background, observations, large-scale structure of universe

---

1. Introduction

The Cosmic Microwave Background Cold Spot (CS) (Vielva *et al.* 2004) could have been originated by a primordial fluctuation on the last scattering surface or by an intervening phenomenon along the line of sight. Contamination from our galaxy or by the Sunyaev-Zeldovich effect from a cluster are quite unlikely (Cruz *et al.* 2006), and cosmic texture hypotheses have already been tested extensively as the origin of the anomaly (Cruz *et al.* 2007). Alternatively, the CS could be imprinted by an intervening supervoid along the line of sight (Inoue and Silk 2007). The supervoid model has been constrained by using radio galaxies of the NVSS survey, Canada-France-Hawaii Telescope (CFHT) imaging of the CS region, redshift survey data using the VIMOS spectrograph on the VLT, and the relatively shallow 2MASS galaxy catalogue. See Finelli *et al.* 2014, and references therein for review. These studies are consistent with an underdensity at $z < 0.3$, but they either run out of objects at low redshift, or have no redshift information to carry out a tomographic imaging of the Cold Spot area. A large area survey reaching up to $z \simeq 0.3$ is needed to map the redshift range unconstrained by previous studies. For this purpose, we use the recently produced WISE-2MASS galaxy catalog (Kovács and Szapudi 2014), with median redshift of $z \simeq 0.14$, and sky coverage of 21,200 square degrees after masking dusty regions. Galaxy samples, however, are biased tracers of the underlying dark matter distribution. We thus measured and modeled the angular power spectrum of WISE-2MASS galaxies, and found a galaxy bias $b_g = 1.41 \pm 0.07$, that we take into account in our further analyses. We show the projected galaxy density profile around the CS in the left panel of Fig. 1. Measurement errors are due to Poisson fluctuations calculated from the expected number of galaxies in a ring or a disk. We detect an underdensity with high significance. At larger radii, the radial profile is consistent with a supervoid surrounded by a gentle compensation that converges to the average galaxy density at $\sim 50°$. The supervoid might also contain a deeper inner void, with its own compensation at around $\simeq 8°$. See Finelli *et al.* 2014 for details.

2. Modeling the supervoid

Next we build a ΛLTB void model for the underdensity (Garcia-Bellido and Haugbølle 2008), characterized by a spatial curvature profile $k(r) = k_0\, r^2 \exp(-r^2/r_0^2)$, written as a linear metric perturbation in ΛCDM,

$$\Phi(\tilde{r}) = \Phi_0\, \tau^2\, e^{-\frac{\tilde{r}^2}{\tilde{r}_0^2}}\,, \tag{2.1}$$

with the LTB radius r related to the co-moving FRW radius through $\tilde{r} = \sqrt{3/4\pi}\, H_0 r$, and τ conformal time. This scalar potential gives rise to a 3D density profile for the void

$$\delta_{3D}(\tilde{r}) = -\delta_0 \left(1 - \frac{2\tilde{r}^2}{3\tilde{r}_0^2}\right) e^{-\frac{\tilde{r}^2}{\tilde{r}_0^2}}\,, \tag{2.2}$$

characterized by two parameters, the co-moving width $\tilde{r}_0$ and the depth δ_0. The 3D density (2.2) is then projected onto the transverse plane, using the WISE window function

$$\delta_{2D}(\theta) = \int_0^\infty \delta_{3D}\left(\tilde{r}(y,\theta)\right)\, y^2 \phi(y) dy\,. \tag{2.3}$$

We also compute the linear Integrated Sachs-Wolfe and the non-linear Rees-Sciama effects on the CMB. For a large compensated void with a profile of Eq. 2.2, the linear ISW effect is dominated by the non-linear Rees-Sciama effect,

$$\delta T(\theta) = -A \left(1 - \frac{28}{13}\frac{\theta^2}{\tilde{\theta}_0^2}\right) e^{-2\frac{\theta^2}{\tilde{\theta}_0^2}}\,, \tag{2.4}$$

where $\tilde{\theta}_0 = \sqrt{3/4\pi}\,\theta_0$, and $\tan\theta \simeq \theta$, is applied. The magnitude of the decrement depends on the parameters of the void,

$$A = 51.0\,\mu\mathrm{K} \left(\frac{r_0}{155.3\, h^{-1}\mathrm{Mpc}}\right)^3 \left(\frac{\delta_0}{0.2}\right)^2\,, \tag{2.5}$$

and $\theta_0 = (180°/\pi)(r_0/d_A(z_0))$, with $d_A(z)$ the angular diameter distance in a flat ΛCDM model ($\Omega_M = 0.3, h = 0.7$), and z_0 the redshift of the center of the void, at co-moving distance $y_0 = y(z_0)$. Note that the LTB model is only used as a general relativistic model for the void dynamics and its effect on the CMB, while the background cosmology is assumed to be standard ΛCDM.

A large LTB supervoid is sufficient to explain a colder spot in the CMB of about half the size, or in other words a shallower supervoid is sufficient to imprint the same temperature depression in the CMB, than is allowed for a supervoid assuming ΛCDM cosmology. Note that the LTB model is only used as a general relativistic model for the void dynamics and its effect on the CMB, while the background cosmology is assumed to be standard ΛCDM. For detailed measurements, a χ^2 statistic is constructed. We perform a simultaneous fit for the projected LTB void in the WISE-2MASS map, and the corresponding temperature depression effect in the CMB data, with parameters δ_0, r_0,

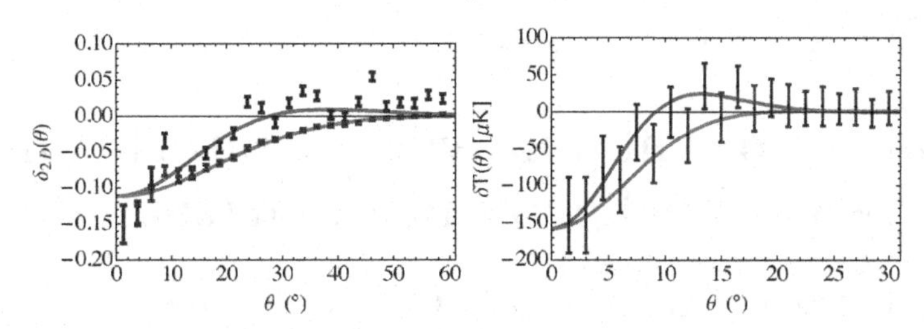

Figure 1. The density profile from WISE-2MASS catalogue compared with the theoretical model for the underdensity (2.3) (left panel). The temperature profile from Planck SMICA map (right panel) (Ade *et al.* 2013) compared with the predicted signal (2.4). The red (blue) lines are the theoretical profiles for rings (disks) and in dark red (blue) are the measurements.

and z_0. The first term corresponds to the χ^2 of the projected LTB void profile (2.3) using uncorrelated Poisson errors, σ_i. The second term is the χ^2 of the CMB profile compared with the LTB prediction (2.4) of the void observed in WISE-2MASS. The best fit LTB void parameters we have found are $\delta_0 = 0.25 \pm 0.10$ (1σ), $r_0 = (195 \pm 35)\mathrm{h}^{-1}\mathrm{Mpc}$ (1σ), and $z_0 = 0.16 \pm 0.04$ (1σ). The LTB model parameter δ_0 is the 3D dark matter density, giving a 12% projected underdensity at the center of the void, i.e. $\delta_{2\mathrm{D}}(\theta = 0) = -0.12$. The angular sizes $\theta_0 = 28.8° \pm 5.2°$, and $\tilde{\theta}_0 = 14.1° \pm 2.5°$ are derived parameters, which correspond to angular scales of the profile on the galaxy map and the CMB, respectively. For later comparison, we calculate the averaged underdensity within the best fit radius $r_0 = 195\,h^{-1}\mathrm{Mpc}$. The 3D top-hat-averaged density from the LTB profile, see Eq. (2.2), is $\bar{\delta} = 3/r_0^3 \int_0^{r_0} r^2 dr \delta(r) = -\delta_0/e$. This finally gives the average void depth $\bar{\delta} = -0.10 \pm 0.03$.

3. Conclusions

We have found a super void aligned with the CS in the WISE-2MASS catalogue and shown for the first time that both the supervoid profile and the CMB profile can be simultaneously fit assuming an LTB profile embedded in an FRW universe located at redshift $z = 0.155 \pm 0.037$ with radius $r_0 = (195 \pm 35)\,h^{-1}\mathrm{Mpc}$, and top-hat-averaged depth $\bar{\delta} = -0.10 \pm 0.03$. Such an LTB supervoid gives a perfect explanation, via a Rees-Sciama effect, of the Cold Spot anomaly, and is strongly preferred (using a Bayesian analysis) over the null hypothesis (statistical fluctuation) or a texture model. Note that these values are in excellent agreement with the findings of Szapudi *et al.* 2014, who combined the WISE-2MASS galaxy data with Pan-STARRS1, optical observations for a direct tomographic imaging of the CS region.

References

Vielva P., Martinez-Gonzalez E., Barreiro R. B., Sanz J. L., & Cayon L., 2004, *Astrophys. J.* **609** 22

Cruz M., Tucci M., Martinez-Gonzalez E., & Vielva P., 2006, *Mon. Not. Roy. Astron. Soc.* **369** 57

Cruz M., Turok N., Vielva P., Martinez-Gonzalez E.,, & Hobson M., 2008, *Mon. Not. Roy. Astron. Soc.* **390** (2008) 913

Inoue, K. T. & Silk, J. 2006, *ApJ*, 648, 23-30

Szapudi, I., Kovács, A., Granett, B. R., *et al.* 2014, *arXiv:1405.1566*

Finelli, F., Garcia-Bellido, J., Kovacs, A., Paci, F., & Szapudi, I. 2014, *arXiv:1405.1555*

Kovács, A. & Szapudi, I. 2014, *arXiv:1401.0156*

Garcia-Bellido, J. & Haugbølle, T. 2008, *JCAP*, 4:3

Ade P. A. R., *et al.* [Planck Collaboration], 2013, arXiv:1303.5072 [astro-ph.CO].

Statistical Challenges in 21st Century Cosmology
Proceedings IAU Symposium No. 306, 2014
A. F. Heavens, J.-L. Starck & A. Krone-Martins, eds.

doi:10.1017/S1743921314013787

Simulation of the analysis of interferometric microwave background polarization data

Emory F. Bunn[1], Ata Karakci[2], Paul M. Sutter[3,4], Le Zhang[5], Gregory S. Tucker[2], Peter T. Timbie[5] and Benjamin D. Wandelt[4]

[1]University of Richmond, USA
email:ebunn@richmond.edu

[2]Brown University, USA

[3]Ohio State University, USA

[4]Institut d'Astrophysique de Paris, France

[5]University of Wisconsin – Madison, USA

Abstract. We present results from an end-to-end simulation pipeline of interferometric observations of cosmic microwave background polarization. We use both maximum-likelihood and Gibbs sampling techniques to estimate the power spectrum. In addition, we use Gibbs sampling for image reconstruction from interferometric visibilities. The results indicate the level to which various systematic errors (e.g., pointing errors, gain errors, beam shape errors, cross polarization) must be controlled in order to successfully detect and characterize primordial B modes and achieve other scientific goals. In addition, we show that Gibbs sampling is an effective method of image reconstruction for interferometric data in other astrophysical contexts.

Keywords. cosmology: cosmic microwave background, techniques: interferometric, methods: statistical

---

Measurement of cosmic microwave background (CMB) anisotropy has become one of the most powerful tools in cosmology. In recent years, researchers have built on the success of these anisotopy measurements by studying linear polarization in the CMB. In particular, considerable attention has been focused on the search for "B-mode" polarization, which has the potential to measure the energy scale of inflation, along with probing cosmology in a variety of other ways (Hu & Dodelson 2002). BICEP2 has measured a B-mode polarization signal in the microwave sky (Ade *et al.* 2014), which if confirmed will represent a major advance in cosmology. The field eagerly awaits other measurements at different frequencies and with different instruments.

Because the B-mode signal is expected to be very faint, control of systematic errors is of paramount importance. An argument can be made (Timbie *et al.* 2006, Bunn 2007) that interferometers provide better control of systematics than imaging telescopes. This is one of the reasons that, for instance, the QUBIC collaboration (Ghribi *et al.* 2013) is constructing an instrument based on the novel approach of bolometric interferometry.

Whether or not interferometers have *better* systematic error properties than imagers, it is clear that they have *different* sensitivity to systematics. Interferometric systematic issues have not received as much attention as systematics in imaging systems. Given the importance of a robust characterization of CMB B modes, it seems worthwhile to study these effects in detail. For these reasons, we have performed detailed simulations of interferometric observations of CMB polarization, in order to characterize the effects of various systematic errors on the reconstruction of the polarization power spectra (Karakci *et al.* 2013a, Karakci *et al.* 2013b, Zhang *et al.* 2013).

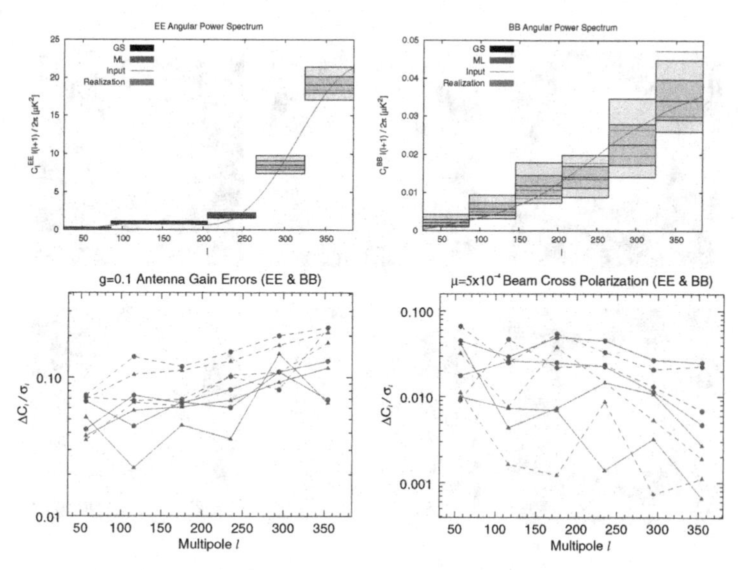

Figure 1. *Top:* Mean posterior band powers obtained by Gibbs sampling (GS) are shown in black. The maximum-likelihood (ML) band powers are shown in blue. Dark and light grey indicate 1σ and 2σ uncertainties on the Gibbs sampling results. Binned power spectra of the signal realization and input power spectra are shown in pink and red. *Bottom*: Examples of the effects of systematic errors, obtained by both ML (triangles) and GS (dots) are shown. Solid and dashed lines correspond to experiments that interfere linear and circular polarization states respectively. Results are shown for the EE (red) and BB (blue) power spectra. For further details, see Karakci *et al.* (2013b).

We have developed tools that generate visibilities for interferometers with arbitrary antenna placement, beam shape, etc., from HEALPix sky maps. We can include the effects of a wide variety of systematic errors, including beam shape and pointing errors, cross-polarization, gain errors, *etc.* We then estimate power spectra from these visibilities in two independent ways, via maximum-likelihood estimation (Hobson & Maisinger 2002) and Gibbs sampling (Larson *et al.* 2007, Sutter *et al.* 2012).

Fig. 1 shows the results of simulations that are described in detail in Karakci *et al.* (2013a), Zhang *et al.* (2013), and Karakci *et al.* (2013b). These results are for a simulated interferometer consisting of a 20×20 close-packed array of feedhorns with a Gaussian beam width of 5° and separation $D = 7.89\lambda$. We include sky rotation for observations from the South Pole, with average noise per visibility of $0.015\,\mu$K. Fig. 1 shows that we correctly reconstruct the power spectra and illustrates the effects of introducing some systematic errors. Order-of-magnitude agreement is found between the results of these simulations and a simple semi-analytic approach (Bunn 2007). Far more detailed results may be found in the papers cited above.

Gibbs sampling provides simultaneous samples of the power spectrum and the signal map. We have shown that these signal map samples provide excellent image reconstruction from visibility data in other (non-CMB) contexts (Sutter *et al.* 2014). Fig. 2 shows some of the results of this work.

On the left of Fig. 2 we show sample images taken from the CASA user guide (Jaeger 2008). These images were then "observed" with a simulated interferometer consisting of 12 randomly-placed antennas, each with a beam size of 0.075 times the image width. We assumed 6 hours of observation with a signal-to-noise ratio of 10 per visibility. The center panel shows the mean reconstructed image calculated via Gibbs sampling, multiplied by the primary beam. The right panel shows the result of ℓ_1 reconstruction in the pixel

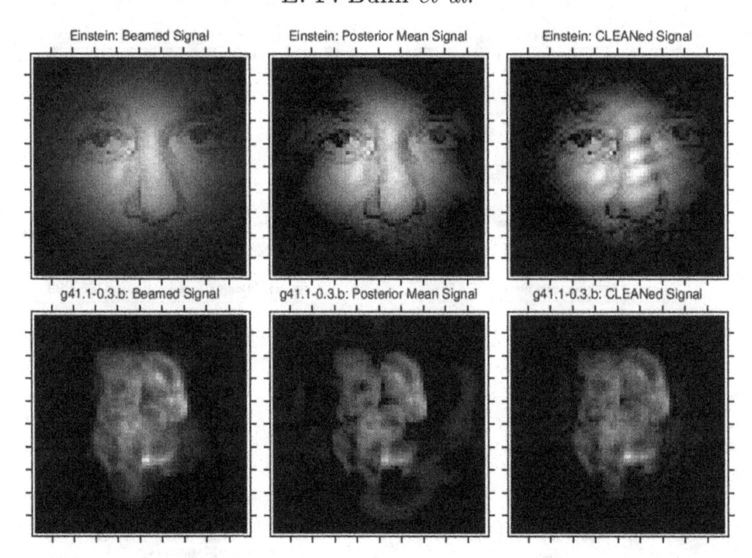

Figure 2. Examples of image reconstruction. The left panel shows test images from the CASA user guide, multiplied by the primary beam of our simulated observations. The center panel shows the result of image reconstruction using Gibbs sampling, and the right shows ℓ_1 reconstruction. Adapted from Sutter *et al.* (2014).

basis, which has been shown (Wiaux *et al.* 2009) to have similar performance to the widely-used CLEAN algorithm (Högbom 1974). By a variety of quantitative measures, Gibbs reconstruction performs better than this proxy for CLEAN reconstruction (Sutter *et al.* 2014).

This research was funded by the following awards from the National Science Foundation: AST-0908902, AST-0908844, AST-0908900, AST-0908855, AST-0927748.

References

Ade, P. A. R., *et al.* 2014, *Phys. Rev. Lett.*, 112, 241101

Bunn, E. F. 2007, *Phys. Rev. D*, 75, 083517

Ghribi, A., *et al.* 2013, *JLTP*, 173

Hobson, M. P. & Maisinger, K. 2002, *MNRAS*, 334, 569

Högbom J. A. 1974, *A&AS*, 15, 417

Hu W. & Dodelson, S. 2002, *ARAA*, 40, 171

Jaeger *et al.* 2008, in Argyle, R. W., Bunclark, P. S., & Lewis J. R., eds, *Astronomical Data Analysis Software and Systems XVII*, Vol. 394 of Astronomical Society of the Pacific Conference Series, The Common Astronomy Software Application (CASA), p. 623

Karakci, A., Sutter, P. M., Zhang, L., Bunn, E. F., Korotkov, A., Timbie, P., Tucker,G. S., & Wandelt, B. D. 2013, *ApJS*, 204, 10

Karakci, A., Zhang, L., Sutter, P. M., Bunn, E. F., Korotkov, A., Timbie, P., Tucker, G. S., & Wandelt B. D. 2013, *ApJS*, 207, 14

Larson, D. L., Eriksen, H. K., Wandelt, B. D., Górski, K. M., Huey, G., Jewell, J. B., & O'Dwyer I. J. 2007, *ApJ*, 656, 653

Sutter, P. M., Wandelt, B. D., & Malu, S. S. 2012, *ApJS*, 202, 9

Sutter P. M., *et al.* 2014, *MNRAS*, 438, 768

Timbie, P. T., *et al.* 2006, *New Astron. Revs*, 50, 999

Wiaux, Y., Jacques, L., Puy, G., Scaife, A. M. M.., & Vandergheynst P. 2009, *MNRAS*, 395, 1733

Zhang, L., Karakci, A., Sutter, P. M., Bunn, E. F., Korotkov, A., Timbie P., Tucker, G. S., & Wandelt, B. D. 2013, *ApJS*, 206, 24

Statistical Challenges in 21st Century Cosmology
Proceedings IAU Symposium No. 306, 2014
A. F. Heavens, J. -L. Starck & A. Krone-Martins, eds.

doi:10.1017/S1743921314013520

Effects of primordial magnetic fields on CMB

Héctor J. Hortúa and Leonardo Castañeda

Grupo de Gravitación y Cosmología, Observatorio Astronómico Nacional,
Universidad Nacional de Colombia, Cra 45 # 26-85, Bogotá D.C., Colombia
email: hjhortua@unal.edu.co

Abstract. The origin of large-scale magnetic fields is an unsolved problem in cosmology. In order to overcome, a possible scenario comes from the idea that these fields emerged from a small primordial magnetic field (PMF), produced in the early universe. This field could lead to the observed large-scales magnetic fields but also, would have left an imprint on the cosmic microwave background (CMB). In this work we summarize some statistical properties of this PMFs on the FLRW background. Then, we show the resulting PMF power spectrum using cosmological perturbation theory and some effects of PMFs on the CMB anisotropies.

Keywords. Cosmic microwave background, Primordial magnetic fields

---

Magnetic fields have been observed in all scales of the universe, from planets and stars to galaxies and galaxy clusters. However, the origin of such magnetic fields is still unknown. Some theories argue magnetic fields we observe today has a primordial origin, indeed, there are some processes in early epoch of the universe that would have created a small primordial magnetic field (PMF). If PMFs really were present, these could have some effect on Nucleosynthesis and would leave imprints in the CMB fluctuation.

The model and statistics for a stochastic PMF. We consider a causal stochastic PMF generated after inflation, thus the maximum coherence lenght for the fields must be not less than the Hubble horizon (Kahniashvili *et al.* 2007). Now, the PMF power spectrum which is defined as the Fourier transform of the two point correlation can be written as

$$\langle B_i(\mathbf{k})B_j^*(\mathbf{k}')\rangle = (2\pi)^3\delta^3(\mathbf{k}-\mathbf{k}')P_{ij}P_B(\mathbf{k}), \tag{1}$$

where $P_{ij} = \delta_{ij} - \frac{\mathbf{k}_i\mathbf{k}_j}{k^2}$ is a projector onto the transverse plane, $P_B(\mathbf{k})$ is the PMF power spectrum. We focus our attention to the evolution of a causally-generated PMF parametrized by a power law with index $n \geqslant 2$, with an ultraviolet cut-off k_D and the dependence of an infrared cutoff, k_m. The power spectrum can be defined as

$$P_B(k) = Ak^n \quad \text{with} \quad A = \frac{B_\lambda^2 2\pi^2\lambda^{n+3}}{\Gamma(\frac{n+3}{2})}; \quad \text{for } k_m \leqslant k \leqslant k_D, \tag{2}$$

where B_λ is the comoving PMF strength smoothing over a Gaussian sphere of comoving radius λ. The energy density of PMF and anisotropic trace-free part are written as

$$\rho_B(k) = \frac{1}{8\pi}\int \frac{d^3k'}{(2\pi)^3}B_l(k)B^l(|\mathbf{k}-\mathbf{k}'|), \tag{3}$$

$$\Pi_{ij}(k) = \int \frac{d^3k'}{2(2\pi)^4}\left[B_i(k')B_j(|\mathbf{k}-\mathbf{k}'|) - \frac{\delta_{ij}}{3}B_l(k')B^l(|\mathbf{k}-\mathbf{k}'|)\right], \tag{4}$$

(in the Fourier space) and we can write the Lorentz force on matter as $\Pi^{(S)} = L^{(S)}(\mathbf{x},\tau_0) + \frac{1}{3}\rho_B(\mathbf{x},\tau_0)$, here $k = |\mathbf{k}|$. Now, we define the two point correlation function as

$$\langle\Xi(k,\tau)\Xi^*(k',\tau)\rangle = (2\pi)^3\left|\Xi(k,\tau)\right|^2\delta^3(\mathbf{k}-\mathbf{k}'), \tag{5}$$

where $\Xi(k,\tau) = \{\rho_B(k,\tau), \Pi(k,\tau), L(k,\tau)\}$, and the cross correlation between them. Now, to calculate the power spectrum, we use the eqs. (1), (3), (4), (5) and the Wick's theorem assuming Gaussian statistics to evaluate the 4-point correlator of the PMF. The power spectrum for $\rho_B(k,\tau)$, $\Pi(k,\tau)$, $L_B(k,\tau)$ are given by

$$|\rho_B(k)|^2 = \frac{1}{256\pi^5}\int d^3k'(1+\mu^2)P_B(k')P_B(|\mathbf{k}-\mathbf{k}'|), \tag{6}$$

$$\left|L^{(S)}(k)\right|^2 = \frac{1}{256\pi^5}\int d^3k'[4(\gamma^2\beta^2-\gamma\mu\beta)+1+\mu^2]P_B(k')P_B(|\mathbf{k}-\mathbf{k}'|), \tag{7}$$

$$\left|\Pi^{(s)}(k)\right|^2 = \frac{1}{576\pi^5}\int d^3k'[4-3(\beta^2+\gamma^2)+\mu^2+9\gamma^2\beta^2-6\mu\beta\gamma]P_B(k')P_B(|\mathbf{k}-\mathbf{k}'|), \tag{8}$$

and for the scalar cross-correlation we have the relations

$$\left|\rho_B(k)L^{(S)}(k)\right| = \frac{1}{256\pi^5}\int d^3k'[1-2(\gamma^2+\beta^2)+2\gamma\mu\beta-\mu^2]P_B(k')P_B(|\mathbf{k}-\mathbf{k}'|), \tag{9}$$

$$\left|\rho_B(k)\Pi^{(S)}(k)\right| = \frac{1}{128\pi^5}\int d^3k'\left(\frac{2}{3}-(\gamma^2+\beta^2)+\mu\gamma\beta-\frac{1}{3}\mu^2\right)P_B(k')P_B(|\mathbf{k}-\mathbf{k}'|), \tag{10}$$

where $\beta = \frac{\mathbf{k}\cdot(\mathbf{k}-\mathbf{k}')}{k|\mathbf{k}-\mathbf{k}'|}$, $\mu = \frac{\mathbf{k}'\cdot(\mathbf{k}-\mathbf{k}')}{k'|\mathbf{k}-\mathbf{k}'|}$, $\gamma = \frac{\mathbf{k}\cdot\mathbf{k}'}{kk'}$. Our results are in agreement with (Kahniashvili *et al.* 2007) and (Finelli *et al.* 2008).

The cutoff dependence and the concern scale. We work with a upper cutoff k_D corresponds to the damping scale due to magnetic energy dissipation into heat through the damping of the Alfvén waves. The upper cutoff of PMF is given by equation (31) in (Hortua *et al.* 2014). The most general scenario at PMFs takes into account a infrared cutoff k_m for low values of k and depends on the generation model of PMF (Yamasaki, 2014). We approximate this infrared cut-off as $k_m = \alpha k_D$ with $0 < \alpha < 1$.

PMF power spectra. In the figure 1(a), we show the magnetic energy density convolution and its dependence with the spectral index ($n = 7/2$ for black and $n = 2$ for gray with points) and the amplitude of PMF at a scale of $\lambda = 1$Mpc (thick for $B = 1nG$, large dashed for $10nG$ and small dashed for $5nG$). In the figure 1(b), the Lorentz force and the anisotropic part spectra are shown with $n = 2$ for dashed lines and $n = 4$ for continue lines and fixed an amplitude of $B_\lambda^2 = 1nG$ at a scale of $\lambda = 1$Mpc. Figure 1(c) shows the cross-correlation between the energy density with Lorentz force and the anisotropic trace-free part ($n = 2$ continue lines and $n = 3$ for dashed lines). We notice that the cross correlation is negative in all range of scales for Lorentz force, while to be negative for values of $k \geqslant 0.05$ (with $n = 3$) and $k \geqslant 0.03$ (with $n = 2$) for anistropic trace-free. The integration domain and the effects of the upper and lower cutoff are described in (Hortua *et al.* 2014).

CMB angular power spectra. If PMFs really were present before the recombination era, these would leave imprints on CMB. Using the total angular momentum formalism, the scalar angular power spectrum of the CMB temperature anisotropy is given by

$$(2l+1)^2 C_l^{\Theta\,\Theta} = \frac{2}{\pi}\int dk k^2 \Theta_l^{(S)\,*}(\tau_0,k)\Theta_l^{(S)}(\tau_0,k), \tag{11}$$

where $\Theta_l^{(S)}(\tau_0,k)$ are the temperature fluctuation multipolar moments. Now, considering only the fluctuation via PMF perturbation, (Kahniashvili *et al.* 2007) found that temperature anisotropy multipole moment becomes $\frac{\Theta_l^{(S)}(\tau_0,k)}{2l+1} \approx \frac{-8\pi G}{3k^2 a_{dec}^2}\rho_B(\tau_0,k)j_l(k\tau_0)$,

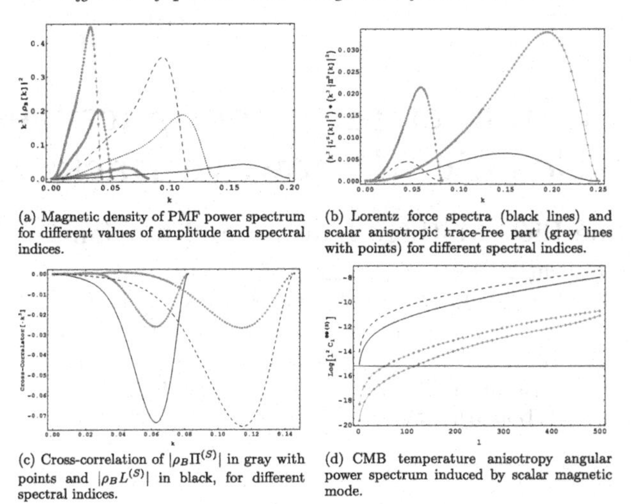

(a) Magnetic density of PMF power spectrum for different values of amplitude and spectral indices.

(b) Lorentz force spectra (black lines) and scalar anisotropic trace-free part (gray lines with points) for different spectral indices.

(c) Cross-correlation of $|\rho_B \Pi^{(S)}|$ in gray with points and $|\rho_B L^{(S)}|$ in black, for different spectral indices.

(d) CMB temperature anisotropy angular power spectrum induced by scalar magnetic mode.

Figure 1. Correlators and power spectrum for CMB anisotropies due to a PMF.

where a_{dec} is the scalar factor at decoupling, G is the Gravitational constant and j_l is the spherical Bessel function. Substituting the last expression in equation (11), the CMB temperature anisotropy angular power spectrum is given by

$$l^2 C_l^{\Theta\,\Theta\,(S)} = \frac{2}{\pi}\left(\frac{8\pi G}{3a_{dec}^2}\right)^2 \int_0^\infty \frac{|\rho_B(\tau_0,k)|^2}{k^2} j_l^2(k\tau_0) l^2 dk, \tag{12}$$

where for our case, the integration is only up to $2k_D$, since it is the range where energy density power spectrum is different from zero. The result of the angular power spectrum induced by scalar magnetic perturbations given by eq. (12) is shown in the figure 1(d). Here, the black thick line defines the power spectrum for $n = 2$ and the black dashed line for $n = 5/2$, both with a strength of $B = 1$nG. The gray thick line with points is for $n = 2$ and the other one is for $n = 5/2$, the gray lines with points are for $B = 10$nG.
Acknowledgments. Héctor J. Hortúa acknowledges the IAU-S306 Grant.

References

Kahniashvili T., Ratra B. 2007, *Phys. Rev. D*, 75, 023002
Finelli F., Paci F., Paoletti D. 2008, *Phys. Rev. D*, 78, 0233510
Yamazaki D. G. 2014, *Phys. Rev. D*, 89, 083528
Hortúa H. J., Castañeda L. 2014, *Phys. Rev. D* 90, 123520 (2014)

Statistical Challenges in 21st Century Cosmology
Proceedings IAU Symposium No. 306, 2014
A. F. Heavens, J.-L. Starck & A. Krone-Martins, eds.

doi:10.1017/S1743921314011028

The impact of superstructures in the Cosmic Microwave Background

Stéphane Ilić[1], Mathieu Langer[2] and Marian Douspis[2]

[1]Institut de Recherche en Astrophysique et Planétologie,
14 avenue Édouard Belin, 31400 Toulouse, France
email: stephane.ilic@irap.omp.eu

[2]Institut d'Astrophysique Spatiale, Université Paris-Sud,
Bâtiment 121, 91405 Orsay Cedex, France
emails: mathieu.langer@ias.u-psud.fr, marian.douspis@ias.u-psud.fr

Abstract. In 2008, Granett *et al.* claimed a direct detection of the integrated Sachs- Wolfe (iSW) effect by a stacking approach of patches of the CMB, at the positions of identified superstructures. However, the high amplitude of their measured signal seems to be at odds with predictions from the standard model of cosmology. However, multiple questions arise from these results and their expected value : I propose here an original theoretical prediction of the iSW effect produced by such superstructures. I use simulations based on GR and the LTB metric to reproduce cosmic structures and predict their exact full theoretical iSW effect. Expected amplitudes are consistent with the measured signal ; however the latter shows non-reproducible features that are hardly compatible with ΛCDM predictions.

Keywords. cosmic microwave background, large-scale structure of universe, dark energy, general relativity

1. Introduction

Dark Energy (DE) is one of the great mysteries of modern cosmology. The integrated Sachs-Wolfe effect (iSW) is an original probe of DE, linked to the large-scale structure of the Universe and the cosmic microwave background (CMB). Indeed, the accelerated expansion stretches gravitational potentials, therefore changing the frequency of the CMB photons that cross them. This effect is integrated along the whole path of the photons and shift the temperature of CMB photons by an amount defined by :

$$\delta_T^{\mathrm{iSW}} = 2\int \dot{\Phi} dt \qquad (\Phi = \text{grav.potential})$$

Consequently, the iSW effect has a direct but very weak impact on the largest scales of the power spectrum of the CMB temperature fluctuations. It therefore requires the use of external data to be detectable. The conventional approach is to correlate the CMB with a tracer of the matter distribution – usually galaxy surveys. Although attempted numerous times (see Dupé *et al.*, 2011 for a review), these have yet to give a definitive and unambiguous detection of the iSW effect. This is mainly due to the shortcomings of current surveys, not deep enough and/or with too small a sky coverage.

In this unclear context, another approach would be to focus on the individual objects expected to leave the biggest imprint on the CMB, i.e. the largest superstructures in the Universe. While their individual imprint is drowned in the primordial CMB fluctuations, we can average patches of the CMB at the locations of many superstructures in order to cancel the random CMB fluctuations while enhancing the iSW signal. In their pioneering work, Granett *et al.* (2008) claimed a 4σ detection of the iSW effect by stacking CMB

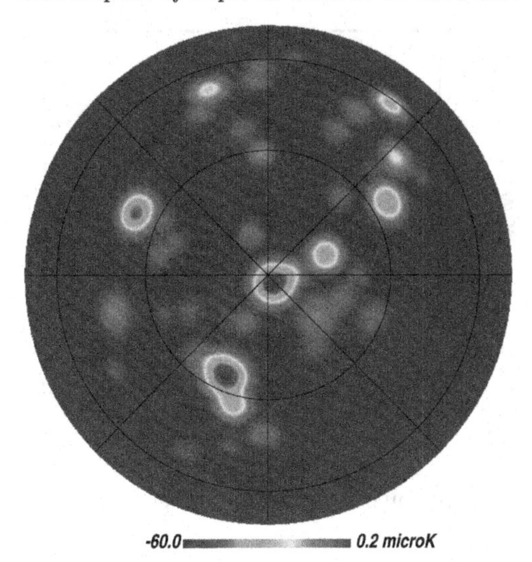

Figure 1. Map of the theoretical temperature shift due to the 50 voids of Granett *et al.* 2008. The map is centred on the galactic north pole.

patches at the positions of 100 superstructures, identified in the SDSS. However, the measured amplitude was reported to be at odds with ΛCDM predictions (e.g. Cai *et al.*, 2013), while some peculiar features were also noted in the signal (Ilić *et al.*, 2013) such as the large hot ring around the cold signal from voids.

2. Simulating a superstructure and its impact on the CMB

The results of the stacking approach raised a number of questions concerning the nature of the expected signal. This led me to propose an original theoretical prediction of the iSW effect produced by superstructures : I modelled each structure and its evolution using the Lemaître-Tolman-Bondi (LTB) metric of GR, the most general metric with a spherical symmetry. Using the data from the Granett *et al.* catalogue, I focused on reproducing their 50 voids and their evolution, and then predict the exact full theoretical iSW effect of these structures by solving geodesic equations for crossing photons.

As expected, the mean effect of these voids is indeed small (about -10μK for photons going through their center). However, using my simulations I showed that in theory, only a fifth of the 50 voids of Granett (as intuited, the largest ones) should contribute for the majority of the predicted signal. However, removing these voids from the real CMB stacking shows that they have almost no contribution to the measured signal. This finding may cast some doubts on whether the measured signal is really due to the iSW effect of the voids.

Going further, I was also able to reconstruct the theoretical “iSW map” associated with these 50 voids (see Fig. 1), i.e. the temperature shift due to the iSW effect associated to each point in the sky, accounting for all the voids present on the corresponding line of sight. Similarly to the analysis on real data, I can perform the stacking procedure on this iSW map, and obtain the expected signal from these voids. However, one should also account for the effect of the CMB primordial fluctuations in this measurement. To do so, I generated many Gaussian realisations of the CMB (with the current best-fit cosmological parameters) that I added on top of (or rather, under) these secondary iSW fluctuations. In the end, I was able to reconstruct both the expected temperature profile

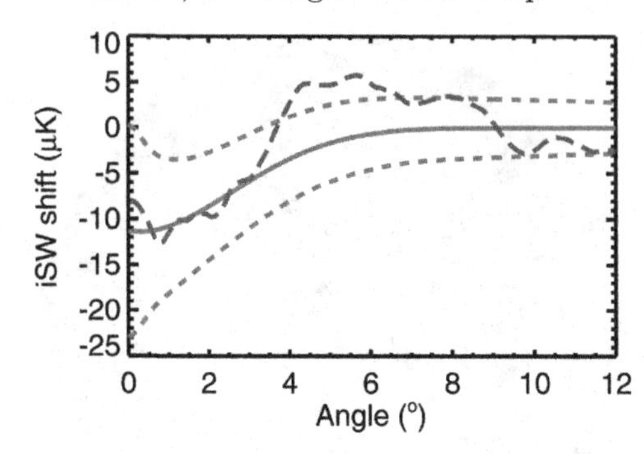

Figure 2. Theoretical iSW temperature profile (solid red) and $1\,\sigma$ limits for the contamination due to primordial CMB fluctuations. The measured temperature profile of the "real" stacked CMB patches is shown in blue.

of the resulting stacked image, but also and foremost its $1\,\sigma$ limits due to CMB noise (see Fig. 2).

The final step in this analysis was to compare those predictions with the measure temperature profile coming from the stacking of real CMB patches (shown also in Fig. 2). Contrary to previous claims in the literature, I show here that the measured signal is compatible with (my) ΛCDM predictions at the $\sim 1\,\sigma$ level. The high reported amplitude seems to be merely due to random (and not particularly rare) fluctuations that artificially enhanced the signal. It is therefore compatible with the expected iSW from such structures.

References

Cai, Y.-C., Neyrinck, M. C., Szapudi, I., Cole, S., & Frenk, C. S. 2014, *ApJ*, 786, 110
Dupé, F.-X., Rassat, A., Starck, J.-L., & Fadili, M. J. 2011, *A&A*, 534, A51
Granett, B. R., Neyrinck, M. C., & Szapudi, I. 2008, *ApJ*, 683, L99-L102
Ilić, S., Langer, M., & Douspis, M. 2013, *A&A*, 556, A51

Statistical Challenges in 21st Century Cosmology
Proceedings IAU Symposium No. 306, 2014
A. F. Heavens, J.-L. Starck & A. Krone-Martins, eds.

doi:10.1017/S174392131401388X

21cm Cosmology

Mario G. Santos[1,2,5], David Alonso[3], Philip Bull[4], Stefano Camera[5,1], and Pedro G. Ferreira[3]

[1]Department of Physics, University of Western Cape, Cape Town 7535, South Africa
[2]SKA SA, 3rd Floor, The Park, Park Road, Pinelands, 7405, South Africa
email: mgrsantos@uwc.ac.za

[3]Astrophysics, University of Oxford, DWB, Keble Road, Oxford OX1 3RH, UK

[4]Institute of Theoretical Astrophysics, University of Oslo,
P.O. Box 1029 Blindern, N-0315 Oslo, Norway

[5]CENTRA, Instituto Superior Técnico, Universidade de Lisboa, Portugal

Abstract. A new generation of radio telescopes with unprecedented capabilities for astronomy and fundamental physics will be in operation over the next few years. With high sensitivities and large fields of view, they are ideal for cosmological applications. We discuss their uses for cosmology focusing on the observational technique of HI intensity mapping, in particular at low redshifts ($z < 4$). This novel observational window promises to bring new insights for cosmology, in particular on ultra-large scales and at a redshift range that can go beyond the dark energy domination epoch. In terms of standard constraints on the dark energy equation of state, telescopes such as Phase I of the SKA should be able to obtain constrains about as well as a future galaxy redshift surveys. Statistical techniques to deal with foregrounds and calibration issues, as well as possible systematics are also discussed.

Keywords. instrumentation: interferometers, methods: data analysis, cosmology: observations, (cosmology:) diffuse radiation, (cosmology:) large-scale structure of universe, radio lines: galaxies.

---

1. Introduction

In recent years we have seen a dramatic improvement in the constraints imposed on the cosmological parameters, entering a phase sometimes called of precision cosmology. One of the most prominent examples of this are the CMB results from WMAP (Hinshaw *et al.* 2013) and more recently the Planck satellite (Planck Collaboration *et al.* 2013b) providing sub-percent constraints on several parameters. The current picture shows an Universe well described by only 6 parameters and a dominant dark energy component very similar to a cosmological constant.

A great effort is now underway to push the boundaries of this standard model and obtain information that can give us new insights into the true nature of these cosmological parameters. Examples are the properties of dark energy, the possibility of modifications to General Relativity or the nature of the primordial fluctuations of the Universe. One of the most accessible methods to probe the energy content and large-scale structure evolution of the Universe is though large galaxy surveys which can be used as tracers of the underlying dark matter distribution. Several surveys are now being done or in preparation, in particular: BOSS (SDSS), DES, and the Euclid satellite†. These surveys are based on imaging of a large number of galaxies at optical or near-infrared wavelengths combined with redshift information to provide a 3-dimensional position of the galaxies.

† www.sdss3.org/surveys/boss.php; www.darkenergysurvey.org; www.euclid-ec.org

Although large radio-galaxy surveys do exist (FIRST, NVSS)‡, they lack direct redshift information from the radio which can seriously impair their usefulness for cosmological applications. The solution will be to use the hydrogen (HI) 21cm line to provide the redshift information from radio surveys (this HI emission comes from the hydrogen spin-flip hyperfine transition at a rest wavelength of 21 cm). HI pervades space from the time of recombination up to the present day. At late times the Universe has reionised, and so the bulk of the neutral hydrogen is thought to reside in comparatively dense gas clouds (damped Lyman-alpha systems) embedded in galaxies, where it is shielded from ionising UV photons. HI is therefore essentially a tracer of the galaxy distribution. At a rest frequency of 1420 MHz, telescopes probing the sky between this and 250 MHz would be able to detect galaxies up to redshift 5 at wavelengths that are mostly immune to obscuration by intervening matter. The problem is that this emission line is usually quite weak: at $z = 1.5$ most galaxies with a HI mass of 10^9 $M_\odot$ will be observed with a flux density of $\sim 1\mu$Jy (see Obreschkow *et al.* 2009). In order to obtain game changing cosmological constraints, experiments with sensitivities better than $\sim 10\mu$Jy will then be required so to provide enough galaxies to beat shot noise at the cosmic variance level (Abdalla *et al.* 2010). Although near term radio telescopes such as ASKAP and MeerKAT should be able to achieve such sensitivities on deep single pointings, it will require a much more powerful telescope such as SKA Phase 2¶, to integrate down to the required sensitivity over the visible sky in a reasonable amount of time. This would imply that one would need to wait until then to use radio telescopes for Cosmology.

Experiments used for galaxy surveys are threshold surveys in that they set a minimum flux above which galaxies can be individually detected. Instead we could consider measuring the integrated 21cm emission of several galaxies in one angular pixel on the sky and for a given frequency resolution. For a reasonably large 3d pixel (angular times frequency) we expect to have several HI galaxies in each pixel so that their combined emission will provide a larger signal. For instance, if we are interested on scales relevant for baryon acoustic oscillations ($\sim$ 150 Mpc), resolutions of around 30 arc min in angle and 1 MHz in frequency are enough. Moreover we can use statistical techniques, similar to what has been applied for instance to CMB experiments, to measure quantities in the low signal to noise regime. By not requiring the detection of each galaxy, the specification requirements imposed on the telescope will be much less demanding. This is what has been commonly called as an intensity mapping experiment. The result is a map of large-scale fluctuations in 21cm intensity, similar to a CMB map, except now the signal is also a function of redshift. Combined with the high frequency (and thus redshift) resolution of modern radio telescopes, this *intensity mapping* (IM) methodology makes it possible to efficiently survey extremely large volumes (Battye *et al.* 2004; McQuinn *et al.* 2006; Chang *et al.* 2008; Mao *et al.* 2008; Loeb & Wyithe 2008; Pritchard & Loeb 2008; Wyithe & Loeb 2008; Wyithe *et al.* 2008; Peterson *et al.* 2009; Bagla *et al.* 2010; Seo *et al.* 2010; Lidz *et al.* 2011; Ansari *et al.* 2012; Battye *et al.* 2013). By not requiring galaxy detections, the intensity mapping technique transfers the problem to one of foreground removal: how to develop cleaning methods to remove everything that is not the HI 21cm signal at a given frequency. This in turn also impacts on the calibration requirements of the instrument. In this paper, we discuss the cosmological applications available with HI IM as well as the statistical challenges that one will have to overcome in order to measure this signal at the required precision level.

‡ sundog.stsci.edu; www.cv.nrao.edu/nvss

¶ www.atnf.csiro.au/projects/askap; www.ska.ac.za/meerkat; www.skatelescope.org

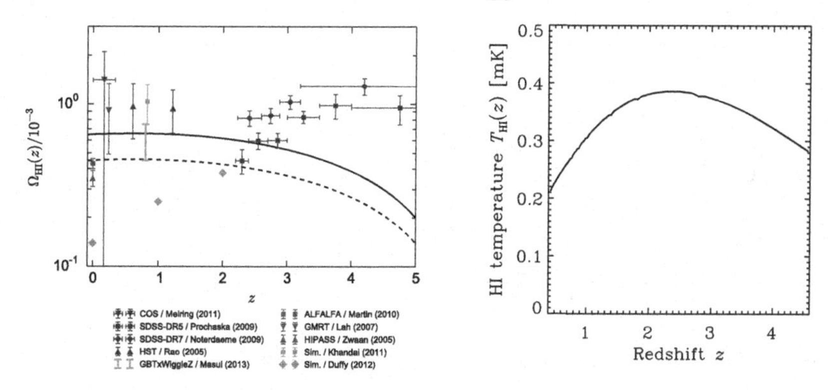

Figure 1. The evolution with redshift of the HI density parameter (left) and the HI temperature signal (right).

2. The HI intensity mapping signal

Radio telescopes measure flux density – the integral of the source intensity, I_ν, over the solid angle of the telescope beam. In the Rayleigh-Jeans limit, this can be converted into an effective brightness temperature, $T_b = c^2 I_\nu / 2k_B\nu^2$, that can be split into a homogeneous part and a fluctuating part, $T_b = \overline{T}_b(1 + \delta_{\rm HI})$, where

$$\overline{T}_b = \frac{3}{32\pi}\frac{hc^3 A_{10}}{k_B m_p \nu_{21}^2}\frac{(1+z)^2}{H(z)}\Omega_{\rm HI}(z)\rho_{c,0}. \tag{2.1}$$

$\Omega_{\rm HI}(z)$ is the comoving neutral hydrogen density in units of the critical density today (see 1) and $\delta_{\rm HI}$ are the fluctuations in the neutral hydrogen mass. At late times, most of the neutral hydrogen content of the Universe is expected to be localised to dense gas clouds within galaxies, where it is shielded from ionising photons. We therefore expect HI to be a biased tracer of the dark matter distribution, just as galaxies are. This allows us to write the HI density contrast as $\delta_{\rm HI} = b_{\rm HI} \star \delta_M$ (where δ_M is the total matter density perturbation, and $\star$ denotes convolution, accounting for the possibility of scale- and time-dependent biasing). Figure 1 shows the expected evolution of the average signal as a function of redshift.

Because the HI intensity is measured as a function of frequency (and thus redshift) rather than comoving distance, we must also account for redshift space distortions (RSDs) caused by the peculiar velocities of the clouds and the galaxies in which they reside. Following Kaiser (1987), we write the (Fourier-transformed) redshift-space HI contrast as

$$\delta_{\rm HI}(\boldsymbol{k}) = (b_{\rm HI} + f\mu^2)\exp\left(-k^2\mu^2\sigma_{\rm NL}^2/2\right)\delta_M(\boldsymbol{k}), \tag{2.2}$$

where $\mu \equiv k_\parallel / k$ and the flat-sky approximation has been used again. We have assumed that the HI velocities are unbiased. The linear growth factor, f, is a key observable, telling us much about the growth of structure on linear scales. The exponential term accounts for the "Fingers of God" effect due to uncorrelated velocities on small scales, which washes out structure in the radial direction past a cutoff scale parametrised by the non-linear dispersion, $\sigma_{\rm NL}$.

One key uncertainty is the behaviour of the HI bias, $b_{\rm HI}$. The bias depends on the size of host dark matter haloes; if a halo is too small, gas clouds would be unable to gain sufficient density to shield themselves and keep the hydrogen neutral. The halo dependence can be modelled using the halo mass function with an appropriate lower mass cutoff (or lower rotation velocity); see (Bagla *et al.* 2010) for example. There are

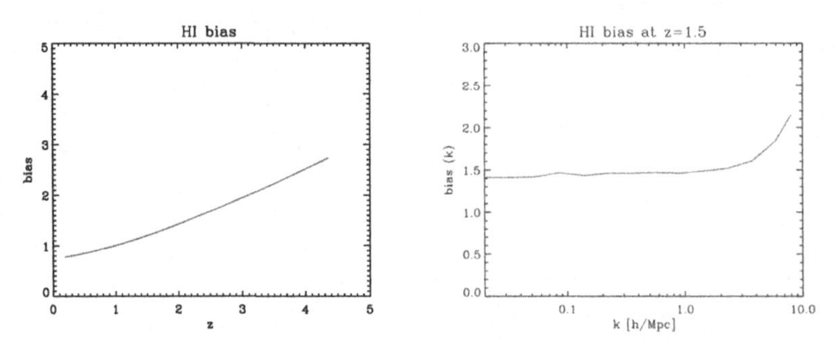

Figure 2. The evolution of the bias as a function of redshift (left) and scale (right).

a few candidate models for the evolution of the bias as a function of redshift that fit the current constraints from observations (Switzer *et al.* 2013) or are calibrated against simulations (Wilman *et al.* 2008), but there is considerable disagreement between them. Unless stated otherwise, we will use a linear bias model for the rest of the paper, and – rather conservatively – marginalise over the value of $b_{\rm HI}$ separately in each redshift bin.

Another major uncertainty is in the HI density fraction, $\Omega_{\rm HI} = \rho_{\rm HI}/\rho_{c,0}$. The current best constraints on the HI fraction come from Switzer *et al.* (2013), who find

$$\Omega_{\rm HI} b_{\rm HI} = 4.3 \pm 1.1 \times 10^{-4}$$

at the 68% confidence level at $z = 0.8$. This constitutes a relatively large uncertainty in the overall amplitude of the HI signal and, correspondingly, the signal-to-noise that can be achieved by a given experiment. For the rest of the paper we will adopt a fiducial value of $\Omega_{\rm HI,0} = 6.5 \times 10^{-4}$. Figure 1 shows the expected HI bias and $\Omega_{\rm HI}$ for a few models as well as current measurements. Given the low resolution of these experiments, it should be a good enough approximation to consider that the total HI mass is a function of the host dark matter halo mass (a redshift dependence can also be included).

Once $M_{\rm HI}(M, z)$ has been specified, we can calculate $\Omega_{\rm HI}$, the HI bias, and HI brightness temperature in a consistent manner. For the mass function, the most straightforward ansatz is to assume that it is proportional to the halo mass – the constant of proportionality can then be fitted to the available data. Even in this case, however, we need to take into account the fact that not all halos contain galaxies with HI mass. Following (Bagla *et al.* 2010), we assume that only halos with circular velocities between 30 – 200 kms^{-1} are able to host HI.

3. Experiments

First attempts at using intensity mapping have been promising, but have highlighted the challenge of calibration and foreground subtraction. The Effelsberg-Bonn survey (Kerp *et al.* 2011) has produced a data cube covering redshifts out to $z = 0.07$, while the Green Bank Telescope (GBT) has produced the first (tentative) detection of the cosmological signal through IM by cross-correlating with the WiggleZ redshift survey (Chang *et al.* 2010; Switzer *et al.* 2013; Masui *et al.* 2013). As probes to constrain cosmological parameters these measurements are, as yet, ineffective, but they do point the way to a promising future.

We can divide the intensity mapping experiments into two types: dish surveys and interferometers. In single dish surveys (auto-correlations) each pointing of the telescope gives us one single pixel on the sky (though more dishes or feeds can be used to increase the field of view). It has the advantage that can give us large scale modes by scanning

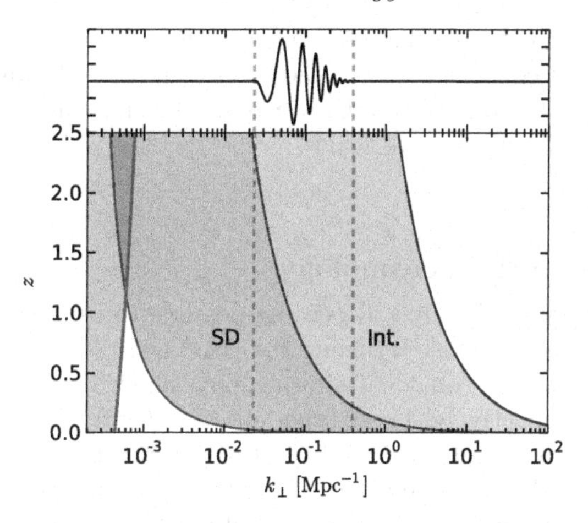

Figure 3. Redshift evolution of the minimum/maximum transverse scales (filled regions) for illustrative interferometer (blue) and single-dish (red) experiments. The BAO are plotted for comparison. The dishes have diameter $D_{\rm dish} = 15$m, the min./max. interferometer baselines are $D_{\rm min} = 15$m and $D_{\rm max} = 1000$m, and the survey has bandwidth $\Delta\nu = 600$ MHz and area $S_{\rm area} = 25,000$ sq. deg. The shaded grey region denotes superhorizon scales, $k < k_H = 2\pi/r_H$.

the sky. Since brightness temperature is independent of dish size we can achieve the same sensitivity with a smaller dish although that will in turn limit the angular resolution of the experiment (a 30 arc min resolution at $z \sim 1$ would require a dish of about 50 m in diameter). One example is the GBT telescope as described above. BINGO (Battye *et al.* 2013) is a proposed 40m multi-receiver single-dish telescope to be set in South America and aimed at detecting the HI signal at $z \sim 0.3$.

Interferometers basically measure the Fourier transform modes of the sky. They have the advantage of easily providing high angular resolution as well being less sensitive to systematics that can plague the auto-correlation power. On the other hand, the minimum angular scale they can probe is set by their shortest baseline which can be a problem when probing the BAO scales. One example of a purpose built interferometer for intensity mapping is CHIME a proposed array, aimed at detected BAO at $z \sim 1$, made up of 20×100m cylinders, based in British Columbia, Canada. The pathfinder has 2 half-length cylinders, and the full experiment has 5 (CHIME Collaboration 2012).

The next generation of large dish arrays can also potentially be exploited for HI intensity mapping measurements. Such is the case of MeerKAT and ASKAP. However, these interferometers do not provide enough baselines on the scales of interest (5m to 80m) so that their sensitivity to BAO will be small. The option is to use instead the auto-correlation information from each dish, e.g. make a survey using the array in single dish mode. Figure 3 shows the scales that can be probed by an array with 15m dishes in single dish and interferometer mode. The large number of dishes available with these telescopes will guarantee a large survey speed for probing the HI signal.

The ultimate example of this approach will be SKA1, the first phase of the SKA telescope, to be built in 2018. This will comprise of 254 single pixel feed dishes to be built in South Africa (SKA1-Mid) and 96 dishes fitted with 36 beam PAFs to increase the field of view to be set in Australia (SKA1-Sur). A HI intensity mapping survey will turn SKA phase 1 into a state of the art cosmological probe. It will allow SKA1 to make detailed measurements of Baryon Acoustic Oscillations and redshift space distortions at several redshifts as well as detect dark matter fluctuations on ultra-large scales past the equality peak and on super-horizon scales. This will make SKA1 capable of addressing crucial questions in Cosmology such as the nature of dark energy and modified gravity, at a level competitive with concurrent experiments as well as push limits on non-standard parameters such as the curvature of the Universe, primordial non-Gaussianity or General Relativistic corrections on ultra-large scales.

4. Forecasts for late time cosmology

Once the BAO are detectable, it will become possible to use IM experiments for precision cosmological measurements. By using the BAO as a 'standard ruler' in the radial and transverse directions, one can constrain the expansion rate, $H(z)$, and angular diameter distance, $D_A(z)$, respectively. These functions of redshift encode crucial information about the energy content and geometry of the Universe; in particular, measurements at $z \lesssim 1.5$ provide a wealth of information about the possible evolution of dark energy, parametrised by its equation of state, $w(z)$. One of the central tasks of observational cosmology over the coming decade will be to determine $w(z)$ to high precision. The spatial curvature of the Universe, Ω_K, can also be pinned down by distance measurements over an even wider redshift range, providing a useful consistency check on the inflationary paradigm.

RSDs, in addition to having a role in separating the radial and transverse BAO, also provide valuable cosmological information in their own right. They are sensitive to the growth rate of structure, $f(z)$, which is a key observable for testing deviations from General Relativity on cosmological scales. Thanks to the high frequency (and thus redshift) resolution of modern radio receivers, experiments like the SKA can be expected to measure RSDs to high precision as well.

Both BAOs and RSDs are now routinely measured in large optical galaxy redshift surveys (e.g. Blake *et al.* (2012); Dawson *et al.* (2013)), which are continually increasing in size and precision. By measuring the same quantities over a wider sweep of redshifts, for a larger fraction of the sky, a HI intensity mapping survey on SKA1 has the potential to provide both competitive constraints on key cosmological parameters like $w(z)$, and vital cross-checks for future optical surveys. Indeed, as surveys begin to come up against the limits imposed by cosmic variance and difficult-to-model systematic effects, it will become particularly important to leverage "multi-tracer" approaches (McDonald & Seljak 2009).

While contingent on the relative importance of foreground contamination and calibration uncertainties, as discussed in subsequent sections, one can nevertheless get some handle on the expected performance of future IM surveys using the Fisher forecasting technique. The basic idea of Fisher forecasting is to derive a Gaussian approximation of the joint likelihood for a set of parameters, given a fiducial model of the expected signal and the noise properties of an experiment. The power of this approach lies in its simplicity; important aspects of measurements such as parameter degeneracies can be taken into account without the need for computationally-expensive simulations (although ultimately these are needed to understand the expected performance characteristics of an experiment in detail).

A Fisher forecasting formalism for intensity mapping was developed in Bull *et al.* (2014), based on a similar method for characterising galaxy redshift surveys. In Fig. 4, we use this method to forecast the expected constraints on the dark energy equation of state parameters and the growth rate for two example SKA1 configurations in single-dish mode. A 10,000 hour IM survey on SKA1 will offer comparable precision to a large multi-year Stage IV galaxy survey such as Euclid or WFIRST in around the same timeframe. Assuming that various statistical challenges can be overcome, intensity mapping therefore

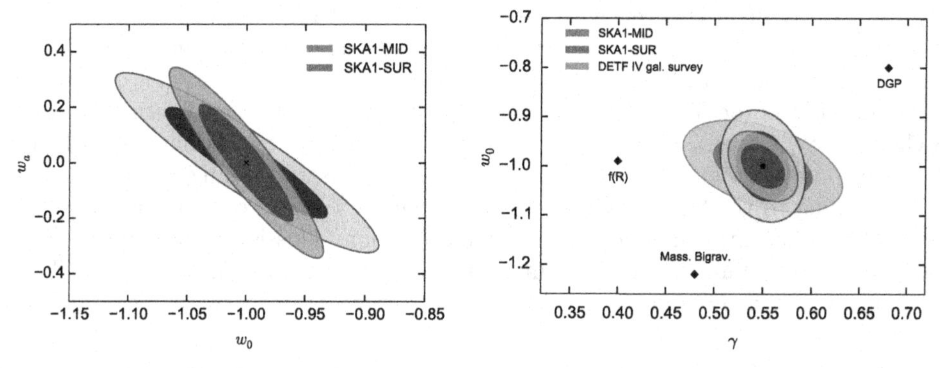

Figure 4. Left: Fisher forecasts for a parametrisation of the dark energy equation of state, $w(z) \approx w_0 + w_a(1-a)$, for two example SKA1 intensity mapping surveys. Right: Forecasts for w_0 and the growth index, γ, (where $f \approx \Omega_M^\gamma(z)$) for the same surveys, with example modified gravity models and a forecast for a DETF Stage IV galaxy redshift survey shown for comparison. Each IM survey is for 10,000 hours over 25,000 sq. deg. of the sky. The redshift ranges depend on the frequency coverage of the instrument; for band 2 of both SKA1-MID and SUR these are $0 \leqslant z \leqslant 0.5$ and $0.2 \leqslant z \leqslant 1.2$ respectively.

has the potential to become a leading precision cosmological probe over the coming decade.

5. Cosmology on ultra-large scales

Besides being a valuable tool for standard cosmological analyses, intensity mapping is incomparable in probing the Universe's largest scales. Constraints on the properties of density perturbations on extremely large cosmic scales can greatly improve our understanding of the very early Universe. Indeed, these scales are well within the linear régime, thus being uncontaminated by non-linear growth of structure. Furthermore, the poorly understood effects of baryons can be safely neglected.

Ultra-large scales are utterly difficult to access with conventional experiments. On the one hand, the potentiality of CMB experiments is wasted by cosmic variance. On the other hand, for large volumes of the Universe we not only need wide fields of view, but also deep observations—in order to probe small perpendicular wavenumbers as well as transverse ones—and this is a problem for usual galaxy redshift surveys. Indeed, it is hard to achieve the required sensitivity at large redshifts over wide areas of the sky. On the contrary, if we forego the identification of individual galaxies, we can greatly speed up the observation and detection of the large-scale structure. Intensity mapping experiments are thus sensitive to density fluctuations at a redshift range observationally difficult to span for standard galaxy surveys (Seo *et al.* 2010).

Amongst the most interesting phenomena occurring on the largest cosmic scales we can list the general relativistic corrections to cosmological observables, modified gravity signatures and large-scale biasing effect entailed to primordial non-Gaussianity. General relativistic effects can significantly deviate from the standard, Newtonian prediction on ultra-large cosmic scales. A measurement of such effects will be a powerful, additional proof of the goodness of Einstein's general relativity. Conversely, it has been argued that modifications of the behaviour of gravity on cosmological distances—which can possibly explain the late-time accelerated expansion of the cosmos with need of neither a cosmological constant nor DE—may hide close to the horizon scale. Thus, it is imperative to scan properly all the cosmological scales in the search for deviation from general relativity. Finally, it is also renown that most models of inflation predict slightly non-Gaussian initial conditions, whose effects on the clustering of galaxies and galaxy clusters are the strongest on ultra-large scales.

Regarding the potentiality of intensity mapping experiments for detecting primordial non-Gaussianity, Camera *et al.* (2013) scrutinised several configurations for both single dish surveys and interferometers. They demonstrated that oncoming experiments such as those leading to the Square Kilometre Array will be capable of reaching such a level of accuracy on the measurement of the primordial non-Gaussianity parameter, $f_{\rm NL}$ to be able to discriminate operatively amongst competing inflationary scenarios. This is particularly true when resorting to large surveys of the sky using single dish mode until $z \sim 3$.

6. Statistical challenges I: Foregrounds

One of the most important challenges facing HI intensity mapping is the presence of foregrounds (both galactic and extra-galactic) with amplitudes several orders of magnitude larger than the signal to be measured. The statistical properties, as well as the frequency dependence of these foregrounds differs significantly from those of the signal, and therefore there is hope that they can be successfully subtracted (Di Matteo *et al.* 2002; Oh & Mack 2003; Santos *et al.* 2005; Morales *et al.* 2006; Wang *et al.* 2006; Gleser *et al.* 2008; Jelić *et al.* 2008; Liu *et al.* 2009; Bernardi *et al.* 2009, 2010; Jelić *et al.* 2010; Moore *et al.* 2013; Wolz *et al.* 2014; Shaw *et al.* 2013). Nevertheless, this foreground subtraction is a potential source of systematic effects that could limit the observational power of intensity mapping for cosmology. Evaluating and modelling these systematics is therefore an essential step in the observational pipeline that requires the use of simulated realisations of these foregrounds.

It has become the norm in the analysis of Cosmic Microwave Background (CMB) data to construct efficient simulations which can then be used to understand the analysis pipeline for any given experiment (Hinshaw *et al.* 2013; Planck Collaboration *et al.* 2013a). By including different foreground contaminants and instrumental systematic effects in the simulation, it is then possible, via Monte Carlo techniques, to accurately estimate the various biases that may enter the final result. In this spirit, Alonso *et al.* (2014) presented a computer code to generate fast mock realizations of the intensity mapping signal, as well as its most relevant foregrounds. The method is similar to those used in galaxy redshift surveys to produce mock catalogs, and is based on generating a lognormal realization of the density field of neutral hydrogen. Through this method it is possible to implement the most important effects (e.g.: the bias of HI with respect to the matter density, the lightcone evolution of the density field, redshift space distortions, frequency decorrelation in the foregrounds, etc.) at a very low computational cost. Five different types of foregrounds have been implemented in the present version of the code: unpolarized and polarized galactic synchrotron, galactic and extragalactic free-free emission and emission from extragalactic point radio sources. Their modelling was based partly on the parametrization of Santos *et al.* (2005) and on the methods used by other groups to simulate radio foregrounds (Jelić *et al.* 2010; Shaw *et al.* 2013, 2014).

The problem of foregrounds has been addressed in the literature mainly for the EoR case, but few studies exist regarding the range of frequencies relevant for intensity mapping. The different algorithms that have been proposed to date can be classified into *blind* (Wang *et al.* 2006; Switzer *et al.* 2013; Chapman *et al.* 2012) and *non-blind* (Liu & Tegmark 2011; Shaw *et al.* 2013) methods, depending on the kind of assumptions made about the nature of the foregrounds (e.g. whether only general properties such as

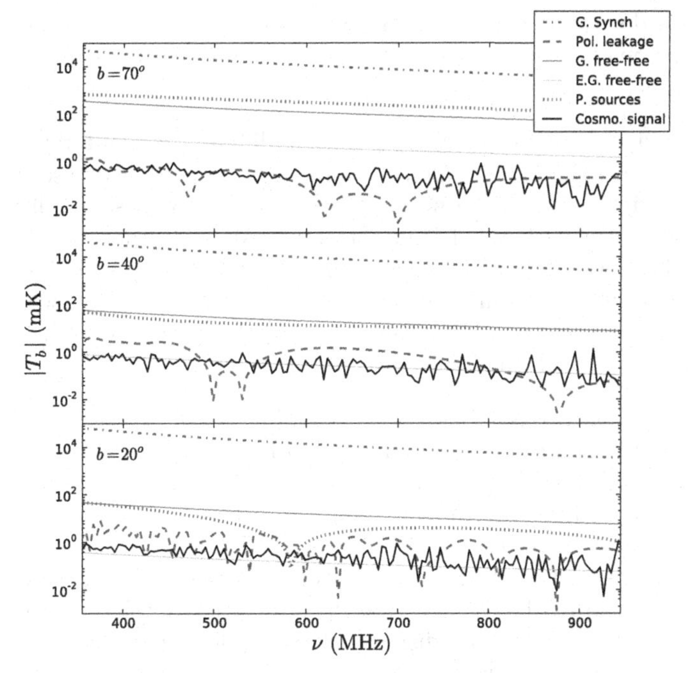

Figure 5. Frequency-dependence of the different foregrounds and the cosmological signal along lines of sight with different galactic latitudes (given in the top right corner of each panel), according to the simulations of Alonso *et al.* (2014). The effect of Faraday decorrelation on the polarized synchrotron emission increases as we approach the galactic plane, making the subtraction of the polarization leakage more challenging.

smoothness and degree of correlation are assumed or whether a more intimate knowledge of the foreground statistics is needed). Recently Wolz *et al.* (2014) studied the effectiveness of the FastICA method for intensity mapping, finding that foreground removal is indeed possible, although it may induce a residual bias on large angular scales that could prevent a full analysis based on the shape of the temperature power spectrum. This result is not surprising: most relevant foregrounds are (fortunately) exceptionally smooth and therefore it should be possible to distinguish them from the much "noisier" cosmological signal. Any foreground residual will probably be dominated by galactic synchrotron emmission, which is most relevant on large angular scales.

Of greater concern is the problem of polarization leakage. Linearly polarized radiation changes its polarization angle in as it traverses the galaxy due to Faraday rotation, an effect that is frequency-dependent and therefore not spectrally smooth. Hence, if part of the polarized synchrotron intensity is leaked into the unpolarized signal due to instrumental issues, it could become an extremely problematic "foreground" (see figure 5). This, together with other frequency-dependent instrumental effects (e.g. a ν-varying beam) could turn out to be greater challenges than the 10^5 times larger foregrounds for intensity mapping.

7. Statistical challenges II: Calibration

The idea of using intensity mapping to reconstruct the large scale structure of the universe bring radio astronomy back to what has been one of its greatest successes-mapping out cosmological diffuse emission. Indeed, tremendous progress has been made in mapping out the cosmic microwave background (CMB) and the hope is that many of the techniques developed there may inform us on how best to proceed. We will now briefly address some of the problems that need to be tackled if we are to move forward with this technique.

For a start, and from figure 3, it is clear that different redshift ranges will require different observation "modes": at high redshift it is preferable to use interferometers while at low redshifts it should be more efficient to work with single dishes (or "auto-correlation" mode in the parlance of radio astronomers). This is not a watertight rule. For example, with the clever use of phase arrays or cylindrical arrays, it should be possible to construct interferometers with short baselines and large fields of view hence accessing larger wavelengths at lower redshifts. But, for now, this separation of scales/redshift is useful in guiding us through the issues.

At low redshifts one needs to perform a classic CMB-like observation which is to raster scan the sky building up a rich set of cross-linked scans that cover as much area as possible with as much depth as is necessary. The key problems are then dealing with the long term drifts in the noise (the $1/f$ noise which is ubiquitous in such experiments) and accurately calibrating the overall signal. Usually, the first problem can be dealt with sufficiently fast scan speed (such that the bulk of the signal is concentrated in frequency in the regime where the $1/f$ has died off and the noise is effectively uncorrelated) but this can be difficult to achieve with large dishes such as the ones that are envisaged in current and future experiments. For some setups, the fortuitous configuration of elevation and location mean that drift scanning may lead to a fast enough scan speed.

With regards to calibrating single dish experiments, this a source of major concern. Major systematic effects to be tackled are spillover and sidelobe pickup as well as gain drifts. Again, these are issues that have been tackled successfully in the analysis of CMB data although novel approaches can be envisaged. So, for example, the BINGO experiment (Battye *et al.* 2013) propose to use a partially illuminated aperture and a fixed single dish, minimising the problems that arise from moving parts. Another intriguing possibility is, for a cluster of single dishes working in autocorrelation mode, to use the cross correlation data for calibrating off known sources. This means that in principle, calibrating the gains should be straightforward using the interferometer data since the high resolution will allow access to a good sky model.

In the case of interferometric measurements, the challenge is to capture as much of the long wavelength modes as possible. The largest wavelength is set by the smallest baseline and in Fig. 3 we can see that arrays with large dishes will not adequately sample BAO scales at low redshift in interferometer mode. To mitigate this problem, one can work with dense aperture arrays – to just use smaller dishes and pack them closer together. This results in smaller baselines and a larger field of view for the interferometer, but reduces its total effective collecting area (and thus its sensitivity). Alternatively, one can use a more novel reflector design– for example long cylindrical reflectors with many closely-spaced receivers installed along the cylinder (Shaw *et al.* 2014). This provides a large number of short baselines, and a primary beam that is $\sim 180^\circ$ in one direction but much narrower along the orthogonal direction. Another possibility is to make interferometric measurements over a number of separate pointings without mosaicing, simply to survey a larger area of sky (White *et al.* 1999). Drift scanning can be seen as a continuous limit of this. The advantage of such a method is that one can greatly reduce the sample variance of the smallest-baseline modes, simply by observing them on several independent patches of the sky.

8. Summary

Neutral hydrogen intensity mapping looks set to become a leading cosmology probe during this decade. Intensity mapping at radio frequencies has a number of advantages over other large scale structure survey methodologies. Since we only care about the large-scale characteristics of the HI emission, there is no need to resolve and catalogue individual objects, which makes it much faster to survey large volumes. This also changes the characteristics of the data analysis problem: rather than looking at discrete objects, one is dealing with a continuous field, which opens up the possibility of using alternative analysis methods similar to those applied (extremely successfully) to the CMB. Thanks to the narrow channel bandwidths of modern radio receivers, one automatically measures redshifts with high precision too, bypassing one of the most difficult aspects of performing a galaxy redshift survey.

These advantages, combined with the rapid development of suitable instruments over the coming decade, look set to turn HI intensity mapping into a highly competitive cosmological probe. One of the key instruments that can be used for this purpose is the Phase I of the SKA. A large sky survey with this telescope should be able to provide stringent constraints on the nature of dark energy, modified gravity models and the curvature of the Universe. Moreover, it will open up the possibility to probe Baryon Acoustic Oscillations at high redshifts as well as ultra-large scales, beyond the horizon size, which can be used to constrain effects such as primordial non-Gaussianity or potential deviations from large-scale homogeneity and isotropy.

Several challenges will have to be overcome, however, if we want to use this signal for cosmological purposes. In particular, cleaning of the huge foreground contamination, removal of any systematic effects and calibration of the system. Foreground cleaning methods have already been tested with relatively success taking advantage of the foreground smoothness across frequency but novel methods need to be explored in order to deal with more complex foregrounds. Other contaminants, such as some instrumental noise bias that shows up in the auto-correlation signal, can in principle be dealt with the same methods. Ultimately, we should deal with the cleaning of the signal and the map making at the same time. This will require even more sophisticated statistical analysis methods and it will be crucial to take on such an enterprise in the next few years in order to take full advantage of this novel observational window for cosmology.

References

Abdalla, F. B., Blake, C., & Rawlings, S. 2010, *MNRAS*, 401, 743
Alonso, D., Ferreira, P. G., & Santos, M. G. 2014
Ansari, R., Campagne, J. E., Colom, P., *et al.* 2012, A. & A., 540, A129
Bagla, J., Khandai, N., & Datta, K. K. 2010, Mon. Not. Roy. Astron. Soc., 407, 567
Battye, R. A., Browne, I. W. A., Dickinson, C., *et al.* 2013, M.N.R.A.S., 434, 1239
Battye, R. A., Davies, R. D., & Weller, J. 2004, Mon. Not. Roy. Astron. Soc., 355, 1339
Bernardi, G., de Bruyn, A. G., Brentjens, M. A., *et al.* 2009, *A&A*, 500, 965
Bernardi, G., de Bruyn, A. G., Harker, G., *et al.* 2010, *A&A*, 522, A67
Blake, C., Brough, S., Colless, M., *et al.* 2012, *MNRAS*, 425, 405
Bull, P., Ferreira, P. G., Patel, P., & Santos, M. G. 2014, arXiv:1405.1452 [astro-ph.CO]
Camera, S., Santos, M. G., Ferreira, P. G., & Ferramacho, L. 2013, Phys. Rev. Lett., 111, 171302
Chang, T.-C., Pen, U.-L., Bandura, K., & Peterson, J. B. 2010, Nature, 466, 463
Chang, T.-C., Pen, U.-L., Peterson, J. B., & McDonald, P. 2008, Phys. Rev. Lett., 100, 091303
Chapman, E., Abdalla, F. B., Harker, G., *et al.* 2012, *MNRAS*, 423, 2518
CHIME Collaboration. 2012, Overview, `http://chime.phas.ubc.ca/CHIME_overview.pdf`
Dawson, K. S., Schlegel, D. J., Ahn, C. P., *et al.* 2013, *AJ*, 145, 10

Di Matteo, T., Perna, R., Abel, T., & Rees, M. J. 2002, *ApJ*, 564, 576
Gleser, L., Nusser, A., & Benson, A. J. 2008, M.N.R.A.S., 391, 383
Hinshaw, G., Larson, D., Komatsu, E., *et al.* 2013, *ApJS*, 208, 19
Jelić, V., Zaroubi, S., Labropoulos, P., *et al.* 2010, *MNRAS*, 409, 1647
Jelić, V. *et al.* 2008, M.N.R.A.S., 389, 1319
Kaiser, N. 1987, Mon.Not.Roy.Astron.Soc., 227, 1
Kerp, J., Winkel, B., Ben Bekhti, N., Flöer, L., & Kalberla, P. M. W. 2011, Astronomische Nachrichten, 332, 637
Lidz, A., Furlanetto, S. R., Oh, S. P., *et al.* 2011, *ApJ*, 741, 70
Liu, A. & Tegmark, M. 2011, *Phys. Rev. D*, 83, 103006
Liu, A., Tegmark, M., Bowman, J., *et al.* 2009, M.N.R.A.S., 398, 401
Loeb, A. & Wyithe, J. S. B. 2008, *Physical Review Letters*, 100, 161301
Mao, Y., Tegmark, M., McQuinn, M., Zaldarriaga, M., & Zahn, O. 2008, *Phys. Rev. D*, 78, 023529
Masui, K. W. *et al.* 2013, Ap. J. L., 763, L20
McDonald, P. & Seljak, U. 2009, JCAP, 0910, 007
McQuinn, M., Zahn, O., Zaldarriaga, M., *et al.* 2006, *ApJ*, 653, 815
Moore, D. F., Aguirre, J. E., Parsons, A. R., Jacobs, D. C., & Pober, J. C. 2013, *ApJ*, 769, 154
Morales, M. F., Bowman, J. D., & Hewitt, J. N. 2006, *ApJ*, 648, 767
Obreschkow, D., Klöckner, H.-R., Heywood, I., Levrier, F., & Rawlings, S. 2009, *ApJ*, 703, 1890
Oh, S. P. & Mack, K. J. 2003, M.N.R.A.S., 346, 871
Peterson, J. B., Aleksan, R., Ansari, R., *et al.* 2009, arXiv:0902.3091 [astro-ph.IM]
Planck Collaboration, Ade, P. A. R., Aghanim, N., *et al.* 2013a, arXiv:1303.5062 [astro-ph.CO]
Planck Collaboration, Ade, P. A. R., Aghanim, N., *et al.* 2013b, arXiv:1303.5076 [astro-ph.CO]
Pritchard, J. R. & Loeb, A. 2008, *Phys. Rev. D*, 78, 103511
Santos, M. G., Cooray, A., & Knox, L. 2005, *ApJ*, 625, 575
Seo, H.-J., Dodelson, S., Marriner, J., *et al.* 2010, Astrophys.J., 721, 164
Shaw, J. R., Sigurdson, K., Pen, U.-L., Stebbins, A., & Sitwell, M. 2013, *ApJ*, 781, 57
Shaw, J. R., Sigurdson, K., Sitwell, M., Stebbins, A., & Pen, U.-L. 2014, arXiv:1401.2095 [astro-ph.CO]
Switzer, E. R., Masui, K. W., Bandura, K., *et al.* 2013, *MNRAS*, 434, L46
Wang, X., Tegmark, M., Santos, M. G., & Knox, L. 2006, *ApJ*, 650, 529
White, M., Carlstrom, J. E., Dragovan, M., & Holzapfel, W. L. 1999, *ApJ*, 514, 12
Wilman, R. J., Miller, L., Jarvis, M. J., *et al.* 2008, M.N.R.A.S., 388, 1335
Wolz, L., Abdalla, F. B., Blake, C., *et al.* 2014, *MNRAS*, 441, 3271
Wyithe, J. S. B. & Loeb, A. 2008, M.N.R.A.S., 383, 606
Wyithe, J. S. B., Loeb, A., & Geil, P. M. 2008, M.N.R.A.S., 383, 1195

Statistical Challenges in 21st Century Cosmology
Proceedings IAU Symposium No. 306, 2014
A. F. Heavens, J. -L. Starck & A. Krone-Martins, eds.

doi:10.1017/S1743921314011053

Deep Source-Counting at 3 GHz

Tessa Vernstrom, Jasper Wall and Douglas Scott

Department of Physics and Astronomy, University of British Columbia
6224 Agricultural Road, Vancouver, Canada V6T 1Z1
emails: tvern@phas.ubc.ca, jvw@phas.ubc.ca, dscott@phas.ubc.ca

Abstract. We describe an analysis of 3-GHz confusion-limited data from the Karl J. Jansky Very Large Array (VLA). We show that with minimal model assumptions, *P(D)*, Bayesian and Markov-Chain Mone-Carlo (MCMC) methods can define the source count to levels some 10 times fainter than the conventional confusion limit. Our verification process includes a full realistic simulation that considers known information on source angular extent and clustering. It appears that careful analysis of the statistical properties of an image is more effective than counting individual objects.

Keywords. source count, confusion limit, *P(D)*

---

1. Introduction

Counts of objects as a function of apparent intensity have been important in understanding the cosmology of our Universe since Herschel's day (Herschel 1785, "On the Construction of the Heavens"). *Evolution* of our Universe was first claimed by Ryle & Scheuer (1955), but the claim was tangled with the simultaneous and unwelcome discovery of *confusion.* Confusion is the integration of the myriad of faint sources into a lumpy background, the "confusion limit" (Fig. 1). The subsequent fight over the interpretation of radio source counts – the Big Bang vs. Steady State controversy – obscured the momentous implications of evolution. In the heat of the battle, Scheuer (1957) showed how confused data could be used to extract information on the source count to levels down to and below this limit. His elegant paper describes *P(D)*, Probability-of-Deflection analyis, now known in terms of one-point statistics.

There are pressing reasons for measuring deep radio source counts. Firstly there is serious disagreement between different direct counts to faint intensities at or near frequencies of 1.4 GHz, as shown in Fig. 2. These are not explicable by sample variance or clustering (Condon 2007). Resolution of this issue is important in view of observations from a balloon experiment (Fixsen *et al.* 2011, ARCADE 2) finding excess radio emission, possibly due to broad-beam integration of discrete sources constituting a newly intruding faint-intensity population (Vernstrom *et al.* 2011; Seiffert *et al.* 2011). *Inter alia*, accurate counts to faint intensities provide powerful data for deriving the cosmic evolution of star-formation. In addition, design of the Square Kilometre Array (SKA) is in progress, as well as SKA pathfinder-telescope surveys. Deep counts, confusion issues and consequent survey plans play into the final design of this billion-dollar experiment.

2. Observations: the Survey Map

To resolve the disagreements Condon *et al.* (2012, CO12) used the VLA in a single-pointing 57h integration at the position of the Owen-Morrison (2008) survey in the Lockman Hole. The 1.4-GHz count from this survey is the most discrepant at faint intensities (Fig. 2). The field is covered in many other wavebands (e.g. *Spitzer, Chandra,*

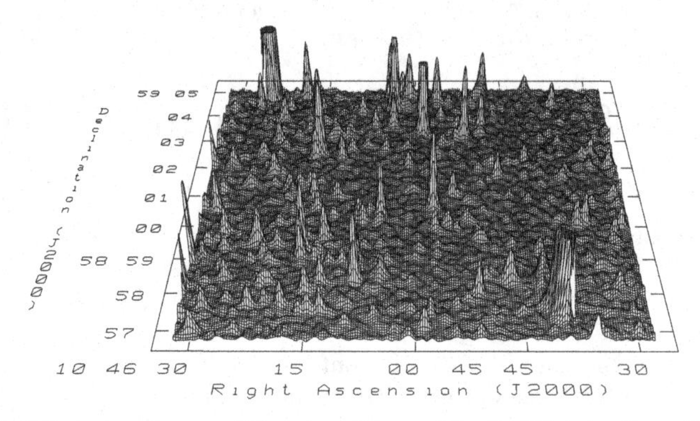

Figure 1. The 3-GHz image from a 57h integration with the VLA in C-configuration (beam 8 arcsec FWHM). The instrumental noise is 1.02 μJy/beam. All features in the image, positive and negative, are on beamwidth scales and these cover the image, indicating that the "confusion limit" has been reached. The large positive deflections represent discrete radio sources which produce the long positive tail of the *P(D)* distribution (Fig. 3).

Herschel, GMRT, CFHT) to allow cross-identifications. The central frequency was 3.02 GHz with the full bandwidth of 2 GHz; the synthesized beam was 8 arcsec FWHM. The survey field is governed by primary beam size, which changes from 22 arcmin FWHM at 2.0 GHz to 11 arcmin at 4 GHz. The broad bandwidth meant rethinking standard analysis procedures and in particular we split the bandwidth into 16 sub-bands and processed these independently. Details of the reduction procedures are in CO12. The resulting survey image (Fig. 1) was confusion-limited as designed, with an instrumental noise of 1.02 μJy/beam.

CO12 used this image with a single-beam *P(D)* analysis (Condon 1974) to estimate the source count at 3 GHz down to 1 μJy. CO12 found a monotonic decrease in the (Eudlidean-normalized) 3-GHz source count, agreeing with predictions from models of luminosity functions and evolution (Massardi *et al.* 2010), and ruling out the appearance of any new population at flux densities above 1 μJy. The CO12 analysis considered only a single short power law over the range 1—10 μJy to describe the counts [best value $\log(dN/dS) = \log(9000) - 1.7 \log(S)$], but the process showed that far more information is present in the *P(D)* distribution. It is the statistical approach to extract this information that we briefly describe here, a process fully described by Vernstrom *et al.* (2014a, V14).

3. Modelling the Source Count

All stages in this detailed analysis became iterative. Instrumental noise must be accurately assessed, as it is of similar central width to the *P(D)* distribution. The *P(D)* distribution itself was compiled with several different sampling procedures, correcting each pixel height for the (frequency-dependent) primary-beam response. There is a trade-off between increasing the field-radius over which *P(D)* is measured to yield more data, while noise rises with radius inversely as the primary-beam response until it overwhelms confusion signal. We performed multiple tests to find the optimum pixel sampling to reduce correlations between binned data points, as well as to test weighting schemes that take into account the radial-dependent noise. Model-selection also folded into this process. We wanted a model that makes minimal assumptions about the form of the count; and eventually – again after trials – we adopted the multi-power-law model of Patanchon *et al.* (2009) in which the count is approximated by a series of short power-law sections, the variables then being the node heights and their selected positions along the

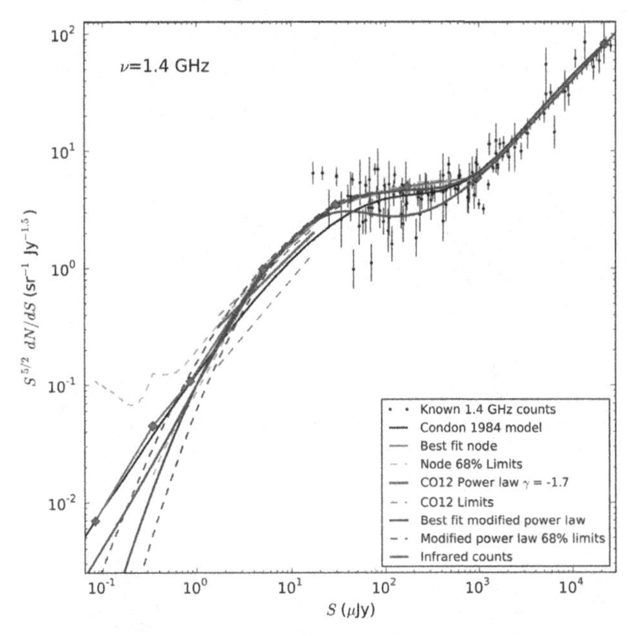

Figure 2. Source counts at 1.4 GHz, shown in relative differential form, i.e counts per bin of intensity divided by the counts expected in a static Euclidean universe, so that a horizontal line indicates an integral power-law count of index -3/2. Points and error bars show the previous badly scattered estimates of the counts. The short straight line plus dashed lines above and below it show results from CO12, the early interpretation of the current deep survey data, while joined points plus dashed error limits show the present results, translated from 3.0 to 1.4 GHz with a spectral index of -0.7. Smooth curves represent counts calculated from models of radio and IR luminosity functions and cosmic evolution.

flux-density axis. Such a model loses no cosmological information provided that the nodes are sufficiently closely spaced in log(S). Uncertainties in all trials were assessed with MCMC sampling, using a generic sampler from the COSMOMC package (Lewis & Bridle 2002). The final noise and *P(D)* distributions used apply to the central 5 arcmin radius of the image (Fig. 3); the noise, weighted following the primary beam response, is accurately Gaussian with $\sigma_n = 1.255$ μJy.

To test our chain of precedures, we used the SKADS simulation (Wilman *et al.* 2008), which computes a source count at the nearby freqeuncy of 1.4 GHz using best estimates of luminosity functions and their evolutions. The simulation data extend to 1 nanoJy, as we require. The simulation includes realistic angular-size and clustering information, and it enabled us to demonstrate retrieval of the exact source count of Wilman *et al.* In the end we fixed the two highest flux-density nodes at positions very well defined by direct source counts (Fig. 2), and floated 6 other nodes down to 1 nanoJy. These two fixed nodes may be considered as priors; in addition we set weak constraints for the node heights to limit MCMC search time, but none of these priors constrained the results significantly. We found it unecessary to iterate the simulation; the Wilman analysis and model count proved to be relatively accurate. Our testing procedure included allowing the noise to be a Gaussian of indeterminate width; the results came back gratifyingly close to our measurement. We tested node frequency and position. The Pearson correlation matrix showed that there is degeneracy between the node heights, telling us that adding any further nodes would not lead to improved count definition.

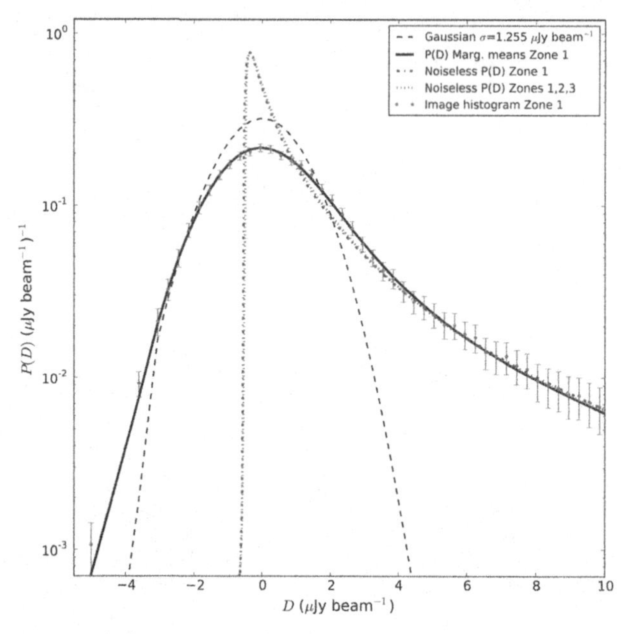

Figure 3. Oversampled *P(D)* distribution for the central 5 arcmin radius of the 3-GHz image (dots and $\sqrt{N}$ error bars), with the best-fitting 8-node model (solid curve) of Fig. 1. The dashed line represents a Gaussian of σ =1.255 μJy, accurately describing the instrumental noise. The dot-dashed line is the noise-free *P(D)* distribution corresponding to the best-fit count model.

4. Results

These are shown in Fig. 2. The *P(D)* distribution contains infomation on the 3-GHz source count down to a level of about 100 nanoJy, at least 10 times below the accepted confusion limit. Integration of this count indicates a source surface density of 2.4×10^5 deg^{-2} at 1.0×10^{-7} Jy. A resolution of 8 arcsec should encompass the extent of most star-forming galaxies at this level, but there remains the possibility of extended-emission sources, such as cluster halos. In a further project (Vernstrom *et al.* 2014b), we have obtained deep data with the Australia Telescope Compact Array (ATCA) in a 7-pointing mosaic at 1.75 GHz, now with a beam on 1 arcmin scales. We are searching for any contribution of extended emission, perhaps low-level AGN emission resolved out with our 8 arcsec VLA beam, or synchrotron radiation from cluster halos, possibilites here including relics and mini-halos (Feretti *et al.* 2012) or WIMP Dark Matter annihilation (Fornengo *et al.* 2011). We have tentative evidence for an excess count corresponding to such low surface-brightness products. If the result stands, the count numbers are nearly a factor of two above the count of Fig. 2. Such a population cannot be from star-formation processes, because the radio – FIR correlation would result in an FIR background well above that observed. Cluster halos from relic radiation or from Dark Matter annihilation are the more likely.

In summary, we have shown how the marriage of confusion-limited imaging, *P(D)*, and modern statistical techniques (Bayesian likelihood, MCMC) yields results on source counts of major astrophysical/cosmological significance. We have looked in detail at statistical *P(D)* issues for broad-band radio interferometry, including noise and sampling. We have shown that the *P(D)* distribution contains information on the source count down to intensities 10 times below the standard "confusion limit". The 3-GHz count (Fig. 2), now defined down to 50 nanoJy, demonstrates a monotonic sinking to below this level.

The results clean up the ambiguities from direct source counts, rule out the intrusion of "new" populations above 1 nanoJy, and indicate that current models of radio emission from star-forming populations require only minor modification. Lastly, we have estimates of source surface densities down to nanoJy levels, vital data for the SKA project.

References

Condon, J. J. 1974, *Astrophys. J.*, 188, 279
Condon, J. J. 2007, in *ASPCS*, eds J. Afonso, *et al.* 380, 189
Condon, J. J., Cotton, W. D., Fomalont, E. B., *et al.* 2012 [CO12], *Astrophys. J.*, 758, 23
Feretti, L., Giovannini, G., Govoni, F., & Murgia, M. 2012, *Astron. Astrophys. Rev.*, 20, 54
Fixsen, D. J., Kogut, A., Levin, S., *et al.* 2011, *Astrophys. J.*, 734, 5
Fornengo, N., Lineros, R., Regis, M., & Taoso, M. 2011, *Phys. Rev. Lett.*, 107, A261302
Herschel, W. 1785, *Phil. Trans. R. Soc. Lond.*, 75, 213
Lewis, A. & Bridle, S. 2002, *Phys. Rev. D*, 66, 103511
Massardi, M., Bonaldi, A., Negrello, M., *et al.* 2010, *Mon. Not. R. astr. Soc.*, 404, 532
Patanchon, G., Ade, P. A. R., Bock, J. J., *et al.* 2009, *Astrophys. J.*, 707, 1750
Ryle, M. & Scheuer, P. A. G. 1955, *Proc. Roy. Soc. Lond. A*, 230, 448
Scheuer, P. A. G. 1957, *Proc. Camb. Phil. Soc.*, 53, 764
Seiffert, M., Fixsen, D. J., Kogut, A., *et al.* 2011, *Astrophys. J.*, 734, 6
Vernstrom, T., Scott, D., & Wall, J. V. 2011, *Mon. Not. R. astr. Soc.*, 415, 3641
Vernstrom, T., Scott, D., Wall, J. V., *et al.* 2014a [V14], *Mon. Not. R. astr. Soc.*, 440, 2791
Vernstrom, T., Norris, R. P., Scott, D., & Wall, J. V. 2014b, submitted
Wilman, R. J., Miller, L., Jarvis, M. J., *et al.* 2008, *Mon. Not. R. astr. Soc.*, 388, 1335

Statistical Challenges in 21st Century Cosmology
Proceedings IAU Symposium No. 306, 2014
A. F. Heavens, J.-L. Starck & A. Krone-Martins, eds.

doi:10.1017/S174392131401343X

The cosmic radio dipole and local structure effects

Matthias Rubart[1], David Bacon[2] and Dominik J. Schwarz[1]
[1]Fakultät für Physik, Universität Bielefeld, Postfach 100131, 33501 Bielefeld, Germany
email: `matthiasr at physik dot uni-bielefeld dot de`
[2]Institute of Cosmology and Gravitation, University of Portsmouth, Burnaby Road, Portsmouth PO1 3FX, United Kingdom

Abstract. We investigate the contribution of a local over- or under-density to linear estimates of the cosmic dipole. We focus on radio continuum surveys. Recently it was shown that the radio dipole amplitude is larger than expected from the corresponding dipole of the CMB. We show that a significant contribution to this excess could come from local structure.

Keywords. Cosmology: observations, large scale structure of the universe, radio surveys, peculiar motion

---

1. Dipole in number counts

The measured CMB dipole is $\Delta T = 3.36 \pm 0.01$ mK in the direction (RA,Dec) = $(168°, -7°)$ and thus the velocity of the Solar system has been inferred to be $v = 369 \pm 1$ km s^{-1} (Hinshaw *et al.* (2009)).

The amplitude of the kinetic radio dipole is given by (Ellis & Baldwin (1984))

$$d = [2 + x(1 + \alpha)]\,(v/c)\,, \tag{1.1}$$

where α is the mean spectral index of radio sources and x is the power law index of the corresponding number counts. This dipole is due to spherical aberration as well as the Doppler effect. Using the inferred CMB dipole velocity we expect $d = (5 \pm 1) \times 10^{-3}$, pointing towards the CMB dipole.

Blake & Wall (2002), Singal (2011), Gibelyou & Huterer (2012) and Rubart & Schwarz (2013) (2D and 3D) attempted to determine the dipole in the NRAO VLA Sky Survey (NVSS), as shown in the table.

The estimator used by Blake & Wall is quadratic, while the other estimators are different implementations of linear estimators, essentially:

$$\vec{D} = \sum_{i}^{N} \hat{r}_i \,. \tag{1.2}$$

Source	Flux > (mJy)	N	RA (°)	Dec (°)	d (10^{-3})
B & W	25	197998	158 ± 30	-4 ± 34	11 ± 3
$\vec{D}_S$	25	184237	159 ± 10	-7 ± 9	22 ± 6
$\vec{D}_{GH}$	15	211487	117 ± 20	$+6 \pm 14$	27 ± 5
$\vec{D}_{2D}$	25	195245	155 ± 14		19 ± 5
$\vec{D}_{3D}$	25	185649	158 ± 19	-2 ± 19	18 ± 6
expected	from CMB:		168	-7	5 ± 1

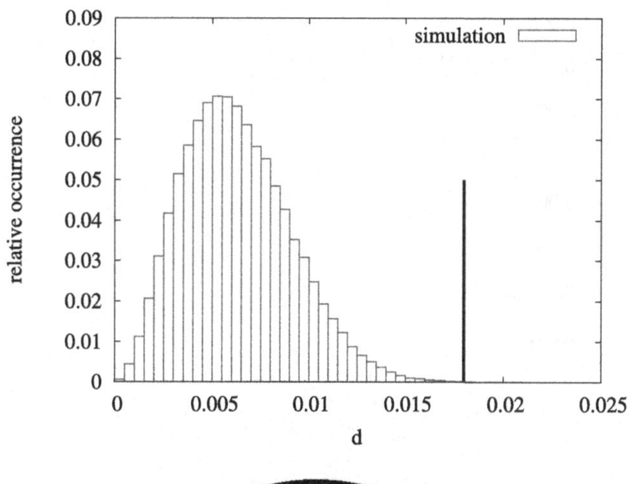

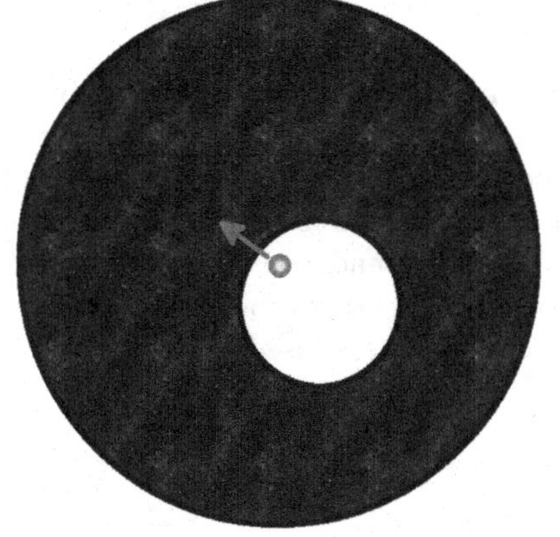

radius of void (redshift)	relative density (in percent)	dipole (10^{-3})
$0.07R_H$	66	2.1 ± 0.1
$0.15R_H$	66	6.9 ± 0.1
$0.11R_H$	40	7.2 ± 0.1

The general direction of the radio dipole coincides with the expectation from the CMB. The amplitudes, especially for the linear estimators, are well above the expectation. In order to investigate the significance of this excess, Rubart & Schwarz (2013) simulated the expected radio sky:

Our simulations show that the observed excess of the amplitude d cannot be explained by shot noise or bias alone.

2. Local structure dipole

There have been studies claiming that the local Universe has an untypically low density of galaxies on 300 Mpc scales, e.g. Keenan *et al.* (2013). We expect to see more galaxies in one direction than in the other, assuming we live in such region. It is likely that the Local Group moves towards this direction, due to a gravitational pull.

The results of Rubart *et al.* (2014) shows a structural dipole component, which affects

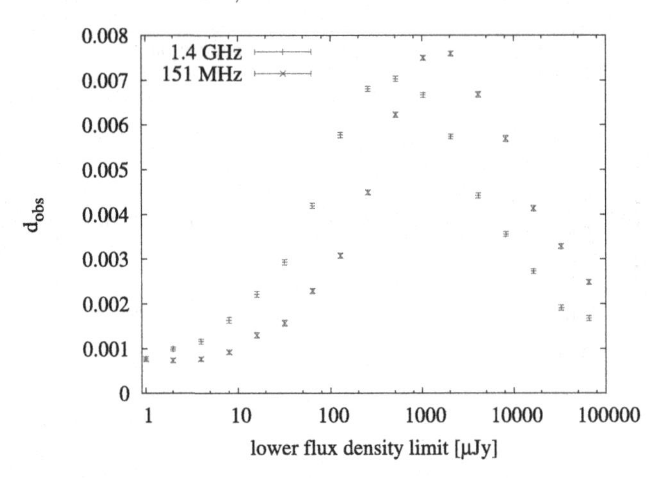

the measured radio dipole. This effect depends on the size of the void and its density contrast.

While the CMB dipole is caused by the motion of the Sun relative to the CMB, the radio dipole can be expected to also receive contributions from the galaxy distribution (structure dipole). Therefore we expect the structural contribution to add up with the velocity dipole resulting in a larger dipole amplitude in radio surveys.

It turns out that there is also a dependence on the lower flux density limit and on the survey frequency band (shown in the graph above for a radius of $0.07R_h$ and $\delta = -0.33$ density contrast). This dependence can be used to distinguish the kinetic dipole from the structural one.

3. Discussion

We find that simple void models have a significant effect on the cosmic radio dipole estimation. The dipole amplitude measured by the linear estimator of a void of radius $0.07R_{\rm H}$ is expected to be $d_{\rm void} = (2.1 \pm 0.1) \times 10^{-3}$. The discrepancy between radio and CMB dipole measurements can be relaxed by such a contribution, but the difference cannot be explained completely. In forthcoming surveys (Lofar MSSS & Tier 1, EMU and SKA surveys), the effect of structures will become more important, due to lower flux limits.

References

Blake, C. & Wall, J. 2002, *Nature*, 416, 150

Ellis, G. F. R. & Baldwin J. E. 1984, *MNRAS*, 206, 377

Gibelyou, C. & Huterer, D. 2012, *MNRAS*, 427, 1994

Hinshaw, G., Weiland, J. L., Hill, R. S., *et al.* 2009, *ApJS*, 180, 225

Keenan, R. C., Barger, A. J., & Cowie, L. L. 2013, *ApJ*, 775, 62

Rubart, M. & Schwarz, D. J. 2013, *A&A*, 555, A117

Rubart, M., Schwarz, D. J., & Bacon, D. 2014, *A&A*, 565, A111

Singal, A. K. 2011, *ApJ*, 742, L23

Statistical Challenges in 21st Century Cosmology
Proceedings IAU Symposium No. 306, 2014
A. F. Heavens, J. -L. Starck & A. Krone-Martins, eds.

doi:10.1017/S1743921314013830

Bayesian Inference for Radio Observations - Going beyond deconvolution

Michelle Lochner[1,2], Bruce Bassett[1,2,3], Martin Kunz[1,4], Iniyan Natarajan[5], Nadeem Oozeer[1,6,7], Oleg Smirnov[6,8] and Jonathan Zwart[9]

[1]African Institute for Mathematical Sciences, 6 Melrose Road, Muizenberg, 7945, South Africa
email: dr.michelle.lochner@gmail.com

[2]Department of Mathematics and Applied Mathematics, University of Cape Town, Rondebosch, Cape Town, 7700, South Africa

[3]South African Astronomical Observatory, Observatory Road, Observatory, Cape Town, 7935, South Africa

[4]Département de Physique Théorique and Center for Astroparticle Physics, Université de Genève, Quai E. Ansermet 24, CH-1211 Genève 4, Switzerland

[5]Astrophysics, Cosmology and Gravity Centre (ACGC), Department of Astronomy, University of Cape Town, Private Bag X3, Rondebosch 7701, South Africa

[6]SKA South Africa, 3rd Floor, The Park, Park Road, Pinelands, 7405, South Africa

[7]Centre for Space Research, North-West University, Potchefstroom 2520, South Africa

[8]Department of Physics and Electronics, Rhodes University, PO Box 94, Grahamstown, 6140, South Africa

[9]Department of Physics & Astronomy, University of the Western Cape, Private Bag X17, Bellville 7535, South Africa

Abstract. Radio interferometers suffer from the problem of missing information in their data, due to the gaps between the antennae. This results in artifacts, such as bright rings around sources, in the images obtained. Multiple deconvolution algorithms have been proposed to solve this problem and produce cleaner radio images. However, these algorithms are unable to correctly estimate uncertainties in derived scientific parameters or to always include the effects of instrumental errors. We propose an alternative technique called Bayesian Inference for Radio Observations (BIRO) which uses a Bayesian statistical framework to determine the scientific parameters and instrumental errors simultaneously directly from the raw data, without making an image. We use a simple simulation of Westerbork Synthesis Radio Telescope data including pointing errors and beam parameters as instrumental effects, to demonstrate the use of BIRO.

Keywords. methods: statistical, methods: data analysis, techniques: interferometric

1. Introduction

The problem of extracting scientific parameters from dirty (dominated by artifacts) interferometric radio images has resulted in many deconvolution algorithms being developed. However, none of these solve the problem of incorporating instrumental errors as a source of uncertainty when making measurements from radio data. Deconvolution algorithms can only produce one image, which the scientist must then assume is correct before extracting any science (for example, a catalogue of source fluxes) from it. Algorithms such as CLEAN (Högbom 1974), the most popular algorithm in use, cannot reliably produce any uncertainties (Junklewitz *et al.* 2013), making it impossible to propagate the uncertainties from instrumental errors to the scientific parameters. Further, if these parameters are correlated, as they likely are, even correcting the data for instrumental errors may lead to biased scientific results, as the measurement of the

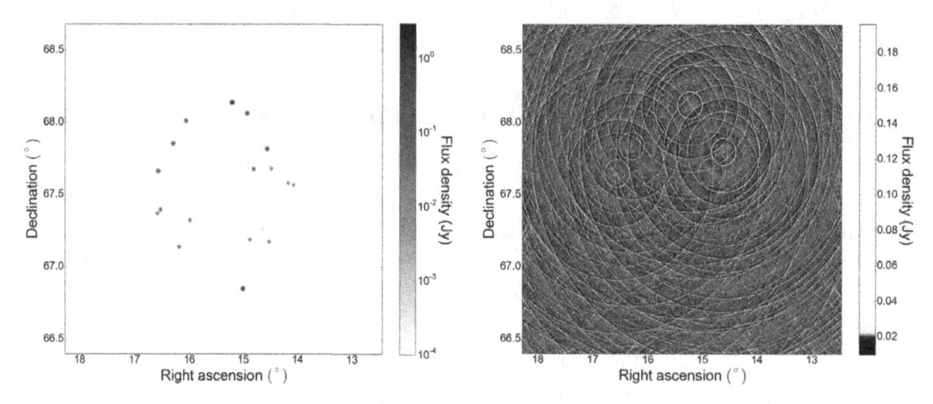

Figure 1. Mock dataset. *Left panel:* The simulated sky model. *Right panel:* The "dirty image", which is how this field appears when convolved with the telescope beam.

instrumental error may be wrong. These instrumental effects will become more important as more sensitive telescopes such as the Square Kilometre Array† come online.

2. BIRO

We propose, in Lochner et al. 2015, a completely different approach, whereby we model the sky and all known sources of instrumental error simultaneously using the radio interferometry measurement equation (RIME) (Hamaker 1996). The RIME can be written as (Smirnov 2011):

$$\mathsf{V}_{pq} = \boldsymbol{J}_{pn}(\ldots(\boldsymbol{J}_{p2}(\boldsymbol{J}_{p1}\mathsf{B}\boldsymbol{J}_{q1}^{H})\boldsymbol{J}_{q2}^{H})\ldots)\boldsymbol{J}_{qm}^{H}, \tag{2.1}$$

where V_{pq} is the visibility matrix (the radio data), $\boldsymbol{J}_{pi}$ is the i'th Jones matrix for antenna p (containing instrumental effects) and B is the brightness matrix (containing the sky model, and hence all scientific parameters). We use the software package MeqTrees (Noordam & Smirnov 2010), which implements the RIME, to model our radio field and any known instrumental effects. We can then estimate the parameters of this model, both scientific and instrumental, in a Bayesian context using a sampling method such as MCMC (Metropolis *et al.* 1953, Hastings 1970). We assume uncorrelated Gaussian noise on the visibilities which leads to a simple Gaussian likelihood for V_{pq}. With this approach, we are able to determine the full posterior for the problem, obtaining not only the best fits for all parameters, but also their uncertainties and correlations.

3. Applying BIRO

Fig. 1 shows the mock WSRT (Westerbork Synthesis Radio Telescope)‡ field which we tested BIRO on. This field, simulated using MeqTrees and based on a real field, consists of 17 point sources. We also applied pointing errors to each antenna as an example of a source of instrumental error, which WSRT (and many other radio telescopes) have had to deal with in the past (Smirnov & de Bruyn 2011). A mispointed antenna will observe a point source through the edge rather than the centre of its beam. Thus, in general, we would expect pointing errors and fluxes to be correlated and we apply our method to determine these correlations, as well as the parameters themselves.

The instrumental parameters for this simulated data consist of the pointing errors (one for each direction for each antenna), the width of the primary beam and the noise on the visibilities, which we assume to be Gaussian, as is widely considered a good approximation. We allow the pointing errors to vary in time as second order polynomial functions,

† Square Kilometre Array, `http://www.skatelescope.org`
‡ WSRT, `https://www.astron.nl/radio-observatory/astronomers/wsrt-astronomers`

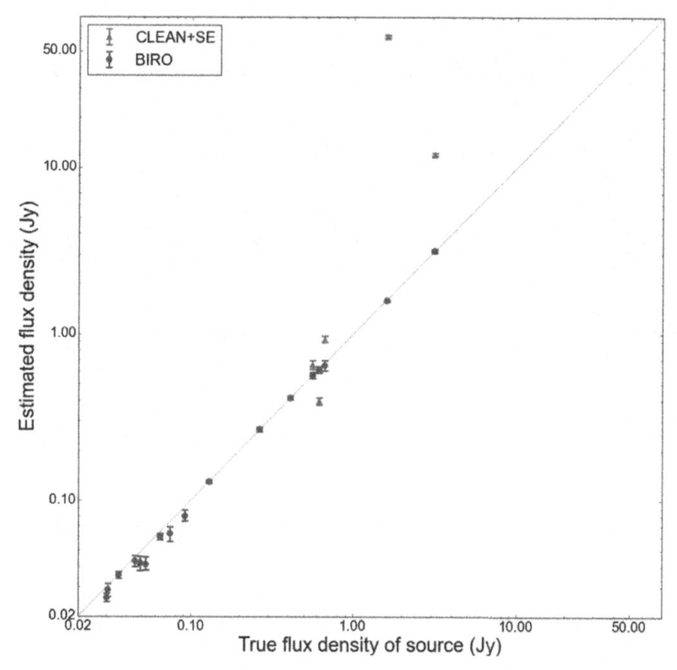

Figure 2. Comparison between BIRO and the fluxes obtained from CLEAN, combined with a source extraction algorithm (CLEAN+SE). While the BIRO results are unbiased, CLEAN+SE misestimates the fluxes by up to 44σ, due to the strong correlations between pointing error and flux.

resulting in 84 pointing error parameters (3 for each direction for each antenna). Our scientific parameters are the flux for each source.

We compared BIRO with the standard, commonly used CLEAN algorithm combined with a source extraction algorithm on the CLEANed image (we call this CLEAN+SE) for this dataset. Fig. 2 illustrates this comparison for the estimated fluxes from BIRO and CLEAN+SE. With no knowledge of the time-varying pointing errors, it is no surprise that CLEAN+SE returns biased fluxes. The danger is that the error bars, estimated using only the surrounding flux of the point source in this source extraction algorithm, do not and cannot take into account the additional source of uncertainty from the instrumental errors. In contrast, the BIRO estimates are unbiased and the error bars are larger, correctly propagating the instrumental errors. We also found that while CLEAN+SE identified many of the point sources as extended (and the only extended source as a point source), BIRO is properly able to determine the type of each source simply by fitting shape parameters to the sources. BIRO is also able to determine the covariance matrix between all the parameters (Fig. 3).

4. Conclusions

We have introduced BIRO, a Bayesian approach to the deconvolution problem of radio interferometry observations. Fig. 3 highlights the importance of fully propagating the uncertainty on instrumental errors, as they can be highly correlated with the scientific parameters and hence bias scientific results. Due to the fully Bayesian nature of BIRO, it allows for very elegant extensions to this simple scenario. For example, the choice of model for the field can be selected for using the Bayesian evidence (see Lochner et al. 2015). BIRO can be useful in any scenario where reliable statistics are required for the science extracted and may be essential to fully exploit the sensitivity of the SKA.

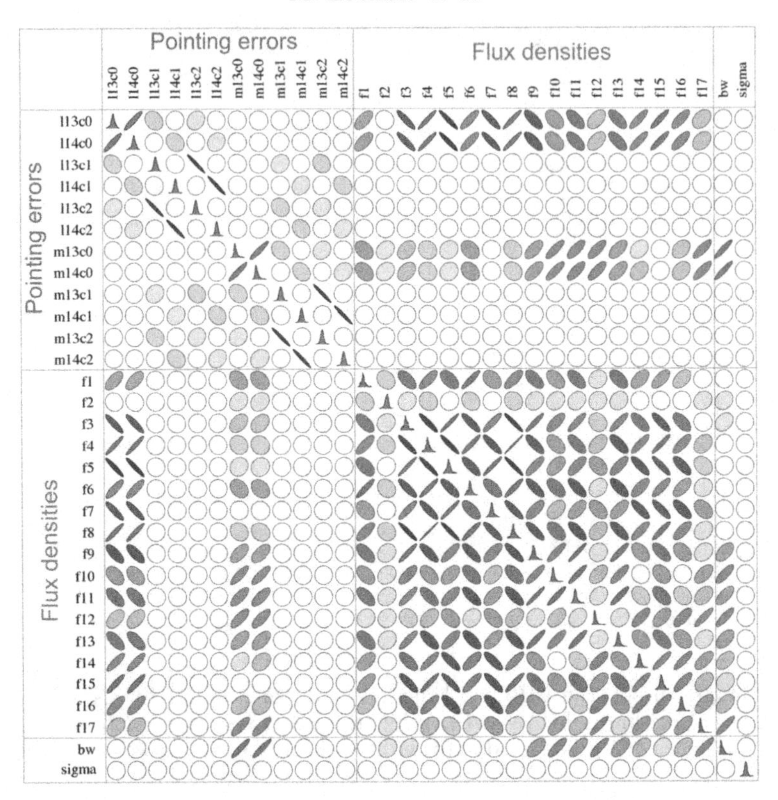

Figure 3. Covariance matrix between a subset of parameters. The parameters are listed on each axis with the correlations between them represented by a coloured ellipse. Highly correlated parameters are red with a thin ellipse angled to the right, whereas anti-correlated parameters have a dark blue ellipse, angled to the left. The diagonal shows the 1D marginalised posterior for each parameter. The correlations in the fluxes arise due to uncertainty in the flux distribution from gaps in the *uv*-plane, while the pointing errors are correlated simply because every pointing error affects every source. Of particular interest is the complex way in which pointing errors correlate with the fluxes, which would be very difficult to determine from first principles.

5. Acknowlegdements

ML and JZ are grateful to the South Africa National Research Foundation Square Kilometre Array Project for financial support. ML acknowledges support from the University of Cape Town and resources from the African Institute for Mathematical Sciences. OS is supported by the South African Research Chairs Initiative of the Department of Science and Technology and National Research Foundation. IN acknowledges the MeerKAT HPC for Radio Astronomy Programme. Part of the computations were performed using facilities provided by the University of Cape Town's ICTS High Performance Computing team: `http://hpc.uct.ac.za`.

References

Hamaker, J. P., Bregman, J. D., & Sault, R. J. 1996, *A&AS*, 117, 137-147
Hastings, W. K. 1970, *Biometrika*, 57, 97-109
Högbom, J. A. 1974, *A&AS*, 15, 417
Junklewitz, H., Bell, M. R., Selig, M., & Enßlin, T. A. 2013, *arXiv:1311.5282*
Metropolis, N. *et al.* 1953, *J. Chem. Phys.*, 21, 1087
Noordam, J. E. & Smirnov, O. M. 2010, *A&A*, 524, A61
Smirnov, O. M. 2011, *A&A*, 527, A106
Smirnov, O. M. & de Bruyn, A. G. 2011, *arXiv:1110.2916*

Statistical Challenges in 21st Century Cosmology
Proceedings IAU Symposium No. 306, 2014
A. F. Heavens, J.-L. Starck & A. Krone-Martins, eds.

doi:10.1017/S1743921314013696

The impact of small absorbers, galactic neutral hydrogen & X-rays on 1-point statistics of the 21-cm line.

C. A. Watkinson[1], J. R. Pritchard[1], A. Mesinger[2] and E. Sobacchi[2]

[1]Department of Physics, Blackett Laboratory, Imperial College, London SW7 2AZ, UK
email: c.watkinson11@imperial.ac.uk

[2]Scuola Normale Superiore, Piazza dei Cavalieri 7, I-56126 Pisa, Italy

Abstract. We discuss a selection of semi-numerical simulations of reionization whose analysis investigates the effect of small absorbing systems, neutral hydrogen within galaxies and the efficiency with which galaxies produce X-rays. We focus on the consequences for both observing the 21–cm 1-point statistics and their interpretation.

Keywords. intergalactic medium, galaxies: high-redshift, cosmology: theory

1. Introduction

Reionization, the process in which the first galaxies ionized the otherwise neutral intergalactic medium (IGM), is one of the least well constrained epochs in the history of our universe. We appear to be observing the final phases of the process at $z \sim 7$ in analysis of the highest known quasar (Bolton *et al.* 2011) and an apparent drop in Lyman-alpha emitters (Ota *et al.* 2010); we also know from the cosmic microwave background (CMB) that the process of reionization was under way by $z \sim 11$ (Planck collaboration 2013). Beyond this we know very little, so it is essential that we understand the effects that the associated uncertainties have on the observables with which we hope to constrain reionization. The 21-cm brightness temperature is an extremely promising such quantity, this temperature describes the intensity of the 21-cm line, a hyperfine transition of neutral hydrogen (H I). The hope is that we can detect this signal with huge radio interferometers such as LOFAR, MWA and SKA to observe evolution in the distribution of hydrogen as we look to $z > 6.5$. Here we overview results of semi-numerical simulations of reionization that allow us to infer the effects on the brightness-temperature 1-point statistics due to: absorbing sinks, H I that remains in galaxies after their local IGM is ionized and the efficiency with which stars produce X-rays.

2. Overview

The 21-cm line is observed through the difference between the transition's brightness temperature and that of the CMB, $\delta T_{\rm b}$. It can in principle be detected whenever the excitation/spin temperature ($T_{\rm s}$) is decoupled from that of the CMB ($T_{\rm cmb}$). The 21-cm brightness temperature is also sensitive to cosmology [$H(z)$, $\Omega_{\rm m}$, $\Omega_{\rm b}$], density contrast (δ), neutral fraction ($x_{\rm HI}$) and peculiar velocities ($\mathrm{d}v_r/\mathrm{d}r$) according to Equation 2.1,

$$\delta T_{\rm b} \approx 27.\, \frac{T_{\rm s} - T_\gamma}{T_{\rm s}}\, x_{\rm HI}(1+\delta) \left[\frac{H(z)/(1+z)}{\mathrm{d}v_{\rm r}/\mathrm{d}r}\right] \left(\frac{1+z}{10}\,\frac{0.15}{\Omega_{\rm m} h^2}\right)^{1/2} \left(\frac{\Omega_{\rm b} h^2}{0.023}\right) \mathrm{mK}\,. \tag{2.1}$$

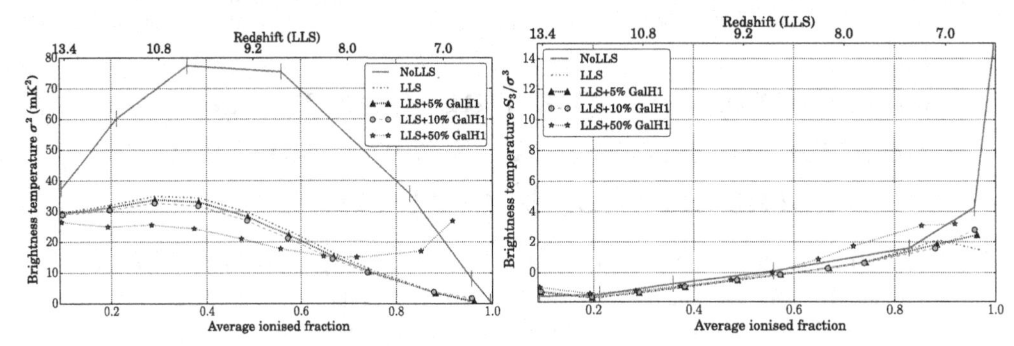

Figure 1. Absorbers & galactic H I: Variance (left) and skewness (right) of δT_b as a function of ionized fraction

C. A. Watkinson, J. R. Pritchard, A. Mesinger & E. Sobacchi 2014 in prep:

Neutral sinks & residual galactic hydrogen (Figure 1)

The effects that ionizing radiation from galaxies has on the IGM has been studied at great length using both numerical and semi-numerical methods; both agree that the collective effect of clustering galaxies will generate large ionized bubbles that exhibit a characteristic size which increases as reionization proceeds. Such simulations, including 21CMFAST (Mesinger *et al.* 2011) labelled as **NoLLS** in Figure 1, only consider homogeneous recombinations, yet we observe strongly absorbent, self-shielded and neutral Lyman-limit systems (**LLS**) at lower redshifts and expect such systems to exist during reionization.

The effect that these systems have on the progress of reionization and the power spectrum was considered by Sobacchi & Mesinger 2014. In this simulation, they use sub-grid physics based on fits to simulations (Miralda-Escude *et al.* 2013, Schaye 2001, Rahmati *et al.* 2013) to self-consistently model localised recombinations. Here we calculate the 1-point statistics of their full simulation (**LLS** in Figure 1) and also consider the impact of remnant galactic H I (**LLS + %GalHI** in Figure 1). The evolution of remnant galactic H I remains completely unconstrained during the final phases of reionization so we assume that a fixed percentage of galaxies (assumed to be well approximated by the extended Press-Schechter collapsed fraction) remains neutral and vary this percentage to understand its impact.

C. A. Watkinson & J. R. Pritchard 2014 in prep:

X-ray efficiency of galaxies (Figure 2)

The spin (excitation) temperature of the 21-cm transition is sensitive to the efficiency with which the first stars produced X-ray radiation which effects the heating of the IGM and causes partial ionizations. Simulations of reionization often assume that the spin temperature has already saturated and so fluctuations in the spin temperature are ignored ($\boldsymbol{T_s}$ **saturated** in Figure 2). Furthermore, the moments of the 21-cm brightness temperature have not been examined in the presence of these fluctuations. We generate a suite of 21CMFAST simulations that model spin-temperature fluctuations. We vary the efficiency of X-ray production by an order of magnitude either side of a fiducial value of 2e57 X-ray photons per solar mass ($\boldsymbol{\xi_x} =$ **2e57** in Figure 2).

3. Implications

Neutral sinks & residual galactic hydrogen (Figure 1): Including small self-shielded neutral absorbers reduces the variance of the 21-cm signal; the inclusion of galactic hydrogen

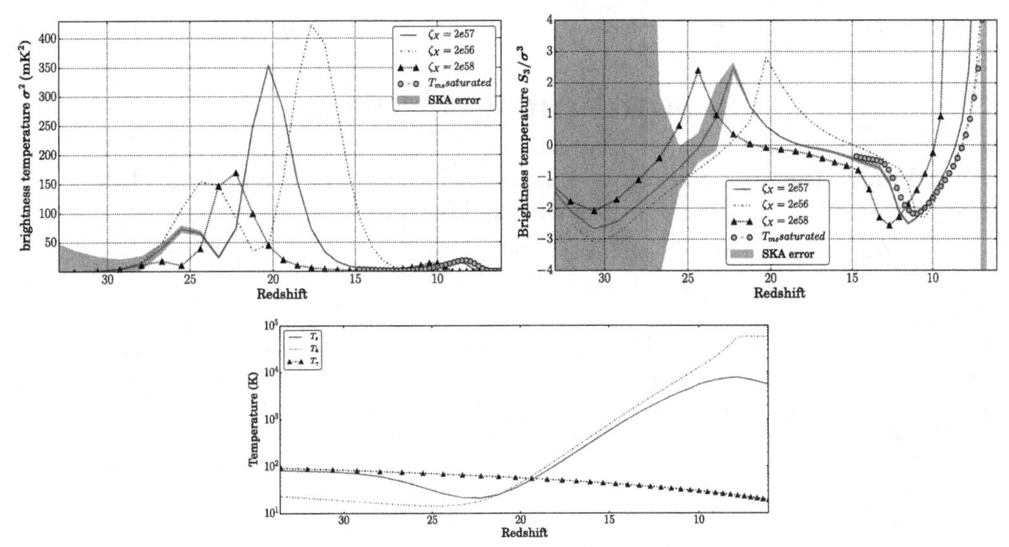

Figure 2. Spin temperature: Variance (top-left) and skewness (top-right) of δT_b as function of redshift; evolution of the spin temperature (red), kinetic temperature (blue dot-dashed) and CMB temperature (black dotted w/triangles) as a function of redshift for $\xi_x = 2e57$ (bottom)

decreases its amplitude further as well as producing a kick up towards the end of reionization. The severity of these two effects is dependent on the percentage of galactic mass that is assigned to remain neutral and can potentially alter the qualitative nature of the variance's evolution. This effect will make the variance more challenging to detect and suppresses a turnover that if observed would provide a signature that reionization was at its mid point. Including LLS also wipes out a late-time signature of reionization in the skewness; this is replaced by a characteristic turnover which is in turn erased when even small amounts of galactic hydrogen remain in galaxies. The amplitude of the variance at the end or reionization can be used to constrain the amount of H I remaining in galaxies.

X-ray efficiency of galaxies (Figure 2): We see that the variance and skewness of the brightness temperature both exhibit distinct turnovers corresponding to the point at which T_s starts increasing with the gas kinetic temperature to which it is becoming coupled. This provides constraints on the timing of X-ray heating processes and the strength of this signature also increases with decreasing ξ_x. We also see the variance is suppressed during reionization due to partial ionizations.

References

Bolton J. S. *et al.* 2011, *MNRAS: Letters*, 416, L70
Mesinger A., Furlanetto S. R., & Cen R., 2011, *MNRAS*, 411, 955
Miralda-Escude J., 2003, *ApJ*, 597, 66
Miralda-Escude J., Haehnelt M., & Rees M. J., 2000, *ApJ*, 530, 1-16
Planck Collaboration: Ade P. A. R.. *et al.*, 2013, *ArXiv preprint*,
Rahmati *et al.* 2013, *MNRAS*, 430: 2427
Robertson B. E. *et al.*, 2013, *ApJ*, 768, 71
Schaye, 2001, *ApJ*, 559: 507
Sobacchi, E. & Mesinger, A. 2014, *MNRAS*, 440, 1662

Statistical Challenges in 21st Century Cosmology
Proceedings IAU Symposium No. 306, 2014
A. F. Heavens, J.-L. Starck & A. Krone-Martins, eds.

doi:10.1017/S1743921315001696

Combining Probes

Anaïs Rassat[1]**, François Lanusse, Donnacha Kirk, Ole Host and Sarah Bridle**

[1]Laboratory of Astrophysics, Ecole Polytechnique Fédérale de Lausanne (EPFL)
email: anais.rassat@epfl.ch

Abstract. With the advent of wide-field surveys, cosmology has entered a new golden age of data where our cosmological model and the nature of dark universe will be tested with unprecedented accuracy, so that we can strive for high precision cosmology. Observational probes like weak lensing, galaxy surveys and the cosmic microwave background as well as other observations will all contribute to these advances. These different probes trace the underlying expansion history and growth of structure in complementary ways and can be combined in order to extract cosmological parameters as best as possible. With future wide-field surveys, observational overlap means these will trace the same physical underlying dark matter distribution, and extra care must be taken when combining information from different probes. Consideration of probe combination is a fundamental aspect of cosmostatistics and important to ensure optimal use of future wide-field surveys.

Keywords. cosmology, methods: statistics

---

1. Introduction

In recent decades, complementary observational cosmological probes have contributed to building our currently accepted standard model of cosmology: a universe dominated by a mysterious dark energy required to explain its acceleration and with most matter being in the form of yet another mysterious component called dark matter. The matter-energy densities ($\Omega_{\rm DE}, \Omega_{\rm DM}$, etc ...) have been studied quantitatively for several decades, often referred to as "precision cosmology" (Dicke *et al.* (1965)). However, this picture of our Universe leaves room for several open questions: 1) on the nature of dark energy and 2) the nature of dark matter, 3) on understanding the initial conditions of the Universe and 4) on testing whether Einstein's theory of General Relativity is the correct prescription for gravity on cosmological scales. Answering these questions requires new physical parameters to be considered.

To answer these open questions, cosmologists have access to two fundamental probes: the expansion history of the Universe and the growth of structures. In practice, these can be observed through several observational probes. At low redshifts, weak gravitational lensing, galaxy clustering (Albrecht *et al.* (2006), Peacock *et al.* (2006)), as well as strong lensing and other observables have been used and are promising tools for further experiments. At high redshift, the cosmic microwave background has been studied extensively with COBE (Bennett *et al.* (1990)), WMAP (Bennett *et al.* (2013)) and Planck (Planck Collaboration (2011)) as well as other smaller scale experiments. Cross-correlation of low and high redshift observables have also been used to further constrain our cosmological model, as for example with the integrated Sachs-Wolfe (ISW) effect (Sachs & Wolfe (1967)).

With the advent of wide-field surveys, observational overlap means that different observational probes will trace the same physical underlying dark matter distribution and extra care must be taken when combining information. This consideration of probe combination methods is a fundamental aspect of reaching precision and accurate cosmology.

2. Forecasting with the Fisher Information Matrix

In order to discuss the constraining power of different observational probes and how best to combine them, we consider the Fisher Information Matrix (FIM).

2.1. *Fisher Information Matrix (FIM)*

It is possible to forecast the precision with which a future experiment will be able to constrain cosmological parameters, by using the FIM (for a detailed derivation of the following see Tegmark *et al.* (1997) or Dodelson (2003)). This method requires only three fundamental ingredients:

• A set of cosmological parameters $\vec{\theta}$ for which one wants to forecast errors, and assumed values of these for a true underlying universe. For example, this could be a set of parameters $\vec{\theta} = (\Omega_{\rm DE}, w_0)$ for which one can assume: $\Omega_{\rm DE}^{\rm true} = 0.75$, $w_0^{\rm true} = -1$."

• A set of n measurements of the data $\vec{x} = (x_1, x_2, ..., x_n)$, say the gravitational weak lensing angular power spectrum $C_{GG}(\ell)$ over a range $\ell = 1...n$, and a model for how the data depend on cosmological parameters, i.e.: $C_{GG}(\ell) = C_{GG}(\ell, \vec{\theta})$

• An estimate of the uncertainty on the data $\Delta C_{GG}(\ell)$, which may depend on the given experiment (instrument noise, shot noise, etc...) as well on the data estimator (e.g., cosmic variance).

The FIM is defined by:

$$F_{\alpha\beta} = \left\langle \frac{\partial^2 \mathcal{L}}{\partial\theta_\alpha \partial\theta_\beta} \right\rangle, \tag{2.1}$$

where $\mathcal{L} = -\ln L$ and $L = L(\vec{x}, \vec{\theta})$ is the likelihood function or the probability distribution of the data $\vec{x}$, which depends on some model parameter set $\vec{\theta}$. If the data are correlated, the Fisher matrix then this might instead be the inverse of the covariance matrix between data points. In this chapter, I only consider uncorrelated data points.

The uncertainty on the parameter θ_α can be estimated directly from the FIM and has been shown to obey:

$$\Delta\theta_\alpha \geqslant \frac{1}{\sqrt{F_{\alpha\alpha}}}, \tag{2.2}$$

if all other parameters are known. This is known as the Cramér-Rao inequality. This is the key strength of the FIM forecast method, in that it places a solid lower limit on the parameter uncertainties, if the underlying probability distribution $L(\vec{x}, \vec{\theta})$ is Gaussian. If the parameters $\vec{\theta}$ also have Gaussian distribution around the fiducial value, then:

$$\Delta\theta_\alpha = \frac{1}{\sqrt{F_{\alpha\alpha}}}, \tag{2.3}$$

when all the other parameters are fixed.

If the vector of parameters $\vec{\theta}$ is allowed to vary, then the uncertainties for each parameter can still be obtained, and obey:

$$\Delta\theta_\alpha \geqslant \sqrt{(F^{-1})_{\alpha\alpha}}, \tag{2.4}$$

where F^{-1} is the inverse of the FIM. In this case, the uncertainty for θ_α has been obtained by implicitly marginalising over the other parameters. An unmarginalised estimate of the uncertainty of θ_α would be $1/\sqrt{F_{\alpha\alpha}}$.

By assuming the errors on the spherical harmonic estimator $C_{GG}(\ell)$ are Gaussian, then:

$$F = \left\langle \frac{\partial \mathcal{L}}{\partial \theta_\alpha \partial \theta_\beta} \right\rangle = \frac{1}{2} \left\langle \frac{\partial^2 \chi^2}{\partial \theta_\alpha \partial \theta_\beta} \right\rangle, \tag{2.5}$$

since $\mathcal{L} = -\ln L = \frac{1}{2}\chi^2$. Using the definition of χ^2:

$$F_{\alpha\beta} \simeq \sum_\ell \frac{1}{(\Delta C_{GG})^2} \frac{\partial C_{GG}}{\partial \theta_\alpha} \frac{\partial C_{GG}}{\partial \theta_\beta}, \tag{2.6}$$

where C_{GG} implicitly depends on ℓ and there is no covariance between the data points.

2.2. *Figure of Merits (FoM)*

In the context of future wide-field survey optimisation, it is often useful to have a single quantity or *Figure of Merit (FoM)* in order to compare the constraining power of different surveys or survey configurations. If one is interested in dark energy parameters, one can use the dark energy FoM:

$$\text{FoM}_{\text{DE}} = \frac{1}{\sigma_{w_0} \sigma_{w_p}}, \tag{2.7}$$

where σ_{w_0} and σ_{w_p} are the marginalised errors, and w_p is the dark energy equation of state parameter at the pivot redshift.

In order to compare the constraining power across different sectors of the cosmological model, one can use a more general FoM which encompasses information from several cosmological parameters, i.e. :

$$\text{FoM}_{\text{TOT}} \equiv \ln \left(\frac{1}{\det (F^{-1})_{\text{cosm}}} \right). \tag{2.8}$$

3. Observables vs. Probes

3.1. *Distinction between an observable and a probe*

It is useful to state the distinction here between an *observable* and a *probe*, with a probe relating to a physical effect and an observable being independent of any model. Often different probes are folded into a single observables, and different observables can fundamentally measure a single probe. As an example, different observables are:

- ϵ: galaxy ellipticities,
- n: galaxy number counts,
- T: temperature.

Each observable can contain information related to several probes, e.g.:

- ϵ: gravitational lensing and intrinsic alignments,
- n: galaxy clustering (including redshift distortions, baryon acoustic oscillations) and cosmic magnification,
- T: primary and secondary temperature fluctuations in the CMB.

Here, we give a few examples of how probe combination within a single observable can have an important impact on observational cosmology or be useful to constrain different sectors of the cosmological model.

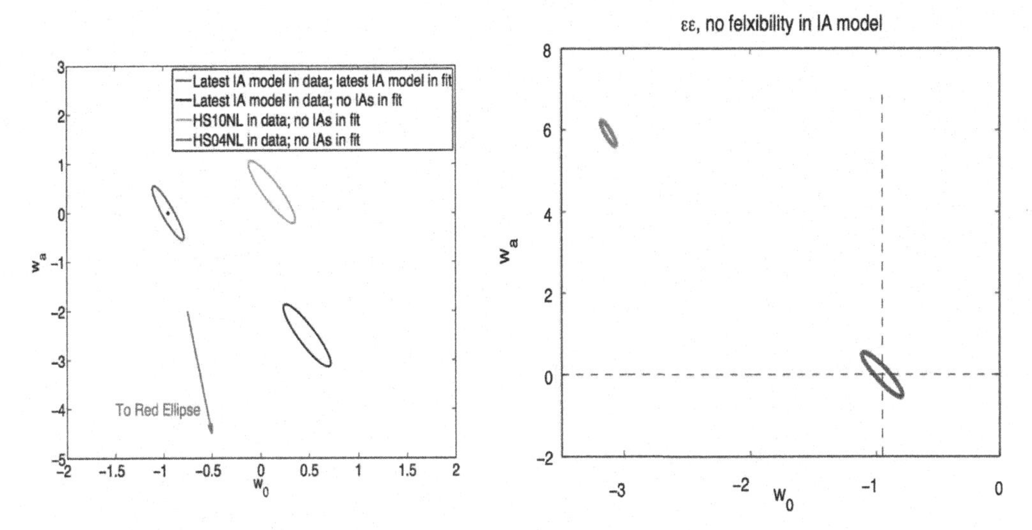

Figure 1. Taken from Kirk *et al.* (2012). *Left*: this figure shows the effect of ignoring intrinsic alignments for different physical models by showing the cosmological bias on dark energy parameters w_0 and w_a (95% confidence limits). *Right*: This figure shows the effect of using the 'wrong' IA model on cosmological parameters. In blue the latest IA model is used in the data and in the fit, whereas in red, the latest IA model is in the data but an older IA model is used in the fit.

3.2. *Intrinsic alignments and galaxy ellipticities*

Weak gravitational lensing has been shown to be one of the most promising probes to study the nature of dark energy. Gravitational lensing can be studied statistically by looking at the correlation function of galaxy elliipticities. One fundamental assumption is that galaxies are randomly oriented before they are affected by gravitational lensing. However, it is known that galaxies can have instrinsic alignments due to tidal forces so that the measured ellipticity is statistically given by:

$$\epsilon = \gamma + I, \tag{3.1}$$

where γ is the galaxy shear and I is the intrinsic alignment component so that the total measured correlation function is given by (for references and further details, see Kirk *et al.* (2012):

$$C_{\epsilon\epsilon} = C_{GG} + C_{II} + C_{GI}. \tag{3.2}$$

The IA terms contain cosmological information through their dependence on the matter power spectrum and is therefore a secondary probe that can help constrain cosmology. not only does IA contain extra cosmological information, but ignoring this effect can lead to a bias on cosmological parameters as shown in the left hand side of figure 1 taken from Kirk *et al.* (2012) showing the importance of performing an analysis including secondary signals from the start.

In the right hand side of figure 1, also taken from Kirk *et al.* (2012), shows the cosmological bias from a different IA model in the analysis than the one contained in the observable, showing again the importance of a good physical model for secondary probes.

3.3. *Galaxy number counts*

Galaxy number counts are used in the literature to measure clustering information in the large scale structure (LSS) by indirectly measure the matter power spectrum. As shown in figure 2, the matter power spectrum itself measures different physical effects or

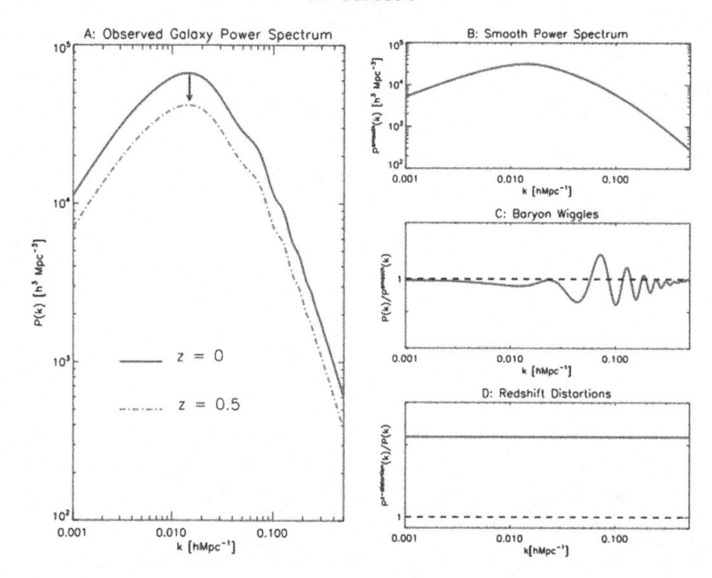

Figure 2. Taken from Rassat *et al.*(2008). The matter power spectrum is shown on the left, and different physical components or probes are shown on the right, including: the broad band power spectrum, baryon acoustic oscillations and redshift distortions.

probes, for e.g.: the broad band power spectrum, baryon acoustic oscillations and redshift distortions. Each of which probe cosmological parameters in different and complementary ways.

4. Probe Combination: Cross-correlations as a probe

Previously, we have briefly discussed how different observables can be studied by considering their auto-correlation functions. Cross-correlating different observables can also be used to study certain physical effects that are different to study with the auto-correlations, such as the integrated Sachs-Wolfe effect, which we discuss in section 4.1. We also use the example of tomographic analysis of LSS to show the importance of using cross-correlations in section 4.2.

4.1. *The integrated Sachs-Wolfe effect*

As photons from the last scattering surface travel towards us, they pass through the gravitational potentials of structures along the line of sight. Photons will gain energy as they enter the gravitational potentials and loose energy on exit. If the potential is unchanged during this travel time, the net effect will be null. However, if the gravitational potential decreases with time (as in universes with dark energy, positive curvature or in some alternative models), the net effect will be that photons will have gained energy by passing through the gravitational potentials. This effect, called the integrated Sachs-Wolfe effect, adds power to the observed CMB temperature-temperature power spectrum, but is difficult to measure due to cosmic variance. Instead, this effect can be measured by cross-correlating tracers of the gravitational potential (e.g. with galaxy surveys) with the CMB (Sachs & Wolfe (1967), Boughn & Crittenden(2002), Rassat *et al.* (2007)).

Measuring the ISW effect can be used as an independent measure of the presence of dark energy, as well as a measure of its equation of state parameters. Though the effect is small, combination of different tracers of large scale structure at different redshifts can also increase the amplitude of the signal (Giannantonio *et al.* (2008)).

Reconstructing the ISW effect (Dupé *et al.* (2011), Rassat & Starck (2013), Rassat *et al.* (2013)Rassat, Starck, & Dupe, Rassat *et al.* (2014)), i.e. by reconstructing a map of the temperature anisotropies due to the ISW effect can be used to study claimed anomalies in the anisotropies of the primordial CMB. The premise is that the most interesting cause of the anomalies would be one resulting from early Universe physics, and that we are therefore interested in studying the primordial CMB instead of the observed CMB, i.e., one free from Galactic emissions, astrophysical and secondary cosmological foregrounds. The observed CMB temperature fluctuations can be written as:

$$\delta_T^{\rm OBS} = \delta_T^{\rm prim} + \delta_T^{\rm ISW} + \delta_T^{\rm other} + \delta_T^{\mathcal{N}}, \tag{4.1}$$

where $\delta_T^{\rm OBS}$ is the total observed CMB, $\delta_T^{\rm prim}$ the primoridal CMB flucutations, $\delta_T^{\rm ISW}$ the signal due to the ISW effect, $\delta_T^{\rm other}$ other effects, and $\delta_T^{\mathcal{N}}$ the noise. On large scales, $\delta_T^{\rm other}$ is expected to include the kinetic Sunyave-Zel'dovich and the kinetic Doppler quadrupole effects, and $\delta_T^{\mathcal{N}}$ should be negligible.

Recently, Rassat *et al.* (2014) analysed six of these claimed anomalies in a new full-sky map of the CMB (provided in Bobin *et al.* (2013)). Analysis of the observed CMB maps showed that only the low quadrupole and quadrupole-octopole alignment seemed significant, but that the planar octopole, Axis of Evil, mirror parity and cold spot were not significant. After subtraction of astrophysical (kinetic Sunyaev-Zel'dovich) and cosmological secondary effects (ISW and the kinetic Doppler quadrupole), only the low quadrupole could still be considered anomalous, meaning the significance of only one anomaly was affected by secondary effect subtraction out of six anomalies considered.

4.2. *Large Scale Structure: tomography vs. 3D analysis*

Unlike the CMB, information in a spectroscopic galaxy survey will be 3-dimensional. It can therefore naturally be split into radial bins. Cosmological information is contained in each bin, but also in the cross-correlation between redshift bins. The question is therefore of how best to combine the information from the galaxy number counts to extract the 3-dimensional information.

This can be done either using tomography, i.e. a spherical harmonic approach including cross-correlation between various bins, or using a full 3D spherical Fourier-Bessel (SFB) approach (Heavens & Taylor (1995)), i.e. by decomposing the field using:

$$f(\vec{r}) = \sqrt{\frac{2}{\pi}} \int {\rm d}k \sum_{\ell m} f_{\ell m}(k) k j_\ell(kr) Y_{\ell m}(\theta, \phi). \tag{4.2}$$

The SFB approach is motivated physically and has natural prescription for selection and physical effects (e.g., redshift distortions).

Figure 3 is taken from Lanusse *et al.* (2014) and shows that while a tomographic reconstruction can recover the information extracted using a SFB analysis, in the case where nuisance parameters are included, the 3D approach should be preferred (see Lanusse *et al.* (2014) for details).

5. Probe Combination: Uncorrelated probes and priors

Different observables can constrain cosmological parameters in complementary ways, for e.g. by having different directions of degeneracy. This complementarity can be used to provide higher precision constraints. For independent probes A and B, the forecasting FIMs can simply be added so that:

$$F_{\rm TOT} = F_A + F_B. \tag{5.1}$$

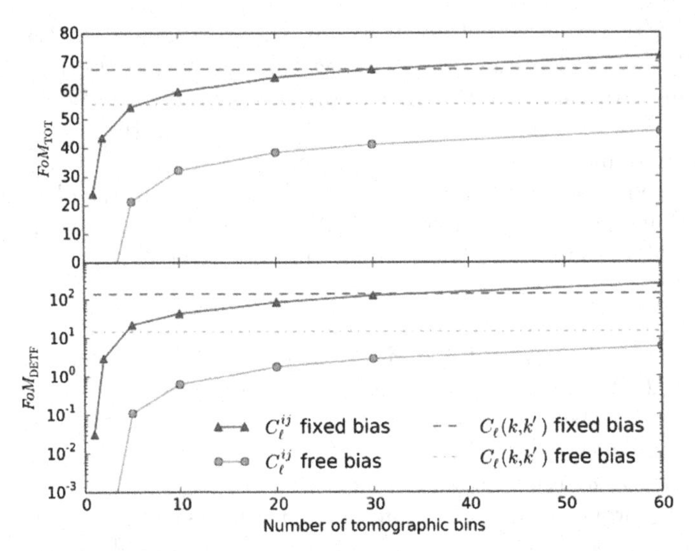

Figure 3. Taken fromLanusse *et al.* (2014). The two different figures of merit for a tomographic vs. SFB analysis, considering only cosmological parameters (top) or including nuisance parameters (bottom).

This method can be applied to different observables of a same experiment (e.g., the expected constraints from gravitational lensing and galaxy number counts for a given wide-field survey), as well as for constraints from different experiments, or to include priors from past experiments.

Combining information in such a way can be used to either provide tighter constraints on cosmological parameters, or expand the parameter space to includ more general parameters. For example, by including CMB Planck priors with gravitational lensing constraints, one can expand the parameters space of the initial conditions sector whilst simultaneously allowing for massive neutrinos (Debono *et al.* (2009)). The parameter space can also be expanded to simultaneously constrain dark energy and modified gravity parameters (Laureijs *et al.* (2011)).

6. Probe Combination: Accounting for correlations

For future wide-field surveys, observational overlap will mean that different observables will often probe the same physical underlying dark matter field, so that these will not in practice be uncorrelated observables. Taking the example of galaxy ellipticities ϵ and galaxy number counts n as before. In section 5, we considered the observables $C_{\epsilon\epsilon}$ and C_{nn} separately. If the observables are correlated, we must consider the $C_{n\epsilon}$ as a new measurement. In this case, we can no longer consider the Fisher matrices to simply add as in equation 5.1, but must calculate a single Fisher matrix for all observables using a covariance matrix which takes into account correlations. The Fisher matrix elements are given by (see Joachimi & Bridle (2010) for full details):

$$F_{\alpha\beta} = \sum_{l=l_{min}}^{l_{max}} \sum_{(i,j),(m,n)} \frac{\partial C_{ij}(l)}{\partial\theta_\alpha} \mathrm{Cov}^{-1}\left[C_{ij}(l), C_{mn}(l)\right] \frac{\partial C_{mn}(l)}{\partial\theta_\beta}. \tag{6.1}$$

Where the power spectra are now combined into a total data vector:

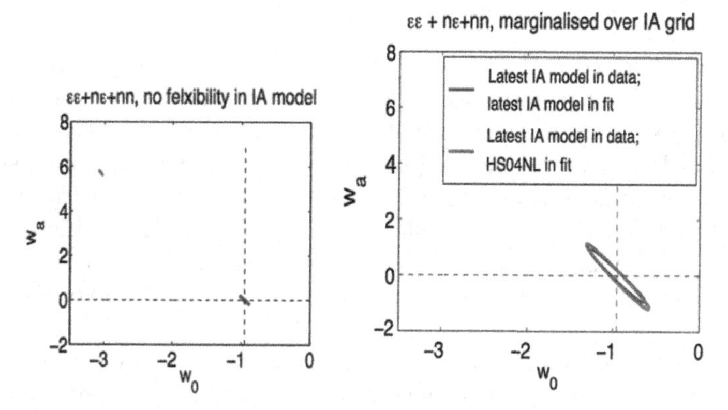

Figure 4. Taken from Kirk *et al.* (2012). The 95% confidence limits on dark energy parameters w_0 and w_a from galaxy ellipticities and number counts, accounting for correlations between both observables, in the context of a future wide-field survey. The blue contours have the same intrinsic alignment model in the fit as in the data, while the red contours have another intrinsic alignment model in the fit. In the left hand side, the Fisher matrix contains only cosmological parameters, while in the right hand side nuisance parameters regarding the galaxy bias and intrinsic alignment amplitude have been marginalised over. In the case where nuisance parameters are considered, the cosmological bias is not longer problematic.

$$\mathcal{D}(\ell) = \Big\{ C_{\epsilon\epsilon}^{(11)}(\ell), .., C_{\epsilon\epsilon}^{(N_{\rm zbin}N_{\rm zbin})}(\ell), C_{n\epsilon}^{(11)}(\ell), .., C_{n\epsilon}^{(N_{\rm zbin}N_{\rm zbin})}(\ell), C_{nn}^{(11)}(\ell), .., C_{nn}^{(N_{\rm zbin}N_{\rm zbin})}(\ell) \Big\} \tag{6.2}$$

for every angular frequency considered. The corresponding covariance, again for every ℓ, reads

$$\mathrm{Cov}(\ell) = \left(\begin{array}{c|c|c} \mathrm{Cov}^{(ijkl)}_{\epsilon\epsilon\epsilon\epsilon}(\ell) & \mathrm{Cov}^{(ijkl)}_{\epsilon\epsilon n\epsilon}(\ell) & \mathrm{Cov}^{(ijkl)}_{\epsilon\epsilon nn}(\ell) \\ \hline \mathrm{Cov}^{(ijkl)}_{n\epsilon\epsilon\epsilon}(\ell) & \mathrm{Cov}^{(ijkl)}_{n\epsilon n\epsilon}(\ell) & \mathrm{Cov}^{(ijkl)}_{n\epsilon nn}(\ell) \\ \hline \mathrm{Cov}^{(ijkl)}_{nn\epsilon\epsilon}(\ell) & \mathrm{Cov}^{(ijkl)}_{nnn\epsilon}(\ell) & \mathrm{Cov}^{(ijkl)}_{nnnn}(\ell) \end{array} \right), \tag{6.3}$$

where the sub-matrices are the usual matrices for each individual observable (i.e., $C_{\epsilon\epsilon}$, C_{nn}, $C_{n\epsilon}$).

In this approach, different probes are combined in several ways. Several probes are folded directly into single observables (e.g., intrinsic alignments and weak gravitational lensing are both included in the physical effects contributing to galaxy ellipticities), while correlations between different observables are accounted for in the covariance matrix.

This "full" calculation returns tighter constraints than using only information from the galaxy ellipiticities (Kirk *et al.* (2013)), as shown in the left-hand side of figure 4. However, there is still evidence of a catastrophic bias on the cosmological parameters when the 'wrong' intrinsic alignment model is used to fit the data, indicating that including all secondary signals is still of utmost importance for precision and accurate cosmology even when correlations between galaxy ellipticities and number counts are accounted for. However, in the right hand side of figure 4, the same calculation is done using a series of nuisance parameters (see Kirk *et al.* 2012 for details).

7. Conclusion

The 20th century saw cosmology transition from a theoretical endeavour to data-driven field. With the advent of future wide-field surveys covering large areas on the sky and wide redshift ranges, this new golden age of data in cosmology will no longer be limited by access to data but by statistical challenges in their processing. Observables such as galaxy ellipticities, galaxy number counts and temperature fluctuations in the microwave sky are each linked to a variety of physical effects, sometimes present in different observables. Future surveys are aiming at high precision cosmology, and correct treatment of all of these effects will be necessary to achieve not only precision cosmology but also accurate cosmology.

These observables can also be cross-correlated in order to study subtle physical effects like the integrated Sachs-Wolfe effect. In addition, correct understanding of these different effects can also help link different sectors of the cosmological model, for example, a correct understanding of secondary cosmological effects in the CMB can help recover the primordial CMB temperature fluctuations.

Together these observables will trace the underlying expansion history and growth of structure in complementary ways and can be combined in order to extract cosmological parameters as best as possible. With future wide-field surveys, observational overlap means these will trace the same physical underlying dark matter distribution, and extra care must be taken when combining information from different probes.

Consideration of probe combination is a fundamental aspect of cosmostatistics and important to ensure optimal use of future wide-field surveys. Probe combination will require combining information from different fields and sub-fields of astrophysics, cosmology and statistics, linking specialists working on different observational probes but that might be considering similar or identical effects, meaning effort within the cosmological community may be duplicated. This effort might be made more efficient by a systematic commitment by all researchers to reproducible research, which includes publication of codes and intermediate results both for existing data analysis and future forecasts.

Acknowledgement

This research is in part supported by the Swiss National Science Foundation (SNSF)

References

Albrecht, A., Bernstein, G., Cahn, R., *et al.* 2006, ArXiv e-prints 0609591

Bennett, C. L., Larson, D., Weiland, J. L., *et al.* 2013, *ApJS*, 208, 20

Bennett, C. L., Smoot, G. F., & Kogut, A. 1990, in Bulletin of the American Astronomical Society, Vol. 22, Bulletin of the American Astronomical Society, 1336–+

Bobin, J., Sureau, F., Paykari, P., *et al.* 2013, *A & A*, 553, L4

Boughn, S. P. & Crittenden, R. G. 2002, *Physical Review Letters*, 88, 021302

Debono, I., Rassat, A., Refregier, A., Amara, A., & Kitching, T. D. 2010, *MNRAS*, 404, 110

Dicke, R. H., Peebles, P. J. E., Roll, P. G., & Wilkinson, D. T. 1965, *ApJ*, 142, 414

Dodelson, S. 2003, Modern Cosmology, Amsterdam (Netherlands): Academic Press. ISBN 0-12-219141-2

Dupé, F.-X., Rassat, A., Starck, J.-L., & Fadili, M. J. 2011, *Astronomy & Astrophysics*, 534, A51

Giannantonio, T., Scranton, R., Crittenden, R., *et al.* 2008, *Physical Review D*, 77, 123520

Heavens, A. F. & Taylor, A. N. 1995, *MNRAS*, 275, 483

Joachimi, B. & Bridle, S. L. 2010, *A&A*, 523, A1

Kirk, D., Lahav, O., Bridle, S., *et al.* 2013, ArXiv e-prints 1307.8062

Kirk, D., Rassat, A., Host, O., & Bridle, S. 2012, *MNRAS*, 424, 1647

Lanusse, F., Rassat, A., & Starck, J. 2014, ArXiv e-prints 1406.5989
Laureijs, R., Amiaux, J., Arduini, S., *et al.* 2011, ArXiv e-prints 1110.3193
Peacock, J. A., Schneider, P., Efstathiou, G., *et al.* 2006, ESA-ESO Working Group on "Fundamental Cosmology", Tech. rep., ArXiV e-prints 0610906
Planck Collaboration, Ade, P. A. R., Aghanim, N., *et al.* 2011, *A&A*, 536, A1
Rassat, A., Land, K., Lahav, O., & Abdalla, F. B. 2007, *MNRAS*, 377, 1085
Rassat, A. & Starck, J.-L. 2013, *A&A*, 557, L1
Rassat, A., Starck, J.-L., & Dupe, F.-X. 2013, *A&A*, 557, A32
Rassat, A., Starck, J.-L., Paykari, P., Sureau, F., & Bobin, J. 2014, JCAP, 8, 6
Sachs, R. K. & Wolfe, A. M. 1967, Astrophys. J., 147, 73
Tegmark, M., Taylor, A. N., & Heavens, A. F. 1997, *ApJ*, 480, 22

Statistical Challenges in 21st Century Cosmology
Proceedings IAU Symposium No. 306, 2014
A. F. Heavens, J. -L. Starck & A. Krone-Martins, eds.

doi:10.1017/S1743921314013647

Cross-correlation between cosmological and astrophysical datasets: the Planck and Herschel case

Federico Bianchini and Andrea Lapi

SISSA, Via Bonomea 265, I-34136, Trieste, Italy
email: fbianchini@sissa.it

Abstract. We present the first measurement of the correlation between the map of the CMB lensing potential derived from the Planck nominal mission data and $z \gtrsim 1.5$ galaxies detected by *Herschel*-ATLAS (H-ATLAS) survey covering about 550 deg^2. We detect the cross-power spectrum with a significance of $\sim 8.5\sigma$, ruling out the absence of correlation at 9σ. We check detection with a number of null tests. The amplitude of cross-correlation and the galaxy bias are estimated using joint analysis of the cross-power spectrum and the galaxy survey auto-spectrum, which allows to break degeneracy between these parameters. The estimated galaxy bias is consistent with previous estimates of the bias for the H-ATLAS data, while the cross-correlation amplitude is higher than expected for a ΛCDM model. The content of this work is to appear in a forthcoming paper Bianchini, *et al.* (2014).

Keywords. cosmology: cosmic microwave background, gravitational lensing, galaxies: high-redshift, methods: data analysis

---

1. Introduction

Cosmological observations carried out in the last two decades have led to the establishment of the standard cosmological model. In this framework observed galaxies form in matter overdensities which are the result of the growth, driven by gravitational instabilities in an expanding Universe, of primordial inhomogeneities generated during an inflationary epoch. A picture of primordial inhomogeneities and the Universe at the beginning of its evolution is provided by observations of the cosmic microwave background (CMB) anisotropy at redshift $z \sim 1100$.

However as CMB photons travel from last scattering surface to the observer, they are gravitationally deflected on the arcminute scale by the intervening matter. The effect of the cosmic web is to induce a small but coherent (on the degree scale) deflections of the observed CMB temperature and polarization anisotropies; as a consequence, these deflections change the statistics of small scale unlensed CMB fluctuations by introducing a correlation between different Fourier modes.

In recent years CMB lensing has been measured in a number of CMB experiments. The first detections were made via cross-correlations with large scale structure probed by galaxy surveys. The higher sensitivity and resolution of recent CMB experiments, such as ACT, SPT and Planck, made possible to detect lensing using CMB data alone.

As already mentioned, the CMB lensing potential is an integrated measure of the matter distribution in the universe, up to the last scattering surface while the galaxies constitute signposts of nonlinear, virialized dark matter structures. In the standard, hierarchical paradigm for structure formation, nonlinear objects are preferentially formed on the peaks of the underlying linear density field, thus a cross-correlation between these two cosmic fields is expected; by measuring this cross-correlation we can determine the linear galaxy bias factor b that relates galaxies δ_g and matter δ fractional overdensities as $\delta_g = b\delta$.

In this work we present the first investigation of the cross correlation between the CMB lensing potential measured by Planck and Herschel-selected galaxies with estimated redshifts $z \gtrsim 1.5$. We adopt the fiducial flat ΛCDM cosmology with best-fit Planck cosmological parameters as provided by the Planck team in Planck Collaboration XVI (2013) (combination of Planck+WP+ highL+lensing datasets).

2. Modeling the expected signal

Both the strength of CMB lensing (encoded in the convergence field κ) and galaxy density fluctuations in a given direction of the sky depend on the projected dark matter density in that direction, so that these fields can be written as

$$X(\hat{\mathbf{n}}) = \int_0^{z_*} dz\, W^X(z)\delta(\chi(z)\hat{\mathbf{n}}, z), \tag{2.1}$$

where $X = \kappa, g$ and $W^X(z)$ is the kernel related to a given field.

The lensing kernel W^κ is

$$W^\kappa(z) = \frac{3\Omega_m}{2c}\frac{H_0^2}{H(z)}(1+z)\chi(z)\frac{\chi_* - \chi(z)}{\chi_*}. \tag{2.2}$$

In this equation, $\chi(z)$ is the comoving distance to redshift z, χ_* is the comoving distance to the last scattering surface at $z_* \simeq 1090$, $H(z)$ is the Hubble factor at redshift z, c is the speed of light, $\Psi(\chi(z)\hat{\mathbf{n}}, z)$ is the 3D gravitational potential at a point on the photon path given by $\chi(z)\hat{\mathbf{n}}$, Ω_m and H_0 are the present-day values of matter density and Hubble parameter respectively.

The galaxy kernel is

$$W^g(z) = \frac{b(z)\frac{dN}{dz}}{\left(\int dz'\,\frac{dN}{dz'}\right)} + \frac{3\Omega_m}{2c}\frac{H_0^2}{H(z)}(1+z)\chi(z)\int_z^{z_*} dz'\left(1 - \frac{\chi(z)}{\chi(z')}\right)(\alpha(z') - 1)\frac{dN}{dz'}. \tag{2.3}$$

The galaxy overdensity kernel is the sum of two terms: the first one is given by the product of the linear bias $b(z)$ by the redshift distribution dN/dz, while the second one takes into account the effect of gravitational magnification on the observed density of foreground sources. This effect depends on the slope, $\alpha(z)$, of their integral counts $(N(>F) \propto F^{-\alpha})$ below the adopted flux density limit.

Since the relevant angular scales are smaller than 1 radian (multipoles $\ell \gtrsim 100$) the theoretical angular cross-correlation can be computed using the Limber approximation:

$$C_\ell^{\kappa g} = \int_0^{z_*} \frac{dz}{c}\frac{H(z)}{\chi^2(z)} W^\kappa(z) W^g(z) P\left(k = \frac{\ell}{\chi(z)}, z\right), \tag{2.4}$$

where $P(k, z)$ is the matter power spectrum evaluated at wavenumber k and redshift z. The strength of the cross-correlation signal is determined by the overlap of the two kernels which are shown in Fig. 1 (left panel) using the H-ATLAS redshift distribution for W^g. The mean redshift probed by this cross-correlation is $\langle z\rangle \simeq 2$.

3. Datasets

We use the publicly released Planck CMB lensing potential map derived from a minimum variance combination of the 143 and 217 GHz temperature anisotropy maps only

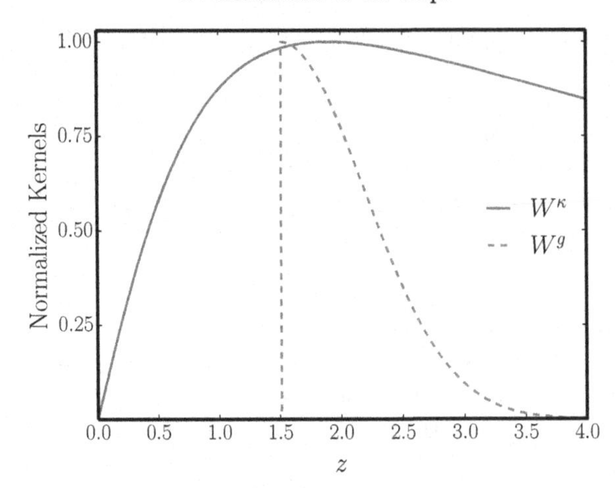

Figure 1. *Left panel*: estimated redshift distribution of the full sample of H-ATLAS galaxies (dashed line) compared with the CMB lensing kernel W^κ (solid line). Both kernels are normalized to a unit maximum. *Right panel*: the CMB convergence - galaxy density cross-spectrum as measured from Planck and Herschel data with best-fit theory lines (see text).

(Planck Collaboration XVII (2013)). To avoid biases in the estimation of the power spectrum for small patches of the sky due to the leakage of the large scale power to small scales, we converted to ϕ map to a κ map.

We have selected our galaxy sample from the catalogue of sources detected in the H-ATLAS fields, covering altogether $\sim$ 550 deg^2 with PACS and SPIRE instruments between 100 and 500 μm. The survey area is divided into five fields: three equatorial fields (GAMA fields, G09, G12 and G15), the North Galactic Pole (NGP) block and the South Galactic Pole (SGP) block.

The $z \lesssim 1$ galaxies detected by the H-ATLAS survey are mostly late-type and starburst galaxies with moderate star formation rates and relatively weak clustering. High-z galaxies are forming stars at high rates ($\gtrsim$ few hundred $M_\odot\,\mathrm{yr}^{-1}$) and are much more strongly clustered, implying that they are tracers of large scale over-densities. Their properties are consistent with them being the progenitors of local massive elliptical galaxies (Lapi, *et al.*(2011)). We aim at correlating high-z H-ATLAS galaxies with the *Planck* CMB lensing map.

To select the high$-z$ population we adopt the criteria developed by Gonzalez-Nuevo, *et al.*(2012): (i) $S_{250\,\mu m} > 35$ mJy; (ii) $S_{350\,\mu m}/S_{250\,\mu m} > 0.6$ and $S_{500\,\mu m}/S_{350\,\mu m} > 0.4$; (iii) $3\,\sigma$ detection at $350\,\mu$m; (iv) photometric redshift $z_{\rm phot} > 1.5$, estimated following Lapi, *et al.*(2011) who showed that the uncertainty is $|z_{\rm phot} - z|/(1+z) \leqslant 20\%$. Our final sample comprises a total of 99823 sources.

4. Methods and Results

We estimate the angular power spectra using a pseudo-C_ℓ estimator, based on MASTER algorithm, using a sky mask leaving out only the regions covered by the H-ATLAS survey. We choose to measure the signal in seven linearly spaced bins of width $\Delta\ell = 100$ in the $100 \leqslant \ell \leqslant 800$ range.

In order to validate the pipeline developed and check that the measured cross- and auto-power spectra estimates are unbiased, we created 500 simulated maps of the CMB convergence field and the galaxy overdensity field with the statistical properties consistent with observations. The analysis pipeline recovers the input power spectra within measured errors. We also verify our pipeline by cross correlating our 500 simulated CMB lensing maps (containing both signal and noise) with the real H-ATLAS galaxy density map and our 500 simulated galaxy maps constructed using $b = 3$ with the true *Planck* CMB convergence map: in both cases no significant signal is detected.

The recovered cross-spectrum is shown in Fig. 1 (right panel). The error bars are estimated cross-correlating 500 Monte Carlo realizations of simulated CMB convergence maps (containing both signal and noise) with the true H-ATLAS galaxy density map. We evaluate the χ^2 of null hypothesis as $\chi^2_{\rm null} = \hat{\mathbf{C}}_L\, {\rm Cov}^{-1}_{\rm LL'}\, \hat{\mathbf{C}}_{\rm L'}$, rejecting no correlation hypothesis with 9σ significance. If all of the signal is indeed due to cosmological origin, the significance of the detection can be evaluated as $\Delta\chi^2 = \sqrt{\chi^2_{\rm null} - \chi^2_{\rm best}} \simeq 8.5\sigma$, where $\chi^2_{\rm best}$ is the chi-squared value of best-fit theoretical spectrum.

We introduce an additional parameter, A, that scales the expected amplitude of the cross-power spectrum, $C_\ell^{\kappa g}$. The best values of the parameters (amplitude A and bias b) have been obtained fitting the function $C_\ell^{\kappa g}$, multiplied by $A \times b$, to the observed cross-power spectrum. The obvious degeneracy between the amplitude A and the bias parameter b can be broken fitting simultaneously also the galaxy auto-correlation that scales as b^2. The best-fit values obtained using only the cross-spectrum data are $b = 5.12^{+3.89}_{-2.86}$ and $A = 1.15^{+1.39}_{-0.52}$ while adding the galaxy auto-spectrum data we are able to break the degeneracy: $b = 2.80^{+0.12}_{-0.11}$ and $A = 2.08^{+0.23}_{-0.23}$.

The amplitude is higher than expected from the standard model and found by cross-correlation analyses with other tracers of the large-scale structure. This tension may be due to magnification bias due to weak gravitational lensing: recent work by Gonzalez-Nuevo, *et al.*(2014) has shown that this effect is substantial for high-z H-ATLAS sources selected with the same criteria as the present sample (the number counts are steep and the slope is $\alpha = 3 \div 5$). The consequence of such effect is to enhance the expected $C_\ell^{\kappa g}$ amplitude, hence decreasing the tension. Nevertheless, even for $\alpha = 5$ the data require $A > 1$. Another systematic effect that can bias our measurement of the CMB convergence-galaxy cross-correlation is the leakage of CIB emission into the lensing map through the temperature maps, as it correlates strongly with the CMB lensing signal. Work is in progress to address these issues.

5. Conclusions

We have presented the first measurement of the correlation between the lensing potential derived from the Planck data and a high$-z$ ($z \gtrsim 1.5$) galaxy catalogue from the H-ATLAS survey, the highest z sample for which the correlation between Planck CMB lensing and tracers of large-scale structure has been investigated so far. We rule out the absence of cross-correlation signal with a significance of 9σ. A joint analysis of the cross-power spectrum and of the auto-power spectrum of the galaxy density contrast yielded a galaxy bias parameter consistent with earlier estimates for H-ATLAS galaxies at similar redshifts (Xia, *et al.*(2012)), although reporting an higher amplitude than expected.

References

F. Bianchini, *et al.* 2014, *in preparation*
A. Lapi, *et al.* 2011, *ApJ*, 742, 24
P. A. R. Ade, *et al.* 2013, arXiv:1303.5076
P. A. R. Ade, *et al.* 2013, arXiv:1303.5077
J.-Q. Xia, *et al.* 2012, *MNRAS*, 422, 1324
J. Gonzalez-Nuevo, *et al.* 2012, *ApJ* 749, 65
J. Gonzlez-Nuevo, *et al.* 2014, arXiv:1401.4094

Statistical Challenges in 21st Century Cosmology
Proceedings IAU Symposium No. 306, 2014
A. F. Heavens, J.-L. Starck & A. Krone-Martins, eds.

doi:10.1017/S1743921314013623

Information Gains in Cosmological Parameter Estimation

Sebastian Seehars[1], Adam Amara, Alexandre Refregier, Aseem Paranjape and Joël Akeret

ETH Zurich, Department of Physics, Wolfgang-Pauli-Strasse 27, 8093 Zurich, Switzerland
[1]email: seehars@phys.ethz.ch

Abstract. Combining datasets from different experiments and probes to constrain cosmological models is an important challenge in observational cosmology. We summarize a framework for measuring the constraining power and the consistency of separately or jointly analyzed data within a given model that we proposed in earlier work (Seehars *et al.* 2014). Applying the Kullback-Leibler divergence to posterior distributions, we can quantify the difference between constraints and distinguish contributions from gains in precision and shifts in parameter space. We show results from applying this technique to a combination of datasets and probes such as the cosmic microwave background or baryon acoustic oscillations.

Keywords. methods: data analysis, cosmology

---

1. Introduction

Observations of cosmological probes such as the cosmic microwave background (CMB) or Type Ia supernovae have established a standard model of cosmology, resting on the concepts of a cosmological constant Λ driving the observed accelerated expansion of the universe at late times and a cold dark matter (CDM) component accounting for about 80% of the matter content in the universe. To learn more about these mysterious dark components, it is crucial to test the ΛCDM model with observations from different astrophysical probes and independent experiments.

We summarize a technique for measuring the progress in constraining the parameters of a cosmological model from different datasets by quantifying the variation between the constraints using the Kullback-Leibler divergence that we proposed in earlier work (Seehars *et al.* 2014). This technique measures both shifts in parameter space and improvements in the precision of the constraints and is hence able to determine the constraining power of a dataset and its consistency with other probes. After formally introducing the method in section 2, we apply it to joint analyses of CMB data from the WMAP mission and measurements of small scale CMB experiments, baryon acoustic oscillations (BAO), and the expansion rate of the universe as published by the WMAP team in Hinshaw *et al.* (2013) as well as to other large scale CMB experiments in section 3. We conclude with a discussion of our findings in section 4.

2. Parameter Estimation and Kullback-Leibler Divergence

We wish to compare two posteriors on the same parameter space in order to measure gains in precision and consistency of the constraints on a given cosmological model. For simplicity, we consider the special case of two complementary datasets which can be analyzed in a sequential manner, i.e. where constraints from one dataset can be used as prior information for analyzing the second. In many applications of cosmology, however, two datasets can be correlated due to cosmic variance, which is the fact that we can observe only a single realization of an inherently stochastic cosmological model. The concepts presented next can also be applied to those scenarios and we refer the reader to Seehars *et al.* (2014) for more details. Returning to the case of two uncorrelated observables $\mathcal{D}_1$ and $\mathcal{D}_2$, we consider two posteriors $p_1(\Theta) \equiv p(\Theta|\mathcal{D}_1)$ and

$$p_2(\Theta) \equiv p(\Theta|\mathcal{D}_1, \mathcal{D}_2) = \frac{\mathcal{L}(\Theta; \mathcal{D}_2)p_1(\Theta)}{\int d\Theta\, \mathcal{L}(\Theta; \mathcal{D}_2)p_1(\Theta)}. \tag{2.1}$$

To measure the difference between p_1 and p_2, we use the Kullback-Leibler divergence (KL-divergence, Kullback & Leibler (1951)):

$$D(p_2||p_1) \equiv \int d\Theta\, p_2(\Theta) \log \frac{p_2(\Theta)}{p_1(\Theta)}. \tag{2.2}$$

The KL-divergence is well suited for our purposes: It is a positive quantity which is zero if and only if p_1 is equal to p_2 almost everywhere and can be interpreted as a pseudo-distance between probability densities (it is not a distance because of the asymmetry in exchanging p_1 and p_2). The KL-divergence is furthermore invariant under invertible parameter transformations and consequently measures differences between the constraints on the model and not only a particular parametrization of the model.

We will distinguish two contributions to the KL-divergence between two posteriors: A contribution from the improvements in precision that are expected even before the actual data is gathered and a second contribution from the shifts in parameter space induced by the observed data. The former can be defined by the expectation value of $D(p_2||p_1)$ when averaging over the prior knowledge on the data:

$$\langle D \rangle \equiv \int d\mathcal{D}_2\, p(\mathcal{D}_2) D(p_2||p_1), \tag{2.3}$$

where $p(\mathcal{D}_2) \equiv \int d\Theta\, \mathcal{L}(\Theta; \mathcal{D}_2)p_1(\Theta)$. The latter is given by the difference between observed and expected KL-divergence $S \equiv D(p_2||p_1) - \langle D \rangle$ and called *surprise* in the following. Similarly, we can consider the expected variations of D around $\langle D \rangle$

$$\sigma^2(D) \equiv \int d\mathcal{D}_2\, p(\mathcal{D}_2)(D(p_2||p_1) - \langle D \rangle)^2. \tag{2.4}$$

In general, we can think of the observed KL-divergence as a realization from the distribution of KL-divergences induced by the prior distribution on $\mathcal{D}_2$ and our knowledge of the likelihood.

Equation (2.2) could be estimated with a numerical Monte Carlo integrator such as nested sampling (Skilling 2004). Equation (2.3), however, is much harder to evaluate numerically. Yet, as the expected KL-divergence is a well-known quantity in Bayesian experimental design (see e.g. Chaloner & Verdinelli (1995)), computational approaches exist. For the applications to constraints on a flat ΛCDM cosmology in section 3 we will take a different approach: As the ΛCDM parameters are tightly constrained by CMB data alone, it is a reasonable approximation to assume a linear model and normal distributions for prior and likelihood when updating these constraints. In this case, equations (2.2)

Table 1. KL-divergence estimates in bits for considered combinations of datasets. In the *data combination* column, WMAP refers to the full WMAP 9 data (Bennett *et al.* 2013). The BAO, H_0, and small scale CMB (eCMB) data are described in Hinshaw *et al.* (2013). The other CMB datasets are from the Boomerang (Jones *et al.* 2006) and Planck (Planck collaboration 2013) teams. The updating column refers to *add* if complementary data is added, *replace* if the dataset is completely replaced, and *part* if parts of the data are replaced. The p-value is an estimate for the prior probability of observing a surprise that is greater or equal (less or equal) than S if S is greater (smaller) than zero. It is an approximation when data is partially replaced.

Data combination	Updating	D	$\langle D\rangle$	S	$\lvert S/\sigma(D)\rvert$	p-value
BOOMERANG → WMAP	replace	22.5	18.4	4.1	1.6	0.07
WMAP → WMAP + eCMB	add	2.1	1.7	0.4	0.5	0.2
WMAP + eCMB → WMAP + eCMB + BAO	add	1.3	1.0	0.3	0.8	0.2
WMAP + eCMB → WMAP + eCMB + H_0	add	0.4	0.3	0.1	0.1	0.3
WMAP + eCMB → WMAP + eCMB + BAO + H_0	add	0.9	1.1	−0.2	0.2	0.6
WMAP → Planck + WP	part	29.8	7.9	21.9	6.5	0.0002

and (2.3) are analytic and given by (Seehars *et al.* 2014):

$$D(p_2||p_1) = \frac{1}{2}\left((\Theta_1 - \Theta_2)^T \Sigma_1^{-1}(\Theta_1 - \Theta_2) + \mathrm{tr}(\Sigma_2\Sigma_1^{-1}) - d - \log\frac{\det\Sigma_2}{\det\Sigma_1}\right), \tag{2.5}$$

$$\langle D\rangle = -\frac{1}{2}\log\frac{\det\Sigma_2}{\det\Sigma_1}, \tag{2.6}$$

$$\sigma^2(D) = \frac{1}{2}\mathrm{tr}\left((\Sigma_1^{-1}\Sigma_2 - \mathbb{1})^2\right), \tag{2.7}$$

with Θ_i and Σ_i being mean and covariance of distribution p_i, d being the dimensionality of the parameter space, and $\mathbb{1}$ being the d-dimensional identity matrix. Equation 2.5 shows that the KL-divergence depends on the ratio of covariance matrices as well as the significance of the shifts in the means Θ_1 and Θ_2. While the former also governs the expected relative entropy (2.6), the latter is driving the surprise contribution. In order to estimate equations (2.5) to (2.7), it is left to estimate mean and covariance matrix of the posteriors from Monte Carlo Markov chains, for example. Note furthermore that the distribution of the KL-divergence induced by prior knowledge on $\mathcal{D}_2$ is a generalized χ^2-distribution (Seehars *et al.* 2014).

3. Application to Data

We apply the results from section 2 to two scenarios: The parameter constraints on a flat ΛCDM cosmology from the joint analyses of the final CMB data release by the WMAP team with other cosmological probes as published by Hinshaw *et al.* (2013) and the constraints from a historical series of CMB experiments. While we use the official Monte Carlo Markov chains from the WMAP team to estimate the information gains in the former application, we use the CosmoHammer framework (Akeret *et al.* 2013) to generate samples for the comparison between CMB datasets. When estimating the p-values of the observed KL-divergences on its generalized χ^2 distribution, we use the R-package `CompQuadForm` by Duchesne & De Micheaux (2010). The results in bits, i.e. when taking the logarithm in equation (2.2) to base two, are shown in Table 1.

Adding external data from small scale CMB experiments, BAO, and H_0 measurements to the WMAP constraints results in an information gain between 0.4 bits for the H_0 prior and 2.1 bits for small scale CMB data. We generally find small surprise compared to the expected variations in D. Hence, a flat ΛCDM model is consistent with all datasets and changes in the constraints are coming from the expected statistical variations in the data.

When analyzing the differences between the constraints from different CMB experiments, we must consider the effects of cosmic variance. This changes the form of equations 2.6 and 2.7, but similar results can be derived when considering the case of replacing correlated data. Results when comparing different CMB experiments are extensively discussed in Seehars *et al.* (2014). Here we focus on replacing CMB data from Boomerang data with WMAP 9 data and replacing the WMAP temperature power spectrum with the results of the Planck team. Both data updates generate large information gains in parameter space as measured by KL-divergences of 22.5 and 29.8 bits, respectively. In these cases, however, it is important to look at the decomposition of D into the expected KL-divergence and the surprise: Comparing the Boomerang constraints with the WMAP 9 results, we find that the KL-divergence is dominated by the increased precision in the constraints as measured by $\langle D \rangle = 18.4$ bits. The Planck update also improves the precision of the constraints ($\langle D \rangle = 7.9$) but furthermore introduces shifts in parameter space that are significantly larger than expected a priori with a surprise of 21.9 ($p = 0.0002$). This significant surprise implies that the model is not able to consistently fit WMAP and Planck temperature data within the errors and hints towards either systematics in the data or physics beyond a flat ΛCDM cosmology.

4. Conclusions

We described a technique for measuring variations in the constraints on cosmological parameters coming from different cosmological probes and datasets that we proposed in earlier work (Seehars *et al.* 2014). It is based on applying the Kullback-Leibler divergence to the posteriors of two separately or jointly analyzed measurements. With this technique, we are able to separate contributions from gains in precision and shifts in parameter space by comparing the observed KL-divergence to a-priori expectations.

By applying these concepts to the constraints on a flat ΛCDM cosmology from CMB, BAO, and H_0 measurements, we show that this technique is able to quantify inconsistencies between the full posteriors that are not immediately apparent from the distributions of the individual parameters. In particular, we find no inconsistencies between the constraints on ΛCDM parameters from joint analyses of WMAP data with small scale CMB, BAO, and H_0 data as published by Hinshaw *et al.* (2013). Comparing the constraints from a joint analysis of Planck temperature and WMAP polarization data with prior expectations from the constraints of WMAP data alone, however, we find significant surprise contributions hinting towards tensions between the datasets. The described method may thus be a valuable tool for quantifying inconsistencies between data from different experiments in order to detect systematics or even signs of new physics.

References

Akeret, J. , Seehars, S., Amara, A., Refregier, A., & Csillaghy, A. 2013 *Astron. Comput.*, 2, 27

C L Bennett, D Larson, J L Weiland, & N Jarosik 2013 *ApJS*, 208(2), 54

K Chaloner, & I Verdinelli 1995 *Statist. Sci.*, 10, 273

P Duchesne, & P L De Micheaux 2010 *Comput. Stat. Data Anal.*, 54(4), 858

G Hinshaw, D Larson, E Komatsu, D N Spergel, C L Bennett, J Dunkley, M R Nolta, M Halpern, R S Hill, N Odegard, L Page, K M Smith, J L Weiland, B Gold, N Jarosik, A Kogut, M Limon, S S Meyer, G S Tucker, E Wollack, & E L Wright 2013 *ApJS*, 208(2), 19

William C Jones, PAR Ade, J J Bock, J R Bond, J Borrill, A Boscaleri, P Cabella, C R Contaldi, B P Crill, & P De Bernardis 2006 *ApJ*, 647(2), 823

S Kullback, & R A Leibler 1951 *Ann. Math. Stat.*, 22(1), 79

Planck collaboration, P A R Ade, N Aghanim, C Armitage-Caplan, M Arnaud, M Ashdown, F Atrio-Barandela, J Aumont, C Baccigalupi, A J Banday, *et al.* 2013 *arXiv*, 1303.5075

S Seehars, A Amara, A Refregier, A Paranjape, & J Akeret 2014 *arXiv*, 1402.3593

J Skilling 2004 *AIP Conference Proceedings*, 735(1), 395

Statistical Challenges in 21st Century Cosmology
Proceedings IAU Symposium No. 306, 2014
A. F. Heavens, J.-L. Starck & A. Krone-Martins, eds.

doi:10.1017/S1743921314013726

Cosmography with high-redshift probes

Vincenzo Vitagliano
CENTRA, Instituto Superior Técnico,
Universidade de Lisboa, Av. Rovisco Pais 1, 1049 Lisboa, Portugal
email: vincenzo.vitagliano@ist.utl.pt

Abstract. I discuss how the cosmographic approach to the determination of cosmological parameters can be implemented with the inclusion of high-redshift data. I argue on the viability of such high-z probes for cosmographic purposes, and resume some statistical issues in finding the most reliable cosmographic truncation.

Keywords. cosmological parameters, cosmology: miscellaneous

---

1. Introduction

Cosmology has recently become one of the most severe referee testing the reliability of theories alternative to General Relativity and models beyond standard particle physics. The analysis of cosmological data and the extraction of consistent results require the definition of a specific underlying theoretical model. The next natural question is then whether or not it is possible to define a scheme with minimal, self-consistent dynamical requests allowing the analysis of cosmological observations while skipping the bias of any theoretical prior.

Cosmography provides one possibility of such tool: starting with the educated guess of a cosmological metric described by the spatially homogeneous and isotropic Friedmann-Lemaitre-Robertson-Walker solution, any relevant distance indicator of an observed object can be then expanded in a power series of a suitable redshift parameter. The coefficients of such powers (evaluated today), casted into a combination of successive weighted derivatives of the scale factor $a(t)$ (the cosmographic parameters), contain the relevant information for a kinematic description of the universe.

It has to be stressed that the ill-behaviour at high redshift of the expansion is known to strongly affect the results. To circumvent the problem one must abandon the standard relation linking the luminosity distance to the ordinary-defined redshift. As already pointed out in Cattoen & Visser (2007), the lack of validity of the Taylor-expanded expression for the luminosity distance could be settled down approximately at $z \sim 1$. In order to avoid problems with the convergence of the series for the highest redshift objects as well as to control properly the approximation induced by truncations of the expansions and the underestimation of the errors, it is useful to recast the luminosity distance as a function of an improved parameter $y = z/(1+z)$ (Visser (2004), Cattoen & Visser (2007)). In such a way, being $z \in (0, \infty)$ mapped into $y \in (0,1)$, one is in principle able to retrieve the right behaviour for series convergence at any distance. The introduction of this new redshift variable will not affect the definition of the cosmographic parameters.

2. The data ensemble

Reaching the highest possible redshift allowed by data is a fundamental condition to disentangle between competing cosmological models. Given that most of the models are built in order to recover Dark Energy at low redshift, their expansion histories are obviously degenerate at late times. To break such a degeneracy, it is required an improvement on the knowledge of the early universe expansion curve: this aim can be achieved only by an accurate determination of the higher order parameters – and higher terms in the cosmographic expansion can be consistently reached only using very high redshift data.

The recent analysis Xia *et al.*(2012) handles the problem of interpreting the whole ensemble of cosmological data sets under a cosmographic perspective. We constrain the cosmographic parameters appearing in the expansions of the characteristic scales associated to Supernovae Type Ia (SNeIa), Gamma Ray Bursts (GRBs), Baryon Acoustic Oscillations (BAOs), the Cosmic Microwave Background (CMB) power spectrum and the determination of the Hubble parameter, estimated from surveys of galaxies (Hub).

Depending on the quantity that has been measured, it could be more appropriate to consider a particular cosmological distant scale than another one. To different distant scales correspond different Taylor expansions whose coefficients will combine the cosmographic parameters in different ways. In our analysis we refer to luminosity distance as the most direct choice for SNeIa and GRBs; volume distance for BAOs; angular diameter distance for the CMB. The inclusion of the direct estimation of the Hubble parameter is pretty peculiar and will be discussed later.

Even though the prominent role of SNe (in the high-z version too) in doing the job is well-known, the potentiality of GRBs as cosmological standard candles has been recently explored as a possible proposal to increase the number of high redshift distance ladders. Data coming from the observations of both SNe and GRBs are used to fit directly the expression for the luminosity distance. The analysis is performed by using Monte Carlo Markov Chains in the multidimensional parameter space to derive the likelihood.

We have also used the BAO (albeit the cosmographic series fitting is only mildly improved): although the physics and the data of BAOs depend on the content in matter of the universe, the impact of spacetime priors on the power spectrum and the volume distances ratio were shown to be only weakly dependent on dynamical feature, leaving the safe possibility to use them as a further constraining tool.

The CMB data account for a very stable and well determined scale. It is worth noting here, anyway, that on the contrary of the other probes, CMB data provide the problem of a lack of universality in the cosmographic approach. Unfortunately, the set of parameters extracts from CMB observations is not truly independent from the dynamics of the underlying gravitational theory. Its definition, in fact, strictly depends on the assumption of a cosmological model that behaves as General Relativity plus a content of matter of arbitrary nature. It is hence impossible to use it straightforwardly within a purely cosmographic analysis which wants to apply also to non-standard cosmologies (based on exotic modified gravity theories). In Xia *et al.*(2012) we proposed CMB data constraints on the cosmographic series by restricting the results to a slightly smaller variety of models (namely models having standard physics up to the decoupling era, and whose eventual new physics after decoupling only modifies the small angle spectrum changing the overall amplitude and the angular diameter distance at the decoupling). A desirable full solution to this problem would be achieved "standardizing" somehow the CMB parameters or alternatively identifying other CMB observables which could be used as standard rulers.

Last probe is the direct determination of the Hubble parameter as determined by the differential ages of galaxies, with a caveat: the coefficient of the n-th y-power in the Taylor expansion already provides a combination of n cosmographic parameters, while the same number of parameters appears only at the $(n + 1)$-th power of the series expansion for the other distant scales. This is due to an extra derivative with respect to time included in the definition of the Hubble parameter. For this reason, and for the different nature of the Hubble data, we initially consider constraints based on standard candles and rulers, and then we add the Hubble data using one order less in the y-power expansion.

3. Statistics selection between two truncations

Higher order powers of the redshift expansions always improve the data fitting, since more free parameters are involved. However, for a given data set, there will be an upper bound on the order which is statistically significant in the data analysis (Vitagliano *et al.*(2010)). An early truncation of the power series also leads to several inconsistencies or artifacts. This justifies the search of some criteria to make a proper choice between two alternative models. The criterion that we use is the so-called F-test. This test compares two nested models (in this case, two different truncations of the Taylor series) in order to find out which is, for a given data set, the most viable approximation of the series. Supposing that the null hypothesis implies the correctness of the first model, the F-test verifies the probability for the alternative model to fit the data as well. If this probability is high, then no statistical benefit comes from the extra degrees of freedom associated to the new model. The less is this probability, the better is the data fitting of the second model against the first one.

The following Table summarizes the estimates obtained in our analysis for the most statistically meaningful (in the F-test sense) term of the series expansion.

Data	SNIa+GRB+BAO+CMB (5th order)				
Parameter	q_0	j_0	s_0	c_0	H_0
Best Fit	-0.17	-6.92	-74.18	-10.58	–
Mean	-0.49 ± 0.29	-0.50 ± 4.74	-9.31 ± 42.96	126.67 ± 190.15	–
$\chi^2_{\rm min}$/d.o.f.	627.61/624				
Data	SNIa+GRB+BAO+CMB (5th order) +Hub (4th order)				
Parameter	q_0	j_0	s_0	c_0	H_0
Best Fit	-0.24	-4.82	-47.87	-49.08	71.65
Mean	-0.30 ± 0.16	-4.62 ± 1.74	-41.05 ± 20.90	-3.50 ± 105.37	71.16 ± 3.08
$\chi^2_{\rm min}$/d.o.f.	639.81/633				

Comparing the parameters against the guess for different cosmological models (see Xia *et al.*(2012) for details), it is interesting to stress the remarkably good performance of ΛCDM, even with respect to a cosmographic expansion with more free parameters. This could be taken as a strong hint in favor of this specific solution. However, we should warn about the (ab)use of a statistical comparison in terms of the derived χ^2. In fact, while this procedure is completely meaningful for a selection between two nested cosmographic expansions, it becomes rather questionable when the comparison is between any fiducial cosmological model and a cosmographic series.

VV is supported by FCT - Portugal through the grant SFRH/BPD/77678/2011.

References

C. Cattoen & M. Visser, 2007 *Class. Quant. Grav.* **24**, 5985.
M. Visser, 2004 *Class. Quant. Grav.* **21**, 2603.
V. Vitagliano, J. -Q. Xia, S. Liberati & M. Viel, 2010 *JCAP* **1003**, 005.
J. -Q. Xia, V. Vitagliano, S. Liberati & M. Viel, 2012 *Phys. Rev. D* **85**, 043520.

Statistical Challenges in 21st Century Cosmology
Proceedings IAU Symposium No. 306, 2014
A. F. Heavens, J.-L. Starck & A. Krone-Martins, eds.

doi:10.1017/S174392131401374X

Cross-correlating spectroscopic and photometric galaxy surveys

Martin B. Eriksen[1] **and Enrique Gaztañaga**[2]

[1]Sterrewacht Leiden, University of Leiden,
NL-2333 CA, Leiden, the Netherlands
email: marberi@strw.leidenuniv.nl

[2]Institut de Ciencies de l'Espai (ICE-IEEC/CSIC), Campus UAB
08193 Bellaterra, Spain
email: gazta@ice.cat

Abstract. Does photometric and spectroscopic survey benefit from overlapping areas? The photometric survey measures 2D Weak Lensing (WL) information from galaxy shape distortions. On the other hand, the higher redshift precision of an spectroscopic survey allows measurements of redshift space distortions (RSD) and baryonic accustic oscillations (BAO) from 3D galaxy counts.

The two surveys are combined using 2D-correlations, using sufficiently narrow bins to capture the radial information. This poster present effects of RSD and intrinsic correlations between narrow redshift bins. In understanding how the effects affects cosmological constrains, we first define two stage-IV and then present forecast for various configurations. When surveys overlap, they benefit from additional cross-correlations and sample variance cancellations from overlapping volumes. For a combined dark energy and growth history figure of merit, the result increase 50% for overlapping surveys, corresponding to 30% larger area.

Keywords. galaxies: distances and redshifts, techniques: spectroscopic surveys

1. Introduction

Galaxy surveys provide important insight to properties of dark matter and modified gravity. Photometric surveys target using galaxy shapes to measure weak gravitational lensing, while spectroscopic surveys can probe redshift space distortions (RSD) and baryonic acoustic oscillations (BAO) from excellent redshift information. The probes are powerful separate, but the combination can break parameter degeneracies. In addition the overlapping surveys allow for cross-correlating the observable. The benefit has been investigated by several groups Bernstein & Cai (2011); Gaztañaga et al. (2012); Cai & Bernstein (2012); Kirk et al. (2013); Font-Ribera et al. (2013); de Putter et al. (2013), but with large disagreement. Here we present our results, which is is a subset of a series of articles (in preparation) dealing with the modeling (paper-I), forecasts (paper-II), galaxy bias (paper-III) and the same-sky benefit (same-sky paper).

2. Modeling

Overdensities of galaxies can be expressed on the form

$$\delta = \delta_I + \delta_r + \delta_\kappa \tag{2.1}$$

where δ_I is the intrinsic overdensity, δ_r is the redshift space contribution and δ_κ is the weak lensing magnification. In this work, we forecast the constraint using cross-correlations in narrow redshift bins. The narrow bins capture the bulk of information

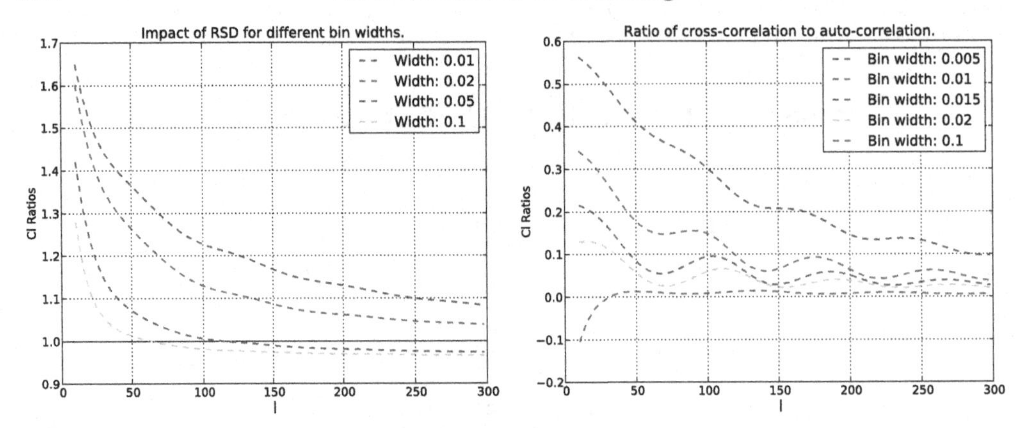

Figure 1. This figure who the effect of RSD and intrinsic correlations for different narrow bins. The left panel show the redshift to real space ratio for an auto-correlation at $z = 0.5$. The right panel shows the ratio of a cross-correlation between an adjacent bin and the auto-correlation, with the first bin starting at $z = 0.5$.

Parameter	Photometric (F)	Spectroscopic (B)
Area [sq.deg]	14000	14000
Redshift uncertainty	0.05(1+z)	0.001 (1+z)
Magnitude limit	i < 24.1	i < 22.5
Bin width	0.07 (1+z)	0.01(1+z)
Density [gal/sq.arcmin]	6.5	0.4

Table 1. The photometric and spectroscopic survey are modeled as two separate galaxy populations, with a brief description given in the table. The redshift uncertainty given is the Gaussian dispersion and both samples are magnitude limited. For redshift binning, the spectroscopic sample is analyzed using 10 times thinner bins, to properly include radial information.

Asorey et al. (2012), while simplifying the covariance between the surveys and the inclusion of redshift uncertainties.

Redshift Space Distortions (2D) and intrinsic cross-correlations.

Fig. 1 show two effects for narrow redshift bins. Decreasing the redshift bin will increase the RSD effect in the auto-correlations. At $l = 10$, $z = 0.5$ and $\Delta z = 0.01$, the RSD increase the signal with 60%. In addition the linear RSD effect is reaching smaller scales for thinner bins. The right panel show a substantial signal of cross-correlations between adjacent redshift bins. These cross-correlations has sufficient signal-to-noise (paper-I) and contribute significantly to the combined constraints (paper-II). Since this effect is zero in narrow redshift bins, it requires performing the exact calculations.

3. Forecasts

3.1. *Assumptions*

The Fisher matrix forecast use 2D-correlations in Fourier space (Cls) as observable and model the two surveys as separate galaxy populations (see Table 1). Observable are limited $l <= 300$, with additional cuts on low redshift to remove non-linear scales (paper-II) and the fiducial parameters are equal to the MICE simulations (Fosalba et al. 2008). To investigate the benefit of overlap, we define

$$\mathrm{FoM}_S \equiv \frac{1}{\det[F_S^{-1}]} \tag{3.1}$$

10^{-3}FoMγw	Fiducial	fixBias	NoLens	NoRSD	NoBAO
FxB-All	31.5	189	5.86	14.7	21.7
F+B-All	20.8	157	4.69	9.22	13.3
Improvement	1.5	1.2	1.2	1.6	1.6
F-All	2.55	38.4	0.031	2.13	1.95
B-All	6.71	44.1	4.14	2.46	4.27
Improvement	1.2	0.99	1.1	1.1	3.4

FoM γ	Fiducial	xBias	No Lens	No RSD	No BAO
FxB-Counts	43	150	38	15	43
F+B-Counts	35	152	34	4.3	34
Improvement	1.2	0.99	1.1	3.4	1.3

Table 2. Subset of the forecast. The rows marked "All" includes galaxy counts and shear, while "Counts" only include galaxy counts. Here FxB and F+B respectively denote combining the surveys on overlapping and separate areas. On the columns, fiducial is the normal forecast, "xBias" fix the galaxy bias, "noLens" does not include weak lensing, "noRSD" perform all calculations in real space, while "noBAO" disable the BAO in the Eisenstein-Hu power spectrum. All numbers include Planck priors.

which extends the dark energy task force (DETF) FoM. The $FoM_{w\gamma}$ with $S = w0, wa, \gamma$ consider both dark energy and growth of structure, while FoM_{γ} with $S = \gamma$ only focus on the growth. When estimating the FoMs, we marginalize over the DETF parameters and the galaxy bias. The bias model is scale-independent and depend linearly on the redshift. For each population, the bias in each bin is treated as a separate variable and without prior knowledge.

3.2. *Results*

Table 2 show as subset of the forecast results (paper-II). Weak lensing is the most important contribution (FxB-All), with RSD and BAO also contributing significantly. In the forecast, the overlapping surveys increasing the FoM$_{w\gamma}$ with 50%, corresponding to 30% larger area. Comparing the first and second column, the benefit is stronger for an unknown bias, but is still present when fixing the bias. The benefit comes from additional cross-correlations and sample variance cancellation (paper-II and same-sky paper). In addition, the same-sky ratio is higher without RSD or BAO, because they constrains parameter combinations where the overlap otherwise would contribute. This highlights the importance of including all effects in the combined forecast.

References

Asorey, J., Crocce, M., Gaztañaga, E., & Lewis, A. 2012, *MNRAS*, 427, 1891
Bernstein, G. M. & Cai, Y.-C. 2011, *MNRAS*, 416, 3009
Cai, Y.-C. & Bernstein, G. 2012, *MNRAS*, 422, 1045
de Putter, R., Doré, O., & Takada M. 2013, ArXiv e-prints
Font-Ribera, A., McDonald, P., Mostek, N., Reid, B. A., Seo, H.-J., & Slosar, A. 2013, ArXiv e-prints
Fosalba, P., Gaztañaga, E., Castander, F. J., & Manera, M. 2008, *MNRAS*, 391, 435
Gaztañaga, E., Eriksen, M., & Crocce, e. 2012, *MNRAS*, 422, 2904
Kirk, D., Lahav, O., Bridle, S., Jouvel, S., Abdalla, F. B., & Frieman, J. A. 2013, ArXiv e-prints

Statistical Challenges in 21st Century Cosmology
Proceedings IAU Symposium No. 306, 2014
A. F. Heavens, J.-L. Starck & A. Krone-Martins, eds.

doi:10.1017/S1743921314013556

Combining cosmological constraints from cluster counts and galaxy clustering

F. Lacasa

ICTP South American Institute for Research & Instituto de Física Teórica - UNESP
Rua Dr. Bento Teobaldo Ferraz 271, Bloco 2 - Barra Funda
01140-070 São Paulo, SP, Brazil
email: fabien@ift.unesp.br

Abstract. Present and future large scale surveys offer promising probes of cosmology. For example the Dark Energy Survey (DES) is forecast to detect ~300 millions galaxies and thousands clusters up to redshift ~1.3. I here show ongoing work to combine two probes of large scale structure : cluster number counts and galaxy 2-point function (in real or harmonic space). The halo model (coupled to a Halo Occupation Distribution) can be used to model the cross-covariance between these probes, and I introduce a diagrammatic method to compute easily the different terms involved. Furthermore, I compute the joint non-Gaussian likelihood, using the Gram-Charlier series. Then I show how to extend the methods of Bayesian hyperparameters to Poissonian distributions, in a first step to include them in this joint likelihood.

Keywords. galaxies: statistics, methods: data analysis, cosmology, joint probes

---

1. Cross-covariance between cluster counts and galaxy clustering

1.1. *Cluster counts and galaxy angular 2-point function*

The number counts in a bin of mass i_M and redshift i_z, can be considered as a monopole of the halo density field :

$$\hat{N}_{\rm cl}(i_M,i_z) = \overline{N}_{\rm cl}(i_M,i_z) + \frac{1}{\Omega_S}\int \mathrm{d}^2\hat{n}\,\mathrm{d}M\,\mathrm{d}z\, r^2\frac{\mathrm{d}r}{\mathrm{d}z}\,\frac{\mathrm{d}^2 n_h}{\mathrm{d}M\mathrm{d}V}\,\delta_{\rm cl}(\mathbf{x}=r\hat{n}|M,z) \quad (1.1)$$

Cluster counts have been shown as a powerful probe of cosmology, e.g. Planck Collaboration XX (2014) has produced constraint on σ_8 and Ω_m with SZ detected clusters.

The study of the clustering of galaxies may be done with the angular correlation function $w(\theta)$ or its harmonic transform C_ℓ, in tomographic redshift bins :

$$C_\ell^{\rm gal}(i_z,j_z) = \frac{2}{\pi}\int k^2\mathrm{d}k\,\frac{\overline{n}_{\rm gal}(z_1)\,\overline{n}_{\rm gal}(z_2)\,\mathrm{d}V_1\,\mathrm{d}V_2}{\Delta N_{\rm gal}(i_z)\Delta N_{\rm gal}(j_z)}\,j_\ell(kr_1)\,j_\ell(kr_2)\,P_{\rm gal}(k|z_1,z_2) \quad (1.2)$$

In the following I use C_ℓ instead of $w(\theta)$ for simpler equations, although they can be related by a simple linear transformation.

1.2. *Cross-covariance derivation with the halo model*

The cross covariance between these two probes involves the halo-galaxy-galaxy angular bispectrum in the squeezed limit (Lacasa & Rosenfeld, in prep.) :

$$\mathrm{Cov}\left(\hat{N}_{\rm cl}(i_M,i_z), C_\ell^{\rm gal}(j_z,k_z)\right) = \int \frac{\mathrm{d}M_1\,\mathrm{d}z_{123}}{4\pi}\,\frac{\mathrm{d}V}{\mathrm{d}z_1}\,\frac{\mathrm{d}^2 n_h}{\mathrm{d}M\,\mathrm{d}V}\bigg|_{M_1,z_1} b_{0\ell\ell}^{hgg}(M_1,z_{123}) \quad (1.3)$$

$b_{0\ell\ell}$ is a projection of the 3D bispectrum, for which we need a non-linear model. In the framework of the halo model + HOD, I have shown that the bispectrum (or higher orders) can be computed with a diagrammatic formalism (Lacasa *et al.*, 2014). See the diagrams for this hgg bispectrum in the left panel of Fig. 1.

1.3. *Current results*

I have shown that the equations for the covariance can be rewritten in terms of effective quantities, e.g. (Lacasa & Rosenfeld, in prep.) :

$$\mathrm{Cov_{2PT}}\left(\hat{N}_{\mathrm{cl}}(i_M, i_z), C_\ell^{\mathrm{gal}}(j_z, k_z)\right) = \frac{\delta_{j_z,k_z}}{4\pi} \int \frac{\overline{n}_{\mathrm{gal}}(z_2)^2\, \mathrm{d}V_1\, \mathrm{d}V_2}{\Delta N_{\mathrm{gal}}(j_z)^2}\, 4F_{\mathrm{sqz}}\, b_1^{\mathrm{gal,eff}}(k_\ell, z_2)^2$$
$$\rho b_1^{\mathrm{halo,eff}}(i_M, z_1)\, P_{\mathrm{DM}}(k_\ell, z_2)\, \Delta_{0,P}(z_1, z_2) \quad (1.4)$$

These intermediate quantities are integrated over the halo mass and contain the HOD and mass function dependency. They can be compared to measurement on data or to other modeling, providing the possibility for some model independence.

I have built a fast and efficient code to compute the different terms of the covariance ; the plots in Fig. 2 illustrate the numerical results. We see that different terms can become important depending on mass and redshift. The code runs in ~ 1 CPU-second, and is thus adequate to be integrated into an MCMC pipeline.

2. Likelihood

2.1. *Joint non-Gaussian likelihood*

Cluster counts follow a Poissonian distribution (up to sample variance), thus one cannot assume that the joint likelihood of $X = (\mathrm{counts}, w_{\mathrm{gal}}(\theta))$ is Gaussian.
To tackle this, I expanded the joint likelihood with the Gram-Charlier series, around a fiducial independent case. I am then able to resum the expansion into (Lacasa & Rosenfeld, in prep.) :

$$\mathcal{L}(X) = \exp\left[-\sum_{i,j} \langle c_i\, w_j\rangle_c\, (\log\lambda_i - \Psi(c_i+1))\, ({}^T w C^{-1} e_j)\right] \mathcal{L}(\mathrm{counts})\, \mathcal{L}(w) \quad (2.1)$$

This analytic form is well-behaved (positive), can be extended straightforwardly to include sample variance, and has correct asymptotic behaviour at large N_{cl} (Gaussian with the correct covariance matrix).

2.2. *Bayesian hyperparameters*

Hyperparameters (HPs) allow to detect over/underestimation of error bars, or inconsistencies between data sets (see e.g. Hobson *et al.* 2002). The method is at the moment only adapted to Gaussian distributions, thus not for Poissonian cluster counts. It is however mathematically impossible to keep the Gaussian properties of HPs in the Poissonian case (that is, rescaling the variance while keeping the mean). However I found a prescription which approximately respect them. On the right panel of Fig. 1 are shown three pdfs, corresponding to three different values of the Bayesian HP α.
This will allow the use of HPs for the cluster counts - galaxy 2-pt combination, after further extension (sample variance, correlation with w_{gal} as treated in Sect. 2.1).

3. Conclusion and perspectives

I sketched how to combine cluster counts and galaxy 2-pt measurements for increased cosmological constraints, from the physical modeling to the likelihood. In the context of the halo model, I introduced a diagrammatic method allowing elegant computation of

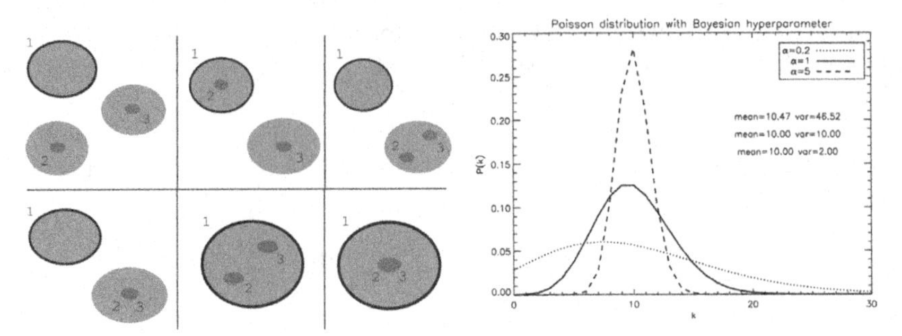

Figure 1. Left : Diagrams for the hgg bispectrum : 3h, 2h_2h, 2h_1h2g, 2h_1h1g, 1h2g and 1h1g. The 3h diagram has two contributions : non-linear evolution of dark matter (2PT), and second-order halo bias (b_2). **Right :** pdfs for a Poisson law and a Bayesian hyperparameter. $\alpha = 1$ corresponds to the original Poisson law (with parameter $\lambda = 10$). Note how the mean is approximately independent of α while the variance scales as $1/\alpha$, as intended.

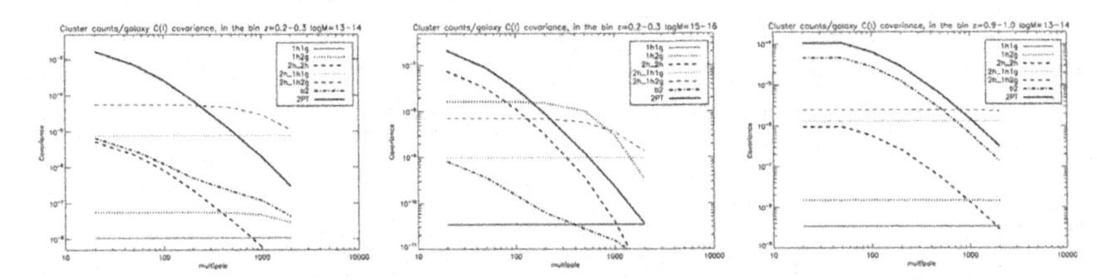

Figure 2. Terms of the covariance for some bins of mass and redshift. **From left to right :** log M=13-14 & z=0.2-0.3 ; log M=15-16 & z=0.2-0.3 ; log M=13-14 & z=0.9-1. The b_2 term can be either positive (high z, galaxies are biased) or negative (low z, galaxies are antibiased). Colored figures are available online as supplementary material.

the equations involved. I derived a non-Gaussian joint likelihood using Gram-Charlier series, and showed how to introduce Bayesian HPs to a Poissonian distribution.

Further work will be necessary to include experimental effects in the covariance and the likelihood : photo-z errors, purity... Next order derivation of the joint likelihood may also be necessary to solve a small bias issue at low $N_{\rm cl}$, and the bayesian hyperparameters method need to be extended to cluster sample variance and correlation with galaxies. In the medium term, I aim to build a full MCMC pipeline of cluster-galaxy combination, for realistic forecasts and application to DES data.

Further details on the model, method, and forecasts will be available in Lacasa & Rosenfeld (in prep.)

4. Figures

Due to the size constraint of this proceeding, I had to cram together the figures below.

References

Planck Collaboration (2014), "Planck 2013 results. XX. Cosmology from Sunyaev-Zeldovich cluster counts", arXiv:1303.5080

F. Lacasa & R. Rosenfeld, "Combining cluster counts and galaxy clustering cosmological constraints", in prep.

F. Lacasa *et al.* (2014), "Non-Gaussianity of the CIB anisotropies - I. Diagrammatic formalism and application to the angular bispectrum", *MNRAS*, Vol 439, p.123-142

M. P. Hobson *et al.*, "Combining cosmological data sets: hyperparameters and Bayesian evidence", *MNRAS*, Vol 335, pp. 377-388

Statistical Challenges in 21st Century Cosmology
Proceedings IAU Symposium No. 306, 2014
A. F. Heavens, J. -L. Starck & A. Krone-Martins, eds.

doi:10.1017/S1743921314011004

How well can the evolution of the scale factor be reconstructed by the current data?

Sandro D. P. Vitenti[1] **and Mariana Penna-Lima**[2]

[1]GReCO, Institut d'Astrophysique de Paris,
UMR7095 CNRS, 98 bis boulevard Arago, 75014 Paris, France
email: dias@iap.fr

[2]Divisão de Astrofísica, Instituto Nacional de Pesquisas Espaciais,
Av. dos Astronautas 1758, 12227-010, São José dos Campos, Brazil
email: mariana.lima@inpe.br

Abstract. Distance measurements are currently the most powerful tool to study the expansion history of the universe without assuming its matter content nor any theory of gravitation. In general, the reconstruction of the scale factor derivatives, such as the deceleration parameter $q(z)$, is computed using two different methods: fixing the functional form of $q(z)$, which yields potentially biased estimates, or approximating $q(z)$ by a piecewise nth-order polynomial function, whose variance is large. In this work, we address these two methods by reconstructing $q(z)$ assuming only an isotropic and homogeneous universe. For this, we approximate $q(z)$ by a piecewise cubic spline function and, then, we add to the likelihood function a penalty factor, with scatter given by σ_{rel}. This factor allows us to vary continuously between the full n knots spline, $\sigma_{rel} \to \infty$, and a single linear function, $\sigma_{rel} \to 0$. We estimate the coefficients of $q(z)$ using the Monte Carlo approach, where the realizations are generated considering ΛCDM as a fiducial model. We apply this procedure in two different cases and assuming four values of σ_{rel} to find the best balance between variance and bias. First, we use only the Supernova Legacy Survey 3-year (SNLS3) sample and, in the second analysis, we combine the type Ia supernova (SNeIa) likelihood with those of baryonic acoustic oscillations (BAO) and Hubble function measurements. In both cases we fit simultaneously $q(z)$ and 4 nuisance parameters of the supernovae, namely, the magnitudes $\mathcal{M}_1$ and $\mathcal{M}_2$ and the light curve parameters α and β.

Keywords. cosmology: miscellaneous, cosmology: observations, cosmology: theory

1. Introduction

The current observational data are still not able to decide between many different models proposed to explain the recent accelerated expansion of the universe. These models include, among others, modified gravitational theories and the addition of an exotic cosmological fluid, in the context of general relativity. As an alternative study, we can make a pure kinematical description of this recent phase of the universe, avoiding the choice of a gravitational theory and matter content. In this description one still models the universe as a metric manifold and uses the cosmological principle restricting the metric to the Friedmann-Lemaître-Robertson-Walker (FLRW) metric. In this case, the deceleration function is given by

$$q(z) = \frac{(1+z)}{H(z)}\frac{\mathrm{d}H(z)}{\mathrm{d}z} - 1, \tag{1.1}$$

where z is the redshift and $H(z)$ is the Hubble function.

Different approaches have been used in the literature in order to reconstruct some kinematic quantities such as the luminosity distance $D_L(z)$, $H(z)$, $q(z)$ and the jerk

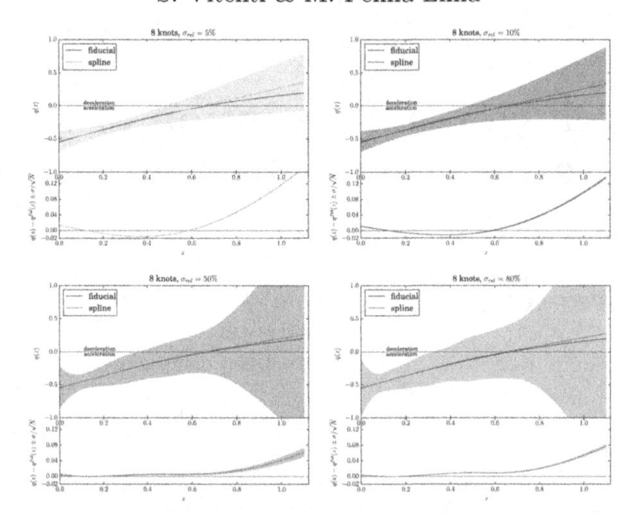

Figure 1. The top part of each panel shows the reconstructed curve of $q(z)$ (spline), computed using the mean values of the estimators, and its 1σ error bar. Each panel displays the result obtained for a given σ_{rel}. The bottom part shows the bias, $q(z) - q^{fid}(z)$, and the 1σ error bar of the mean curve, i.e., $\sigma/\sqrt{N}$.

function. One of these methods assumes *a priori* a functional form of the scale factor $a(t)$ $(1 + z = a_0/a)$, or equivalently $D_L(z)$, $H(z)$ and so on (Riess *et al.* 2004, Visser 2004, Rapetti *et al.* 2007, Lu *et al.* 2011, Shafieloo 2012). This parametric reconstruction is strongly model dependent, so, besides having small variances the results present large biases.

In order to minimize the assumptions on the fitted curve, a second approach is to approximate the kinematic quantity by a piecewise nth-order polynomial function. This "non-parametric" reconstruction is dominated by the over-fitting feature, but it provides small biased estimators (see Daly *et al.* 2008 and Lazkoz *et al.* 2012).

2. Methodology

In this work, we describe $q(z)$ by a piecewise cubic polynomial function, also known as *cubic spline*. Imposing continuous second derivatives, the only free parameters of $q(z)$ are its values at the knots, i.e., $q(z_i) = q_i$, where i runs from 0 to $n-1$ (being n the number of knots). We then address both parametric and non-parametric methods including a penalty factor $P_i(\sigma)$ in the likelihood L, namely,

$$F = -2\ln(L) + \sum_{i=2}^{n-1} P_i(\sigma), \tag{2.1}$$

where the penalty factor is given by

$$P_i(\sigma) = \left((\bar{q}_i - q_i)/(\sigma_{abs} + \bar{q}_i\sigma_{rel})\right)^2, \tag{2.2}$$

$\bar{q}_i = (q_{i-1} + q_{i+1})/2$ and $\sigma_{abs} = 10^{-5}$. Varying the value of σ_{rel}, see Figs. 1 and 2, we are able to recover a full n knots spline (over-fitting dominated), for large σ_{rel}, and a single linear function in the entire redshift interval, for small σ_{rel}, where q_i's are better constrained but they can be biased (if the assumed functional form significantly differs from the true one).

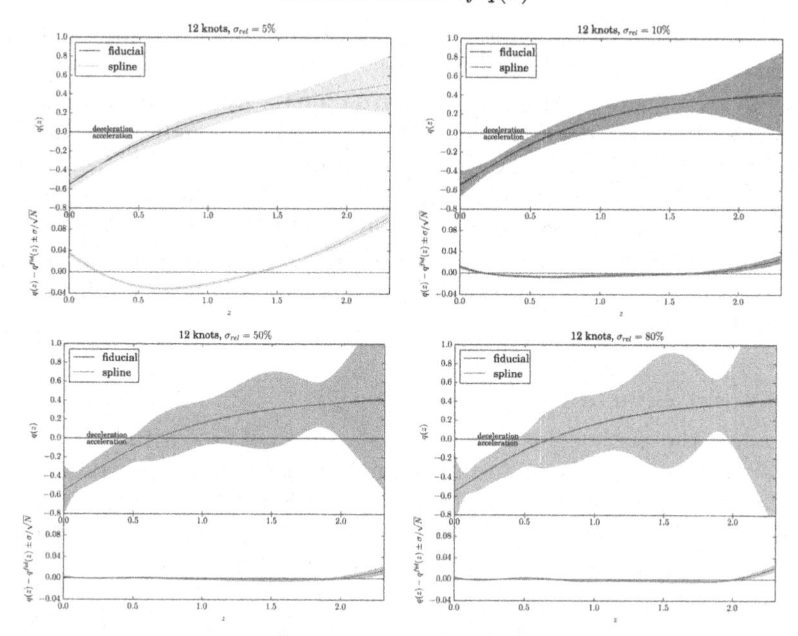

Figure 2. Equivalent to the caption of Fig. 1. These results were obtained considering Eq. 3.1 and 12 knots.

We use the SNeIa sample from SNLS3 (Conley *et al.* 2011 and Sullivan *et al.* 2011) and, thus, the likelihood is build as

$$-2\ln(L_{SNIa}) = \Delta\vec{m}^T \mathsf{C}_{SNIa}^{-1}\Delta\vec{m} + \ln(\det(\mathsf{C}_{SNIa})), \tag{2.3}$$

where $\mathsf{C}_{SNIa} = \mathsf{C}_{stat}(\alpha,\beta) + \mathsf{C}_{sys}$ is the data covariance and

$$\Delta m_i = 5\log_{10}(D_L(z_i^{cmb}, z_i^{hel})) - \alpha(s_i - 1) + \beta\mathcal{C}_i + M_{h_i} + 5\log_{10}(c/H_0) + 25 - m_{Bi}. \tag{2.4}$$

The SNIa phenomenological model contains four parameters α, β, M_1 and M_2, where the first two are related to the stretch-luminosity and colour-luminosity, respectively, and M_1 and M_2 are absolute magnitudes.

We use data from Baryon Acoustic Oscillation (BAO) surveys such as WiggleZ, SDSS and 6dFGRS as described in Hinshaw *et al.* (2013). The BAO likelihood is given by

$$-2\ln(L_{BAO}) = \Delta\vec{d_v}^T \mathsf{C}_{BAO}^{-1}\Delta\vec{d_v}, \tag{2.5}$$

where C_{BAO} is the BAO constant covariance matrix and

$$\Delta d_{vi} = d_{vi} - \frac{r_{rec}}{D_v(z_i)}. \tag{2.6}$$

The BAO depends on the comoving sound horizon at recombination r_{rec} and the geometric estimate of the effective distance $D_v(z_i)$. The last is calculated directly rewriting the distance as an integral of $q(z)$, however, r_{rec} have to be treated as an additional parameter since our model does not define r_{rec}.

The Hubble function likelihood is build as

$$-2\ln(L_{Hubble}) = \Delta\vec{H}^T \mathsf{C}_{Hubble}^{-1}\Delta\vec{H}, \tag{2.7}$$

where C_{Hubble} is the diagonal constant covariance matrix and $\Delta H_i = H_i - H(z_i)$. This likelihood includes the $H(z)$ measurements which were obtained independently of BAO.

The data is summarize, for example, in Zheng *et al.* (2014). We also include the H_0 ($z = 0$) measurement from Riess *et al.* (2011).

We apply the Monte Carlo (MC) approach to obtain the estimators of q_i's, α, β, M_1 and M_2. For this, we first define a fiducial model from which N random realizations will be generated. In particular, we assume a flat universe ($k = 0$) and a ΛCDM cosmological model with $H_0 = 73$ km s^{-1} Mpc^{-1}, $\Omega_m = 0.3$ and $\Omega_\Lambda = 0.7$. For each realization, the best-fitting values of the parameters are obtained by minimizing Eq. (2.1). At each step, i.e., for each resample, the arithmetic mean and the variance of the estimators are computed. This loop ends when the required precision is achieved. In this work, the number of realizations needed varies between 10^4 and 2×10^5.

3. Results and Concluding Remarks

We first apply the methodology considering a spline of $q(z)$ with 8 knots within the redshift interval $[0.01, 1.4]$. We perform this analysis using only the SNLS3 data, i.e., its covariance matrix to generate the realizations, and Eqs. (2.5) and (2.1) to obtain the means and variances of the 9 q_i estimators, α, β, M_1 and M_2. We consider 4 different values of $\sigma_{rel} = 0.05$, 0.1, 0.5 and 0.8. In Fig. 1 we show the reconstructed $q(z)$ function, computed using the mean of the estimators, and its 1σ error bar.

In this second case we consider 12 knots within the redshift interval $[0, 2.3]$. We perform this analysis using the joint likelihood:

$$-2\ln(L) = -2\ln(L_{SNIa}) - 2\ln(L_{BAO}) - 2\ln(L_{Hubble}). \tag{3.1}$$

Fig. 2 displays the results for those 4 different σ_{rel} values.

It is worth emphasizing that the parameters α, β, M_1, M_2 and H_0 are recovered in all cases, i.e. ML estimators for these quantities have negligible biases.

In both figures we plotted the threshold line, $q(z) = 0$, which indicates the deceleration/acceleration regions. In the 8 cases studied, we obtain the indication of an accelerated expansion with more than 1σ confidence level in the interval $0 \leqslant z \lesssim 0.4$.

In this work, we used ΛCDM as the fiducial model, in which $q(z)$ is almost linear. Therefore, in order to obtain a more conservative reconstruction of $q(z)$, it is necessary to apply this study for fiducial models with different functional forms of $q(z)$. Once we get suitable number of knots and σ_{rel}, which work well for all fiducial models, we can apply the method using real data and recover a conservative estimate of the recent expansion history of the universe (Vitenti & Penna-Lima, in prep.).

References

Conley, A., Guy, J., Sullivan, M., Regnault, N., & Astier, P., *et al.* 2011, *ApJS*, 192, 1
Daly, R. A., Djorgovski, S. G., Freeman, K. A., Mory, M. P., & O'Dea, C., *et al.* 2008, *ApJ*, 677, 1
Hinshaw, G., Larson, D., Komatsu, E., & Spergel, D. N. *et al.* 2013, *ApJS*, 208, 19
Lazkoz, R., Salzano, V., & Sendra, I. 2012, *Eur. Phys. J. C*, 72, 2130
Lu, J., Xu, L., & Liu, M. 2011, *Phys. Lett. B*, 699, 246
Rapetti, D., Allen, S. W., Amin, M. A., Blandford, R. D. 2007, *MNRAS*, 375, 1510
Riess, A. G., Strolger, L-G., Tonry, J., Casertano, S., Ferguson, H. C., *et al.* 2004, *ApJ*, 607, 665
Riess, A. G., Macri, L., Casertano, S., Lampeitl, H., Ferguson, H. C., *et al.* 2011, *ApJ*, 730, 119
Shafieloo, A. 2012, *JCAP*, 8, 2
Sullivan, M., Guy, J., Conley, A., Regnault, N., Astier, P., *et al.* 2011, *ApJ*, 737, 102
Turner, M. S. & Riess, A. 2002, *ApJ*, 569, 18
Visser, M. 2004, *Classical Quant. Grav.*, 21, 2603
Zheng, W., Li, H., Xia, J. Q. *et al.* 2014, *Int. J. Mod. Phys. D*, 23, 50051

Statistical Challenges in 21st Century Cosmology
Proceedings IAU Symposium No. 306, 2014
A. F. Heavens, J.-L. Starck & A. Krone-Martins, eds.

doi:10.1017/S1743921314013593

Precision cosmology, Accuracy cosmology and Statistical cosmology

Licia Verde[1,2]

[1] ICREA & Instituto de ciencias del Cosmos, Iniversitat de Barcelona ICC-UB IEEC
[2] Institute of Theoretical Astrophysics, University of Oslo, 0315 Oslo, Norway
email: liciaverde@icc.ub.edu

Abstract. The avalanche of data over the past 10-20 years has propelled cosmology into the "precision era". The next challenge cosmology has to meet is to enter the era of accuracy. Because of the intrinsic nature of studying the Cosmos and the sheer amount of data available now and coming soon, the only way to meet this challenge is by developing suitable and specific statistical techniques. The road from precision Cosmology to accurate Cosmology goes through statistical Cosmology. I will outline some open challenges and discuss some specific examples.

Keywords. methods: data analysis, methods: statistical, Cosmology: cosmological parameters, cosmology: large-scale structure of universe

---

1. Introduction

Cosmology in the past twenty years or so has made the transition to *precision cosmology*. Cosmological parameters that were known only within an order of magnitude are now measured with percent precision. This transition was brought about by the avalanche of data provided by massive large-scale structure surveys and ambitious mapping of the Cosmic Microwave Background (CMB) radiation. Thanks to this, over the past twenty years, cosmology has made the transition from a data-starved science to a data driven science. As a result, cosmology has now a standard (base, or ΛCDM) model. The standard model for cosmology requires only a handful of parameters to describe the origin, composition and evolution of the entire Universe. However, there is a big difference between modelling and understanding. While the standard cosmological model works extremely well it is highly unsatisfactory, as it has many ingredients we do not understand and we know it is incomplete. This is driving virtually all the experimental efforts in the field of the near future. From the modelling point of view, there are many plausible extensions of the base model, where in practice one or more parameters are being added. In these extended models the "precision" even of the base parameters gets significantly degraded.

In this Cosmology is special among the experimental sciences, but the peculiarity of Cosmology goes deeper than that.

2. Cosmology is special

As an experimental science Cosmology has a peculiarity in that we cannot make controlled experiments. When we mention experiments we should always bear in mind that we can only make observations†. And we only have one observable Universe. This is what I call the curse of cosmology. There are two important consequences of this limitation. One is that all we can do is to fit models to observations. To be more precise we constrain

† I have seen this limitation only in disciplines like the study of the effect of nuclear fallout from "unlikey" (and unlucky) events such as accidents with power plants and bombs.

numerical values of the model's parameters using observations; very few quantities are direct measurements: therefore any statement is model-dependent. To make matters worst, non-linearities (which are very hard to model) and poorly known astrophysical processes (*gastrophysics*, for aficionados) get in the way. As a result, different observations are more or less "clean" and robust and more or less trustable. It is however somewhat a question of personal taste which I like to compare to the Standard & Poor's credit rating for countries (enough said).

Cosmological results (and constraints on cosmological parameters) therefore depend not only on the adopted cosmological model but also on the data sets one is (willing to) consider. Therefore there is not one, unique, determination from cosmology of a given quantity (cosmological parameter). This can be quite confusing at first for those not working in the field.

Like every cloud has a silver lining, the curse of cosmology is also a blessing: we might have only one observable Universe and we might not perform controlled experiments on it, but we can –at least in principle– observe all there is to see. This is what I call "ultimate experiment" in cosmology. The big development for the near future (which has however already started in the past few years) is that advances in observational techniques have made possible to perform ultimate experiments and the next generation of large-scale structure surveys will be a collection of "ultimate experiments". For example the *Planck* satellite (Planck collaboration, 2013a, Planck collaboration, 2013b) has provided us with the "ultimate experiment" for the Cosmic Microwave Background *primary* temperature anisotropies. A mission like the proposed *COrE+*(COrE collaboration, 2012; PRISM collaboration, 2014) will provide the ultimate experiment for CMB lensing and polarisation – both E modes and B modes– signal.

The other important consequence of having only one observable Universe is more subtle. In Cosmology we believe that the observed Universe is only one part of the entire Universe, which is seen as an *ensemble* of all possible observ-ed/able Universes of which the observed one is a random draw. Ambitiously, from the limited observations we can gather, we want to make *inferences* about the properties of the entire Universe. In this sense, statistics is very much ingrained in cosmology: any theoretical model will mostly predict the statistical properties of the Universe. The observations are intrinsically a statistical quantity. There is therefore a fundamental error floor for any measurement given by cosmic variance. No wonder many cosmologists are Bayesian!

3. Challenges

The avalanche of data of the past twenty years has brought about challenges, which the community was able to address and solve. The next generation of surveys will push the field in to the "big data" era. Undoubtably this will bring in "big challenges" but, given that so far the community has a good track record on this, I will not focus on this aspect here.

I will instead dwell on the fact that with more data available, making the statistical error shrink, the systematic errors must be kept under exquisite control. Here we are entering uncharted territory. While there is a at least well defined framework to deal with statistical errors, there is no systematic way to address systematic errors.

There are several types of systematic errors, and here is my personal view (paraphrasing Rumsfeld).

- There are things that we know that we know. But what if what we think we know isn't true? Recall that any statement in cosmology is model dependent. "Essentially, all models are wrong, but some are useful" (Box and Draper, 1987).

- Known unknowns. These are the systematic effects that are probably easier to deal with: at least we know what we have to watch out for. Below I will give an example of how to deal with these.

- Unknown unknowns. Here the general wisdom is to try to measure the same thing in different, independent ways (e.g., different data) and compare. It might be a very expensive route, but at the moment it is the only option.

To summarise, after precision cosmology, cosmology should strive to become accurate. This is not news, see Peebles (2002). Systematic effects may be in the data but may also be in the model used for their interpretation. In what follows I will present a small selection of examples where the application of statistical techniques can help in the transition from precision to accurate cosmology. This selection is not meant to be exhaustive or representative, it just cover some of the problems I have been working on with my collaborators over the past couple of years.

3.1. *Example 1: Being Bayesian with non-Gaussianity*

The search for deviations from primordial non-Gaussianity is a very active research subject: any deviation from the simplest implementation of slow-roll inflation implies primordial non-Gaussianity, possibly at a detectable level. In addition, even if the initial conditions were Gaussian, non-linear gravitational evolution creates non-Gaussianity. Despite the fact that Bayesian methods are routinely employed in cosmology, the state-of-the-art non-Gaussianity analyses are done in the Frequentist framework: a bispectrum estimator is employed which has been shown to be optimal but only in the limit of vanishing non-Gaussianity.

The problem in applying the Bayesian method in this context is that a full, analytic probability distribution function does not exist even for the simplest non-Gaussian model. The available literature on this has concentrated on the *local* type of non-Gaussianity where the field of interest Φ is expressed in terms of a Gaussian auxiliary field ϕ and a dimensionless non-Gaussianity parameter $f_{\rm NL}$:

$$\Phi(\mathbf{x}) = \phi(\mathbf{x}) + f_{\rm NL}\left[\phi^2(\mathbf{x}) - \langle\phi^2(\mathbf{x})\rangle\right] . \tag{3.1}$$

Recall that $\langle\phi^2(\mathbf{x})\rangle = \sigma_\phi^2$. Here we also consider this model. The goal is to find the posterior distribution of the amplitude of non-Gaussianities given the data, $P(f_{\rm NL}|d)$; this includes information from all correlation orders not just the bispectrum. This has been attempted by Elsner and Wandelt (2010), Elsner *et al.* (2010), Ensslin *et al.* (2008), here I follow the approach of Verde *et al.* (2013). I will only outline the philosophy of the approach and explain the implications of the findings, the detailed expressions and equations may be found in Verde *et al.* (2013).

To understand the set up, let us warm up by re-deriving known results in the Gaussian case.

Let us define a field A, which is the observable field and can be the temperature of the CMB or the a_ℓ^m or the matter overdensity field etc. The value of the observed field at any spatial point i is related to the underlying potential Φ via

$$A(\mathbf{x}) = A_i = \int d^3y \mathcal{M}(\mathbf{x},\mathbf{y})\Phi(\mathbf{y}) \equiv (\mathcal{M},\Phi) . \tag{3.2}$$

The full information about the Gaussian field ϕ is given by the Gaussian generating functional $\mathcal{P}[\phi]$; for example the joint multivariate probability of $A_1, ..., A_n$ is

$$\mathcal{P}(A_1, A_2, ...A_n) = \int [\mathcal{D}\phi]\, \mathcal{P}(\phi) \prod_{i=1}^{n} \delta^D\left[A_i - (\mathcal{M},\Phi(\phi))\right] . \tag{3.3}$$

Here,

$$\mathcal{P}(\phi) = \frac{\exp[-1/2(\phi, K_0, \phi)]}{\int D[\phi] \exp[-1/2(\phi, K_0, \phi)]} \tag{3.4}$$

and we have used the shorthand notation

$$\int d^3y\, d^3z\, \phi(\mathbf{y}) K(\mathbf{y}-\mathbf{z}) \phi(\mathbf{z}) \equiv (\phi, K, \phi), \tag{3.5}$$

with K_0 defined by the relation to the 2-point correlation function of ϕ, ξ^ϕ:

$$\int d^3y K_0(|\mathbf{x}-\mathbf{y}|) \xi^\phi(|\mathbf{y}-\mathbf{z}|) = \delta(|\mathbf{x}-\mathbf{z}|). \tag{3.6}$$

This is a very powerful approach although sometimes it is tricky to deal with the Dirac delta function which Fourier representation involves highly oscillatory functions. On the other hand, observations are often discretized/pixelized, so we discretize these integrals to simplify their analytic calculation, then, if needed, one take the continuous limit at the end. All the above convolutions then become just matrix operations.

For example it is easy to see that if $\mathcal{M}$ is invertible and Φ is Gaussian then A is also Gaussian with covariance given by the covariance of A: $\hat{K}_0 = \left(\mathcal{M}^{-1}\right)^T K_0 \mathcal{M}^{-1}$ (matrix multiplication) and $[\hat{K}_0^{-1}]_{ij} = \xi^A_{ij}$.

If $\mathcal{M}$ is not invertible (say Φ is the three dimensional gravitational potential and the observable field A is the CMB temperature spherical harmonic coefficients a_ℓ^m) it is still possible to split it in two blocks $\mathcal{M} = (\aleph \quad \mathcal{N})$ where the $\aleph$ block is invertible. Then standard linear algebra yields the known results i.e., that the A (i.e., a_ℓ^m) field is Gaussian with correlation function given by $\mathcal{M}^T \xi^\Phi \mathcal{M}$.

The moment a nonzero $f_{\rm NL}$ is introduced, things complicate. It is instructive to first consider the case where $\mathcal{M}$ is the identity matrix. Then it is well known that in this case one can solve the Dirac delta function in Eq. (3.3) by finding its roots. Neglecting the exponentially suppressed root (which is an excellent approximation over virtually all the support of the Probability Density Function, PDF) it is possible to write down the exact multi-variate PDF, $\mathcal{P}(\Phi)$.

Of course in the more general case for $\mathcal{M}$, once one has the expression for $\mathcal{P}(\Phi)$, one could simply say:

$$\mathcal{P}(\vec{A}|f_{\rm NL}) = \int \mathcal{D}[\Phi] \mathcal{P}(\Phi) \delta^D(A - \int M\Phi)\,. \tag{3.7}$$

This then must be integrated numerically via Monte-Carlo methods, as for example explored in Elsner *et al.* (2010), Elsner and Wandelt (2010). This is computationally extremely intensive, as the dimensionanity of the integrand is given by Φ it is not that of the observable field. We can attempt to proceed further analytically. Of course if $\mathcal{M}$ is invertible (e.g., A is the matter over density field) one could do a suitable transformation on the observables:

$$A_b \longrightarrow \sum_a (\mathcal{M}^{-1})_{ba} A_a \equiv \tilde{A}_b\,, \tag{3.8}$$

and get a nice closed expression. In the general case when $\mathcal{M}$ is not invertible it is possible to further simplify the resulting expressions, to yield a (still) highly-dimensional integral but where the dimensionality is that of the observed field A. What remains to perform numerically is still computationally intensive especially for applications such as mega-pixel CMB maps. Still, compared to Eq. 3.7, the dimensionality has been dramatically reduced, for example, by a factor $\mathcal{O}(10^6)$, for a full sky CMB map from current experiments!

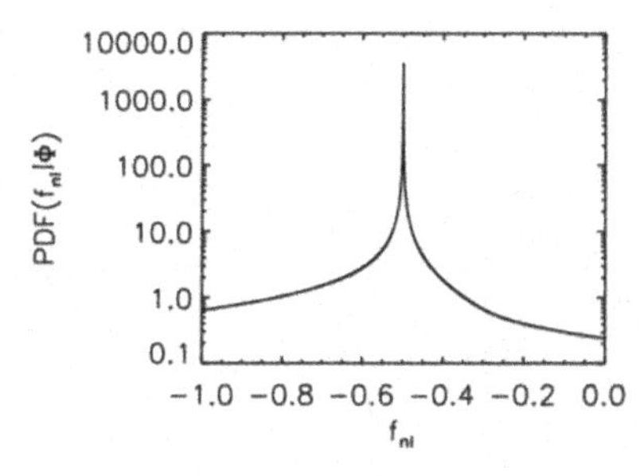

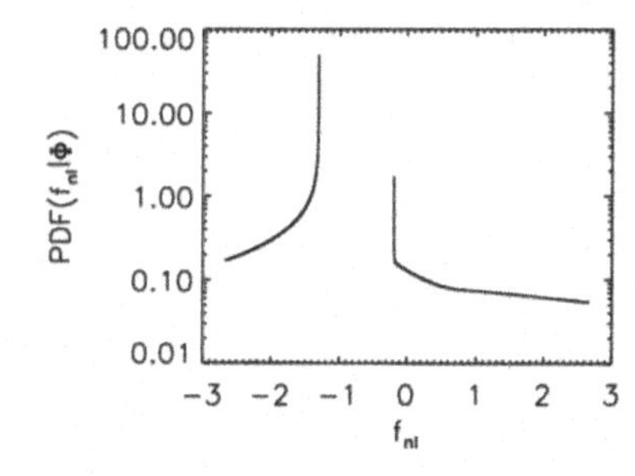

Figure 1. PDF for the non-Gaussianity parameter $f_{nl} = f_{\rm NL}\sigma_\phi$ given a value of the field Φ: left panel $\Phi = 1$, right panel $\Phi = 1.5$. Note the discontinuity on the left panel and, on the right panel, the sharp divergencies at the edge of a region where the PDF is zero.

Before proceeding further it is important to understand the following warning and its consequences. So far we have been aiming at constructing $\mathcal{P}(\Phi|f_{\rm NL})$. Assuming a uniform prior on $f_{\rm NL}$, this can be interpreted as $\mathcal{P}(f_{\rm NL}|\Phi)$. However $\mathcal{P}(f_{\rm NL}|\Phi)$ has discontinuities which prevent any perturbative expansion. Only well away from these these discontinuities the moments are defined and the central limit theorem applies. In particular while in $\mathcal{P}(\Phi|f_{\rm NL})$ $f_{\rm NL}$ is fixed and can be taken to be small, and maybe there one might find that perturbative expansions work well, in $\mathcal{P}(f_{\rm NL}|\Phi)$ the values of $f_{\rm NL}$ are not bounded and can (and do) get very large. Fig. 1 shows an example of a 1-dimensional PDF for $f_{nl} = f_{\rm NL}\sigma_\phi$, for different values of Φ.

Discontinuities are clearly visible. The support of the PDF depends on the value of the parameter $f_{\rm NL}$: the PDF is zero for $\Phi < -f_{nl} - 1/(4f_{nl})$. This creates many problems. For example, one could try to find a maximum likelihood estimator (rather than working with the full PDF). But because of this the Cramer-Rao bound is invalid. This sharp discontinuity will also prevent any truncated expansion of the PDF being a good approximation.

The above discussion is valid for a single pixel, but it has consequences when many pixels (measurements) are combined. The full posterior when n independent pixels can be computed analytically and is shown in the left panel of Fig. 2: the PDF is zero for $f_{nl} > f_{nl}^{true}$. This sharp discontinuity will also prevent any truncated expansion of the PDF being a good approximation.

Under some conditions however, the chances of reaching the "excluded" regions can be made vanishingly small, especially for small f_{nl}, for example because of sampling or in the presence of noise as shown in the right panel of Fig. 2. The cutoff disappears also if we now consider that for a sample of finite size the tails of the distribution are not well sampled. For example for for $f_{nl} = 0.05$ the "excluded" region is not sampled if the sample is smaller than about 1.7×10^6. This explains why for small values of $f_{\rm NL}$ and for realistic surveys the popular Non-Gaussianity estimator (which is a maximum likelihood estimator only if $f_{\rm NL}$ is zero) works. This also explains why, at the end of the day, perturbative expansions could work. Note however that a truncation to first order will yield to a PDF that does not have a maximum, which therefore cannot be a good approximation. At the minimum to have a maximum the approximated PDF must be truncated at second order or higher in $f_{\rm NL}$.

Encouraged by this finding, we proceed in finding a useful approximation to the full PDF based on a second-order Edgeworth expansion (i.e., for small deviations from Gaussianity). The full expression can be found in Eqs. 4.28 and 5.2 of Verde *et al.* (2013), and we will not report it here in full. It will suffice to say that it involves the three- and four-point functions and that, when written explicitly for the CMB a_ℓ^m, one can recog-

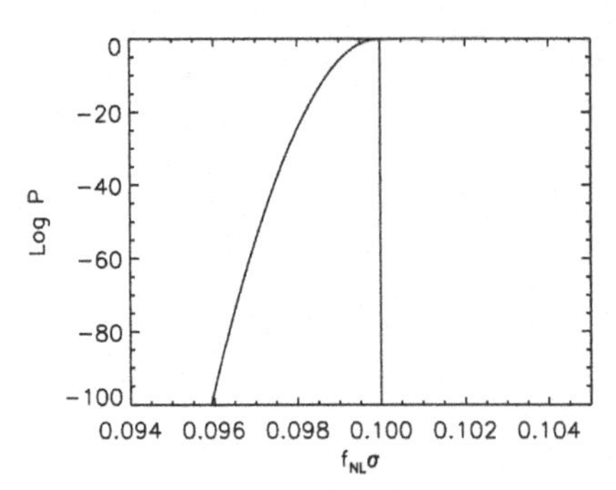

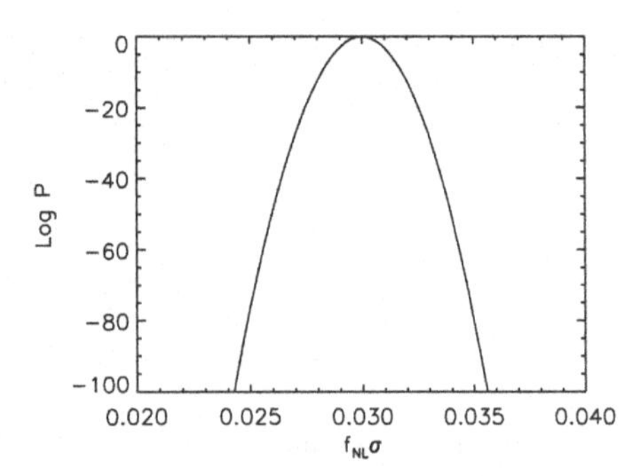

Figure 2. Left panel: Log(Posterior) for f_{nl}, with true $f_{nl}^{true} = 0.1$ and 10^6 pixels but ignoring sampling issues and noise. $f_{nl} > f_{nl}^{true}$ is excluded, as the true PDF is non-zero for some values of Φ where the trial PDF is zero. Right panel: Log(Posterior) for f_{nl} given 10^6 pixels, and $f_{nl}^{true} = 0.03$. A Gaussian noise of rms $0.01\sigma_\phi$ is added. The noise makes all values of Φ reachable in principle, and so the posterior is always non-zero. This removes the strict cutoff apparent in the left panel.

nise the optimal bispectrum estimator (see e.g., Mangilli and Matarrese's contributions) and the trispectrum estimator of Regan *et al.* (2010). The expression is valid for non-Gaussianities more general than the local form as long as departures from Gaussianity is small. This opens up the possibility to extent the work beyond the CMB and possibly to galaxy surveys where, however, non-Gaussianity is not strictly small so the applicability of the expansion would need to be verified.

3.2. *Example 2: Are two (or more) measurements in agreement?*

Let us imagine we have performed two measurements (A and B) of cosmologically interesting quantities in the form of a two or higher dimensional posterior distribution. In the Bayesian framework, how would one quantify whether these two measurements are or not in agreement (tension)? (see Fig.3).

In other words, if the null hypothesis is that the two measurements are sampled from the base model adopted, when should the null hypothesis be abandoned? If the answer is that the two measurements are in tension, Bayesian model selection can be used to study extensions to the base model adopted and select which is the favoured model. Alternatively the detected tension might indicate the presence of unaccounted for, residual systematic errors (e.g., "unknown unknowns"). Possible options at this point are: discredit the measurement most likely affected by systematics or artificially increase its errors. If instead no tension is detected, the measurements can be combined to perform, for example, joint parameter estimation. This has been investigated for example in Verde *et al.* (2013b) which is the approach I will outline here but see also Marshall *et al.* (2016), March *et al.* (2011). For each measurement A, B, we produce a posterior $P_{A,B}(\theta|D_{A,B})$ where θ represents the parameters of the model and $D_{A,B}$ represents the data from experiments A, B respectively. Let us also assume that for producing both posteriors we have used the same, uniform priors, π, over the same support, x, i.e., $\pi_A = \pi_B = \pi$, $\pi = 1$ or 0 and therefore $\pi_A\pi_B = \pi$. Let H_1 be the (null) hypothesis that both experiments measure the same quantity, the models are correct and there are no unaccountable errors. In this case, the two experiments will produce two posteriors, which, although can have different (co)variances, and different distributions, have means that are in agreement. The alternative hypothesis, $H_{\neg 1}$ is when the two experiments, for some unknown reason, do not agree, either because of systematic errors or because they are effectively measuring different things or the model (parameterization) is incorrect. In

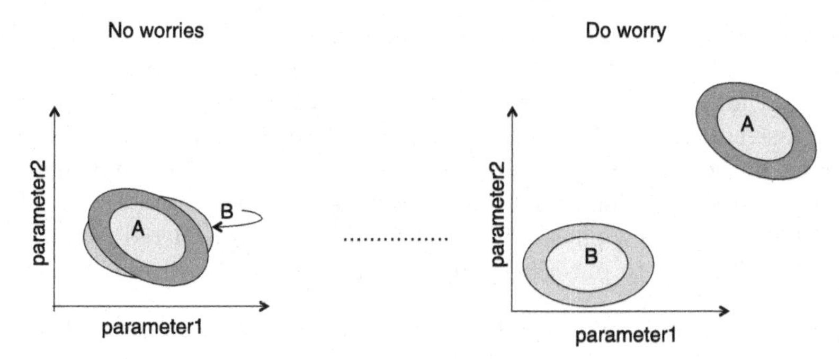

Figure 3. Schematic representation of the set up. Although the contours are reminiscent of Gaussians 1 and 2 σ confidence regions, the argument applies to any form for the distributions. The left hand side situation can be taken as the reference ("straw man") null hypothesis, a "just so" scenario. On the right hand side a situation where clearly the two measurements are not in agreement, but there is a continuum between these two cases, and we would like to find a quantitative scale for it (and know at what point the two measurements are not consistent).

this case, the two experiments will produce two posteriors with two different means and different variances.

To distinguish the two hypothesis we use the Bayes factor, that is the ratio of the Bayesian Evidences (see D. Mortlock contribution and references there). In any practical application, the absolute normalization of the posteriors is often unknown, but we can still work as follows. We define:

$$\int P_A P_B dx = \lambda \int \mathcal{L}_A \mathcal{L}_B \pi_A \pi_B dx = \lambda \int \mathcal{L}_A \mathcal{L}_B \pi dx = \lambda E = \mathcal{E} \,, \tag{3.9}$$

where $\mathcal{L}$ denotes the likelihood and $\lambda^{-1} = \int \mathcal{L}_A \pi_A dx \int \mathcal{L}_B \pi_B dx'$. E is the Bayesian Evidence for the joint distribution, thus $\mathcal{E}$ is akin to an unnormalized Evidence.

However we can consider a "straw man" null hypothesis where the maxima of the two distributions coincide. For example, imagine that we perform a *translation* (shift) of (one or both of) the distributions in x and let us define $\bar{P}_A$ the shifted distribution. Eq. (3.9) becomes

$$\int \bar{P}_A \bar{P}_B dx = \bar{\mathcal{E}}|_{\max A = \max B} \,. \tag{3.10}$$

This translation changes the location of the maximum but does not change the shape or the width of the distribution. Clearly the Evidence ratio for the (null) hypothesis E_1 is $\mathcal{E}/\bar{\mathcal{E}}|_{\max A = \max B}$, as the normalization factors λ cancel out, and the Evidence ratio for the alternative $H_{\neg 1}$ is its reciprocal. We therefore introduce:

$$\mathcal{T} = \frac{\bar{\mathcal{E}}|_{\max A = \max B}}{\mathcal{E}} \,, \tag{3.11}$$

which denotes the degree of tension that can be interpreted in the widely used Jeffrey's (Jeffreys, 1973; Kass and Raftery, 1995) scale. $\mathcal{T}$ indicates the odds: 1 : $\mathcal{T}$ are the chances for the null hypothesis. In other words, a large tension mens that the null hypothesis ($\max A = \max B$) is unlikely. $\mathcal{E}$ can be seen as the evidence for the joint distribution where we interpret one data set as the prior. Then this is normalised by a "just so" scenario: $\bar{\mathcal{E}}|_{\max A = \max B}$.

Of course, and as already mentioned, the obvious application of this is in searching for either new physics or systematics of the "unknown unknowns" type. But, is there a way to distinguish new physics from systematics? As far as I know there is no systematic treatment for that but some insights can be garnered by considering more than two measurements. Let us take for example the recent collection of claims of non-zero neutrino mass that have appeared in the literature and follow the approach of Leistedt *et al.* (2014). The authors argue that the need for extra parameters describing new physics beyond the base cosmological model, yielding a new cosmological concordance can only be convincing if the combined datasets are in tension in the minimal model, and in agreement in extended model. If the tension remains in the extended model it is due to systematic effects (or to new physics different from the one considered). In my view more work remains to be done to put all this into a consistent framework.

3.3. *Example 3: Bias (clustering of peaks of a Gaussian field)*

Most theoretical models predict the statistical properties of the distribution of dark matter. Unfortunately most of the observations rely on objects that "light up" (such as galaxies) which may not be faithful tracers of the dark matter distribution. This is known as (galaxy) bias: the clustering properties of the observed (tracers) field are not the same as those of the dark matter. For most applications the bias is assumed to be scale-independent or to be very slowly varying with scale. A strongly scale-dependent bias would be very problematic for the interpretation of large-scale structure data. But galaxies are expected inhabit dark matter halos which are believed to correspond to high peaks of the *initial* density field. The initial distribution is expected to be very close to Gaussian, thus modelling the clustering properties of peaks of a Gaussian distribution is of great interest. Attempts to address this go back to Otto Politzer and Wise (1986). We can start by considering the expression for the (joint) N points probability distribution of peaks above a threshold t of a Gaussian field $\phi(\mathbf{r})$ which can be written using the Gaussian path integral:

$$P(\mathbf{r}_1,\ldots,\mathbf{r}_N) = \int [d\phi(\mathbf{r})]P[\phi(\mathbf{r})] \times \tag{3.12}$$
$$\prod_{j=1}^{N}\left[\int dw_{(j)}|\det w_{(j)}|\delta^3(\nabla\phi(\mathbf{r}_j))\delta^6(\nabla\nabla\phi(r_j) - w_{(j)})\theta(\phi(\mathbf{r}_j) - t)\right].$$

In this equation $P[\phi]$ is the Gaussian probability distribution function, θ denotes the Heaviside step function and $w_{(j)}$ in three spatial dimensions is the symmetric 3×3 matrix of the second derivatives of ϕ at position $\mathbf{r}_j$ and δ^3, δ^6 denote Dirac delta functions. In Eq. (3.13) the integration on $dw_{(j)}$ has to be extended only over negative definite eigenvalues, in order to identify local maxima. Then the correlation function ξ_N is given by $1+\xi_n = P(\mathbf{r}_1..\mathbf{r}_N)/P^N(\mathbf{r})$ where $P(\mathbf{r})$ is known (e.g., Bardeen *et al.*, 1986). Eq. 3.13 might look innocuous but it is extremely complicated, no exact analytic solution has been found beyond one spatial dimension, despite extensive studies in the literature over a number of years, staring with the pioneering works of Rice(1944), Rice(1945), Adler(1981), Bardeen *et al.* (1986), Peacock and Heavens (1985), Kaiser(1984), Jensen & Szalay(1986). Here I follow the approach of Verde *et al.* (2014) and report their findings. They argue (and show on simulations) that virtually all extrema are peaks above a not too high threshold, therefore one could simply compute the properties of extrema rather than peaks without introducing significant systematics. This gets rid of the (immensely complicated to deal with) six dimensional Dirac delta function.

For a high threshold the value of the integral above the threshold can be approximated by the value of the function at the threshold. Thus the integral involving the threshold can be dropped (or performed eventually numerically if needed).

At this point the residual hurdle is the presence of the $|\det w|$ in the integrand. However this could be regarded as a weight and so one could simplify it (or drop it altogether) if

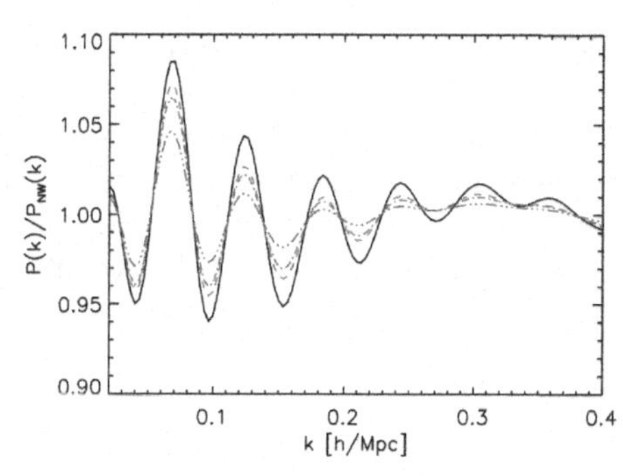

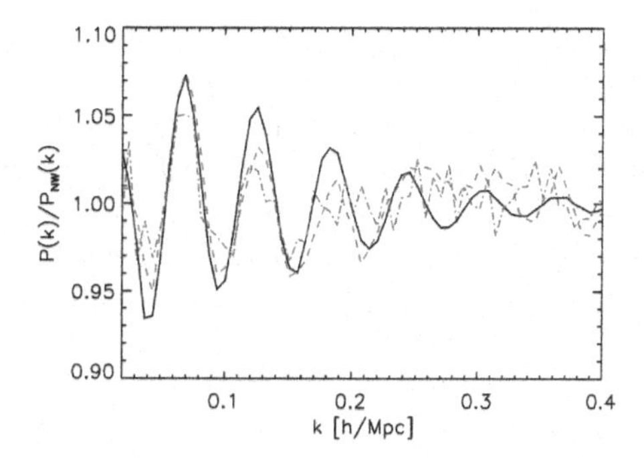

Figure 4. The BAO feature as $P(k)/P_{\mathrm{NW}}$. The left panel shows the theory prediction obtained by Fourier transforming eq.3.13, the right panel shows the mean of 20 Gaussian realisations. Solid/Black for matter, green/dashed for 2σ peaks, red/dot-dashed for 3σ peaks; purple/dashed-dot-dot-dot for 4σ peaks.

the data could be suitably weighted to compensate. To begin with, $|\det w|$ was simply substituted by unity, leaving to forthcoming work to find a weighting scheme for the observations that can compensate for this.

With these simplifications it is possible to find an analytic expression at least for the two-point function:

$$P'(\mathbf{r}_1, \mathbf{r}_2, m_1, m_2) = \frac{2\pi}{\det X'} \frac{(2\pi)^{1/2}}{(\det H')^{1/2}} \times \frac{2\pi}{\det(\xi - Q^T H'^{-1} Q)} \exp\left[-\frac{1}{2} m^T (\xi - Q^T H'^{-1} Q)^{-1} m\right], \quad (3.13)$$

where m_1 and m_2 denote the thresholds, and Q, H' and X' are matrices that can be computed from the two point function of the dark matter, see Verde *et al.* (2014) for the explicit expressions. This might look ugly, but it is not. An important implication of this result is that, since the matter power spectrum is not a featureless power law the tracers' bias as function of scale $b(r)$, defined from the ratio of the correlation functions of tracers to that of matter, is not scale independent, even at linear scales. In particular, a new scale-dependent feature is found in the bias (a "bump") which is located very near the Baryon Acoustic Oscillation feature. The BAO feature corresponds to the (local) minimum of $b(r)$ at around $r = 110$ Mpc/h. The presence of the BAO signal in the matter power spectrum introduces a changing first and second derivative of the correlation function which are responsible for non-negligible effects. In other words, selecting peaks of a Gaussian field is a highly non-linear operation which creates a highly non-Gaussian peaks field. The essence of non-Gaussianity is mode coupling which, by moving power across scales, tends to move around, distort or even erase localized features. In the BAO case the localised scale-dependence of the bias does not move the BAO feature but reduces its amplitude. The smoothing is more marked for higher thresholds as seen in Fig. 4.

The figure shows the ratio of the power spectrum divided by a power spectrum without the BAO feature as a function of the threshold. Note that the BAO feature is smoothed and the smoothing increases with the threshold height. While it is reassuring that this effect does not change the location of the BAO feature, it might have nevertheless important practical implications. The signal-to-noise for measurements that depend on the BAO location is usually computed adopting a model with a fixed BAO smoothing parameter interpreted as the one generated by non-linearities. Should this be underestimated, as the effective one is a combination of the effect of non-linearities and this new effect, then the signal-to-noise would be overestimated.

There are consequences also for survey selection considerations: highly biased tracers are preferentially targeted to beat shot noise when designing a BAO survey –in technical terms, to maximise nP where n denotes the tracer number density–. In fact given a finite amount of observing time and a finite aperture telescope the selection of bright, highly biased objects yields the best signal-to-noise. However highly biased tracers will have a reduced BAO feature: it may be advantageous to select less rare and less biased tracers if they carry a more pronounced signal.

3.4. *Example 4: Cancelling out systematics*

Suppose we have N observables (data points) O_i, $i = 1, .., N$ that depend on m interesting quantities μ_i and n nuisance quantities ν_i which can possibly introduce systematic errors on μ_i if fixed at incorrect values. The usual approach is to treat systematics as nuisance parameters and marginalise. However this might not be satisfactory especially if we are ignorant about the mean values and the errors of the nuisance parameters. There are some example in the literature on how to deal with this situation but not a lot of work on a systematic approach. Here I outline the approach proposed by Noreña *et al.* (2012).

Our goal should be to find combinations of the observables O_i that are insensitive to the nuisance parameters ν_i. In other words we must find a set of $M = N - n$ functions

$$f_k = f_k(O_1, .., O_N) \tag{3.14}$$

where $k = 1, .., M$, so that

$$\frac{df_k}{d\nu_i} = \sum_{j=1}^{N} \frac{\partial f_k}{\partial O_j}\frac{\partial O_j}{\partial \nu_i} = 0 \text{ for } i = 1, .., n\,. \tag{3.15}$$

A natural interpretation of this is the *renormalisation group* equation (Wilson & Kogut,1974). Of course one must know how O_i depend on ν_i so this can apply only to "known unknowns". Let us consider for example a power-law dependence on nuisance parameters ν_i:

$$O_i - \widehat{O}_i = g(\vec{\mu})\prod_{j=1}^{n}(\nu_j - \widehat{\nu}_j)^{\alpha_{ij}} \tag{3.16}$$

where the $\widehat{}$ symbol denotes the true value and g is some function of $\vec{\mu}$, a vector containing all other quantities on which the observables depend. Then the solution for the system of differential equations, Eq. 3.15, is of the form:

$$f_k = \prod_{i=1}^{N}(O_i - \widehat{O}_i)^{b_i^k} \tag{3.17}$$

which gives a system of linear algebraic equations for the unknown b_i^k. If $M = N - n > 0$, i.e., there are more data than nuisance parameters, then there are m non -trivial solutions:

$$\sum_{i=1}^{M} \alpha_{ij} b_i^k = 0\,. \tag{3.18}$$

There are therefore $N - n$ combinations of data that cancel out the effect of the systematics (nuisance parameters). A similar solution can be found in other cases, e.g, linear dependence on the ν_i etc.

Obviously it follows that it is not possible to do better than marginalisation, but with marginalisation the true value and the dispersion of the nuisance parameters must be known or assumed. Also one needs more data than nuisance parameters otherwise exact solutions do not exist. However, even for more nuisance parameters than data it is still possible in some cases to find *approximate* solutions. A particular case of interest is when observables have similar, but not identical, dependences on some of the nuisance parameters. In this case it is possible to minimise (rather than cancel completely) the effects of systematics. It turns out that one must minimise this:

$$\mathcal{L}_k = \sum_{j=1}^{n} \left(\frac{df_k}{d\nu_j} \right)^2 \Delta^2_{\nu_j} - \lambda_k \left(\sum_i (b_i^k)^2 - A_k^2 \right) \tag{3.19}$$

where λ_k is a Lagrange multiplier to be solved for, A_k is the norm of the vector b_i^k , and $\Delta\nu_j$ is the uncertainty on the j nuisance parameter. For example for the power law case discussed above we obtain the eigenvalue equations:

$$\sum_{l=1}^{N} \mathcal{M}_{il} = \sum_{l=1}^{N} \left[f_k^2 \sum_{j=1}^{n} \left(\frac{\Delta\nu_j}{\nu_j} \right) \alpha_{ij}\alpha_{lj} \right] b_l^k = \lambda_k b_l^k \,. \tag{3.20}$$

This formulation is general and include the case discussed above. Note that the eigenvalues of the matrix $\mathcal{M}$ measure how much the solution is affected by the nuisance parameters. The eigenvectors correspond to independent combinations of observables, and, if we are interested in minimizing the impact of nuisance parameters, we should choose those eigenvectors corresponding to eigenvalues which are small with respect to typical entries of the matrix. If an eigenvalue is zero, there is an independent non-trivial solution which is unaffected by changes of the nuisance parameters, as n the previous set-up.

For an example of a possible application, let us consider the case of Baryon Acoustic Oscillations which rely on the CMB to determine the "standard ruler": the sound horizon at radiation drag r_s. Since the determination of r_s is model-dependent, using an incorrect model (e.g., adiabatic model in the presence of isocurvature or non-standard neutrino properties in the presence of new physics in the neutrino sector) can introduce unwanted systematic errors in the interpretation of the results. For N redshift bins and therefore $2N$ measurements (radial and tangential BAO scale) there are $2N-1$ independent combinations (i.e., $N-1$ tangential, $N-1$ radial and one tangential vs radial relative BAO measurements) which cancel out completely the dependence on r_s. For more than ~ 10 redshift bins this does not increase significantly the statistical errors on the recovered cosmological parameters, but makes the measurement much more robust.

4. Summary and conclusions

I hope I have motivated why one cannot study Cosmology without being fluent in statistical techniques. Moreover, the forthcoming challenges that Cosmology faces cannot be addressed without the development of suitable statistical techniques. I have concentrated on dealing with systematic errors (rather than statistical errors), because systematic errors will likely be the major limitation in the near future and there is no systematic framework to deal with them. I have presented four specific examples, which are not representative of exhaustive, but I hope they serve to illustrate the point. The route from precision cosmology to accurate cosmology goes trough statistical cosmology.

Acknowledgments

I am supported by the European Research Council under the European Communitys Seventh Framework Programme grant FP7- IDEAS-Phys.LSS and I acknowledge Mineco grant FPA2011-29678- C02-02. I would like to thank all my collaborators in the works mentioned here: Luis Alvarez-Gaumé, Cesar Gomez, Alan Heavens, Raul Jimenez, Boris Leistedt, Sabino Matarrese, Jorge Noreña, Hiranya Peiris, Carlos Peña-Garay, Pavlos Protopapas, Fergus Simpson.

References

Planck Collaboration, *et al.*, 2013, arXiv, arXiv:1303.5076

Planck Collaboration, *et al.*, 2013, arXiv, arXiv:1303.5062

The COrE Collaboration, Armitage-Caplan, C., Avillez, M., *et al.* 2011, arXiv:1102.2181

André, P., Baccigalupi, C., Banday, A., *et al.* 2014, JCAP, 2, 6

Box, George, E. P., Norman, R. Draper (1987). *Empirical Model-Building and Response Surfaces*, p. 424, Wiley.

Peebles, P. J. E. 2002, arXiv:astro-ph/0208037

Elsner, F. & Wandelt, B. D. , *Astrophys. J.* **724**, 1262 (2010) [arXiv:1010.1254 [astro-ph.CO]].

Elsner, F., Wandelt, B. D. & Schneider, M. D. , *Astron. Astrophys.* **513**, A59 (2010) [arXiv:1002.1713 [astro-ph.CO]].

Ensslin, T. A. , Frommert, M., & Kitaura, F. S., *Phys. Rev. D* **80** (2009) 105005 [arXiv:0806.3474 [astro-ph]].

Verde, L., Jimenez, R., Alvarez-Gaume, L., Heavens, A. F., & Matarrese, S. 2013a, *JCAP*, 6, 23

Regan, D. M., Shellard, E. P. S., & Fergusson, J. R. 2010, *PRD*, 82, 023520

Marshall, P., Rajguru, N., & Slosar, A., *Phys. Rev. D* **73**, 067302 (2006)

March, M. C. , Trotta, R., Amendola, L., & Huterer, D., *Mon. Not. Roy. Astron. Soc.* **415**, 143 (2011)

Verde, L., Protopapas, P., & Jimenez, R. 2013b, *Physics of the Dark Universe*, 2, 166

Jeffreys, H., 1973. Scientific Inference. Cambridge University Press.

Kass, R. E. & Raftery, A. E., 1995, *Bayes factors. JASA* 90, 430, 773-795.

Leistedt, B., Peiris, H. V., & Verde, L., 2014, arXiv:1404.5950

Otto, S., Politzer, H. D., & Wise, M. B., 1986, *Physical Review Letters*, 56, 1878

Rice, S. O., 1944, *Bell Systems Tech. J.*, Volume 23, p. 282-332, 23, 282

Rice, S. O., 1945, *Bell Systems Tech. J.*, Volume 24, p. 46-156, 24, 46

Adler, R. J., 1981, *The Geometry of Random Fields.* Chichester: Wiley

Bardeen, J. M., Bond, J. R., Kaiser, N., & Szalay, A. S., 1986, *ApJ*, 304, 15

Jensen, L. G. & Szalay, A. S., 1986, *ApJLett*, 305, L5

Kaiser, N., 1984, *ApJL*, 284, L9

Peacock, J. A. & Heavens, A. F., 1985, *MNRAS*, 217, 805

Noreña, J., Verde, L., Jimenez, R., Peña-Garay, C., & Gomez, C. 2012, *MNRAS*, 419, 1040

Wilson, K. G. & Kogut, J., 1974, *PhR*, 12, 75

Statistical Challenges in 21st Century Cosmology
Proceedings IAU Symposium No. 306, 2014
A. F. Heavens, J.-L. Starck & A. Krone-Martins, eds.

doi:10.1017/S1743921314010783

Optimal observables in galaxy surveys

Julien Carron and István Szapudi

Institute for Astronomy, University of Hawaii,
2680 Woodlawn Drive, Honolulu, HI 96822,USA
email: carron@ifa.hawaii.edu

Abstract. The sufficient statistics of the one-point probability density function of the dark matter density field is worked out using cosmological perturbation theory and tested to the Millennium simulation density field. The logarithmic transformation is recovered for spectral index close to -1 as a special case of the family of power transformations. We then discuss how these transforms should be modified in the case of noisy tracers of the field and focus on the case of Poisson sampling. This gives us optimal local transformations to apply to galaxy survey data prior the extraction of the spectrum in order to capture most efficiently the information encoded in large scale structures.

Keywords. cosmology: large-scale structure of universe, cosmology: theory, methods: statistical, methods: analytical

1. Motivations

Among the principal motivations behind this study of the cosmological information within the matter density field are (i) the improved quality and size of modern and future cosmological data sets, that should allow understanding of more than the traditional and well-understood yet fairly crude descriptor the power spectrum. Its statistical power is known to be limited beyond the linear regime due to the tri-spectrum including beat-coupling (Rimes & Hamilton (2006)), or super-sample covariance (Takada & Hu (2013)) caused by large scales modes (ii) the very specific type of non-Gaussianity induced by gravity, which is characterised by extreme events, that renders mainstream tools designed for mildly non-Gaussian fields, higher order N-point functions, inadequate, as showed by Carron & Neyrinck (2012) (iii) the correct generalization of non-linear transformations (such as those of Neyrinck *et al.* (2009) or Seo *et al.* (2011)) to noisy tracers of the fields. This is important as noise modifies the statistical properties of the data and the optimal statistics or transform must take this into account in order to be efficient.

2. Overview

We use the fact that for any PDF p and parameter α of interest, the observable defined as $\partial_\alpha \ln p$ always captures the entire Fisher information content F of the PDF. It is therefore a 'sufficient statistics'. Our starting point is the formal Edgeworth series expansion for the logarithm of the PDF (here for one variable),

$$\ln p(\nu) = -\frac{\nu^2}{2} - \frac{1}{2}\ln\sigma^2 + \sum_{n=1}^{\infty}\sigma^n g_{n+2}(\nu), \qquad \nu = \frac{\delta}{\sigma}. \tag{2.1}$$

In that equation $g_k(\nu)$ is a polynomial of degree k given by combinations of Hermite polynomials and the cumulants. Terms proportional to some power ν^k of the field enters only with power of σ^{k-2} and higher. This simple form of $\ln p$ allows us to see through the structure of the information within the moments of the field : in the expansion of F in powers of σ^2, it is always possible to capture the k first terms with a polynomial of order $k+2$. The variance δ^2 captures the leading Gaussian information, a degree three polynomial in the field will capture the next to leading term, etc. This gives us in the next section the Taylor expansion of the sufficient statistic of the PDF. The same structure holds of course for the hierarchy of N-point functions associated to the multivariate PDF, giving us the multivariate Taylor expansion of the optimal observables. This is a generalization left for future work.

The sufficient observable of the matter field one-point PDF.
Next we go further with the one-point PDF as a function of scale in details. A useful reference is Carron & Szapudi (2013). The log-variance $\ln \sigma^2$ can be chosen as the relevant parameter. By reorganizing the series in power of δ rather than ν, the sufficient statistic has a complicated form involving different functions of δ,

$$f_0(\delta) + \sigma^4 f_4(\delta) + \sigma^6 f_6(\delta) + \cdots , \tag{2.2}$$

Nevertheless, it is found that the leading function $f_0(\delta)$ completely dominates the information. Besides, $f_0(\delta)$ depends only on the leading order cumulants. Writing f_0 as a power series $f_0 = \delta^2 + a_3\delta^3 + \cdots$, we gather that the coefficients are the leading coefficients of the polynomials g_k in Eq. (2.1). These coefficients are explicitly given by

$$a_n = \frac{2}{n!} \sum_{\mathbf{k}} (-1)^{|\mathbf{k}|} \left(n - 2 + |\mathbf{k}|\right)! \prod_{i \geqslant 3} \frac{S_i^{k_i}}{(i-1!)^{k_i}\, k_i!}, \tag{2.3}$$

where the sum runs over all vectors of positive integers $\mathbf{k} = (k_3, k_4, \cdots)$ of any dimension such that $\sum_i ik_i = n-2$, and where $|\mathbf{k}|$ stands for $\sum_i k_i$. Given the well known values of the leading cosmological cumulants calculated by Bernardeau (Bernardeau (1994)), we can obtain explicitly the first few coefficients for power-law power spectra $P(k) \propto k^n$. Two simple functions provide an almost perfect match for any value of n of interest. First the square of the power transformation ω_n^2 (Box-Cox transformation)

$$\omega_n(n) = \frac{(1+\delta)^{(n+1)/3} - 1}{(n+1)/3}, \quad \tau(n) = \frac{3}{2}\left(1+\delta\right)^{(n+3)/6}\left[(1+\delta)^{-2/3} - 1\right]. \tag{2.4}$$

and second the squared linear density contrast τ^2 from spherical collapse. Note that in the former case we recover precisely the logarithmic transform $\ln(1+\delta)$ of Neyrinck *et al.* (2009) for $n=-1$.

Test to the Millennium simulation density field. We tested our results and transforms ω_n and τ_n using the ΛCDM, $z=0$ matter density field from the Millennium simulation by Springel *et al.* (2005). We extracted from the $500h^{-1}$ Mpc box the one-point PDF on scales $i \times 1.95^{-1}$Mpc, with $i = 1, \cdots, 29$, corresponding roughly to $\sigma^2 \sim 10 - 0.1$. Poisson noise is negligible on all these scales. We then obtained straightforwardly the derivatives of the PDF with respect to $\ln \sigma^2$ using finite differences. The spectral index was estimated according to $n = -3 - \partial_{\ln R} \ln \sigma^2$ at each scale and lies between -0.8 and -1.2 at all scales. This gives us then both the total information content of the PDF as well as the efficiency of the statistics introduced above. Fig. 1 shows F as the crosses, as a function of σ^2. The three upper lines almost indistinguishable from F show the efficiency of ω_n^2, τ^2 as well as the logarithmic mapping $\ln^2(1+\delta)$. They are efficient over the full range. The two lower solid lines show the efficiency of the variance and that of the variance and third moment jointly. They show the very same behavior than in lognormal fields,

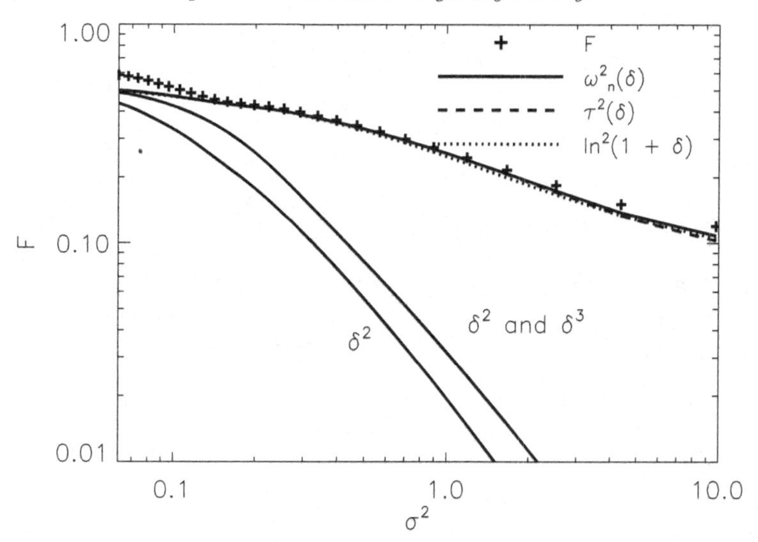

Figure 1. The information content of various statistics in the Millennium density field simulation for different smoothing scales shown as a function of the variance. The crosses show the total information of the one-point PDF. The moments (lower lines) perform poorly as for log-normal field statistics. The statistics derived in this work (upper lines) are maximally efficient over the full range.

becoming rather dramatically poor quite quickly. This is not very surprising given the non-analytic form of the three transforms for $n \sim -1$ with Taylor expansions breaking down quickly for moderate values of δ. At this point we can only speculate that the same happens for higher order moments as well, as the finite volume of the simulations did not allow us to obtain the PDF sufficiently accurately for this purpose.

Optimal transforms in the presence of observational noise. We show now how to adapt our statistics in the presence of noise. For galaxies sampling the density field, we can write very generically for the probability of observing numbers N of galaxies in cells (in a one-dimensional notation for clarity)

$$P(N|\alpha) = \int_{-\infty}^{\infty} dA\, p_A(A|\alpha)P(N|A), \quad \text{with } A = \ln(1+\delta). \tag{2.5}$$

After differentiating under the integral sign and using Bayes theorem the sufficient statistics for α becomes

$$\partial_\alpha \ln P(N|\alpha) = \int_{-\infty}^{\infty} dA\, p(A|N)\partial_\alpha \ln p_A(A,\alpha) = \langle \partial_\alpha \ln p_A(A,\alpha)\rangle_{AIN} \,. \tag{2.6}$$

Note that the weight function is now the posterior probability for A given the observations N. In the case that the data constrains well the signal the sufficient statistics of the observation becomes simply the sufficient statistics of the signal evaluated at its value favored by the data $A^*(N)$. Alternatively one can apply a saddle-point approximation to the above integral effectively treating the posterior as a Gaussian. Due to the above we can safely use a lognormal signal, to which we add Poisson sampling. This gives the following non-linear equation to solve for the saddle point,

$$A^*(N) + \bar{N}\sigma_A^2 e^{A^*(N)} = \sigma_A^2\,(N - 1/2)\,. \tag{2.7}$$

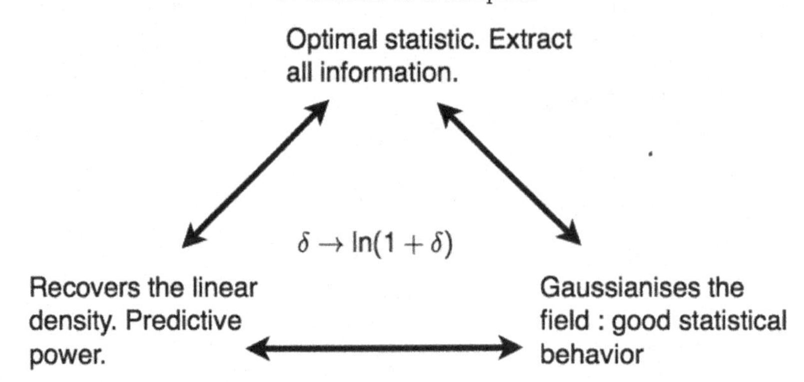

Figure 2. Three different perspectives on the logarithmic mapping of the matter density field. By way of the arguments of this paper, they are roughly equivalent. This is only because of Gaussian initial conditions linking notably the lower two corners and the fact that $n \sim -1$ on the scales of interest.

Further the mean and variance and $A^*(N)$ capture the entire information of $P(N)$, see Carron & Szapudi (2014).

3. Discussion

Our rigorous derivation of the sufficient statistics of the one-point PDF from cosmological perturbation theory points towards the well-known logarithmic transform as the optimal local transformation. In fact, this work unifies three different facets of that particular transform, illustrated in Fig. 2 : the capture of the entire information (because $n \sim -1$), the undoing of the non-linear dynamics and the Gaussianization of the PDF (because of the Gaussian initial conditions). One key aspect of the methods introduced is the systematic way the sufficient statistics and transforms are adapted to the presence of noise. The use of non-linear transforms was until now mostly useful in cosmology in simulations of noise-free fields. This opens the door to the analysis of actual data with efficient non-linear transformations. The analysis of the projected Canada-France-Hawaii Telescope Large Survey (CFHTLS†) data using the power spectrum of the A^* non-linear transform will be exposed by Wolk *et al.* (2014).

References

Carron, J. & Neyrinck, M. C. 2012, *ApJ*, 750,28
Carron, J. & Szapudi, I. 2014, *MNRAS*, 439, L11-L15
Carron, J. & Szapudi, I. 2013, *MNRAS*, 434, 2961-2970
Neyrinck, M. C. & Szapudi, I, Szalay, A. S. 2009, *ApJ*, 698, L90-L93
Springel, V., *et al.* 2015, *Nature*, 435, 629-636
Bernardeau, F. 1994, *A&A*, 291, 697-712
Seo,H. J., Sato, M., Dodelson, S. Jain, B., & Takada, M. 2011, *ApJ*, 729, L11+
Rimes, C. D. & Hamilton, A. J. S. 2006, *MNRAS*, 371, 1205-1215
Takada, M. & Hu, W. 2013, *Phys. Rev. D*, 87,12, 123504
Wolk, M., *et al.* 2014, *In preparation*

† `http://www.cfht.hawaii.edu/Science/CFHLS/`

Statistical Challenges in 21st Century Cosmology
Proceedings IAU Symposium No. 306, 2014
A. F. Heavens, J.-L. Starck & A. Krone-Martins, eds.

doi:10.1017/S1743921314010709

Morpho-statistical characterization of the cosmic web using marked point processes

Radu S. Stoica

Université Lille 1 - Laboratoire Paul Painlevé
Observatoire de Paris - Institut de Mécanique Céleste et de Calcul des Éphémérides
email: radu.stoica@univ.lille1.fr

Abstract. The cosmic web is the intricate network of filaments outlined by the galaxies positions distribution in our Universe. One possible manner to break the complexity of such an elaborate geometrical structure is to assume it made of simple interacting objects. Under this hypothesis, the filamentary network can be considered as the realization of an object or a marked point process. These processes are probabilistic models dealing with configurations of random objects given by random points having random characteristics or marks. Here, the filamentary network is considered as the realization of such a process, with the objects being cylinders that align and connect in order to form the network. The paper presents the use of marked point processes to the detection and the characterization of the galactic filaments.

Keywords. methods: statistical, large-scale structure of universe

---

1. Introduction

The marked point processes were already used in cosmology to describe the galaxies distribution and the different correlations of their corresponding characteristics (Martinez & Saar 2002). The galaxies are not spread uniformly within the observed Universe. Their positions form an intricate network made of complex geometrical objects such as filaments, walls and clusters. During the last years, the filamentary network was studied under the assumptions that the filaments are made of small segments (cylinders) that connect and align forming the network. From a mathematical point of view, this hypothesis allows to consider the cosmic network as the realisation of a marked point process. Under this assumption, the network was detected and characterized from both, morphological and statistical, points of view (Stoica *et al.* 2005b; Stoica *et al.* 2007; Stoica *et al.* 2010; Tempel *et al.* 2013; Tempel *et al.* 2014).

2. Modelling the filamentary network

Let the observation window be a measure space $(K, \mathcal{B}, \nu)$, with $K \subset \mathbb{R}^d$, $\mathcal{B}$ the Borel $\sigma-$algebra and $0 < \nu(K) < \infty$ the Lebesgue measure. A point process in K is a finite random configuration of points leaving in K. To each point, marks or characteristics may be attached. For this purpose, the marks probability space $(M, \mathcal{M}, \nu_M)$ is considered. A marked point process is a random sequence $\boldsymbol{x} = \{x_n = (k_n, m_n)\}$ such that the points k_n are a point process in K and m_n are the marks corresponding for each k_n. A random configuration of cylinders is a marked point process given by the centres positions of the cylinders and their corresponding random geometrical shapes attached to each point position. For further studying and reading on marked point processes we recommend the excellent monographs (van Lieshout 2000; Møller and Waagepetersen 2004).

The most known marked point process is the stationary Poisson point process of unit intensity with i.i.d. marks. This process works as it follows : choose a number of points according to a Poisson law of intensity parameter $\nu(K)$, spread the chosen points uniformly in K and to each point attach an independent mark using the distribution ν_M. Due to the independence property, the Poisson point process exhibits no structure. Nevertheless, it is perfectly possible to build models taking into account objects interactions, hence forming structures. Such a process can be described by a probability density with respect to the reference measure given by the unit intensity Poisson point process. Here, the probability density of the filamentary network has the general expression

$$p(\boldsymbol{x}|\theta) = \frac{\exp\left[-U(\boldsymbol{x}|\theta)\right]}{Z(\theta)}, \tag{2.1}$$

where $Z(\theta)$ is the normalising constant, θ is the model parameters vector, and $U(\boldsymbol{x}|\theta)$ is the (Gibbs) energy function of the system.

Our main hypotheses are that the galaxies forming filaments group together inside rather small cylinders, and that these small cylinders may connect and align to build the network.

Following these ideas, the energy function in (2.1) can be written as the sum :

$$U(\boldsymbol{x}|\theta) = U_{\boldsymbol{d}}(\boldsymbol{x}|\theta) + U_i(\boldsymbol{y}|\theta), \tag{2.2}$$

where $U_{\boldsymbol{d}}(\boldsymbol{x}|\theta)$ is the data energy and $U_i(\boldsymbol{x}|\theta)$ is the interaction energy. The data energy is related to the first assumption, that is the position of the cylinders in the galaxy field, whereas the interaction energy is related to the second hypothesis, that is the alignment and the connection of the cylinders forming the filamentary pattern. The data term is constructed using different statistical tests to verify that the galaxies are aligned along the main axis of the cylinder and that from a statistical point of view, the number of the galaxies inside of the cylinder is higher than outside of it. The interaction term is given by the Bisous model, a marked point process able to simulate random configurations of connected objects. All the details can be found in (Tempel *et al.* 2014).

Neither the filaments nor their model parameters are known. Therefore, under the Bayesian framework, the model is completed with a prior density $p(\theta)$ for the model parameters. This allows to write the joint estimator of the filamentary pattern and the parameters as

$$\begin{aligned}(\widehat{\boldsymbol{x}}, \widehat{\theta}) &= \arg\max_{\Omega\times\Psi} p(\boldsymbol{x}, \theta) = \arg\max_{\Omega\times\Psi} p(\boldsymbol{x}|\theta)p(\theta) \\ &= \arg\min_{\Omega\times\Psi} \left\{\frac{U_{\boldsymbol{d}}(\boldsymbol{x}|\theta) + U_i(\boldsymbol{x}|\theta)}{Z(\theta)} + \frac{U_p(\theta)}{Z_p(\theta)}\right\},\end{aligned} \tag{2.3}$$

with $p(\theta) = \exp[-U_p(\theta)]/Z_p(\theta)$ the prior law for the model parameters, Ω the cylinders configurations space and Ψ the model parameters space.

3. Simulation and statistical inference

The normalizing constant of the models (2.1) does not have always a precise closed analytical form. This implies that the simulation of the model should be done using elaborate methods. Several Markov chains Monte Carlo (MCMC) techniques are available to simulate marked point processes (van Lieshout 2000;Møller and Waagepetersen 2004). Here, we have adopted the Metropolis-Hastings algorithm with tailored to the model transition kernels (van Lieshout and Stoica 2003;Stoica *et al.* 2005a). The classical Metropolis-Hastings randomly adds and deletes a cylinder from a configuration.

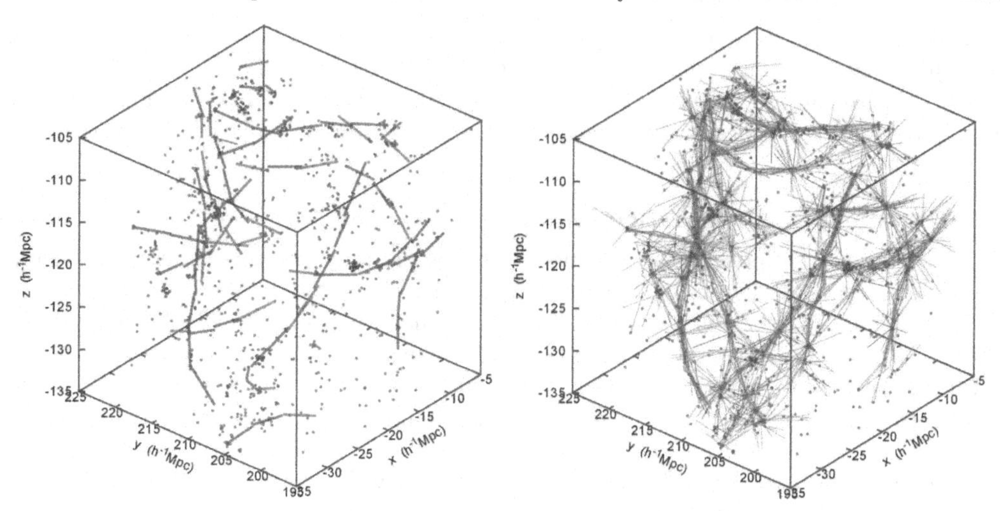

Figure 1. a) Detection result using the simulated annealing algorithm ; b) Superposition of several detection results.

This algorithm uses uniform proposals for its transition kernel, in order to guarantee the theoretical convergence to the desired equilibrium distribution. Nevertheless, whenever models exhibiting complex interactions are simulated, this convergence is very slow in practice. This is also the case for the filamentary network model. The solution is to build an algorithm using not only uniform proposals. More precisely, this means to not propose only uniformly spread cylinders, but also cylinders that tend to connect in order to form a network. As showed in (van Lieshout and Stoica 2003) such an algorithm has the required convergence properties.

The estimator in (2.3) is computed using the simulated annealing algorithm. This global optimization technique iteratively samples from $p(\boldsymbol{x}, \theta)^{1/T}$ while the temperature parameter T goes slowly to 0. The output of the algorithm converges in law towards the uniform distribution over the configuration sub-space made of the solutions of (2.3). The sampling mechanism at fixed temperature is a tailored Metropolis-Hastings. In order to ensure the algorithm convergence, the cooling schedule of the temperature should be slow (Stoica *et al.* 2005a). The solution given by the simulated annealing is not unique. Therefore, we are interested in averaging the obtained solution. This can be done using visit maps or level sets (Heinrich *et al.* 2012). Fig. 1 shows a detection result of filaments in a sample of cosmological catalogue. The superposition of several results is shown as well. Based on these results, visit maps and spines can be computed for the filaments (Tempel *et al.* 2014). All these together gave a rather robust detector of filaments.

A configuration of cylinders is made of cylinders that are connected at both extremities, the double connected cylinders, that are connected at one extremity only, the single connected cylinders, and that are not connected at all, the free cylinders. The sufficient statistics of the model are given by the number of these three types of cylinders, respectively. These statistics determine entirely the model, together with the mark distribution and the different interactions ranges. Furthermore, these statistics allow to characterize different galaxy catalogues from a morphological point of view. The authors in (Stoica *et al.* 2010) compare mock catalogues with real observation. It was stated, that the real filaments are longer and less fragmented. The sufficient statistics were also used to test the general method. This test was based on a bootstrap procedure, and it certifyied that the method does not detect filaments by chance, but only because these filamentary structures do really exist in the observed data (Stoica *et al.* 2007). Once in possession of a map of filaments, interesting cosmological questions can be answered, as well. As an example, we mention (Tempel *et al.* 2013), where the authors found correlations between the filaments orientations and the spin alignment for spiral and elliptical galaxies.

4. Conclusions and perspectives

The different components of our model are simple and intuitive, allowing a visual definition and construction of a filament. The simplicity characteristic is adopted on purpose. Since, the objects we are looking for are not observed, the simpler hypotheses should formulated and tested first. Under these circumstances, our results are comparable with all the other methods. All these techniques use a local gradient or score for the filaments together with a smoothing or regularization procedure. The strong points of our approach are the versatility of the model and its probabilistic general framework. Clearly, the model can be and should be improved. If more knowledge is available concerning the local definition of a filament and the general topology of the network, this knowledge must be integrated in the model. The probabilistic framework allows the computation of integrals, average quantities and characteristics of the filaments, leading to a reliable description of one the most intriguing pattern in our observed Universe.

Aknowledgements

This paper is based on very recent work done together with E. Saar, E. Tempel, V. J. Martinez, L. J. Liivamägi and G. Castellan. E. Tempel provided also the Fig. 1.

References

P. Heinrich, R. S. Stoica, & V. C. Tran. *Spatial Statistics*, 2:47–61, 2012.

M. N. M. van Lieshout. Imperial College Press, London, 2000.

M. N. M. van Lieshout & R. S. Stoica. *Statistica Neerlandica*, 57:1–30, 2003.

V. J. Martinez & E. Saar. Chapman and Hall, 2002.

J. Møller & R. P. Waagepetersen. Chapman and Hall/CRC, Boca Raton, 2004.

R. S. Stoica, P. Gregori, & J. Mateu. *Stochastic Processes and their Applications*, 115:1860–1882, 2005.

R. S. Stoica, V. J. Martinez, J. Mateu, & E. Saar. *Astronomy and Astrophysics*, 434:423–432, 2005.

R. S. Stoica, V. J. Martinez, & E. Saar. *Journal of the Royal Statistical Society. Series C (Applied Statistics)*, 55:189–205, 2007.

R. S. Stoica, V. J. Martinez, & E. Saar. *Astronomy and Astrophysics*, 510(A38):1–12, 2010.

E. Tempel, R. S. Stoica, & E. Saar. *Monthly Notices of the Royal Astronomical Society*, 428:1827–1836, 2013.

E. Tempel, R. S. Stoica, E. Saar, V. J. Martinez, L. J. Liivamägi, & G. Castellan. *Monthly Notices of the Royal Astronomical Society*, 438: 3465-3482, 2014.

Statistical Challenges in 21st Century Cosmology
Proceedings IAU Symposium No. 306, 2014
A. F. Heavens, J. -L. Starck & A. Krone-Martins, eds.

doi:10.1017/S1743921314013891

Measuring the clustering of photometric quasars through blind mitigation of systematics

Boris Leistedt, Hiranya V. Peiris, Nina Roth

Department of Physics and Astronomy,
University College London, London WC1E 6BT, U.K.
boris.leistedt.11@ucl.ac.uk, h.peiris@ucl.ac.uk, n.roth@ucl.ac.uk

Abstract. We present accurate measurements of the large-scale clustering of photometric quasars from the Sloan Digital Sky Survey. These results, detailed in Leistedt & Peiris (2014), rely on a novel technique to identify and treat systematics when measuring angular power spectra, using null-tests and analytical marginalisation. This approach can be used to maximise the extraction of information from current and future galaxy or quasar surveys. For example, it enables to robustly constrain primordial non-Gaussianity (PNG), which modifies the bias of galaxies and quasars on large scales – the most sensitive to observational systematics. The constraints on PNG obtained with the quasar power spectra are detailed in Leistedt, Peiris & Roth (2014); these are the most stringent constraints to date obtained with a single tracer of the large-scale structure.

Keywords. quasars: general, cosmology: large-scale structure of universe, early universe.

---

1. Motivation

Quasars are bright, highly biased tracers of the large scale structure (LSS) of the universe, and are useful for testing cosmological models over large volumes and extended redshift ranges. In particular, photometric quasar surveys can be used to tightly constrain primordial non-Gaussianity (PNG) of local type, parameterised by $f_{\rm NL}$, which is predicted to enhance the bias of LSS tracers on large scales (*e.g.,* Dalal *et al.* 2008; Matarrese & Verde 2008; LoVerde *et al.* 2008). However, this requires accurate auto- and cross-correlation power spectrum measurements, which are complicated by the presence of significant systematics in the data. These systematics can be intrinsic (*e.g.,* dust extinction), observational (*e.g.,* seeing, airmass), or instrumental (*e.g.,* instrument calibration), and affect the properties of the data in complex ways, creating spurious spatial correlations in quasar catalogues. Previous studies of the clustering of quasars from the Sloan Digital Sky Survey (SDSS) exhibited power excesses on large and small scales (*e.g.,* Myers *et al.* 2007), particularly significant in the DR6 and DR8 photometric quasar catalogues (Richards *et al.* 2009; Bovy *et al.* 2012). Cuts to the data were not sufficient to remove this excess power, apparently pointing to significant levels of PNG (Xia *et al.* 2009, 2011; Giannantonio *et al.* 2012; Karagiannis, Shanks & Ross 2014; Giannantonio *et al.* 2014). However, recent work has demonstrated that the excess power was in fact due to systematics (Pullen & Hirata 2013; Ho *et al.* 2013; Agarwal *et al.* 2014) and could be eliminated using mode projection (Leistedt *et al.* 2013). The latter consists of an analytical marginalisation of the spurious clustering using templates of systematics, and is a generalisation of the correction techniques introduced in Ross *et al.* (2011, 2012); Ho *et al.* (2012); Agarwal *et al.* (2014).

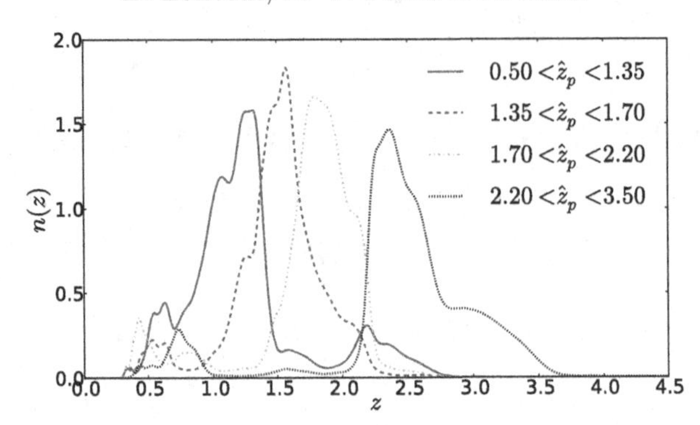

Figure 1. Estimated redshift distributions of the four quasar samples used in this analysis.

2. Data samples

We consider the XDQSOz catalogue of photometric quasars (Bovy *et al.* 2012), which contains quasar candidates selected from point sources detected in the SDSS DR8 imaging data, covering $\sim 10^4$ deg^2 of the southern and northern Galactic sky. We consider all objects with $P_{\rm QSO} > 0.8$, and exclude regions of the sky with PSFWIDTH$_i > 2.0$, SCORE $>$ 0.6, and $E(B-V) > 0.08$, to remove the most dusty regions in the South Galactic Cap. XDQSOz also includes photometric redshift estimates $\hat{z}_p$, defined as the highest peak in the posterior distribution of the individual photo-z estimates. We separate this catalogue into four samples by selecting objects with photometric redshifts $\hat{z}_p$ in top-hat windows ranging $[0.5, 1.35]$, $[1.35, 1.7]$, $[1.7, 2.2]$ and $[2.2, 3.5]$, which have comparable numbers of objects, and thus analogous shot noise. The redshift distributions of these samples, which are needed to produce theoretical predictions for the clustering of quasars, are shown in Figure 1, and were estimated by stacking the posterior distributions of the redshift of the individual objects.

3. Power spectrum measurements

We use a quadratic maximum likelihood estimator (QML) to simultaneously estimate the angular power spectra of these four quasar samples. The resolution of the data and the settings of the estimator are fixed to obtain accurate power spectra for $\ell \in [2, 140]$ (smaller scales are dominated by shot noise), following the rules detailed in Leistedt *et al.* (2013). In particular, we estimate band powers in multipole bins $\Delta\ell = 15$ and 21.

We create a set of 220 templates of potential systematics, using external data (*e.g.*, maps of extinction by dust and stellar density), and numerous calibration and observing condition quantities, presented in detail in Leistedt & Peiris (2014). The red data points in Figure 2 show the estimated band-power spectra when projecting out the modes of this base set of templates ('basic mp'). The black lines denote the theoretical power spectra corresponding to a *Planck* ΛCDM cosmology (Planck Collaboration 2013), $f_{\rm NL} = 0$, and fiducial linear quasar bias $b(z) = 1 + [(1+z)/2.5]^5$. The shaded band indicates the zone spanned by the theory curves when varying $-50 < f_{\rm NL} < 50$.

Projecting out these templates in the estimator is equivalent to marginalising their linear effect on the quasar maps, therefore removing their influence on the measured power spectra to first order. However, the latter still suffer from excess power on large scales when compared to the theoretical prediction. This could indicate significant non-linear effects of the systematics, which cannot be addressed by marginalising over the base templates.

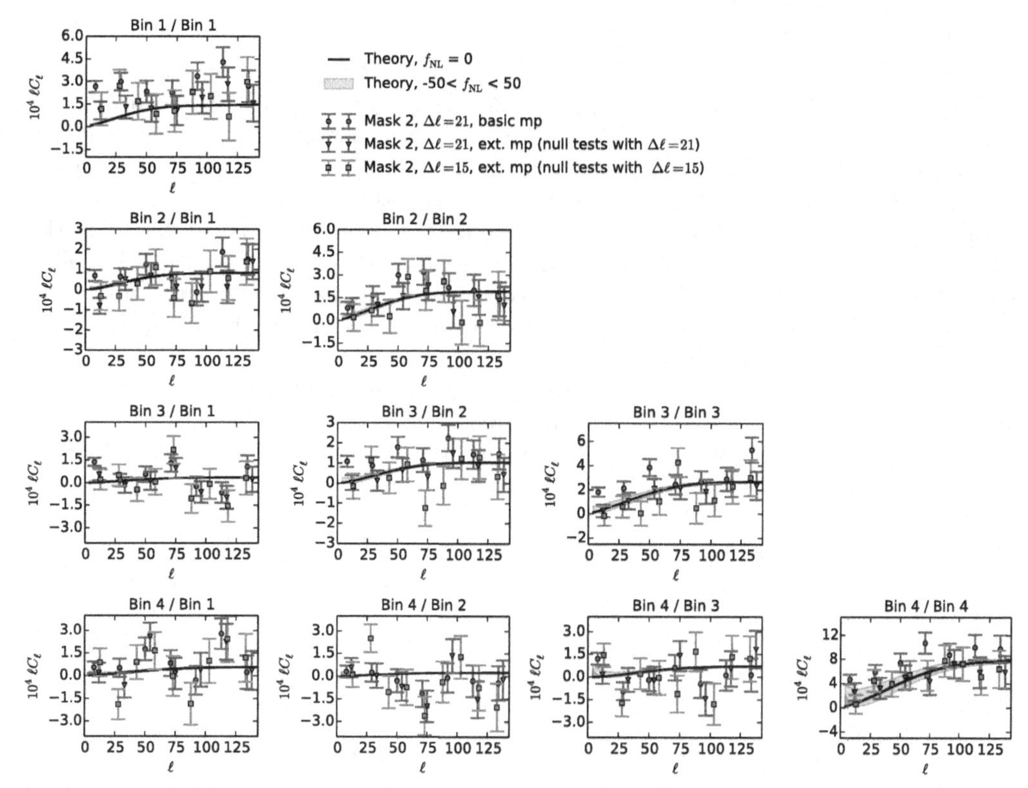

Figure 2. Angular power spectra of the four XDQSOz quasar samples. The effects of systematics are mitigated using either simple mode projection (linear contamination model of ~ 200 templates), or the novel extended mode projection approach (quadratic model with $\sim 20,000$ templates, decorrelated on the sky and cross-correlated with the data). See text for more details.

4. Extended mode projection

To enable a more complex treatment of systematics and cleaner power spectrum measurements, we exploit a novel technique of extended mode projection, introduced in Leistedt & Peiris (2014). This involves having a more complete set of systematics templates but only projecting out the ones which are significantly correlated with the data. For this purpose, we complement the base set of systematics with pairs of products of templates, yielding a set of $\sim 20,000$ templates. We then decorrelate them on the sky, by applying a principal component analysis and discarding the modes with eigenvalues consistent with noise or numerical precision, leading to $\sim 3,700$ uncorrelated templates (replacing the $\sim 20,000$ input templates). We then cross-correlate each template with the four redshift samples in order to obtain cross-angular power spectra which can be used as null tests. We only marginalise (through mode projection) over the modes which fail the null tests, *i.e.*, where the cross-power spectra are not consistent with zero, as measured by the reduced χ^2 using a simple Gaussian likelihood. The blue and grey data points in Figure 2 show the results of this approach with two multipole resolutions $\Delta\ell = 15$ and $\Delta\ell = 21$. In both cases, the excess power due to systematics is eliminated, and the measurements are in good agreement with the theoretical predictions, while not showing further evidence for contamination by systematics (*i.e.*, their effects lie within the statistical errors).

5. Outlook

It was previously shown that effects of systematics on angular power spectrum measurements could be robustly marginalised through mode projection. However, quasar surveys are subject to strong non-linear systematics which require a more complicated treatment, for example involving non-linear contamination models with large numbers of templates. Extended mode projection can deal with this setting by decorrelating the templates and only projecting out the ones which are significantly correlated with the data. This approach is entirely based on mapping potential systematics and performing null tests with the data, and is therefore a robust way to handle systematics in a blind fashion. More details on extended mode projection and the power spectra of the XDQSOz data can be found in Leistedt & Peiris (2014).

Figure 2 also demonstrates that the measured power spectra reach a sufficient precision and robustness to constrain PNG, which modifies the bias of quasars on large scales. The constraints on PNG from the XDQSOz power spectra are presented in Leistedt, Peiris & Roth (2014), and are the most stringent constraints to date obtained with a single tracer of the large-scale structure. The power of the extended mode projection approach can be applied to extract the maximum information from future galaxy and quasar surveys, which will suffer from numerous observational systematics, most of which can be mapped on to the sky.

References

Agarwal, N., Ho, S., Myers, A. D., Seo, H.-J., Ross, A. J. *et al.* 2014, *JCAP*, 1404, 007, arXiv:1309.2954

Bovy, J., *et al.* 2012, *ApJ*, 749, 41, arXiv:1105.3975

Dalal, N., Doré, O., Huterer, D., & Shirokov, A. 2008, *Phys. Rev. D.*, 77, 123514, arXiv:0710.4560

Giannantonio, T., Crittenden, R., Nichol, R., & Ross, A. J. 2012, *MNRAS*, 426, 2581, arXiv:1209.2125

Giannantonio, T., Ross, A. J., Percival, W. J., Crittenden, R., Bacher, D., Kilbinger, M., Nichol, R., & Weller, J. 2014, *Phys. Rev. D*, 89, 023511, arXiv:1303.1349

Ho, S., *et al.* 2013, *arXiv preprint*, arXiv:1311.2597

Ho, S., *et al.* 2012, *ApJ*, 761, 14, arXiv:1201.2137

Karagiannis, D., Shanks, T., & Ross, N. P. 2014, *MNRAS*, 441, 486, arXiv:1310.6716

Leistedt, B. & Peiris, H. V. 2014, *arXiv preprint*, arXiv:1404.6530

Leistedt B., Peiris H. V., Mortlock D. J., Benoit-Lévy A., & Pontzen, A. 2013, *MNRAS*, 435 , 1857, arXiv:1306.0005

Leistedt, B., & Peiris, H. V., & Roth, N. 2014, *arXiv preprint*, arXiv:1405.4315

LoVerde, M., Miller, A., Shandera, S., & Verde, L. 2008, *ApJ (Letters)*, 4, 14, arXiv:0711.4126

Matarrese, S. & Verde, L. 2008, *ApJ (Letters)*, 677, L77, arXiv:0801.4826

Myers, A. D., Brunner, R. J., Richards, G. T., Nichol, R. C., Schneider, D. P., & Bahcall, N. A. 2007, *ApJ*, 658, 99, arXiv:arXiv:astro-ph/0612191

Planck Collaboration, 2013, *arXiv preprint*, arXiv:1303.5076

Pullen, A. R. & Hirata, C. M. 2013, *PASP*, 125, 705, arXiv:1212.4500

Richards, G. T., *et al.* 2009, *ApJ (Supp.)*, 180, 67, arXiv:0809.3952

Ross, A. J., *et al.* 2011, *MNRAS*, 417, 1350, arXiv:1105.2320

Ross, A. J., *et al.* 2012, *MNRAS*, 424, 564, arXiv:1203.6499

Xia, J. -Q., Baccigalupi, C., Matarrese, S., Verde, L., & Viel, M. 2011, *J. Cosmol. Astropart. Phys.*, 8, 33, arXiv:1104.5015

Xia, J. -Q., Viel, M., Baccigalupi, C., & Matarrese, S. 2009, *J. Cosmol. Astropart. Phys.*, 9, 3, arXiv:0907.4753

Statistical Challenges in 21st Century Cosmology
Proceedings IAU Symposium No. 306, 2014
A. F. Heavens, J.-L. Starck & A. Krone-Martins, eds.

doi:10.1017/S1743921314013817

Internal errors: a valid alternative for clustering estimates?

Pablo Arnalte-Mur and Peder Norberg

Institute for Computational Cosmology, Durham University
South Road, Durham DH1 3LE, United Kingdom
email: pablo.arnalte-mur@durham.ac.uk

Abstract. We investigate the validity of internal methods to estimate the uncertainty of the galaxy two-point correlation function. We consider the jackknife and bootstrap methods, which are based on re-sampling sub-regions of the original data. These are cheap computationally, and do not depend on the accuracy of external simulations. We test the different methods over a large range of scales using a set of 160 mock catalogues from the LasDamas set of simulations. Our results show that the standard bootstrap method significantly overestimates the true uncertainty at all scales. We try two possible generalisations of the bootstrap, but find them not to be robust. Regarding jackknife, we obtain that this method provides an unbiased estimation of the error at small and intermediate scales, up to $\sim 40\,h^{-1}$ Mpc. At larger scales, it typically overestimates the error by a $\sim 13\%$.

Keywords. methods: data analysis, methods: statistical, galaxies: statistics, large-scale structure of universe

---

1. Introduction

The study of galaxy clustering in large redshift surveys through the measurement of the two-point correlation function (2PCF) provides valuable information about the global properties of the Universe and about the process of formation and evolution of galaxies. It is important to understand the different methods available to accurately estimate the uncertainty of these measurements.

One option is to use an external estimate: generate a large number of simulated realisations of the data, and compute the uncertainty from the distribution of the measurements in them (Dodelson & Schneider 2013; Taylor *et al.* 2013). An alternative is to use internal error estimations, based on the re-sampling of the data itself. Different methods exist, such as jackknife and bootstrap methods (Efron & Tibshirani 1993). These methods are very cheap computationally and, as they use the data directly, do not depend on the accuracy of external simulations. However, their bias and variance are not very well understood for the case of clustering statistics.

In this work, we test the validity of some of these internal methods for a typical galaxy sample. We focus on the errors of the 2PCF over a large range of scales.

2. Resampling methods

We focus on the study of the redshift-space 2PCF estimated using the Landy & Szalay (1993) estimator, which is based on counting pairs of galaxies in both the data sample and in an auxiliary Poisson catalogue containing the same selection effects:

$$\hat{\xi}(s) = 1 + \frac{DD(s)}{RR(s)} - 2\frac{DR(s)}{RR(s)}, \tag{2.1}$$

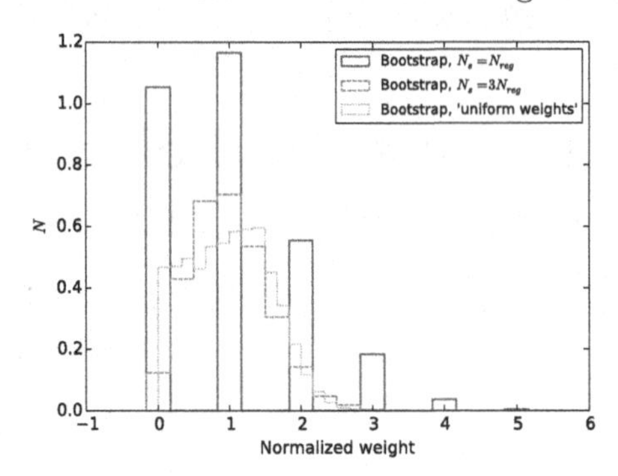

Figure 1. Distribution of weights w_i^k for different bootstrap methods, for the case $N_r = 10$. The weights are normalised such that $\sum_i^{N_r} w_i^k = 1$ for each realisation k.

where the different symbols correspond to the suitably-normalised pair counts in the data catalogue ($DD(s)$), in the random catalogue ($RR(s)$), and to the cross-pairs between both catalogues ($DR(s)$), as function of pair separation.

The internal error methods used in Cosmology (see Norberg *et al.* 2009 for a review) are based on dividing the data in a set of N_r non-overlapping regions or 'blocks'. New 'mock realisations' are created by resampling these blocks, the 2PCF for each realisation, ξ^i, is computed using eq. (2.1) and the uncertainty is computed from the distribution of ξ^i. The resampling is based on blocks instead of individual data points to take into account the fact that the points themselves are correlated.

In the standard bootstrap method, each realisation is created by selecting randomly N_r blocks with replacement from the data. The uncertainty is estimated directly from the variance of the ξ^i. In the jackknife method, N_r realisations are created by omitting in each case one of the blocks from the data. The uncertainty is now obtained by multiplying the variance of the ξ^i by a factor of $(N_r - 1)$, to take into account the large overlap between realisations constructed in this way.

Generalisation of the bootstrap method. Norberg *et al.* (2009) noted that the realisations generated using the standard bootstrap method encompass only $\sim 60\%$ of the total volume covered by the data, inducing a systematic overestimation of the error. They proposed to generalise the bootstrap method by selecting in each realisation N_s blocks from the N_r available, with $N_s > N_r$. An equivalent approach is to regard the bootstrap procedure as a re-weighting of the 'partial pair counts' $DD_{ij}(s)$: the number of pairs of galaxies with separation s in which one of the galaxies is in region i and the other in region j. The total pair counts for a given bootstrap realisation k is then obtained as

$$DD^k(s) = \sum_i^{N_r} \sum_j^{N_r} w_i^k w_j^k DD_{ij}(s) , \qquad (2.2)$$

where w_i^k is the normalised weight assigned to region i in bootstrap realisation k. This same procedure is repeated for DR^k and RR^k and ξ^k is estimated using eq. (2.1).

Fig. 1 shows the distribution of weights that result from applying the bootstrap method with either $N_s = N_r$ (the standard case) or $N_s = 3N_r$. This approach allows us to investigate the possibility of generating bootstrap realisations using different weight distributions. As an example of this possibility, we test the case in which the weights for each block are drawn randomly from a uniform distribution. The resulting distribution of normalised weights w_i^k is also shown in Fig. 1.

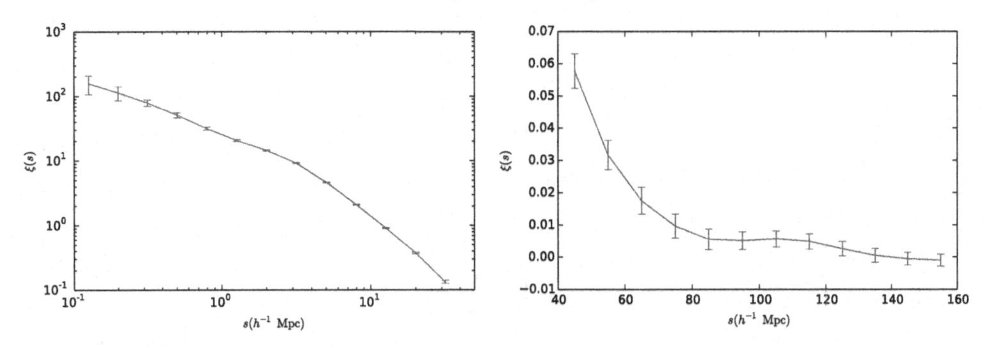

Figure 2. Mean 2PCF $\xi(s)$ measured from the 160 mock LasDamas catalogues at small (left) and large (right) scales. The error bars show the standard deviation of the measurements $\sigma_T(s)$, and correspond to the uncertainties for a single realisation.

3. Design of tests

We tested the different methods using a set of mock galaxy catalogues from the public LasDamas simulations† (McBride *et al.* in prep.). We used 160 independent realisations of their Luminous Red Galaxies (LRGs) sample, which reproduces the properties of the SDSS-DR7 catalogue ($V = 6.6 \times 10^8 \mathrm{Mpc}^3\, h^{-3}$, $N_{gal} \simeq 75000$).

We study the 2PCF for these catalogues using logarithmically-spaced bins in the range $0.1 < s < 40\, h^{-1}$ Mpc, and linearly-spaced bins in $40 < s < 160\, h^{-1}$ Mpc. From the different realisation's results, we compute the standard deviation of the $\xi(s)$ values in each bin, $\sigma_T(s)$, which we take as our 'true' uncertainty to use as reference for the internal error estimates. We plot the 2PCF of the mock catalogues in Fig. 2

For our resampling tests, we divide the survey in $N_r = 100$ regions, using a 10×10 grid in the RA, DEC coordinates. We have tested that our results do not change significantly with the number or geometry of the regions considered.

4. Results

We computed the 2PCF for each of the 160 mock realisations, and estimated its uncertainty in each case using the four methods described above: jackknife, standard bootstrap (with $N_s = N_r$), bootstrap with a larger number of resampled blocks ($N_s = 3N_r$) and bootstrap with uniformly distributed weights. Fig. 3 shows the estimation of the error σ_ξ we obtain in each case, as function of scale, compared with our reference value σ_T.

We find that the standard bootstrap method overestimates the error at all scales by $\sim 40 - 60\%$, in agreement with previous results (e.g. Norberg *et al.* 2009). We tested two generalisations of the bootstrap method that aimed at correcting this effect. At first sight, these work better than the standard bootstrap, reducing the bias in the estimation to $7 - 25\%$. However, they end up underestimating the errors in most cases. Moreover, these results are not robust to changes in the N_s used or the continuous distribution used for the weights.

Jackknife is clearly the method giving the best results. At small scales ($s < 40\, h^{-1}$ Mpc), the bias in the estimated error is very small ($\sim 0.5\%$). This can be explained by the fact that at these scales most of the pairs considered correspond to galaxies in the same region, so we are close to the conditions in which the standard jackknife was shown to be valid (Efron & Tibshirani 1993). We note, however, that even in this case there is an important dispersion ($\sim 10\%$) of the values obtained for different realisations.

† http://lss.phy.vanderbilt.edu/lasdamas/

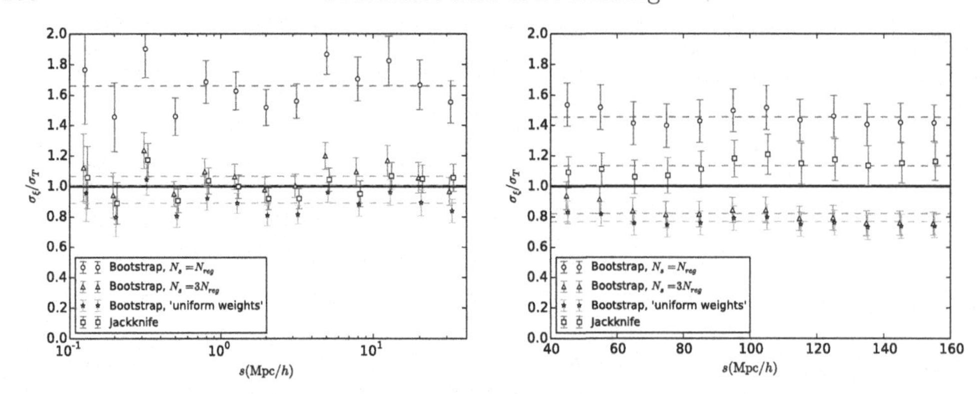

Figure 3. Errors estimations obtained using different internal methods applied to the LasDamas mocks, divided by the 'true' uncertainty (obtained from the variance of the realisations). The points and errorbars correspond to the mean and standard deviation of the errors estimated from the 160 realisations.

At larger scales, the estimator is biased, overestimating the error by a $\sim$ 13%, in agreement with the results of Guo *et al.* (2013). This could be caused by the significant fraction of 'crossed pairs' coming from galaxies in different regions at these scales. The fraction of 'auto' and 'cross' pairs as function of scale can be directly computed from the data, and thus could be used as a way of assessing the range of validity of the jackknife method.

5. Conclusions

We studied the performance of different internal methods to estimate the uncertainty on the galaxy 2PCF on a set of 160 realistic simulations. We also introduced a new approach to generalise the bootstrap method, based on the weighting of the different blocks.

Our results show that the standard bootstrap systematically overestimates the 2PCF uncertainty. The generalised method is not reliable, as it depends on an arbitrarily chosen distribution of weights. We found that the jackknife method provides an unbiased estimation of the error at small scales, while slightly overestimating the error at large scales. At these small scales, cosmological simulations could have more difficulties to reproduce the galaxy clustering pattern, so the external error estimation methods are less reliable. We consider therefore that the jackknife method can be very useful for clustering studies in this regime.

References

Dodelson, S. & Schneider, M. D. 2013, *Phys. Rev. D*, 88, 063537

Efron, B. & Tibshirani, R. J. 1993, *An Introduction to the Bootstrap.* Chapman & Hall/CRC, Boca Raton

Guo, H., *et al.* 2013, *ApJ*, 767, 122

Landy, S. D. & Szalay, A. S. 1993, *ApJ*, 412, 64

Norberg, P., Baugh, C. M., Gaztañaga, E., & Croton, D. J. 2009, *MNRAS*, 396, 19

Taylor, A., Joachimi, B., & Kitching, T. 2013, *MNRAS*, 432, 1928

Statistical Challenges in 21st Century Cosmology
Proceedings IAU Symposium No. 306, 2014
A. F. Heavens, J.-L. Starck & A. Krone-Martins, eds.

doi:10.1017/S1743921314013702

Transformationally decoupling clustering and tracer bias

Mark C. Neyrinck

Department of Physics and Astronomy, The Johns Hopkins University, Baltimore, MD 21211
email: `neyrinck@pha.jhu.edu`

Abstract. Gaussianizing transformations are used statistically in many non-cosmological fields, but in cosmology, we are only starting to apply them. Here I explain a strategy of analyzing the 1-point function (PDF) of a spatial field, together with the 'essential' clustering statistics of the Gaussianized field, which are invariant to a local transformation. In cosmology, if the tracer sampling is sufficient, this achieves two important goals. First, it can greatly multiply the Fisher information, which is negligible on nonlinear scales in the usual δ statistics. Second, it decouples clustering statistics from a local bias description for tracers such as galaxies.

Keywords. large-scale structure of universe, cosmology: theory, methods: statistical

1. Transformations and Information

Cosmologists have been trained to look at the world through linear two-point statistics: the power spectrum and correlation function of the overdensity field, $\delta = \rho/\bar{\rho} - 1$. This is for good reason: linear perturbation theory is naturally expressed in terms of the power spectrum of δ, which sources gravity. The raw power spectrum works well for the nearly Gaussian cosmic microwave background (CMB) as well, and the galaxy and matter density fields on large scales. Also, δ has the benefit that the power at a given scale is largely invariant if the resolution is increased. But the usual correlation function and power spectrum dramatically lose constraining power in a non-Gaussian field such as the matter or galaxy density field on small scales, so to reach the highest-possible precision in cosmology, other approaches are necessary.

Illustrating the nature of non-Gaussianity encountered in cosmology, Fig. 1 shows a dark-matter density slice from the Millennium simulation, together with the results after two operations: PDF (probability density function) Gaussianization, and randomizing Fourier phases. The original field and the phase-randomized field look staggeringly different, but have the same δ two-point statistics. The fact that the power spectrum does not distinguish these fields has been used to demonstrate the need to go to higher-point statistics. But I assert that this is the wrong way to go, to added complexity and difficulty of analysis.

The main difference by eye is simply in the PDF, shown at bottom left. But even more differences are extractable between the two starkly different fields at 'lower order' ($N \leqslant 2$). Applying a nonlinear mapping to a field changes its higher-point statistics. For example, a Gaussian field, with higher-point ($N > 2$) functions identically zero, will sprout nonzero higher-point functions at all orders if a nonlinear function is applied to it (Szalay 1988). I assert that it is useful to define Gaussianized clustering statistics ([$N > 1$]-point functions), in which first, the field is PDF-Gaussianized (i.e. a mapping is applied to give a field with as Gaussian a 1-point PDF as possible), and then clustering statistics are measured.

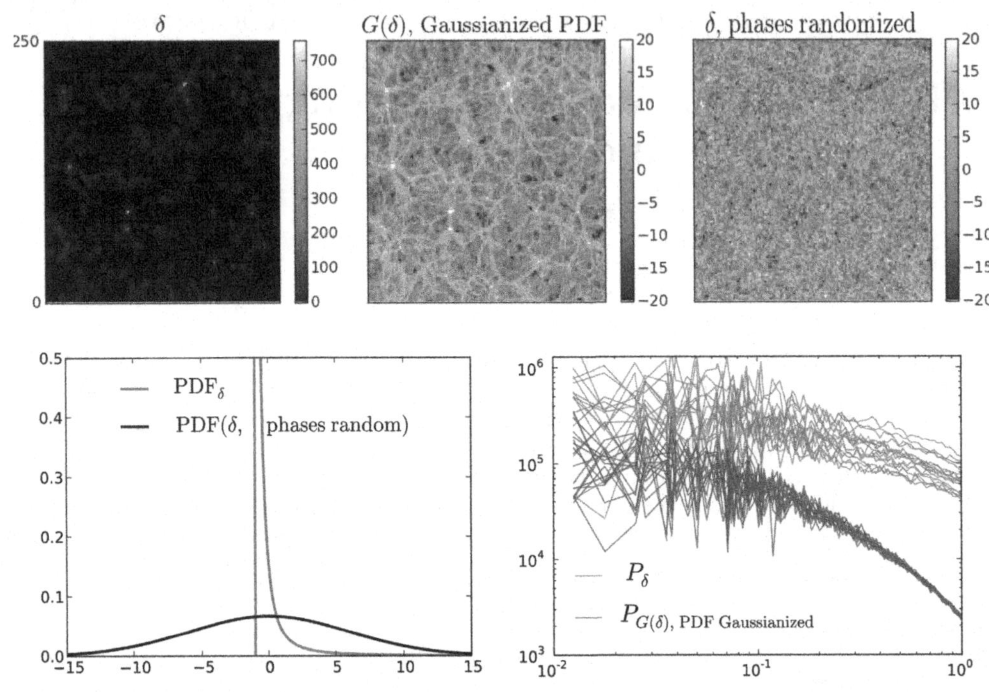

Figure 1. ***Upper left***: a quadrant of the the overdensity $\delta = \rho/\bar{\rho} - 1$ in a $2\,h^{-1}$ Mpc-thick slice of the $500\,h^{-1}$ Mpc Millennium simulation (Springel *et al.* 2005), viewed with an unfortunate linear color scale. ***Upper middle***: the same slice, rank-order-Gaussianized, i.e., applying a function on each pixel to give a Gaussian PDF. ***Upper right***: the δ field with its Fourier phases randomized. ***Lower left:*** PDFs of the upper-left and upper-right 3D fields. ***Lower right:*** P_δ and $P_{G(\delta)}$ of δ (red) and $G(\delta)$ (blue), measured from each of several 2D slices such as those shown in the upper-left and middle panels. The wild, coherent fluctuations in P_δ from slice to slice illustrate its high (co)variance, absent in $P_{G(\delta)}$, which has nearly the covariance properties as in a Gaussian random field.

The phase randomization already imparts a Gaussian PDF to the upper-right panel, so Gaussianization does nothing. Gaussianization greatly changes the upper-left panel, though, into the upper-middle panel. A mapping was applied to give a Gaussian PDF of variance $\mathrm{Var}[\ln(1+\delta)]$ for each slice. This changes the (2-point) power spectra in slices from the red to blue curves, as shown at bottom right in Fig. 1. This produces a change in both the mean, and, crucially, the covariance of the power spectrum. The covariance reduction can be seen qualitatively in the lower-right panel of Fig. 1, which shows the density power spectra P_δ and Gaussianized-density power spectra $P_{G(\delta)}$ for several Millennium Simulation slices. There are wild fluctuations (i.e. variance, or covariance) on small scales of P_δ that are not present in $P_{G(\delta)}$. These wild fluctuations show up as a drastic reduction in the cosmological-parameter Fisher information in P_δ, i.e. an enlargement in error bars (e.g. Meiksin and White 1999; Rimes and Hamilton 2005; Neyrinck *et al.* 2006; Neyrinck and Szapudi 2007; Takahashi *et al.* 2009). Analyzing $P_{G(\delta)}$, on the other hand, enhances cosmological parameter constraints, in principle by a factor of several (Neyrinck 2011).

My proposal is to measure the 1-point PDF and Gaussianized clustering statistics together; the Gaussianization step not only reduces covariance in the power spectrum itself, but also the covariance between the power spectrum and 1-point PDF. A mathematical reason to analyze the complete 1-point PDF (not simply its moments) is that, if it is sufficiently non-Gaussian, analyzing even arbitrarily high moments does not give all of its information, as has long been known in the statistical literature (e.g. Aitchison and Brown 1957). This was pointed out in a cosmological context by Coles and Jones (1991), and Carron (2011) more recently emphasized the mistakenness of the conventional wisdom that measuring all N-point functions gives all spatial statistical information in cosmology, and the consequences for constraining parameters.

2. Tracer bias

From a statistical point of view, measuring the 1-point PDF along with Gaussianized clustering statistics is a superior approach to measuring just the raw power spectrum. One might be fearful of additional irritants from galaxy-vs-matter bias in Gaussianized clustering statistics, but in fact Gaussianization automatically provides a natural framework to incorporate bias issues, potentially *simplifying* analysis greatly. Suppose the tracer δ_g, and matter density δ_m are related by any invertible function. Then mapping both PDF's onto the same function (for example, a Gaussian) will give precisely the same fields. This fact has long been exploited in the field of genus statistics (e.g. Rhoads *et al.* 1994; Gott *et al.* 2009); any local monotonic density transformation will leave Gaussianized statistics unchanged in a local-bias approximation. Gaussianization was first applied for power spectra in cosmology by Weinberg (1992). Unfortunately, it does not seem to reconstruct the initial conditions perfectly on small scales, as was the original aim, although it does largely restore the initial power spectrum's shape (Neyrinck *et al.* 2009). One way to understand this shape restoration is that whereas in P_δ, power smears only from large to small scales, power in $P_{G(\delta)}$ migrates rather evenly both upward and downward in scale (Neyrinck and Yang 2013). This is because P_δ is mainly sensitive to overdense regions where fluctuations contract, and a sort of one-halo shot noise appears from sharp spikes (Neyrinck *et al.* 2006). $P_{G(\delta)}$, on the other hand, is rather equally sensitive to underdense regions as well, where fluctuations expand.

Fig. 2, taken from Neyrinck *et al.* (2014), shows what Gaussianizing does for different tracers explicitly. It uses the MIP (*multim in parvo*) ensemble of N-body simulations (Aragón-Calvo 2012), in which 225 realizations were run with the same initial large-scale modes (with $k < 2\pi/(4h^{-1}\,\mathrm{Mpc})$), but differing small-scale modes. So each simulation gives a different realization of haloes in the same cosmic web. For Fig. 2, we averaged together the halo and matter density fields over the realizations, and measured the δ and $G(\delta)$ power spectra. In the ensemble, there is a rather clean mapping between mean matter density and mean halo density, a power law with an exponential cutoff at low density; see Neyrinck *et al.* (2014) for details. The correspondence in the Gaussianized power spectra is impressive.

However, in this discussion, we have ignored an important caveat: galaxy discreteness. If empty, zero-density pixels exist, this makes a naive logarithmic transform inapplicable. Also, if there are multiple pixels with the same density, then any assumed mapping from a perfect Gaussian to δ is not invertible. In this case, δ can be rank-ordered, and for a δ appearing multiple times, $G(\delta)$ can be set to the average of all $G(\delta)$ that would map to that range of δ; see Neyrinck (2011) for more details. This problem can be negligible, as in the cases of the two figures above, or it can be substantial, in the high-discreteness limit. A rule of thumb is to use pixels large enough to contain on average several galaxies. As long as this scale is in the non-linear regime, it will be fruitful to Gaussianize. A promising new alternative is an optimal transform for a pixelized Poisson-lognormal field (Carron and Szapudi 2014). This gives the maximum posterior density from a single pixel in a lognormal-Poisson Bayesian reconstruction, as in Kitaura *et al.* (2010).

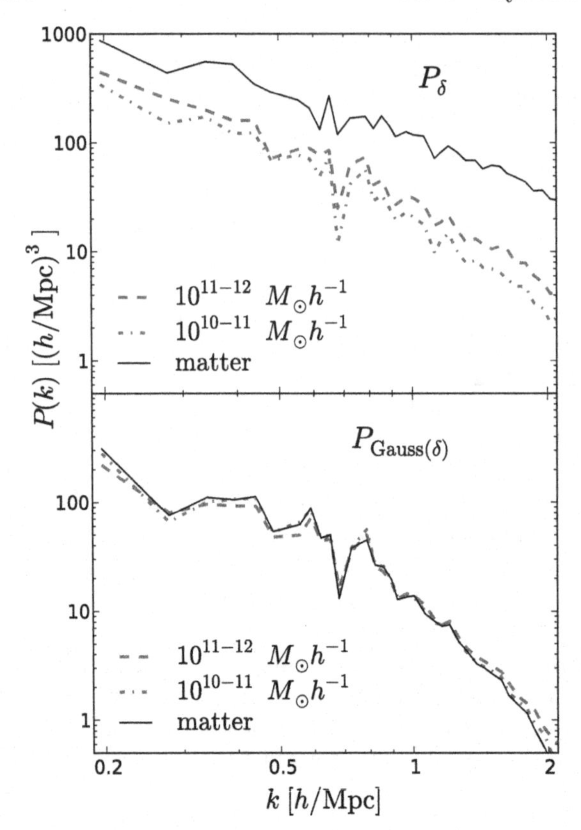

Figure 2. Power spectra of matter and two mass ranges of haloes in the MIP ensemble-mean fields. The Gaussianized-density power spectra $P_{\mathrm{Gauss}(\delta)}$ show substantially less difference among the various density fields than the raw density power spectra P_δ, supporting the hypothesis that a local, strictly-increasing density mapping captures the mean relationship between matter and haloes.

The usual δ clustering statistics have large statistical error bars on nonlinear scales, which can swamp errors from sub-optimal measurement. But Gaussianized clustering statistics have great statistical power; with that power comes great responsibility to measure them accurately, which is what we plan to do in future work.

References

Aitchison, J. & Brown, J. 1957, *The Lognormal Distribution*, Department of Applied Economics Monographs, Cambridge University Press

Aragón-Calvo, M. A. 2012, *MNRAS, submitted*

Carron, J. 2011, *ApJ* **738**, 86

Carron, J. & Szapudi, I. 2014, *MNRAS* **439**, L11

Coles, P. & Jones, B. 1991, *MNRAS* **248**, 1

Gott, J. R., Choi, Y.-Y., Park, C., & Kim, J. 2009, *ApJL* **695**, L45

Kitaura, F.-S., Jasche, J., & Metcalf, R. B. 2010, *MNRAS* **403**, 589

Meiksin, A. & White, M. 1999, *MNRAS* **308**, 1179

Neyrinck, M. C. 2011, *ApJ* **742**, 91

Neyrinck, M. C., Aragón-Calvo, M. A., Jeong, D., & Wang, X. 2014, *MNRAS* **441**, 646

Neyrinck, M. C. & Szapudi, I. 2007, *MNRAS* **375**, L51

Neyrinck, M. C., Szapudi, I., & Rimes, C. D. 2006, *MNRAS* **370**, L66

Neyrinck, M. C., Szapudi, I., & Szalay, A. S. 2009, *ApJL* **698**, L90

Neyrinck, M. C. & Yang, L. F. 2013, *MNRAS* **433**, 1628

Rhoads, J. E., Gott, III, J. R., & Postman, M. 1994, *ApJ* **421**, 1

Rimes, C. D. & Hamilton, A. J. S. 2005, *MNRAS* **360**, L82

Springel, V., White, S. D. M., Jenkins, A., Frenk, C. S., Yoshida, N., Gao, L., Navarro, J., Thacker, R., Croton, D., Helly, J., Peacock, J. A., Cole, S., Thomas, P., Couchman, H., Evrard, A., Colberg, J., & Pearce, F. 2005, *Nature* **435**, 629

Szalay, A. S. 1988, *ApJ* **333**, 21

Takahashi, R., Yoshida, N., Takada, M., Matsubara, T., Sugiyama, N., Kayo, I., Nishizawa, A. J., Nishimichi, T., Saito, S., & Taruya, A. 2009, *ApJ* **700**, 479

Weinberg, D. H. 1992, *MNRAS* **254**, 315

Statistical Challenges in 21st Century Cosmology
Proceedings IAU Symposium No. 306, 2014
A. F. Heavens, J.-L. Starck & A. Krone-Martins, eds.

doi:10.1017/S1743921314013866

Constraints on Growth Index from LSS

Athina Pouri[1,2], Spyros Basilakos[1] and Manolis Plionis[3,4,5]

[1]Academy of Athens, Research Center for Astronomy and Applied Mathematics, Soranou Efesiou 4, 11527, Athens, Greece

[2]Faculty of Physics, Department of Astrophysics - Astronomy - Mechanics, University of Athens, Panepistemiopolis, Athens 157 83

[3]Physics Dept., Sector of Astrophysics, Astronomy & Mechanics, Aristotle Univ. of Thessaloniki, Thessaloniki 54124, Greece

[4]Instituto Nacional de Astrofísica Óptica y Electronica, 72000 Puebla, México

[5]IAASARS, National Observatory of Athens, P.Pendeli 15236, Greece

emails: **athpouri@phys.uoa.gr**, **svasil@academyofathens.gr**, **mplionis@auth.gr**

Abstract. We utilize the clustering properties of the Luminous Red Galaxies (LRGs) and the growth rate data in order to constrain the growth index (γ) of the linear matter fluctuations based on a standard χ^2 joint likelihood analysis between theoretical expectations and data. We find a value of $\gamma = 0.56 \pm 0.05$, perfectly consistent with the expectations of the ΛCDM model, and $\Omega_{m0} = 0.29 \pm 0.02$, in very good agreement with the latest Planck results. Our analysis provides significantly more stringent growth index constraints with respect to previous studies as indicated by the fact that the corresponding uncertainty is only $\sim 0.09\gamma$.

Keywords. cosmological parameters, dark matter, large-scale structure of universe

---

1. Introduction

One of the most intriguing problems in Cosmology is to explain the fact that the universe is in a phase of accelerated expansion. In order to deal with this problem, one has to see Dark Energy (DE) either as a new field in nature or as a modification of General Relativity (see for review Copeland *et al.* 2006, Caldwell & Kamionkowski 2009, Amendola & Tsujikawa 2010). An interesting approach to discriminate between the aforementioned is to use the evolution of the linear growth of matter perturbations. Specifically, a useful tool in this kind of studies is the so called growth rate of clustering, which is defined as $f(a) = \frac{d\ln D}{d\ln a} \simeq \Omega_m^\gamma(a)$, where $a(z) = (1+z)^{-1}$ is the scale factor of the universe, $\Omega_m(a)$ is the dimensionless matter density parameter, γ is the growth index and $D(a) = \delta_m(a)/\delta_m(a=1)$ is the linear growth factor usually scaled to unity at the present time (Peebles 1993, Wang & Steinhardt 1998). Notice that γ may in general vary with redshift; $\gamma \equiv \gamma(z)$. Theoretically speaking, it has been shown that for those DE models which are within the framework of GR and have a constant Equation of State parameter, the growth index γ is equal to $\gamma \simeq 6/11$ for the ΛCDM model.

Since gravity reflects, via gravitational instability, on the nature of clustering (Peebles 1993) it has been proposed to use the clustering/biasing properties of the mass tracers in constraining cosmological models (see Matsubara 2004, Basilakos & Plionis 2005, Basilakos & Plionis 2006, Krumpe *et al.* 2013) as well as to test the validity of GR on extragalactic scales (Basilakos *et al.* 2012, for a recent review see Bean *et al.* 2013). Using the above notations, the aim of the current study is to place constraints on the (Ω_m, γ) parameter space using the joint analysis of the measured two-point angular correlation function (ACF) of LRGs (Sawangwit *et al.* 2011) and the growth rate of clustering data as collected by Basilakos *et al.* (2013) (see Table 1 of Basilakos *et al.* 2013).

2. Angular Correlation Function

Considering a spatially flat Friedmann-Lemaître-Robertson-Walker (FLRW) geometry, we can easily relate via the Limber's inversion equation the ACF, $w(\theta)$, with the two point spatial correlation function $\xi(r,z)$: $w(\theta) = 2\frac{H_0}{c}\int_0^\infty (\frac{1}{N}\frac{dN}{dz})^2 E(z)dz \int_0^\infty \xi(r,z)du$, where $1/N\ dN/dz$ is the normalized redshift distribution of the LRGs.

The redshift distribution for this case is given by the relation: $\frac{dN}{dz} \propto (\frac{z}{z_\star})^{(a+2)} e^{-(\frac{z}{z_\star})^\beta}$, where $(a, \beta, z_\star) = (-15.53, -8.03, 0.55)$ and $z_\star$ is the characteristic depth of the subsample studied. The spatial correlation function of the mass tracers is given by: $\xi(r,z) = b^2(z)\xi_{DM}(r,z)$, where $b(z)$ is the evolution of the linear bias, and ξ_{DM} is the corresponding correlation function of the underlying mass distribution which is written as $\xi_{DM}(r,z) = \frac{1}{2\pi^2}\int_0^\infty k^2 P(k,z)\frac{\sin(kr/a)}{(kr/a)}dk$. Concerning $P(k,z)$, it is $P(k,z) = D^2(z)P(k)$ with $P(k) = P_0 k^n T^2(k)$ denoting the CDM power spectrum of the matter fluctuations. Note that $T(k)$ is the CDM transfer function of Eisenstein & Hu (1998) or this of Bardeen *et al.* (1986) and $n \simeq 0.9671$ following the recent reanalysis of the Planck data by Spergel *et al.* (2013). The variable r corresponds to the physical separation between two sources having an angular separation, θ (in steradians) and $E(z) = H(z)/H_0$, is the normalized Hubble parameter. Non linear effects have been also taken into account through the slope of the power spectrum at the relevant scales: $n_{\rm eff} = d\ln P/d\ln k$ (Peacock & Dodds 1994, Smith *et al.* 2003, Widrow *et al.* 2009).

3. The evolution of linear bias, $b(z)$

The linear bias is defined as the ratio of density perturbations in the mass-tracer field to those of the underling total matter field: $b = \delta_{tr}/\delta_m$. In this analysis we use the bias evolution model of Basilakos *et al.* (2011) and Basilakos *et al.* (2012) which is valid for any DE model (scalar or geometrical) and it is given by:

$$b(z) = 1 + \frac{b_0 - 1}{D(z)} + C_2\frac{\mathcal{J}(z)}{D(z)} \tag{3.1}$$

with $\mathcal{J}(z) = \int_0^z \frac{(1+y)}{E(y)}dy$. The constants b_0 (the bias at the present time) and C_2 depend on the host dark matter halo mass M_h (see Basilakos *et al.*2012).

4. Fitting Theoretical Models to the data

A standard χ^2 minimization statistical analysis is implemented in order to provide constraints in the (Ω_{m0}, γ) parameter space. It is defined as follows:

$$x_t^2 = \sum_{i=1}^{n}\left[\frac{X_{obs} - X_{th}}{\sigma_i^2}\right]^2$$

where the index t stands for the clustering data of LRGs or the growth data, X is the angular correlation function or the $f\sigma_8$ respectively and σ_i is the corresponding uncertainty. We perform a joint likelihood analysis of the two cosmological probes using:

$$x^2 = x_{LRGs}^2 + x_{gr}^2$$

The joint result of the analysis for the case of Eisenstein & Hu (1998) transfer function is $(\Omega_{m0}, \gamma) = (0.29 \pm 0.02, 0.56 \pm 0.05)$, which is the strongest (to our knowledge) joint constraint appearing in the literature. The results are shown in Table 2. For a more

Table 1. Results in the $(\Omega_{m0}, \gamma, M_h, n_{\rm eff})$ parameter space for the different $T(k)$ and σ_8.

$T(k)$	Ω_{m0}	γ	$M_h/10^{13}M_\odot$	$n_{\rm eff}$	$\chi^2_{t,\min}/df$
$\sigma_8 = 0.797\,(0.30/\Omega_{m0})^{0.26}$ Hajian *et al.* 2013					
Eisenstein & Hu (1998)	0.29 ± 0.02	0.56 ± 0.05	1.90 ± 0.10	0.10 ± 0.20	$16.36/23$
Bardeen et al (1986)	0.29 ± 0.01	0.56 ± 0.10	1.80 ± 0.30	$-0.10^{+0.30}_{-0.10}$	$16.56/23$
$\sigma_8 = 0.818\,(0.30/\Omega_{m0})^{0.26}$ Spergel *et al.* 2013					
Eisenstein and Hu (1998)	$0.29^{+0.03}_{-0.02}$	$0.58^{+0.02}_{-0.06}$	1.70 ± 0.20	0.30 ± 0.20	$15.90/23$
Bardeen et al (1986)	$0.29^{+0.02}_{-0.03}$	0.56 ± 0.10	1.60 ± 0.4	$0.0^{+0.10}_{-0.20}$	$16.13/23$

detailed analysis, as well as for the case of a varying growth index with redshift $\gamma = \gamma(z)$, see Pouri *et al.* (2014).

Acknowledgements

Athina Pouri is greatly thankful to the Scientific and Local Organizing Committee of the IAU Symposium 306 for the financial support.

References

Copeland, E. J., Sami, M., & Tsujikawa, S. 2006, *Int. J. of Mod. Phys. D.*, 15, 1753

Caldwell, R. R. & Kamionkowski, M. 2009, *Ann. Rev. Nucl. Part. Sci.*, 59, 397

Amendola, L. & Tsujikawa, S. 2010, *Dark Energy: Theory and Observations, Cambridge University Press, Cambridge UK*

Peebles, P. J. E. 1993, *"Principles of Physical Cosmology", Princeton University Press, Princeton, New Jersey*

Wang, L. & Steinhardt, P. J. 1998, *ApJ*, 508, 483

Matsubara, T. 2004, *ApJ*, 615, 573

Basilakos, S. & Plionis, M. 2005, *MNRAS*, 360, L35

Basilakos, S. & Plionis, M. 2006, *ApJ*, 650, L1

Krumpe, M., Miyaji, T., & Coil, A. L. 2013, *arXiv:1308.5976*, 650, L1

Basilakos, S., Dent, J. B., Dutta, S., Perivolaropoulos, L., & Plionis, M. 2012, *Phys. Rev. D.*, 85, 123501

Bean, R., *et al.* 2013, *arXiv:1309.5385*

Sawangwit, U., *et al.* 2011, *MNRAS*, 416, 3033

Basilakos, S., Nesseris, S., & Perivolaropoulos, L. 2013, *Phys. Rev. D.*, 87, 123529

Eisenstein, D. J. & Hu, W. 1998, *ApJ*, 496, 605

Bardeen, J. M., Bond, J. R., Kaiser, N., & Szalay, A. S. 1986, *ApJ*, 304, 15

Spergel, D., Flauger, R., & Hlozek, R. 2013, *arXiv:1312.3313*

Peacock, J. A. & Dodds, S. J. 1994, *MNRAS*, 267, 1020

Smith, R. E., *et al.* 2003, *MNRAS*, 341, 1311

Widrow, L. M., Elahi, P. J., Thacker, R. J., Richardson, M., & Scannapieco, E. 2009, *MNRAS*, 397, 1275

Basilakos, S., Plionis, M., & Pouri, A. 2011, *Phys. Rev. D.*, 83, 123525

Hajian, A., Battaglia, N., Spergel, D. N., Bond, J. R., Pfrommer, C., & Sievers, J. L. 2013, *JCAP*, 11, 64

Pouri, A., Basilakos, S., & Plionis, M. 2014, *arXiv:1409.0964*

Nesseris, S., Basilakos, S., Saridakis, E. N., & Perivolaropoulos, L. 2013, *Phys. Rev. D.*, 88, 103010

Statistical Challenges in 21st Century Cosmology
Proceedings IAU Symposium No. 306, 2014
A. F. Heavens, J.-L. Starck & A. Krone-Martins, eds.

doi:10.1017/S1743921314010904

Non-Gaussian inference from non-linear and non-Poisson biased distributed data

Metin Ata, Francisco-Shu Kitaura and Volker Müller

Leibniz Institute for Astrophysics (AIP),
An der Sternwarte 16, 14482 Potsdam
email: mata@aip.de

Abstract. We study the statistical inference of the cosmological dark matter density field from non-Gaussian, non-linear and non-Poisson biased distributed tracers. We have implemented a Bayesian posterior sampling computer-code solving this problem and tested it with mock data based on N-body simulations.

Keywords. cosmology: large-scale-structures, methods: statistical

---

1. Introduction

The distribution of galaxies poses a challenging multivariate statistical problem. Galaxies are biased tracers of the underlying three-dimensional dark matter density field. To accurately infer such fields one needs to account for the non-Gaussian, non-Poisson and non-linear biased distribution of galaxies. We show that this is possible within the Bayesian formalism by explicitly writing down the posterior distribution including these effects.

2. Method

We rely on the Bayesian framework to express the posterior distribution function (likelihood weighted prior) of matter fields given a set of tracers, a biasing and a structure formation model. In particular we consider a negative binomial distribution for the likelihood modeling the data (the galaxy field). This permits us to model the over-dispersed galaxy counts (see Kitaura *et al.* 2014 and references therein). The expected galaxy number density is related to the dark matter density through a nonlinear scale-dependent expression extracted from N-body simulations (see Cen & Ostriker 1993; de la Torre & Peacock 2013; Kitaura *et al.* 2014; Neyrinck *et al.* 2014). In this way we extend the works based on the Poisson and linear bias models (Kitaura & Enßlin 2008; Kitaura *et al.* 2010; Jasche & Kitaura 2010; Jasche & Wandelt 2013) following the ideas presented in Kitaura 2012 and Kitaura *et al.* 2014. In particular, we implement these improvements in the ARGO Hamiltonian-sampling code able to jointly infer density, peculiar velocity fields and power-spectra (Kitaura, Gallerani & Ferrara 2012). For the prior distribution describing structure formation of the dark matter field we use the lognormal assumption (Coles & Jones 1991). We note however, that this prior can be substituted by another one, e. g. based on Lagrangian perturbation theory (see Kitaura 2013; Kitaura *et al.* 2012; Jasche & Wandelt 2013; Wang *et al.* 2013; Heß, Kitaura & Gottlöber 2013). Alternatively, one can extend the lognormal assumption in an Edgeworth expansion to include higher order correlation functions (Colombi 1994; Kitaura 2012). We calculate the posterior distribution on a grid of N_C cells dividing our observed volume. The lognormal distribution for the dark matter field $\delta_{\rm M}$ can be written as a function $\mathcal{P}(\delta_{\rm M}|\mathbf{S}(\{p_C\}))$, with $\{p_C\}$ being a set of cosmological parameters that go into the covariance matrix **S**. For the likelihood we consider the negative binomial distribution (NB), so that the probability of observing N_i galaxies in a voxel given the expectation value $\lambda^i = f_{\bar{N}} w^i (1+\delta^i_{\rm M})^\alpha \exp[-(\frac{1+\delta^i_{\rm M}}{\rho_\epsilon})^\epsilon]$ is given in the following form: $\mathcal{L}(\mathbf{N}|\lambda,\beta)$, with the parameter β modeling the deviation of Poissonity. For $\beta \to \infty$ or for low expectation values (low λ) the NB tends towards the Poisson distribution function. We note that deviations from Poissonity to model the galaxy distribution have been considered in previous works (see e. g. Saslaw & Hamilton 1984; Sheth 1995).

The expected number counts λ^i is constructed from the mean number per voxel $\bar{N}$, the completeness w^i, and the density $(1+\delta^i_{\rm M})^\alpha \exp[-(\frac{1+\delta^i_{\rm M}}{\rho_\epsilon})^\epsilon]$ in the voxel. $f_{\bar{N}}$ is the normalisation factor: $f_{\bar{N}} = \frac{\bar{N}}{\langle (1+\delta^l_{\rm M})^\alpha e^{[-(\frac{1+\delta^l_{\rm M}}{\rho_\epsilon})^\epsilon]}}\rangle$, where $\langle ... \rangle$ denotes the ensemble average over the whole volume. The biasing parameters are given by $\{\alpha, \beta, \epsilon, \rho_\epsilon\}$. Combining these terms to a posterior function gives a full description of the desired probability to infer the dark matter field from the observed galaxy distribution $P(\delta_M|\mathbf{N}, \mathbf{S}(\{p_C\})) \propto \mathcal{P}(\delta_{\rm M}|\mathbf{S}(\{p_C\})) \times \mathcal{L}(\mathbf{N}|\lambda,\beta)$:

$$P(\delta_M|\mathbf{N}, S(\{p_C\})) = \frac{1}{\sqrt{(2\pi)^{N_C}\det(\mathbf{S})}} \prod_{l=1}^{N_C} \frac{1}{1+\delta^l_M} \tag{2.1}$$

$$\times \exp\Big(-\frac{1}{2}\sum_{ij}\Big[(\ln(1+\delta^i_M) - \mu^i) S^{-1}_{ij} (\ln(1+\delta^j_M) - \mu^j)\Big]\Big)$$

$$\times \prod_{l=1}^{N_C} \left(\frac{f_{\bar{N}} w^l (1+\delta^l_{\rm M})^\alpha e^{\left[-\left(\frac{1+\delta^l_{\rm M}}{\rho_\epsilon}\right)^\epsilon\right]} \Gamma(\beta + N^l)}{N^l!\,\Gamma(\beta)\left(\beta + f_{\bar{N}} w^l (1+\delta^l_{\rm M})^\alpha e^{\left[-\left(\frac{1+\delta^l_{\rm M}}{\rho_\epsilon}\right)^\epsilon\right]}\right)^{N^l} \left(1 + \frac{f_{\bar{N}} w^l (1+\delta^l_M)^\alpha e^{\left[-\left(\frac{1+\delta^l_{\rm M}}{\rho_\epsilon}\right)^\epsilon\right]}}{\beta}\right)^\beta} \right).$$

Note that for the field $\Phi = \ln(\varrho) - \langle\varrho\rangle = \ln(1+\delta_{\rm M}) - \mu$ our prior is exactly a Gaussian prior. Therefore, our method yields at the same time the optimal logarithmic Gaussianised density field (Neyrinck *et al.* 2009, Carron & Szapudi 2014).

3. Validation

We employ our recently developed Hamiltonian Markov Chain Monte Carlo based computer code for Bayesian inference from the posterior distribution shown in Eq. 2.1 (see Ata *et al.*in prep). To test our method we take a subsample of $2 \cdot 10^5$ halos from the halo catalog of the Bolshoi dark matter only N-body simulation at redshift zero comprising the halo masses between 10^9 and 10^{14} $M_\odot$ (Klypin *et al.* 2011). We run two independent chains: the first one with the classical Poisson-Lognormal model, and the second one with the novel NB-Lognormal model. Our results demonstrate that our new model is able to recover the dark matter density field yielding unbiased power-spectra (within 5%) in the k-range of 0.02 to 0.6 h Mpc^{-1} (see solid blue and bashed dotted light blue lines in Fig. 1). However, the Hamiltonian sampling run with the Poisson-Lognormal model including the same nonlinear deterministic bias produces biased reconstructions with power-spectra deviating about 20 % at $\sim k = 0.4$ h Mpc^{-1} (see red dashed line in Fig. 1). The superior three-dimensional resamblance between the original dark matter field from the N-body ($8 \cdot 10^9$ dark matter particles) and the reconstruction (based on

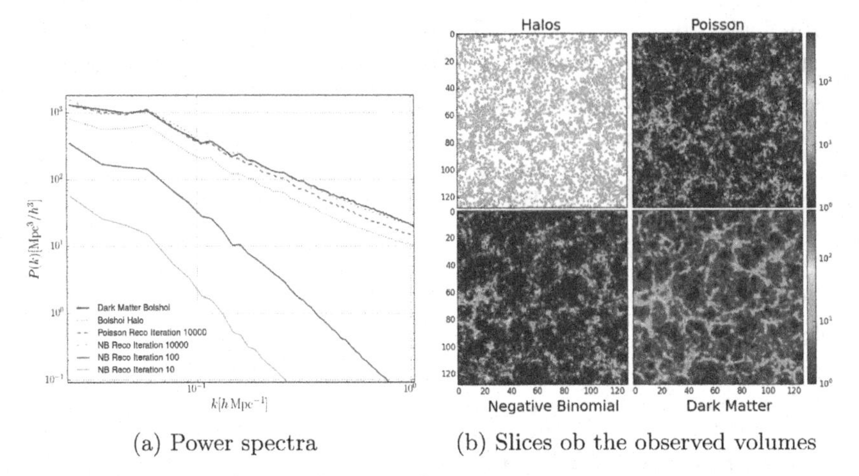

(a) Power spectra (b) Slices ob the observed volumes

Figure 1. Power spectra
Slices ob the observed volumes

$2 \cdot 10^5$ halos), using the NB likelihood as compared to the Poisson likelihood, is apparent in the slice cuts shown in Fig. 1.

4. Conclusions

We have introduced a detailed posterior distribution within the Bayesian framework accurately modeling the statistical nature of the distribution of galaxies. Moreover, we have implemented a Hamiltonian sampling code to infer the corresponding matter density fields. We validated our method against realisitic mock data for which the underlying dark matter density field is known. Our numerical tests emphasize the importance of a scale-dependent nonlinear bias and the deviation from Poissonity.

References

Carron, J. & Szapudi, I., 2014, *MNRAS* (Letters), 439, L11
Cen, R. & Ostriker, J. P., 1993, *ApJ*, 417, 415
Coles, P. & Jones, B. J. T., 1991, *MNRAS*, 248, 1
Colombi, S., 1994, *ApJ*, 435, L536
de la Torre, S. & Peacock, J. A., 2013, *MNRAS*, 435, 743
Heß, S., Kitaura, F.-S.& Gottlöber, S., 2013, *MNRAS*, 435, 2065
Jasche, J. & Kitaura, F.-S., 2010a, *MNRAS*, 407, 29
Jasche, J.& Wandelt, B. D., 2013, *ApJ*, 779, 15
Kitaura, F.-S., 2012, *Springer Series in Astrostatistics*, 143
Kitaura, F.-S., 2012, *MNRAS*, 420, 2737
Kitaura, F.-S., 2013, *MNRAS* (Letters), 429, 1, L84
Kitaura, F.-S. & Enßlin, T. A., 2008, *MNRAS*, 389, 497
Kitaura, F.-S., Erdogdu, P., Nuza, S. E., Khalatyan, A., Angulo, R. E., Hoffman, Y.& Gottlöber, S., 2012, *MNRAS* (Letters), 427, L35
Kitaura, F.-S., Gallerani, S., & Ferrara, A., 2012, *MNRAS*, 420, 61
Kitaura, F.-S., Jasche J. & Metcalf R. B., 2010, *MNRAS*, 403, 589
Kitaura, F.-S., Yepes, G.& Prada, F., 2014, *MNRAS* (Letters), 439, L21
Klypin, A. A., Trujillo-Gomez, S., & Primack, J., 2011, *ApJ*, 740, 102
Neyrinck, M. C., Aragón-Calvo, M. A., Jeong, D., & Wang, X. 2014, *MNRAS*, 441, 646

Neyrinck, M. C. and Szapudi, I., & Szalay, A. S., 2009, *ApJ* (Letters), 698L, 90N
Saslaw, W. C. & Hamilton, A. J. S.., 1984, *ApJ*, 276, 13
Sheth, R. K., 1995, *MNRAS*, 274, 21
Wang, H., Mo H. J., Yang X. & van den Bosch F. C., 2012, *MNRAS*, 420, 1809

Statistical Challenges in 21st Century Cosmology
Proceedings IAU Symposium No. 306, 2014
A. F. Heavens, J.-L. Starck & A. Krone-Martins, eds.

doi:10.1017/S1743921314011041

A Monte Carlo study of cosmological parameter estimators from galaxy cluster number counts

Mariana Penna-Lima[1], Martín Makler[2] and Carlos A. Wuensche[1]

[1]Divisão de Astrofísica, Instituto Nacional de Pesquisas Espaciais,
Av. dos Astronautas 1758, 12227-010, São José dos Campos, Brazil
email: mariana.lima@inpe.br

[2]Centro Brasileiro de Pesquisas Físicas,
Rua Dr. Xavier Sigaud 150 22290-180, Rio de Janeiro, Brasil
email: martin@cbpf.br
email: ca.wuensche@inpe.br

Abstract. Models for galaxy clusters abundance and their spatial distribution are sensitive to cosmological parameters. Present and future surveys will provide high-redshift sample of clusters, such as Dark Energy Survey ($z \leqslant 1.3$), making cluster number counts one of the most promising cosmological probes. In the literature, some cosmological analyses are carried out using small cluster catalogs (tens to hundreds), like in Sunyaev-Zel'dovich (SZ) surveys. However, it is not guaranteed that maximum likelihood estimators of cosmological parameters are unbiased in this scenario. In this work we study different estimators of the cold dark matter density parameter Ω_c, σ_8 and the dark energy equation of state parameter w_0 obtained from cluster abundance. Using an unbinned likelihood for cluster number counts and the Monte Carlo approach, we determine the presence of bias and how it varies with the size of the sample. Our fiducial models are based on the South Pole Telescope (SPT). We show that the biases from SZ estimators do not go away with increasing sample sizes and they may become the dominant source of error for an all sky survey at the SPT sensitivity.

Keywords. (cosmology:) cosmological parameters, galaxies: clusters: general

1. Introduction

Galaxy cluster abundance and their spatial distribution as functions of redshift and mass provide strong constraints on the matter density parameter Ω_m and the amplitude σ_8 of the power spectrum. For high redshift surveys ($z > 1.0$), clusters are also powerful tools to study dark energy (DE) since they depend on the linear growth of density perturbations and the comoving volume (see Allen *et al.* 2011 and references therein).

Usually the cluster likelihood function is built taking into account the following error sources: the uncertainties of the mass-observable relations and the photometric redshifts, sample-variance and shot-noise. However, in general, it is assumed that the cosmological parameter estimators themselves do not induce any extra uncertainty.

The Maximum Likelihood (ML) method is extensively used to estimate the values of parameters given a data set. ML estimators are usually consistent, i.e., the expected value of a parameter tends to its true value when the size of the data sample (n) is sufficiently large, and asymptotically efficient, i.e., the variance of the estimator attains the minimum variance bound as $n \to \infty$. However, ML estimators are not necessarily unbiased.

In this work, we study some estimators of Ω_c, σ_8 and w_0 in the context of SZ surveys, specifically from the SPT. For the redshift and mass ranges of SPT catalogs, sample variance is negligible. Therefore, we build the likelihood only for the cluster number counts and do not include the spatial clustering.

2. Methodology

The likelihood for cluster number counts is

$$\ln \mathcal{L}(\{\theta_j\}, \{\xi_i, z_i\}) = \sum_{i=1}^{n} \ln \left(\frac{d^2 N(\xi_i, z_i, \{\theta_j\})}{dz d\xi} \right) - N(\{\theta_j\}) - \ln n!, \tag{2.1}$$

where n is the total number of clusters, ξ is the detection significance (Vanderlinde *et al.* 2010 and Benson *et al.* 2013) and

$$\frac{d^2 N(\xi_i, z_i, \{\theta_j\})}{dz d\xi} = \int d\ln M \int d\zeta \frac{d^2 N(M, z_i, \{\theta_j\})}{dz d\ln M} P(\xi_i|\zeta)\, P(\ln\zeta|\ln M). \tag{2.2}$$

The normalization factor $N(\{\theta_j\})$ is the expected number of clusters with $0.3 \leqslant z \leqslant 1.1$ and $\xi_{min} \geqslant 5$. We assume that the catalog is complete above this threshold and in the entire redshift interval.

Estimators for a given set of parameters $(\{\hat{\theta}_j\})$ are obtained by minimizing $-2\ln \mathcal{L}(\{\theta_j\}, \{\xi_i, z_i\})$ with respect to these parameters. The bias of $\hat{\theta}_j$ is defined as

$$b_{\hat{\theta}_j} = \langle \hat{\theta}_j \rangle - \theta_j^0, \tag{2.3}$$

where $\langle \hat{\theta}_j \rangle$ is the expected value and θ_j^0 is the true value (fiducial). We cannot compute $\langle \hat{\theta}_j \rangle$ analytically. Therefore, we use the Monte Carlo (MC) method, i.e., we generate a set of realizations of the input fiducial model and compute the best-fitting value of $\hat{\theta}_j$ for each realization. Then we compute the expected value using as an estimate

$$\langle \hat{\theta}_j \rangle \approx \bar{\hat{\theta}}_j \pm \sigma\left(\bar{\hat{\theta}}_j\right), \qquad \bar{\hat{\theta}}_j = \sum_{l=1}^{m} \frac{\hat{\theta}_{jl}}{m}, \quad \sigma\left(\bar{\hat{\theta}}_j\right) = \frac{\sigma(\hat{\theta}_j)}{\sqrt{m}}, \tag{2.4}$$

where m is the number of realizations and $\hat{\theta}_{jl}$ is the best-fitting value for the l-th realization. In cases where $b_{\hat{\theta}_j} \neq 0$, the significance of this bias is ensured computing $\sigma(\bar{\hat{\theta}}_j)$. In particular for our purposes, it was necessary to generate 10,000 realizations for each choice of the fiducial model.

To obtain a realization, we first randomly generate the number of objects n using $N(\{\theta_j^0\})$ as the mean of a Poisson distribution. Then, we apply the inverse transform sampling (ITS): giving the probability distribution of finding a cluster with detection significance greater than ξ_{min} and in $[z, z+dz]$, i.e.,

$$\mathcal{P}(z)dz = \frac{1}{N(\{\theta_j\})} \left(\frac{dN}{dz} \right) dz, \tag{2.5}$$

we define the cumulative

$$f(z) = \int_0^z \mathcal{P}(z')dz'. \tag{2.6}$$

Given that $f(z)$ is a monotonically increasing function, whose image is $[0, 1]$, and a random variable with uniform distribution over $[0, 1]$, we generate n random numbers $\{u_i\}$ from a uniform distribution and invert Eq. 2.6 obtaining $z(f)$ and, therefore,

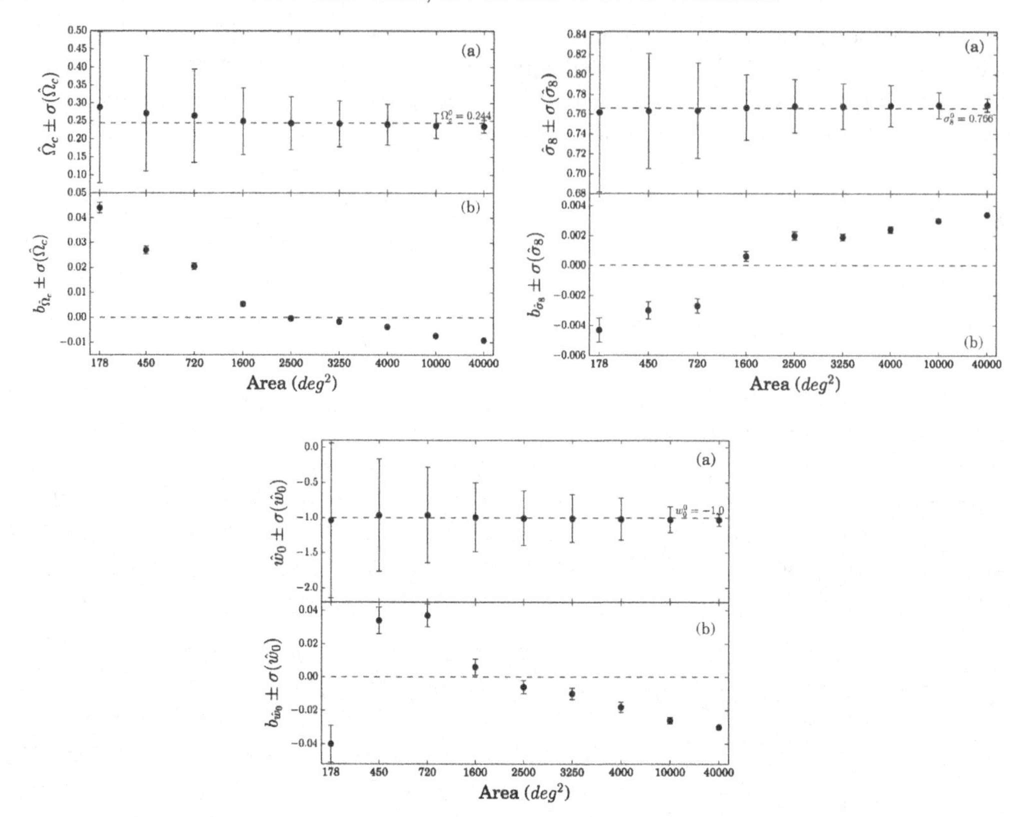

Figure 1. The expected values of the estimators (dots) are displayed with the error bars of the estimators (a) and of their means (b) for different survey areas. This analysis was done in the three-dimensional parametric space $(\Omega_c, \sigma_8, w_0)$, keeping all other parameters fixed.

$\{z_i\} = \{z(u_i)\}$. Analogously, we obtain a value ξ_i from the conditional distribution of obtaining a cluster with detection significance in the range ξ and $\xi + d\xi$ given z_i,

$$P(\xi|z_i) = \frac{P(\xi, z_i)}{\mathcal{P}(z_i)}. \tag{2.7}$$

Then, we apply the ITS method, given the cumulative $f(\xi|z_i)$, obtaining $\{\xi_i, z_i\}$.

3. Results and Concluding Remarks

The fiducial model corresponds to the best-fit values obtained in Benson *et al.* (2013) by combining cluster abundance, Hubble parameter and Big-Bang nucleosynthesis measurements, assuming a ΛCDM model: $\Omega_c = 0.244$, $\Omega_b = 0.0405$, $H_0 = 73.9$, $n_s = 0.966$, $\sigma_8 = 0.766$, $w_0 = -1$, $A_{SZ} = 5.31$, $B_{SZ} = 1.39$, $C_{SZ} = 0.9$ and $D_{SZ} = 0.21$. Giving this fiducial model and varying only the angular survey area, we apply the MC approach and compute Eqs. (2.3) and (2.4).

Fig. 1 shows that almost all estimators of $\hat{\Omega}_c$, $\hat{\sigma}_8$ and $\hat{w}_0$ are biased. Both $b_{\hat{\theta}}$ and the relative biases, $B_{\hat{\theta}_j} \equiv b_{\hat{\theta}_j}/\sigma(\hat{\theta}_j)$, depend on the survey area (sample size). $B_{\hat{\theta}}$ for $\Delta\Omega = 10,000$ and $40,000\, deg^2$ are not negligible, namely, for the last, $B_{\hat{\Omega}_c} = 52\%$, $B_{\hat{\sigma}_8} = 51\%$ and $B_{\hat{w}_0} = 33\%$. These results in the full-sky limit evince that cluster abundance estimators using SPT ξ-mass relation are not consistent.

Besides the results presented, we performed a comprehensive study in Penna-Lima *et al.* (2014), considering, for example, photometric redshift uncertainty and a combined likelihood of clusters and the distance priors of the cosmic microwave background. To perform this work, we developed fast, accurate, and adaptable codes for cluster counts in the framework of the Numerical Cosmology library (www.nongnu.org/numcosmo).

References

Allen, S. W., Evrard, A. E., & Mantz, A. B. 2011, *ARAA*, 49, 409
Benson, B. A., de Haan, T., Dudley, J. P., Reichardt, C. L., Aird, K., *et al.* 2013, *ApJ*, 763, 147
Penna-Lima, M., Makler, M., & Wuensche, C. A. 2014, *JCAP*, 5, 39
Vanderlinde, K., Crawford, T. M., de Haan, T., Dudley, J., Shaw, L., *et al.* 2010, *ApJ*, 722, 1180

Statistical Challenges in 21st Century Cosmology
Proceedings IAU Symposium No. 306, 2014
A. F. Heavens, J.-L. Starck & A. Krone-Martins, eds.

doi:10.1017/S1743921314013738

Understanding Cosmological Measurements with a large number of mock galaxy catalogues.

M. Manera[1]†, W. J. Percival[2], Ashley Ross[2], R. Tojeiro[3], L. Samushia[2], C. Howlett[2] M. Vargas-Magaña[4] and A. Burden[2], for the SDSS-III BOSS Galaxy Working Group.

[1]Department of Physics & Astronomy, University College London,Gower St, London WC1E 6BT, UK

[2]Institute of Cosmology and Gravitation, Dennis Sciama Building, University of Portsmouth, Burnaby Road, Portsmouth, PO1 3FX, UK

[3]School of Physics & Astronomy, St Andrews University, North Haugh St Andrews, KY16 9SS, UK

[4]Department of Physics, Carnegie Mellon University, 5000 Forbes Avenue, Pittsburgh, PA 15213, USA

Abstract. Mock galaxy catalogues are essential to the error analysis of cosmological measurements from big galaxy surveys covering thousands of square degrees in the sky, like BOSS, WiggleZ, DES, or Euclid. The PTHalos mock galaxy catalogues were used in the BOSS survey to analyse the BAO measurement from the CMASS ($z \sim 0.57$) and LOWZ ($z \sim 0.32$) galaxy samples, which provided the best estimate to date of the cosmic distance scale from galaxy surveys at these redshifts. We review the PTHalos mocks galaxy catalogues and their key contributions to these analyses.

Keywords. cosmology: large scale structure of the universe, distance scale; methods: simulations, statistical

---

1. The BAO scale measurement from BOSS galaxy samples

The Baryon Acoustic Oscillation Survey (BOSS) is a spectroscopic survey that targets galaxies from SDSS-III (Eisenstein *et al.* 2011) imaging. BOSS targets two galaxy samples, CMASS ($z \sim 0.57$) and LOWZ ($z \sim 0.32$), which respectively cover 8,976 and 7,341 square degrees in the Data Release 11 (DR11).

BOSS has detected and accurately measured the Baryon Acoustic Oscillations (BAO) in the two-point correlation functions and the power spectra of both the CMASS and LOWZ DR11 galaxy samples (Anderson *et al.* 2014). The BAO scale is given by the distance the sound waves travel before the coupling between baryons and radiation breaks down, r_d, which is 103.8 h^{-1} Mpc in Anderson *et al.* (2014) fiducial flat LCDM cosmology.

The cosmic distance measurement, D_V is related to the observed BAO scale, and is a combination of the angular distance, D_A, and the Hubble parameter $H(z)$,

$$D_V = [cz(1+z)^2 D_A^2(z) H^{-1}(z)]^{1/3} \ . \tag{1.1}$$

The corresponding cosmic distance measurement for the CMASS galaxy sample is given by $D_V\, r_{d,fid} = \alpha\, D_{V,fid}\, r_d = (2056 \pm 20)\, r_d$ (which is at 1% accuracy); and for the LOWZ galaxy sample by $D_V r_{d,fid} = (1264 \pm 25)\, r_d$ (which is sub 2% accuracy), where α parametrizes the position of the BAO peak.

† email:m.miret@ucl.ac.uk

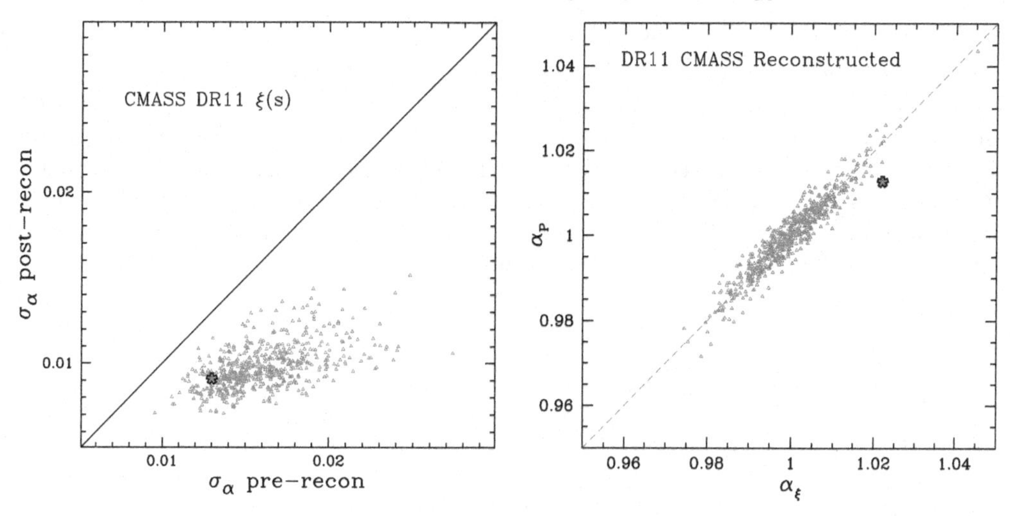

Figure 1. Left. Comparison between the errors of the BAO peak, α, before and after applying reconstruction. Right. Comparison between the BAO peak position α estimated from the correlation function $\xi(s)$ and that estimated from the power spectrum $P(k)$. Both panels show results from the 600 CMASS PThalos mock catalogues. The results from data are shown as stars. For more details see Anderson *et al.* 2014.

2. The creation of PTHalos mock galaxy catalogues

The 600 (CMASS) and 1,000 (LOWZ) PTHalos mock galaxy catalogues used to analyse the clustering of BOSS galaxy samples have been constructed in four main steps (Manera *et al.* 2013 & 2014):

- First, dark matter particle fields are created using 2nd order Lagrangian Perturbation Theory (2LPT). This step is what makes this method fast.
- Second, using a Friends of Friends (FoF) algorithm (with the appropriate linking length from theory) halos are found.
- Third, these halos are then populated with galaxies using a Halo Occupation Distribution prescription that fits the data
- Fourth, finally the observational masks are applied.

3. PTHalos mocks and the BAO measurement analysis

The reconstruction technique (see eg Padmanabhan *et al.* 2012) moves galaxies back in time in order to recover a more linear correlation function, and thus a more prominent BAO peak, i.e, one less smoothed by the non-linear evolution. The PThalos mock catalogues have been used to characterize the gain in accuracy when using this technique (Anderson *et al.* 2014). The results for the CMASS sample are shown in the left panel of Fig 1; points show the mocks, and the star shows the data.

The BAO scale can be measured using different estimators. The PTHalos mocks have been used to characterize the correlation between them (Anderson *et al.*2014), enabling their combination. This is shown in the right panel of Fig 1. Mock galaxy catalogues have also been essential to understand systematics (Ross *et al.* 2012, Vargas-Magaña *et al.* 2014) and to choose the best binning for the BAO measurement (Percival *et al.* 2014).

4. Finite number of mock catalogues

Having a large number of mock galaxy catalogues allows for the computation of covariance matrices. However, as the number of mock catalogues is finite, an accurate estimation of the cosmic distance measurement error requires correcting for the error in the inverse covariance matrix caused by the finite number of mocks available (Taylor *et al.*2012), and for the derived parameter error. Given a covariance matrix, C, which has been computed with n_s mock catalogues for correlation function or power spectrum with n_b bins, the unbiased estimator of the inverse covariance Ψ is given by:

$$\Psi = \frac{n_s - n_b - 2}{n_s - 1} C^{-1} \tag{4.1}$$

Furthermore, the error corrections for the derived parameters depend on the way in which the parameter error is determined, and are different if this is calculated from the distribution of recovered values from the mocks, or the likelihood surface (Percival *et al.* 2014). The correction factor to the estimate of the variance of a parameter given by the standard method of integrating over the likelihood

$$c_1 = \frac{1 + B(n_s - n_p)}{1 + A + B(n_p + 1)} \tag{4.2}$$

where $A = [0.5(n_s - n_b - 1)(n_s - n_b - 4)]^{-1}$ and $B = 0.5A(n_s - n_b - 2)$ If the variance is computed from the distribution of the best fit values of the mocks catalogues that have also been used to compute the covariance matrix, then the correction of the variance requires an extra factor (Percival *et al.* 2014) of $c_2 = \frac{n_s - 1}{n_s - n_b - 2}$.

5. Acknowledgements

SDSS-III is managed by the Astrophysical Research Consortium for the Participating Institutions of the SDSS-III Collaboration including the University of Arizona, the Brazilian Participation Group, Brookhaven Na- tional Laboratory, University of Cambridge, Carnegie Mellon University, University of Florida, the French Participation Group, the German Participation Group, Harvard University, the Instituto de Astrofisica de Canarias, the Michigan State/Notre Dame/JINA Participation Group, Johns Hopkins University, Lawrence Berkeley National Laboratory, Max Planck Institute for Astrophysics, Max Planck Institute for Extraterrestrial Physics, New Mexico State University, New York University, Ohio State University, Pennsylvania State University, University of Portsmouth, Princeton University, the Spanish Participation Group, University of Tokyo, University of Utah, Vanderbilt University, University of Virginia, University of Washington, and Yale University.

References

Anderson, *et al.* 2014, *MNRAS* 441, 24 BOSS collaboration paper
Eisenstein, D. J., Weinberg, D. H., Agol, E., *et al.* 2011, *AJ*, 142, 72
Manera, *et al.* 2013 *MNRAS* 428, 1036
Manera, *et al.* 2014, arXiv 1401.4171
Padmanabhan, *et al.* 2012, *MNRAS*,427,2132
Percival, *et al.* 2014, *MNRAS* 439,2531
Ross, *et al.* 2012 *MNRAS*,424,564
Taylor, *et al.* 2012, *MNRAS*,432,1928
Vargas-Magana, *et al.* 2014, arXiv:1312.4996

Statistical Challenges in 21st Century Cosmology
Proceedings IAU Symposium No. 306, 2014
A. F. Heavens, J.-L. Starck & A. Krone-Martins, eds.

doi:10.1017/S1743921314013684

Supervoid Origin of the Cold Spot in the Cosmic Microwave Background

András Kovács[1,2], István Szapudi[3], Benjamin R. Granett[4], Zsolt Frei[1,2], Joseph Silk[5], Will Burgett[3], Shaun Cole[6], Peter W. Draper[6], Daniel J. Farrow[6], Nicholas Kaiser[3], Eugene A. Magnier[3], Nigel Metcalfe[6], Jeffrey S. Morgan[3], Paul Price[7], John Tonry[3], Richard Wainscoat[3]

[1] Institute of Physics, Eötvös Loránd University, 1117 Pázmány Péter sétány 1/A Budapest, Hungary
[2] MTA-ELTE EIRSA "Lendület" Astrophysics Research Group, 1117 Pázmány Péter sétány 1/A Budapest, Hungary
[3] Institute for Astronomy, University of Hawaii 2680 Woodlawn Drive, Honolulu, HI, 96822, USA
[4]INAF OA Brera, Via E. Bianchi 46, Merate, Italy
[5]Department of Physics and Astronomy, The Johns Hopkins University, Baltimore MD 21218, USA
[6]Department of Physics, Durham University, South Road, Durham DH1 3LE, UK
[7]Department of Astrophysical Sciences, Princeton University, Princeton, NJ 08544

Abstract. We use a WISE-2MASS-Pan-STARRS1 galaxy catalog to search for a supervoid in the direction of the Cosmic Microwave Background Cold Spot. We obtain photometric redshifts using our multicolor data set to create a tomographic map of the galaxy distribution. The radial density profile centred on the Cold Spot shows a large low density region, extending over 10's of degrees. Motivated by previous Cosmic Microwave Background results, we test for underdensities within two angular radii, 5°, and 15°. Our data, combined with an earlier measurement by Granett *et al.* 2010, are consistent with a large $R_{\rm void} = (192 \pm 15)\mathrm{h}^{-1}\mathrm{Mpc}$ (2σ) supervoid with $\delta \simeq -0.13 \pm 0.03$ centered at $z = 0.22 \pm 0.01$. Such a supervoid, constituting a $\sim 3.5\sigma$ fluctuation in the ΛCDM model, is a plausible cause for the Cold Spot.

Keywords. cosmic microwave background, observations, large-scale structure of universe

1. Introduction

The Cosmic Microwave Background (CMB) Cold Spot (CS) is an exceptionally cold area centered on on $(l, b) \simeq (209°, -57°)$ and was first detected in the Wilkinson Microwave Anisotropy Probe (Bennett *et al.* 2012) at $\simeq 3\sigma$ significance using wavelet filtering (Vielva *et al.* 2003). The CS is perhaps the most significant among the anomalies recently confirmed by *Planck* (Planck 2013 results. XXIII.). These physical anomalies are significant enough to motivate further studies. Some of the models include exotic physics, e.g., cosmic textures (Cruz *et al.* 2008), while e.g. Inoue and Silk 2007 claim that the CS, and possibly other anomalies, are caused by the Integrated Sachs-Wolfe effect (ISW) of the decaying gravitational potentials, which in turn is caused by Dark Energy. The latter explanation would require of a large, $\gtrsim 200\mathrm{h}^{-1}\mathrm{Mpc}$ supervoid centered on the CS, readily detectable in large scale structure surveys.

The supervoid model has been constrained by using radio galaxies of the NVSS survey, Canada-France-Hawaii Telescope (CFHT) imaging of the CS region, redshift survey data using the VIMOS spectrograph on the VLT, and the relatively shallow 2MASS galaxy catalogue. See Szapudi *et al.* 2014, and references therein for review. Although these works report a possible underdense region at redshifts $z < 0.3$, they either run out of objects at low redshift, or have no redshifts for tomographic imaging. Note that although no void has been found that could fully explain the CS anomaly, there is strong, $\gtrsim 4.4\sigma$, statistical evidence that superstructures imprint on the CMB as cold and hot spots (Granett *et al.* 2008).

2. Analysis of WISE-2MASS-PS1 galaxy counts

Next we describe some of our methods and procedures, while Szapudi *et al.* 2014 and Finelli *et al.* 2014 can be consulted for further details. We probe the $z \simeq 0.3$ redshift range, unconstrained by previous studies, using the WISE-2MASS galaxy catalog (Kovács and Szapudi 2014) with $z_{\rm med} \simeq 0.14$, and 21,200 square degrees of coverage. The catalog contains sources to flux limits of $W1_{WISE} \leqslant 15.2$ mag and $J_{2MASS} \leqslant 16.5$ mag, resulting in a galaxy sample deeper than 2MASS (Skrutskie *et al.* 2006) and more uniform than WISE (Wright *et al.* 2010).

In this projected catalog, we detect an underdensity centred on the Cold Spot. In particular, we find signal-to-noise ratios $S/N \sim 12$ for rings at our pre-determined radii, 5° and 15°, and an extended underdensity to $\sim 20°$ with $\gtrsim 5\sigma$ significance. At larger radii, the radial profile is consistent with a supervoid surrounded by a gentle compensation that converges to the average galaxy density at $\sim 50°$. The supervoid appears to contain an extra depression in its centre, with its own compensation at around $\simeq 8°$. The projected underdensity in WISE-2MASS is modeled with a ΛLTB model (Garcia-Bellido and Haugbølle 2008) by Finelli *et al.* 2014.

We then set out to create a detailed map of the CS region in three dimensions. Therefore, WISE-2MASS galaxies have been matched with Pan-STARRS1 (Kaiser 2004, PS1) objects within a $50° \times 50°$ area centred on the CS, except for a Dec $\geqslant -28.0$ cut to conform to the PS1 boundary. We used PV1.2 reprocessing of PS1 in an area of 1,300 square degrees. For PS1, we required a proper measurement of Kron and PSF magnitudes in $g_{\rm P1}$, $r_{\rm P1}$ and $i_{\rm P1}$ bands that were used to construct photometric redshifts (photo-z's) with a Support Vector Machine algorithm. The training and control sets were created matching WISE-2MASS, PS1, and the Galaxy and Mass Assembly Driver *et al.* 2011 (GAMA) redshift survey. Our three dimensional catalog of WISE-2MASS-PS1 galaxies contains photo-z's with an estimated error of $\sigma_z \approx 0.034$. We count galaxies as a function of redshift in disks centred on the CS using the two pre-determined angular radii, $R = 5°$, and 15°. The galaxy density calculated from the average redshift distribution is shown in the upper panel of Fig. 1. Note that the larger disk is not fully contained in our photo-z catalog due to the limited PS1 footprint. In photo-z bins of width of $\Delta z = 0.035$, we found $S/N \sim 5$ and $S/N \sim 6$ for the deepest under-density bins, respectively. The measurement errors are due to Poisson fluctuations in the redshift bins. To extend our analysis to higher redshifts, we add a previous measurement (Granett *et al.* 2010) in a photo-z bin centred at $z = 0.4$.

For a rudimentary understanding of our counts, we build toy models from top-hat voids in the z direction convolved with the photo-z errors. This model has three parameters, redshift ($z_{\rm void}$), radius ($R_{\rm void}$), and depth (δ_g). We carry out the χ^2-based maximum likelihood parameter estimation for $R = 15°$, using our first 10 bins combined with the extra bin measured by Granett *et al.* 2010 for $n = 11$ data points. We find $\chi_0^2 = 92.4$ for the null hypothesis of no void. The marginalized results for the parameters of our toy model are $z_{\rm void} = 0.22 \pm 0.01$ (2σ), $R_{\rm void} = (192 \pm 15)h^{-1}$Mpc ($2\sigma$), and $\delta_g = -0.18 \pm 0.01$ (2σ) finding $\chi^2_{min} = 5.97$. We take the $b_g = 1.41 \pm 0.07$ galaxy bias of the catalog into

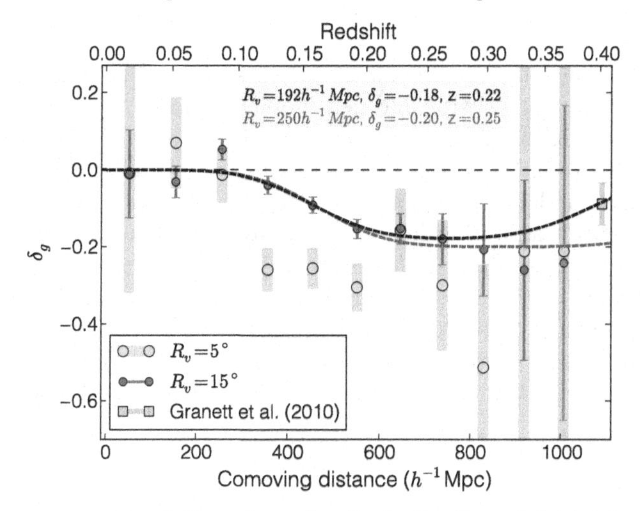

Figure 1. Our measurements of the galaxy density in the line-of-sight using the $\Delta z = 0.035$ photo-z bins we defined. We detected a significant depression in galaxy density in $r = 5°$ and 15° test circles. We used our simple modeling tool to examine the effects of photo-z errors, and test the consistency of simple top-hat voids with our measurements. The last bin (shown in grey) was excluded from the analysis, and a data point by Granett *et al.*(2010) accounts for the higher redshift part of the measurement. See text for details.

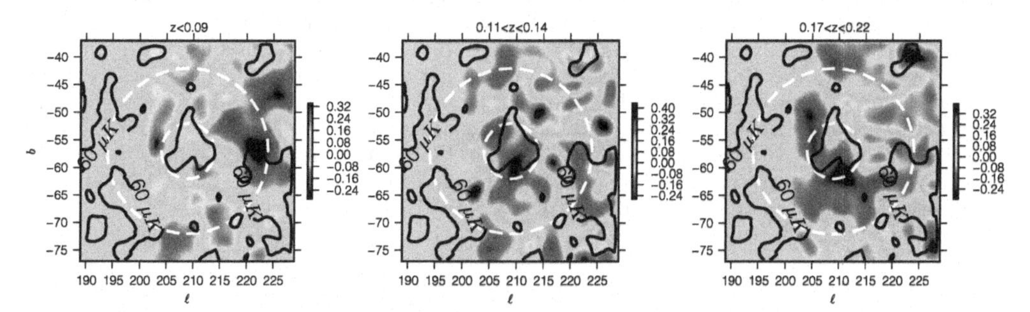

Figure 2. Tomographic view of the CS region. Below $z < 0.09$ appears to show the front compensation (higher density area) of the large void, although there is an under dense structure to the right. The slice at $0.11 < z < 0.14$ cuts into front of the supervoid, and the ring round represents a slice of the compensation. Finally the slice $0.17 < z < 0.22$ cuts into the front half of the supervoid with compensation around it. Note that the left side of the images reflects the mask of the PS1 data set.

account finding an average depth of $\delta = \delta_g/b_g \simeq -0.13 \pm 0.03$ (2σ). While the toy model quantifies the properties of the void, careful inspection of the Figures reveals a complex structure. There appears to be a compensation in front of the supervoid at around $z \simeq 0.05 - 0.08$, and the significantly deeper counts at the smaller radii show that the void is deeper at the centre. An approximate tomographic imaging of the CS region is presented in Fig. 2. in three slices of $z < 0.09$, $0.11 < z < 0.14$, $0.17 < z < 0.22$ and smoothed at 2° scales. The Planck SMICA CMB temperature map is over-plot with contours. While we characterized the profile in radial bins, the geometry of the supervoid is more complex. It is noteworthy that deepest part of the void is close to the centre of the CS in the middle slice. Given the enormous size, the rich structure is expected: the supervoid contains voids, their compensations, filaments, and appears to have its own compensation.

3. Conclusions

We have found strong statistical evidences for a supervoid at redshift $z = 0.22 \pm 0.01$ with radius $R = (192 \pm 15)\, h^{-1}$Mpc and depth of $\delta = -0.13 \pm 0.03$ (Szapudi *et al.* 2014). These parameters are in excellent agreement with the projected density analysis results by Finelli *et al.* 2014, with the redshift slightly closer. We estimated the linear ISW effect due to such an underdensity from a simple approximation. We found that it could significantly affect the CMB, of order -20 μK, falling short of a full explanation of the CS anomaly. However, a non-linear LTB supervoid based on the projected profile in the WISE-2MASS catalog matches well the profile observed on the CMB (Finelli *et al.* 2014). The observed alignment and morphological similarity to a compensation is noteworthy. We estimate that the supervoid we detected corresponds to a rare, at least $\gtrsim 3.5\sigma$, density fluctuation in ΛCDM; thus chance alignment with another rare structure, the CS, is negligible.

Acknowledgments

We acknowledge the support of NASA grants NNX12AF83G and NNX10AD53G. AK and ZF acknowledge support from OTKA through grant no. 101666. In addition, AK acknowledges support from the Campus Hungary fellowship program. We make use of HEALPix Gorski *et al.* 2005 software in our project. The Pan-STARRS1 Surveys (PS1) have been made possible through contributions by the Institute for Astronomy, the University of Hawaii, the Pan-STARRS Project Office, the Max-Planck Society and its participating institutes, the Max Planck Institute for Astronomy, Heidelberg and the Max Planck Institute for Extraterrestrial Physics, Garching, The Johns Hopkins University, Durham University, the University of Edinburgh, the Queen's University Belfast, the Harvard-Smithsonian Center for Astrophysics, the Las Cumbres Observatory Global Telescope Network Incorporated, the National Central University of Taiwan, the Space Telescope Science Institute, and the National Aeronautics and Space Administration under Grant No. NNX08AR22G issued through the Planetary Science Division of the NASA Science Mission Directorate, the National Science Foundation Grant No. AST-1238877, and the University of Maryland, and Eotvos Lorand University (ELTE).

References

Inoue, K. T. & Silk, J. 2006, *ApJ*, 648, 23-30
Szapudi, I., Kovács, A., & Granett, B. R., *et al.* 2014, *arXiv:1405.1566*
Finelli, F., Garcia-Bellido, J., Kovacs, A., Paci, F., & Szapudi, I. 2014, *arXiv:1405.1555*
Kovács, A. & Szapudi, I. 2014, *arXiv:1401.0156*
Garcia-Bellido, J. & Haugbølle, T. 2008, *JCAP*, 4:3
K. M. Gorski, E. Hivon, *et al.* 2005, *Apj*, 622:759–771
E. L. Wright, *et al.* 2010, *AJ*, 140:1868–1881
B. R. Granett, M. C. Neyrinck, & I. Szapudi 2008, *ApjL*, 683:L99–L102
B. R. Granett, I. Szapudi, & M. C. Neyrinck 2010, *Apj*, 714:825–833
M. F. Skrutskie, R. M. Cutri, R. Stiening, *et al.* 2006, *AJ*, 131:1163–1183
C. L. Bennett, D. Larson, J. L., *et al.* Weiland. 2012 *ArXiv e-prints*
P. Vielva, *et al.* 2004, *Apj*, 609:22–34.
M. Cruz *et al.* 2008, *MNRAS*, 390:913–919
S. P. Driver, D. T. Hill, *et al.* 2011 *MNRAS*, 413:971–995
Planck 2013 results. XXIII. 2013 *ArXiv e-prints*
N. Kaiser. *SPIE Conference Series*, 2004.

Statistical Challenges in 21st Century Cosmology
Proceedings IAU Symposium No. 306, 2014
A. F. Heavens, J.-L. Starck & A. Krone-Martins, eds.

doi:10.1017/S1743921314010886

Towards a better understanding of galaxy clusters

Pedro T. P. Viana[1,2]

[1]Centro de Astrofísica da Universidade do Porto,
Rua das Estrelas, 4150-762 Porto, Portugal

[2]Departamento de Física e Astronomia, Faculdade de Ciências, Universidade do Porto,
Rua do Campo Alegre, 687, 4169-007 Porto, Portugal
email: viana@astro.up.pt

Abstract. Observational data on clusters of galaxies holds relevant information that can be used to determine the relative plausibility of different models for the large-scale evolution of the Universe, or estimate the joint posterior probability distribution function of the parameters that pertain to each model. Within the next few years, several surveys of the sky will yield large galaxy cluster catalogues. In order to make use of the vast amount of information they will contain, their selection functions will have to be properly understood. We argue this, as well as the estimation of the full joint posterior probability distribution function of the most relevant cluster properties, can be best achieved in the framework of bayesian statistics.

Keywords. galaxies: clusters: general, surveys, methods: statistical

1. Introduction

Clusters of galaxies are the largest gravitationally collapsed structures in the Universe. The hierarchical growth of large-scale structure also ensures they are the rarest. Further, although not all important physical processes involved in the assembly of galaxy clusters are well known, they are relatively simple astrophysical objects. All of these characteristics make clusters of galaxies excellent probes of the growth of structure in the Universe and of its large-scale evolution.

Within the next few years, several surveys of the sky will detect many thousands of galaxy clusters. The largest catalogue is expected to result from the Euclid mission (*www.euclid-ec.org*). However, the accuracy with which we will be able to distinguish, for example, competing hypothesis for the cause of the present accelerating expansion of the Universe, will be limited by our understanding of the galaxy cluster catalogue selection function, systematic errors and the uncertainty in the galaxy cluster mass determinations. The combination of Euclid data with that obtained through other galaxy cluster surveys, based on the detection of the galaxy cluster signal on the X-rays (XCS, *www.xcs-home.org*, and eROSITA, *www.mpe.mpg.de/eROSITA*, surveys) and due to the Sunyeav-Zel'dovich effect in the mm/sub-mm (SPT, *www.pole.uchicago.edu/spt*, and Planck, *www.sci.esa.int/planck*, surveys), can help in this respect, by enabling the cross-calibration of the selection functions of those surveys and the mass-observable relations, unearthing possible systematic errors in the process.

2. Overview

Let $\xi = \{\xi_{\alpha,i}\}$ be the set of all galaxy cluster quantities of interest, each associated with a different α, for all clusters in some catalogue, each identified by a given i. We would like to characterise the probability distribution for any $\xi_{\alpha,i}$ given the information contained in some dataset, D, and any relevant prior information, I_0,

$$P(\xi_{\alpha,i}|D,I_0) = \int P(\xi|D,I_0)d\xi_{\beta,j} \tag{2.1}$$

where the integral runs over all possible combinations of β and j that are different from the combination of α and i. Using Bayes Theorem, we can now write

$$P(\xi_{\alpha,i}|D,I_0) \propto \int P(\xi|I_0)P(D|\xi,I_0)d\xi_{\beta,j} \tag{2.2}$$

where the proportionality or normalisation constant is equal to the inverse of the model evidence, $P(D|I_0)$. The estimation of the likelihood, $P(D|\xi,I_0)$, is usually difficult given that the data available can be heterogenous, always consisting of a finite number of measurements affected by heteroscedastic noise and possibly correlated.

The problem simplifies considerably if the data pertaining to each galaxy cluster, D_i, is acquired independently. We can thus write

$$P(\xi_{\alpha,i}|D_i,I_0) \propto \int P(\xi|I_1)P(D_i|\xi,I_1)d\xi_{\beta,i} \tag{2.3}$$

Assuming that all galaxy clusters in the catalogue belong to the same statistical population, the prior probability inside the integral becomes independent of i. And it can be inferred from the rest of the data, that we will continue to call D,

$$P(\xi|I_1) = P(\xi|D,I_2) \propto P(\xi|I_2)P(D|\xi,I_2) \tag{2.4}$$

The most general way to evaluate the likelihood in the previous expression is by using a non-parametric or semi-parametric density estimation procedure (e.g. Bovy, Hogg & Roweis 2011, Sarkar *et al.* 2014). The prior probability can be minimally informative, or include any assumptions that may follow from what we believe were the physical conditions under which galaxy clusters assembled. Presently, the most credible prior assumptions follow from the so-called standard cosmological model (e.g. Lahav & Liddle 2014). Among them, it is particularly relevant the halo mass function, $n(M,z)$, describing the number density of gravitationally collapsed structures as a function of their mass, M, and redshift, z. If the data being considered contains information about the masses and redshifts of at least some of the galaxy clusters in the catalogue, then we can write

$$P(\xi|I_2) = P(\xi_\gamma|M,z,I_3)P(M,z|I_3) = P(\xi_\gamma|M,z,I_4)\frac{n(M,z)}{\int n(M,z)dMdz} \tag{2.5}$$

where γ identifies all cluster quantities of interest except for mass and redshift, while $P(\xi_\gamma|M,z,I_4)$ is a minimally informative prior. Otherwise, assumptions about relations that connect cluster mass and redshift to some other cluster property, information about which is contained in the dataset D, will have to be included in the prior assumptions.

The halo mass function depends on the values taken by some parameters in the standard cosmological model, whose set we will denote by θ, like the mean matter density in the Universe. Thus, in fact

$$P(\xi|I_2) = P(\xi_\gamma|M,z,I_4)\int P(\theta|I_4)\frac{n(M,z,\theta)}{\int n(M,z,\theta)dMdz}d\theta \tag{2.6}$$

where $P(\theta|I_4)$ represents the prior distribution of the cosmological parameters.

In the previous expression, the full joint posterior distribution was marginalised over the cosmological parameters, θ. If we had marginalised instead over ξ, we would have obtained the posterior distribution of those parameters given the dataset, D,

$$P(\theta|D, I_0) \propto \int P(\xi|\theta, I_0)P(\theta|I_0)P(D|\theta, \xi, I_0)d\xi \tag{2.7}$$

where $P(\theta|I_0)$ is a minimally informative prior. In an analogous manner to expression (2.5), we could then write

$$P(\xi|\theta, I_0) = P(\xi_\gamma|M, z, I_4)\frac{n(M, z, \theta)}{\int n(M, z, \theta)dMdz} \tag{2.8}$$

The density estimation procedure needed to evaluate $P(D|\xi, I_2)$ or $P(D|\theta, \xi, I_0)$ can be computationally very intensive. This problem can be alleviated if so-called galaxy cluster scaling relations are used to describe how the quantities ξ_α relate to each other. These are often assumed to be simple power-laws (e.g. Kelly 2007, Maughan 2014), that can be linearised by changing to $\log(\xi_\alpha)$, with some associated (so-called intrinsic) scatter. This is usually taken to be normally distributed with respect to $\log(\xi_\alpha)$, and independent from the values that $\log(\xi_\alpha)$ may take. If we label as S the set of parameters that define such relations, then expression (2.4) becomes

$$P(\xi|I_1) = \int P(\xi, S|D, I_5)dS \propto \int P(\xi|S, I_6)P(S|I_6)P(D|\xi, S, I_6)dS \tag{2.9}$$

where $P(\xi|S, I_6)$ and $P(S|I_6)$ can be minimally informative priors. In this case, had we marginalised instead over ξ, we would have obtained the posterior distribution of the linear regression parameters, S, given the dataset, D,

$$P(S|D, I_0) \propto \int P(\xi|S, I_7)P(S|I_7)P(D|S, \xi, I_7)d\xi \tag{2.10}$$

Taking into account prior information about the cosmological model would imply using expressions (2.5) or (2.6) in both expressions (2.9) and (2.10). Further, expressions (2.7) and (2.10) can be combined to infer the joint posterior distribution function of θ and S.

Finally, it should be remembered that all prior probabilities have to include the effects of the selection procedures followed in the assembly of the galaxy cluster catalogues considered. These depend on at least one galaxy cluster property, and most often also on the assumed cosmological model (e.g. Lloyd-Davies *et al.* 2011).

Acknowledgements

This work is being carried out with financial support from Fundação para a Ciência e a Tecnologia (FCT), through project PTDC/FIS/111725/2009.

References

Bovy, J., Hogg, D. W., & Roweis, S. T. 2011, *Annals of Applied Statistics*, 5, 2B, 1657
Sarkar, A., Pati, D., Mallick, B. K., & Carroll, R. J. 2014, *arXiv:1404.6462*
Lahav, O. & Liddle, A. R. 2014, *The Review of Particle Physics 2014*, ch. 23
Kelly, B. C. 2007, *ApJ*, 665, 1489
Maughan, B. J. 2014, *MNRAS*, 437, 1171
Lloyd-Davies, E. J., *et al.* (XCS collaboration), 2011, *MNRAS*, 418, 14

Statistical Challenges in 21st Century Cosmology
Proceedings IAU Symposium No. 306, 2014
A. F. Heavens, J.-L. Starck & A. Krone-Martins, eds.

doi:10.1017/S1743921314010679

A new test of uniformity for object orientations in astronomy

V. Pelgrims

IFPA, AGO Dept., University of Liège, Allée du 6 Août, 17, B4000 Liège, Belgium
email: `pelgrims@astro.ulg.ac.be`

Abstract. We briefly present a new coordinate-invariant statistical test dedicated to the study of the orientations of transverse quantities of non-uniformly distributed sources on the celestial sphere. These quantities can be projected spin-axes or polarization vectors of astronomical sources.

Keywords. method: statistics, method: data analysis, polarization, quasars: general

1. Introduction

In past decades, large amounts of data have been recorded in various fields of astronomy, propelling poorly known and little understood phenomena into the area of precision science. The cosmic microwave background is one of the most striking example. Such full- and or deep-sky surveys are facing on the problem of huge dataset treatment. However, aside from these huge observational campaigns, some catalogues contain relevant information for only sparse and non-uniformly scattered sources on the sky. Appropriate and robust statistical tests have to be dedicated to these small datasets. To the best of our knowledge, there was a lack of dedicated tests capable of analysing transverse data in astronomy, such as polarizations or projected-spin axes of galaxies or quasars. Indeed, these quantities are always defined in the plane orthogonal to the line of sight pointing to their corresponding sources through angles relative to one of the basis vectors. Therefore, if we consider sources that are scattered on the celestial sphere, the results of any statistical test performed on these angles will depend on the coordinate system in which the source positions are reported. When numerous point sources are uniformly scattered the problem is overcome through the use of spin 2 spherical harmonics (in the case of CMB polarization, for instance). Nevertheless, as soon as sources are rare and non-uniformly distributed over the sky, what is recurrent in astronomy, the problem of tackling the coordinate-system dependence appears. Forgetting this fact could, in some cases, lead to a misinterpretation of observational data.

In Pelgrims & Cudell (2014), we developed a dedicated tool to process this type of sparse data. In the following, we introduce this new statistical test and present an application to the sample of optical polarization measurements of 355 quasars non-uniformly scattered on the sky.

2. The new method

The basic idea of our method is to consider the quantity which is really measured. When studying polarization or other transverse data, this quantity is three-dimensional and, the measured axis is defined by the direction of the line of sight $\vec{s}$, where the source is observed, and by the position angle (PA) (defined modulo π radians when axial) which indicates the orientation in the plane orthogonal to $\vec{s}$. In order to study the uniformity of orientations of transverse quantities in a sample, we move from the usual circular data treatment to a spherical data treatment (in the sense of Fisher (1993) and Fisher *et al.* (1993), respectively). Transporting the axes to the origin of the coordinate system, we study their relative orientations through the study of density of points on a unit 2-sphere, where points are the intersections of the axes with the sphere. Simple spherical data analysis, such as those presented in Fisher *et al.* (1993), are not applicable in the case of transverse quantities since points are constrained to lie on great circles embedded in planes having the lines of sight as normal vectors. One can nevertheless evaluate the density of points at each location on the unit 2-sphere by adopting Kamb-like methods (e.g. Vollmer 1995).

We found convenient to use equal-area spherical caps. Indeed, under the assumption of uniformity of orientations, this particular shape allows us to predict in a simple way the density of points at each location, evaluating densities through a standard step function. Namely, the probability $\ell^{(i)}$ that the point attached to the source i, located in $\vec{s^{(i)}}$, falls inside the cap of half aperture angle η and centred in $\vec{c}$ is given by:

$$\ell^{(i)} = \begin{cases} \frac{2}{\pi}\,\mathrm{acos}\left(\frac{\cos\eta}{\sin\tau^{(i)}}\right) & \text{if } \sin\tau^{(i)} \geqslant \cos\eta \\ 0 & \text{otherwise} \end{cases} . \tag{2.1}$$

where $\cos\tau^{(i)} = \vec{c}\cdot\vec{s^{(i)}}$. In fact, adopting a step function, the probability is simply the ratio of the arc length of the great circle embedded inside the cap to that of the great circle itself (i.e. π, as axial data are under consideration here). The case $\ell^{(i)} = 0$ is well understood through geometrical argument. It is clear that this probability is independent of the system of coordinates.

Now, considering a sample of sources, one obtains a set of individual probabilities $\{\ell^{(i)}\}$ for each location on the sphere. Then, at each location, the Poisson-binomial probability distribution of the density of points is built from the $\ell^{(i)}$'s using a recursive algorithm presented in Pelgrims & Cudell (2014), Chen & Liu (1997) and Howard (1972). Then, we are able to test the hypothesis of random orientations at each location of the sphere by evaluating the p-value of the observed density.

For a given sample of sources, the direction of the strongest over-density defines the alignment direction and the corresponding p-value, denoted by p_{min}, gives the likelihood of this over-density in this direction. This value is a *local* probability, it is obtained semi-analytically and it is coordinate-invariant. A *global* probability is also computed, answering the question of likelihood of such an over-density in arbitrary direction. This probability is evaluated through Monte Carlo simulations by keeping source positions fixed while randomly varying PA according to a uniform distribution.

3. Application to real data

In Pelgrims & Cudell (2014), we applied this new statistical test to the sample of 355 optical polarization measurements of quasars for which unexpected large-scale correlations have been previously reported in Hutsemékers (1998) and Hutsemékers *et al.* (2005). Performing an analysis on the data sample with our test of density, we confirmed these reported large-scale anomalies and, we proceeded to a precise identification of the quasar regions showing aligned polarization vectors. Illustration of the new test for one of these regions is shown in Fig. 1. Local and global probabilities for this region of quasars are found to be 1.9×10^{-6} and 1.0×10^{-5}, respectively.

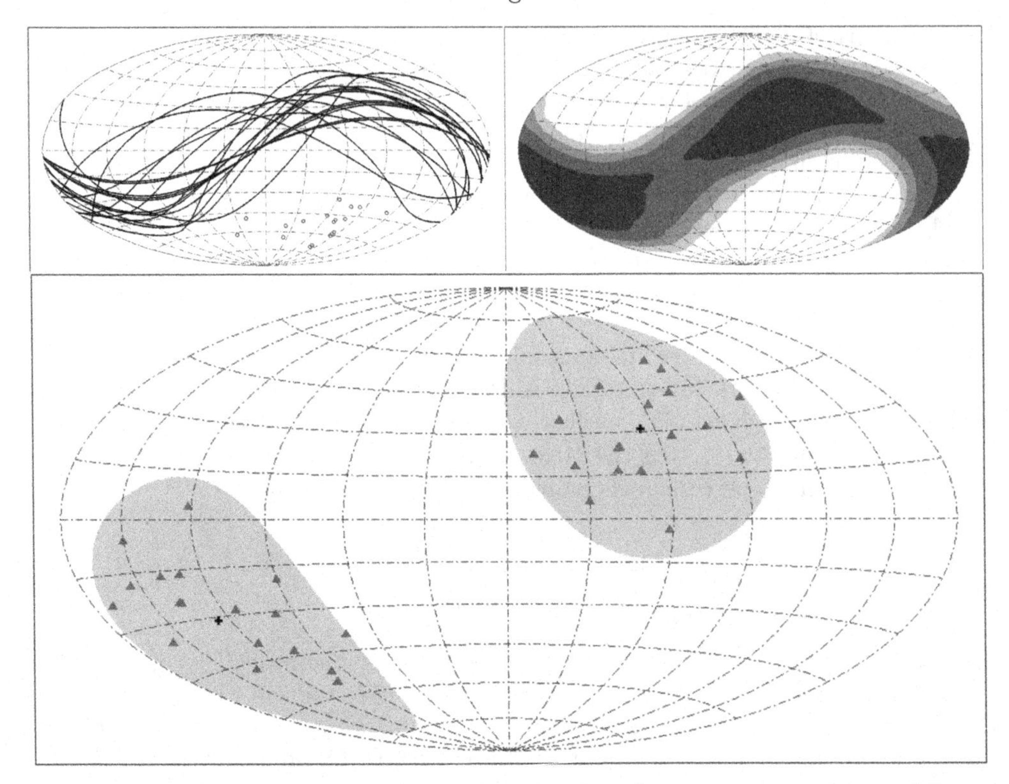

Figure 1. The new method at work. Top left: Small circles are source positions while continuous lines are geometrical loci of corresponding polarization points. Note the symmetry as polarization are axial quantities. The probability distributions are computed at each location from the arclength of geometrical loci embedded in the caps. Top right: We build expected density contours from these distributions, taking the mean densities. Dark shades indicate higher density. Bottom: Observed densities are evaluated by counting the number of polarization points (triangles) falling in each cap. Reporting the observed densities to their corresponding distribution, the alignment direction (crosses) is defined as the centre of the cap (transparent patches) showing the most unexpected over-density. Hammer-Aitoff projected maps are centred on the Galactic centre with positive Galactic latitude at the top and increasing longitude to the right. The half aperture angle being used is $\eta = 45°$.

4. Acknowledgements

V.P. thanks J.R. Cudell and D. Hutsemékers for text enhancement. This work was supported by the Fonds de la Recherche Scientifique - FNRS under grant 4.4501.05.

References

Chen, S. X. & Liu, J. S. 1997, *Stat. Sin.*,7, 875

Fisher, N. I. 1993, *Statistical Analysis of Circular Data*, Cambridge University Press, Cambridge

Fisher, N. I., Lewis, T., & Embleton, B. J. J. 1987, *Statistical Analysis of Spherical Data*, Cambridge University Press, Cambridge

Howard, S. 1972, *J. R. Stat. Soc. Ser. B*, 34, 2012

Hutsemékers, D. 1998, *A&A*, 332, 410

Hutsemékers, D., Cabanac, R., Lamy, H., & Sluse, D. 2005, *A&A*, 441, 915

Pelgrims, V. & Cudell, J. R. 2014, *MNRAS*, 442, 1239

Vollmer, F. W. 1995, *Computers & Geosciences*, 21, 31

Statistical Challenges in 21st Century Cosmology
Proceedings IAU Symposium No. 306, 2014
A. F. Heavens, J.-L. Starck & A. Krone-Martins, eds.

doi:10.1017/S1743921314013672

Machine-learning in astronomy

Michael Hobson[1], Philip Graff[2], Farhan Feroz[1] and Anthony Lasenby[1]

[1]Astrophysics Group, Cavendish Laboratory, J.J. Thomson Avenue, Cambridge, CB3 0HE, UK
email: mph@mrao.cam.ac.uk, ff235@mrao.cam.ac.uk, anthony@mrao.cam.ac.uk

[2]Gravitational Astrophysics Laboratory, NASA Goddard Space Flight Center, 8800 Greenbelt Rd., Greenbelt, MD 20771, USA
email: philip.b.graff@nasa.gov

Abstract. Machine-learning methods may be used to perform many tasks required in the analysis of astronomical data, including: data description and interpretation, pattern recognition, prediction, classification, compression, inference and many more. An intuitive and well-established approach to machine learning is the use of artificial neural networks (NNs), which consist of a group of interconnected nodes, each of which processes information that it receives and then passes this product on to other nodes via weighted connections. In particular, I discuss the first public release of the generic neural network training algorithm, called SkyNet, and demonstrate its application to astronomical problems focusing on its use in the BAMBI package for accelerated Bayesian inference in cosmology, and the identification of gamma-ray bursters. The SkyNet and BAMBI packages, which are fully parallelised using MPI, are available at http://www.mrao.cam.ac.uk/software/.

Keywords. methods: data analysis, methods: statistical, cosmological parameters, gamma rays: bursts

1. Introduction

In astrophysics and cosmology, one is faced with analysing large, complicated and multidimensional data sets. Such analyses typically include tasks such as: data description and interpretation, pattern recognition, prediction, classification, compression, inference and many more. One way of performing such tasks is through the use of machine-learning methods (see, e.g., MacKay (2003), Ball & Brunner (2010) and Way *et al.* (2012)).

In supervised learning, the goal is to infer a function from labeled training data, which consist of a set of training examples. Each example has known 'input' quantities whose values are to be used to predict the values of the 'outputs'. Thus, the function to be inferred is the mapping from input to outputs. Once learned, this mapping can be applied to datasets for which the values of the outputs are not known. Supervised learning is usually further subdivided into *classification* and *regression*.

An intuitive and well-established approach to machine learning is based on the use of artificial neural networks (NNs), which are loosely inspired by the structure and functional aspects of a brain. They consist of a group of interconnected nodes, each of which processes information that it receives and then passes this product on to other nodes via weighted connections. In this way, NNs constitute a non-linear statistical data modeling tool, which may be used to model complex relationships between a set of inputs and outputs. Many machine-learning applications can be performed using only feed-forward NNs: an input layer of nodes passes information to an output layer via zero, one, or many 'hidden' layers in between. Moreover, a universal approximation theorem Hornik *et al.* (1990) assures us that we can approximate any reasonable mapping with a NN of a given form. A useful introduction to NNs can be found in MacKay (2003).

In astronomy, feed-forward NNs have been applied to various machine-learning problems for over 20 years (see, e.g., Way *et al.* (2012), Tagliaferri *et al.* (2003)). Nonetheless, their more widespread use in astronomy has been limited by the difficulty associated with standard techniques, such as backpropagation, in training networks having many nodes and/or numerous hidden layers (i.e. 'large' and/or 'deep' networks), which are often necessary to model the complicated mappings between numerous inputs and outputs in modern astronomical applications. We therefore introduce SkyNet Graff *et al.* (2014), an efficient and robust neural network training algorithm that is capable of training large and/or deep feed-forward networks.

An important recent application of regression supervised learning in astrophysics and cosmology is the acceleration of the Bayesian analysis (both parametet estimation and model selection) of large data sets in the context of complicated models. At each point in parameter space, Bayesian methods require the evaluation of a 'likelihood' function describing the probability of obtaining the data for a given set of model parameters. For some problems each such function evaluation may take up to tens of seconds. Substantial gains in performance can thus be achieved if one is able to speed up the evaluation of the likelihood, and a NN is ideally suited for this task. The blind accelerated multimodal Bayesian inference (BAMBI) algorithm Graff *et al.* (2012) uses SkyNet for training such NNs, and combines them with a nested sampling Skilling (2004) approach that efficiently calculates the Bayesian evidence (also referred to as the marginal likelihood) for model selection and produces samples from the posterior distribution for parameter estimation. It particular, BAMBI employs the MultiNest algorithm Feroz & Hobson (2008), Feroz *et al.* (2009), Feroz *et al.* (2013), which is a generic implementation of nested sampling, extended to handle multimodal and degenerate distributions, and is fully parallelised.

This paper will discuss results from applying SkyNet within the BAMBI algorithm to accelerate Bayesian inference in cosmology, and also to the identification of gamma-ray bursters.

2. Network structure

A multilayer perceptron feed-forward neural network is the simplest type of network and consists of ordered layers of perceptron nodes that pass scalar values from one layer to the next. The perceptron is the simplest kind of node, and maps an input vector $\mathbf{x} \in \Re^n$ to a scalar output $f(\mathbf{x}; \mathbf{w}, \theta)$ via

$$f(\mathbf{x}; \mathbf{w}, \theta) = \theta + \sum_{i=1}^{n} w_i x_i, \tag{2.1}$$

where $\mathbf{w} = \{w_i\}$ and θ are the parameters of the perceptron, called the 'weights' and 'bias', respectively. In a feed-forward NN, the value at each node is computed by applying an 'activation function', g, to the scalar value calculated in Equation 2.1. The activation function used for nodes in hidden layers is $g(x) = 1/(1 + e^{-x}) = \mathrm{sig}(x)$, the sigmoid function. For output layer nodes, $g(x) = x$. The non-linearity of g for the hidden layers is essential to allowing the network to model non-linear functions. The weights and biases are the values we wish to determine in our training (described in Section 3). As they vary, a huge range of non-linear mappings from inputs to outputs is possible. By increasing the number of hidden nodes, we can achieve more accuracy at the risk of overfitting to our training data.

3. Network training

In training a NN, we wish to find the optimal set of network weights and biases that maximise the accuracy of predicted outputs. However, we must be careful to avoid over-fitting to our training data at the expense of making accurate predictions for input values on which the network has not been trained. The set of training data inputs and outputs, $\mathcal{D} = \{\mathbf{x}^{(k)}, \mathbf{t}^{(k)}\}$, is provided by the user. Approximately 75 per cent should be used for actual NN training and the remainder retained as a validation set that will be used to determine convergence and to avoid overfitting. This ratio of 3:1 gives plenty of information for training but still leaves a representative subset of the data for checks to be made.

3.1. *Network objective function*

Let us denote the network weights and biases collectively by the vector $\mathbf{a}$. SkyNet considers the parameters $\mathbf{a}$ to be random variables with a posterior distribution given by the likelihood multiplied by the prior. The likelihood, $\mathcal{L}(\mathbf{a}; \boldsymbol{\sigma})$, encodes how well the NN, characterised by a given set of parameters $\mathbf{a}$, is able to reproduce the known training data outputs. This is modulated by the prior $\mathcal{S}(\mathbf{a}; \alpha)$, which is assumed to have the form of a Gaussian centered at the origin with multiplicative factor α inside the exponential. α determines the relative importance of the prior and the likelihood. The prior here acts as a regularization function; in a full Bayesian treatment it may have a different functional form.

The form of the likelihood depends on the type of network being trained. For regression problems, SkyNet assumes a log-likelihood function for the network parameters $\mathbf{a}$ given by the standard χ^2 misfit function, but including hyperparameters $\boldsymbol{\sigma}$ that describe the standard deviation (error size) of each of the outputs. For classification problems, SkyNet again uses continuous outputs (rather than discrete ones), which are interpreted as the probabilities that a set of inputs belongs to a particular output class. This is achieved by applying the *softmax* transformation to the output values, so that they are all non-negative and sum to unity. The classification likelihood is then given by the *cross-entropy* of the targets and softmaxed output values MacKay (2003).

3.2. *Initialisation and pre-training*

The training of the NN can be started from some random initial state, or from a state determined from a 'pre-training' procedure. In the former case, the network training begins by setting random values for the network parameters, sampled from a normal distribution with zero mean and variance of 0.01. SkyNet can also make use of the pre-training approach developed by Hinton *et al.* (2006), Hinton & Salakhutdinov (2006), which obtains a set of network weights and biases close to a good solution of the network objective function.

3.3. *Optimisation of the objective function*

Once the initial set of network parameters have been obtained, either by assigning them randomly or through pre-training, the network is then trained by iterative optimisation of the objective function. NN training proceeds using an adapted form of a truncated Newton (or 'Hessian-free' Martens (2010)) optimisation algorithm, to calculate the step, $\delta\mathbf{a}$, that should be taken at each iteration. The required step is obtained using a stable and efficient procedure applicable to NNs that avoids explicit calculation and calculation of the Hessian Schraudolph (2002), Pearlmutter (1994). Following each such step, adjustments to α and $\boldsymbol{\sigma}$ may be made before another step is calculated (using formulas derived from the MemSys software package Gull & Skilling (1999)). The combination of the above methods makes practical the use of second-order derivative information even for large networks.

3.4. *Convergence*

Following each iteration of the optmisation algorithm, the likelihood, posterior, correlation, and error squared values are calculated both for the training data and for the

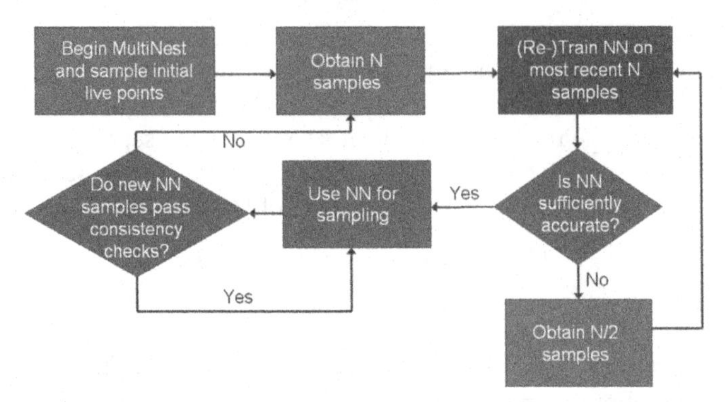

Figure 1. A flowchart depicting the transitions between sampling and NN training within BAMBI. N is given by `updInt` from MultiNest.

validation data. When these values begin to diverge and the predictions on the validation data no longer improve, it is deemed that the training has converged and optimization is terminated. This helps to prevent excessive over-fitting to the training data.

4. SkyNet Regression and BAMBI: learning likelihoods

The BAMBI algorithm combines neural networks with nested sampling. After a specified number of new samples from MultiNest have been obtained, BAMBI uses these to train a regression network on the log-likelihood function using the SkyNet algorithm in order to perform likelihood interpolation. Using the network reduces the log-likelihood evaluation time from seconds to microseconds, allowing MultiNest to complete analysis much more rapidly. The user also obtains a network or set of networks that are trained to easily and quickly provide more log-likelihood evaluations near the peak if needed, or in subsequent analyses. Other methods exist for likelihood interpolation and extrapolation, as well as the fast estimation of the Bayesian evidence, best-fit parameters and errors. See Higden *et al.* (2012), Rasmussen (2003), Bliznyuk *et al.* (2008) and references therein for more information.

4.1. *The structure of BAMBI*

The flow of sampling and training within BAMBI is demonstrated by the flowchart given in Figure 1. After convergence to the optimal NN weights, we test that the network is able to predict likelihood values to within a specified tolerance level. If not, sampling continues using the original log-likelihood until enough new samples have been made for training to be resumed. Once a network is trained that is sufficiently accurate, its predictions are used in place of the original log-likelihood function for future samples in MultiNest. Consistency checks are made to ensure the NN is not making predictions outside the range of data on which it was trained.

The optimal network possible with a given set of training data may not be able to predict log-likelihood values accurately enough, so an additional criterion is placed on when to use the trained network. This requirement is that the RMSE of log-likelihoods predictions is less than a user-specified tolerance, `tol`. When the trained network does not pass this test, then BAMBI will continue using the original log-likelihood function to replace the older half of the samples to generate a new training data set. Network training will then resume, beginning with the weights that it had found as optimal for the previous data set. Since samples are generated from nested contours and each new data set contains half of the previous one, the saved network will already be able to produce reasonable predictions on this new data.

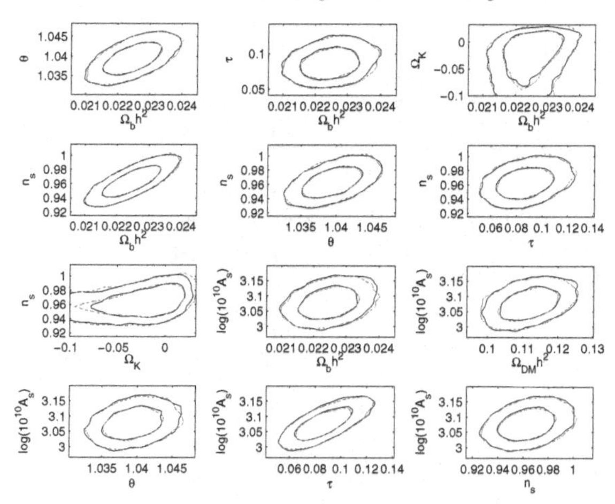

Figure 2. Marginalised 2D posteriors for the non-flat model (ΛCDM+Ω_K) using only CMB data. The 12 most correlated pairs are shown. MULTINEST is in solid black, BAMBI in dashed blue. Inner and outer contours represent 68% and 95% confidence levels, respectively.

Once a NN is in use in place of the original log-likelihood function its evaluations are taken to be the true log-likelihoods. Checks are made to ensure that the network is maintaining its accuracy and will re-train when these fail. To re-train the network, BAMBI first substitutes the original log-likelihood function back in and gathers the required number of new samples from MULTINEST. Training then commences, resuming from the previously saved network. These criteria ensure that the network is not trusted too much when making predictions beyond the limits of the data it was trained on.

4.2. *Accelerated cosmology inference using BAMBI*

We implement BAMBI within the standard COSMOMC code Lewis & Bridle (2002). Bayesian parameter estimation in cosmology requires evaluation of theoretical temperature and polarisation CMB power spectra (C_l values) using codes such as CAMB Lewis *et al.* (2000). These evaluations can take on the order of tens of seconds depending on the cosmological model. The C_l spectra are then compared to observations. Considering that thousands of these evaluations will be required, this is a computationally expensive step and a limiting factor in the speed of any Bayesian analysis. BAMBI has the benefit of not requiring a pre-computed sample of points as in COSMONET Auld *et al.* (2008a), Auld *et al.* (2008b) INTERPMC Bouland *et al.* (2011), PICO Fendt & Wandelt (2007), and others, which is particularly important when including new parameters or new physics in the model.

We use a standard set of eight cosmological parameters, each with a uniform prior distribution. A non-flat cosmological model incorporates all of these parameters, while we set $\Omega_K = 0$ for a flat model. We use $w = -1$ in both cases. The flat model thus represents the standard ΛCDM cosmology. We use two different data sets for analysis: (1) CMB observations alone and (2) CMB observations plus Hubble Space Telescope constraints on H_0, large-scale structure constraints from the luminous red galaxy subset of the SDSS and the 2dF survey, and supernovae Ia data.

Analyses with MULTINEST and BAMBI were run on all four combinations of models and data sets. In Figure 2 we show the recovered two-dimensional marginalised posterior probability distributions for the non-flat model using the CMB-only data set. We see very close agreement between MULTINEST (in solid black) and BAMBI (in dashed blue) across all parameters. Using the full data set one finds similar correspondence in the

Table 1. Time per likelihood evaluation in a follow-up analysis and speed-up factors with respect to MultiNest.

Method	Model	Data	$t_{\mathcal{L}}$ (ms)	factor
MultiNest	flat	CMB	2775	—
		all	3813	—
	non-flat	CMB	12830	—
		all	10980	—
BAMBI	flat	CMB	0.2077	13360
		all	0.2146	17770
	non-flat	CMB	0.08449	151900
		all	0.2032	54040

posterior probability contours. The posterior probability distributions for the flat model with either data set are extremely similar to those of the non-flat flodel with setting $\Omega_K = 0$, as expected. Moreover, the estimated log-evidences obtained by MultiNest and BAMBI for each case are consistent to within the (chosen) statistical error of 0.1 units.

An important comparison is the running time required. The analyses were run using MPI parallelisation on 48 processors. We recorded the time required for the complete analysis, not including any data initialisation prior to initial sampling. We then divide this time by the number of likelihood evaluations performed to obtain an average time per likelihood. Therefore, time required to train the NN is still counted as a penalty factor. If a NN takes more time to train, this will hurt the average time, but obtaining a usable NN sooner and with fewer training calls will give a better time since more likelihoods will be evaluated by the NN. We are able to obtain a significant decrease in running time of order 40%, while adding in the bonus of having a NN trained on the likelihood function.

A major benefit of BAMBI is that following an initial run the user is provided with a trained NN, or multiple ones, that model the log-likelihood function. These can be used in any subsequent analysis to obtain much faster results. This is a comparable analysis to that of CosmoNet Auld *et al.* (2008a), Auld *et al.* (2008b), except that the NNs here are a product of an initial Bayesian analysis where the peak of the distribution was *not* previously known. No prior knowledge of the structure of the likelihood surface was used to generate the networks that are now able to be re-used. When multiple NNs are trained and used in the initial BAMBI analysis, we must determine which network's prediction to use in the follow-up. We decide this by using a computationally cheap method of estimating the error on predictions suggested by MacKay (1995).

To demonstrate the speed-up potential of using the NNs in follow-up analyses, we repeated each of the four analyses above, but set the prior ranges to be uniform over the region defined by $\mathbf{x}_{\max(\log(\mathcal{L}))} \pm 4\boldsymbol{\sigma}$, where $\boldsymbol{\sigma}$ is the vector of standard deviations of the marginalised one-dimensional posterior probabilities. Table 1 shows the average time taken per log-likelihood function evaluation when this follow-up was performed with MultiNest and with BAMBI. In all cases, the BAMBI evidences matched the MultiNest values to within statistical error and the posteriors agreed closely. We can thus obtain accurate posterior distributions and evidence calculations orders of magnitude faster than originally possible.

5. SkyNet Classification: identifying gamma-ray bursters

Long gamma-ray bursts (GRBs) are almost all indicators of core-collapse supernovae from the deaths of massive stars. The ability determine the intrinsic rate of these events

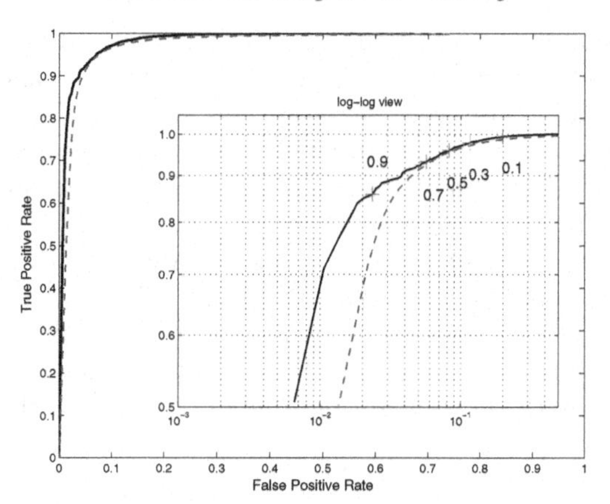

Figure 3. Actual (black solid) and expected (dashed red) ROC curves for a NN classifier that predicts whether a GRB will be detected by Swift.

as a function of redshift is essential for studying numerous aspects of stellar evolution and cosmology. The Swift space telescope is a multi-wavelength detector that is currently observing hundreds of GRBs Gehrels *et al.* (2004). However, Swift uses a complicated combination of over 500 triggering criteria for identifying GRBs, which makes it difficult to infer the intrinsic GRB rate.

To investigate this issue, a recent study by Lien *et al.* (2012) performed a Monte Carlo analysis that generated a mock sample of GRBs, using the GRB rate and luminosity function of Wanderman & Piran (2010), and processed them through an entire simulated Swift detection pipeline.

5.1. *Form of the classification problem*

Our goal here is to replace the simulated Swift trigger pipeline with a classification NN, which (as we will see) can determine in just a few microseconds whether a given GRB is detected. We use as training data a pre-computed mock sample of 10,000 GRBs from Lien *et al.* (2012). In particular, we divide this sample randomly into ~4000 for training, ~1000 for validation, and the final ~5000 as a blind test set on which to perform our final evaluations. For each GRB we have 13 inputs that describe the GRB and how it is seen by the detector. The two outputs correspond to the probabilities that the GRB is or is not detected.

5.2. *Results*

We can investigate the quality of the classification as a function of a threshold probability, $p_{\rm th}$, required to claim a detection. As discussed in Feroz *et al.* (2008), we can compute the *expected* number of total GRB detections, correct detections, and false detections, as well as other derived statistics as a function of $p_{\rm th}$, *without* knowing the true classification. From these, we can compute the completeness ϵ (fraction of detected GRBs that have been correctly classified; also referred to as the efficiency) and purity τ (fraction of all GRBs correctly classified as detected).

Values of the actual and expected completeness and purity are nearly identical. Thus, without knowing the true classifications of the GRBs as detected or not, we can set $p_{\rm th}$ to obtain the desired completeness and purity levels for the final sample.

With this information, we can also plot the actual and expected receiver operating characteristic (ROC) curves (see, e.g., Fawcett (2006)). The ROC curve plots the true positive rate (identical to completeness) against the false positive rate (also known as contamination). In general, the larger the area underneath a ROC curve, the more powerful the classifier. From Figure 3, we can see that the expected and actual ROC curves for a NN classifier are very close, with deviations occuring only at very low false positive rates. We conclude that $p_{\rm th} = 0.5$, the original naive choice, is a near-optimal threshold value. The curves also indicate that this test is quite powerful at predicting which GRBs will be detected by Swift.

Using $p_{\rm th} = 0.5$, we now wish to determine how well the GRB detection rate with respect to redshift is reproduced, since this relationship is key to deriving scientific results. The detected GRB event counts as a function of redshift for both our NN classifier and the original Swift pipeline agree very well. When histogrammed, the error between the counts differs by less than the Poissonian counting uncertainty on the true values inherent in this type of process.

6. Conclusion

We present an efficient and robust neural network training algorithm, called SkyNet. This generic tool is capable of training large and deep feed-forward networks and may be applied to machine-learning tasks in astronomy. We describe the application of SkyNet to the regression problem of learning likelihoods, in combination with MultiNest to create the BAMBI algorithm for accelerated Bayesian inference. This is a generic tool and requires no pre-processing. We apply BAMBI to the problem of cosmological parameter estimation and model selection. By calculating a significant fraction of the likelihood values with the NN instead of the full function, we are able to reduce the running time by $\sim 40\%$ on an initial run, and subsequent analyses using the resulting trained networks enjoy speed-up factors of $O(10^{4-5})$. We also demonstrate the application of SkyNet to the classification problem of identifying gamma-ray bursters.

References

Auld, T., Bridges, M., Hobson, M. P., & Gull, S. F. 2008, *MNRAS*, 376, L11
Auld, T., Bridges, M., & Hobson, M. P. 2008, *MNRAS*, 387, 1575
Ball, N. M. & Brunner, R. J. 2010, *Int. J. Mod. Phys.*, 19, 1049
Bliznyuk, N., *et al.* 2008, *J. Comput. Graph. Statist.*, 17, 270
Bouland, A., Easther, R., & Rosenfeld, K. 2011, *J. Cosmol. Astropart. Phys.*, 5, 016
Fawcett, T. 2006, *Pattern Recogn. Lett.*, 27, 861
Fendt, W. A. & Wandelt B. D. 2007, *ApJ*, 654, 2
Feroz, F. & Hobson, M. P. 2008, *MNRAS*, 384, 449
Feroz, F., Marshall, P. J., & Hobson M. P. 2008, arXiv:0810.0781 [astro-ph]
Feroz, F., Hobson, M. P., & Bridges, M. 2009, *MNRAS*, 398, 1601
Feroz, F., Hobson, M. P., Cameron, E., & Pettitt A. N. 2013, arXiv:1306.2144 [astro-ph.IM]
Gehrels, N., *et al.* 2004, *ApJ*, 611, 1005
Graff, P. Feroz, F., Hobson, M. P., & Lasenby, A. N. 2014, *MNRAS*, 441, 1741
Graff, P., Feroz, F., Hobson, M. P., & Lasenby, A. N. 2012, *MNRAS*, 421, 169
Gull, S. F. & Skilling, J. 1999, *Quantified Maximum Entropy: MemSys5 Users' Manual* (Maximum Entropy Data Consultants Ltd. Bury St. Edmunds, Suffolk, UK. http://www.maxent.co.uk/)
Higdon, D., Lawrence, E., Heitmann, K., & Habib S. 2012, in: E.D. Feigelson & G.J. Babu (eds.), *Statistical Chanllenges in Modern Astronomy V* (New York: Springer), p. 41
Hinton, G. E., Osindero, S., & Teh, Y.-W. 2006, *Neural Comput.*, 18, 1527
Hinton, G. E. & Salakhutdinov, R. R. 2006, *Science*, 313, 504
Hornik, K., Stinchcombe, M., & White, H. 1990, *Neural Networks*, 3, 359
Lewis, A. & Bridle, S. 2002, *Phys. Rev. D*, 66, 103511,

Lewis, A., Challinor, A., & Lasenby, A. N. 2000, *ApJ*, 538, 473
Lien, A., Sakamoto, T., Gehrels, N., Palmer, D., & Graziani, C. 2012, *Proceedings of the International Astronomical Union*, 279, 347
MacKay, D. J. C. 1995, *Network: Computation in Neural Systems*, 6, 469
&MacKay, D. J. C., 2003, *Information Theory, Inference, and Learning Algorithms* (Cambridge: CUP)
Martens, J. 2010, in: J. Fürnkranz & T. Joachims (eds.), *Proc. 27th Int. Conf. Machine Learning* (Haifa: Omnipress), p. 735
Pearlmutter, B. A. 1994, *Neural Comput.*, 6, 147
Rasmussen, C. E. 2003, in: J.M. Bernardo, M.J. Bayarri, J.O. Berger, A.P. Dawid, D. Heckerman, A.F.M. Smith, & M. West (eds.), *Bayesian statistics 7* (New York: OUP), p. 651
Schraudolph, N. N. 2002, *Neural Comput.*, 14, 1723
Skilling, J. 2004, *AIP Conference Series*, 735, 395
Tagliaferri, R. *et al.* 2003 *Neural Networks*, 16, 297
Wanderman, D. & Piran, T. 2010, *MNRAS*, 406, 1944
Way, M. J., Scargle, J. D., Ali, K. M., & Srivastava, A. N. 2012, *Advances in Machine Learning and Data Mining for Astronomy* (CRC Press)

Statistical Challenges in 21st Century Cosmology
Proceedings IAU Symposium No. 306, 2014
A. F. Heavens, J.-L. Starck & A. Krone-Martins, eds.

doi:10.1017/S1743921314013842

Machine Classification of Transient Images

Lise du Buisson[1]†, Navin Sivanandam[2]‡, Bruce A. Bassett[1,2,3]¶ and Mathew Smith[4]

[1]Department of Mathematics and Applied Mathematics, University of Cape Town, Cross Campus Rd, Rondebosch 7700, South Africa

[2]African Institute for Mathematical Sciences, 6–8 Melrose Rd, Muizenberg 7945, South Africa

[3]South African Astronomical Observatory, Observatory Rd, Observatory 7925, South Africa

[4]University of the Western Cape, Robert Sobukwe Rd, Bellville 7535, South Africa

Abstract. Using transient imaging data from the 2nd and 3rd years of the SDSS supernova survey, we apply various machine learning techniques to the problem of classifying transients (e.g. SNe) from artefacts, one of the first steps in any transient detection pipeline, and one that is often still carried out by human scanners. Using features mostly obtained from PCA, we show that we can match human levels of classification success, and find that a K-nearest neighbours algorithm and SkyNet perform best, while the Naive Bayes, SVM and minimum error classifier have performances varying from slightly to significantly worse.

Keywords. methods: data analysis, methods: statistical, techniques: image processing, techniques: photometric, surveys.

---

1. Introduction

Next-generation of survey experiments, like the Large Synoptic Survey Telescope (LSST) and the Dark Energy Survey (DES) will produce a huge photometric data deluge that will require new statistical and machine learning techniques for processing. This data deluge create problems in the supernovae (SNe) detection pipeline. Difference images are scanned for objects of interest (e.g. potential SNe) and artefacts are removed to avoid overwhelming spectroscopic follow-up capabilities with an unnecessary flood of data.

This classification of images into real objects and artefacts are carried out by human hand scanners, leading to large amounts of images having to be scanned each night. Additionally, human classifiers have different sets of biases that change with mood and circumstances, something that cannot be characterised in a systematic way and leads to irreproducible results.

Using machine learning techniques instead of human scanners for this classification is therefore an important next step in achieving the goals of transient surveys in the next decade. We here compare various machine learning algorithms in order to see if a desired classification accuracy can be achieved. It should be noted that this contribution is a condensed version of work that is soon to be published (du Buisson *et al.* 2015).

2. Data

We use g, i and r-band difference images from the 2nd and 3rd years of the Sloan Digital Sky Survey (SDSS) supernova survey (Frieman *et al.* 2008, Sako *et al.* 2014). The survey's transient detection algorithms consisted of producing difference images, which were then

† Email: lisedubuisson@gmail.com
‡ Email: navin.sivanandam@gmail.com
¶ Email: bruce.a.bassett@gmail.com

Table 1. The three hierarchical classification systems and how they relate to one another. A description of the different original classes (Frieman *et al.* 2008) can be found in Bassett *et al.* (2005).

Original class	Visual class	Real/Not-Real
Artefact	Artefact	Not-Real
Dipole	Dipole/Saturated	Not-Real
Saturated Star	Dipole/Saturated	Not-Real
Moving	Real	Real
Variable	Real	Real
Transient	Real	Real
SN Other	Real	Real
SN Bronze	Real	Real
SN Silver	Real	Real
SN Gold	Real	Real

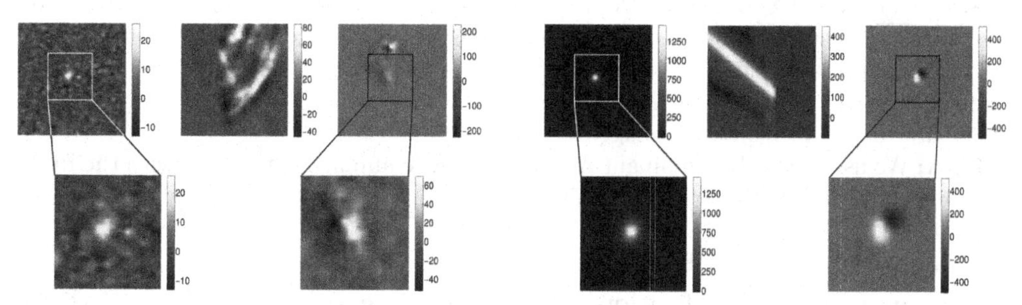

Figure 1. The three visual classes in the r-band. **(Left)** Representative (SNR < 20) examples of (left to right) real, artefact and dipole/saturated with zooms. **(Right)** High-quality (SNR $>$ 40) examples of (left to right) real, artefact and dipole/saturated with zooms.

labelled as candidates for human hand scanning if they passed the necessary threshold cuts. Because this process is not perfect, it introduced a large number of artefacts into the candidate group.

Human hand scanners then sorted each candidate object into one of ten original classes. We decided to re-group the original classes into three visual classes (see Fig. 1) based on the similarities in appearance between some of the original classes, and used these three classes to carry out single-class PCA (see section 4). Ultimately, we are concerned with being able to correctly predict whether an object is real or not, and this is therefore the main classification used by the classifiers. To see how these hierarchical classification systems relate to one another, see Table 1.

3. Testing and Performance Measures

Testing: 25% of our data set were held apart for final testing after the various classifiers have been trained using the remaining 75% of the data, of which a further 30% were set apart for preliminary testing and parameter optimisation of the learners.

Performance Measures: For preliminary testing, optimisation and final testing we used two common performance measures, namely the accuracy (A) and recall (R), defined in terms of true/false positives/negatives (t_p, t_n, t_p and f_p) as follows:

$$A = \frac{t_p + t_n}{t_p + t_n + f_p + f_n} \qquad R = \frac{t_p}{t_p + f_n} \tag{3.1}$$

A Receiver Operating Characteristic (ROC) curve shows a binary classifier's performance as its discrimination threshold is varied. It is a plot of the true positive rate (TPR), or recall, versus the false positive rate (FPR) (Eq. 3.2) at various different threshold values.

$$FPR = \frac{f_p}{f_p + t_n} \tag{3.2}$$

The "Area Under Curve" (AUC) is often used for classifier comparison, and is equal to the probability that a classifier will correctly classify a randomly chosen instance (Hanley & McNeil 1982). We used the AUC to compare classifiers.

4. Feature Extraction

Multi-class PCA: We carry out Principal Component Analysis (PCA) on the full training set and then characterise individual objects by including a certain number of principal components (PCs) in their feature vectors.

Single-class PCA: We apply PCA independently on the three different visual classes (see Table 1), resulting in each class having a unique eigenspace. An image's features is obtained by reconstructing it using the PCs of each individual class in turn, and then finding the error between each reconstructed image and the original. The three calculated errors are then features for an image.

LDA: We use Linear Discriminant Analysis (LDA) components together with the PCA components as features.

5. Machine Learning Techniques

Minimum Error Classification (MEC): MEC takes only the three error features (section 4) as input, and classifies an object to be of the class corresponding to the minimum error.

Naïve Bayes (NB): NB is a practical Bayesian classifier, resting on the assumption that, given the class, the feature values are conditionally independent.

K-Nearest Neighbours (KNN): KNN finds the K nearest training objects in feature space to a test item, averages their classes (with a uniform or distance weight) and classifies the test item accordingly.

Support Vector Machine (SVM): An SVM is a maximum margin classifier – it aims to find a decision boundary in such a way that the largest possible separation between classes is obtained. The kernel trick can be applied to transform from the linear decision boundary to a non-linear one.

Artificial Neural Network (ANN): ANNs have interconnected nodes that processes information and passes it on to other nodes along weighted connections that are directed from an input layer of nodes to an output one. An ANN is trained to learn a mapping between the input and output layer, and it then predicts the outputs for new input data. We use SkyNet as our ANN (Graff *et al.* 2014).

6. Results and Discussion

The performance results of the various classifiers upon final testing can be seen in Table 2, and the ROC curve for each of the classifiers can be seen in Fig. 2. It is interesting that the simple KNN performs better than (or as well as) the most intricate classifiers, like SkyNet. Furthermore, it can be seen that KNN's, SkyNet's and MEC's classification performance on fake SNe equals that of human scanners (0.956 ± 0.010 fake SN-tag recall (Kessler 2007)). In order to determine how much overlap there is between the performance of the different classifiers on the test set we calculated the Cohen's Kappa

Table 2. The classification performance of the various classifiers in final testing (see section 3) together with the performance on the fake SNe in the dataset.

Machine Learning Technique	AUC	Accuracy	Recall	Fake SN-tag Recall
KNN	0.94	0.89	0.90	0.96
SkyNet	0.94	0.88	0.89	0.96
SVM	0.93	0.86	0.90	0.94
MEC	0.90	0.84	0.92	0.96
NB	0.81	0.78	0.77	0.85

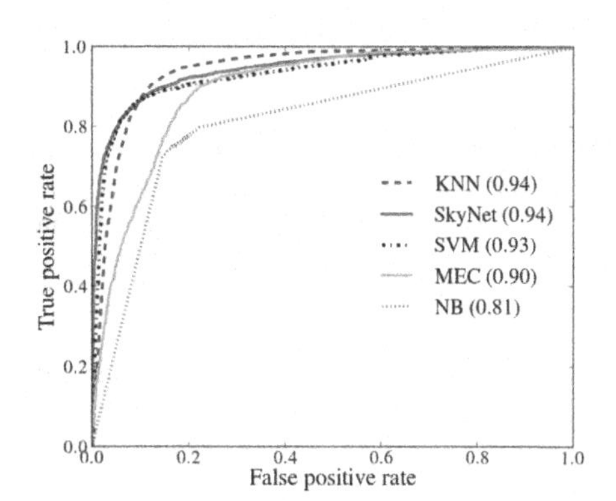

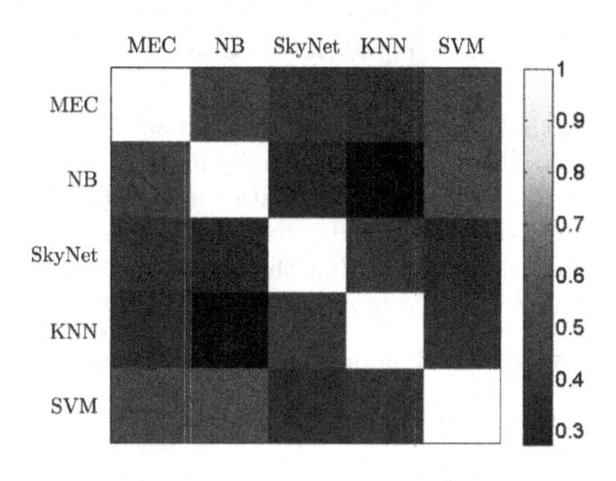

Figure 2. **(Left)** ROC curve for the five classifiers, listed in order of performance based on the AUC statistic (section 3), given in brackets. **(Right)** The Cohen's Kappa value (κ) for pairs of classifiers. Agreement: $0.0 < \kappa \leqslant 0.2$ – Slight; $0.2 < \kappa \leqslant 0.4$ – Fair; $0.4 < \kappa \leqslant 0.6$ – Moderate; $0.6 < \kappa \leqslant 0.8$ – Substantial; $0.8 < \kappa \leqslant 1.0$ – Almost Perfect (Landis & Koch 1977).

statistic κ (Cohen 1960) for each pair of classifiers. Fig. 2 shows that no pair of classifiers is highly correlated, which suggests that there may be significant gain from combining classifiers. Further improvements will also likely come from adding host galaxy and multi-epoch information, studies that we leave to the future.

Acknowledgements

LdB is supported by an NRF/SKA scholarship, NS is funded by a Claude Leon Foundation fellowship and BB and MS is funded by the SKA and the NRF.

References

Bassett, B. A., Richmond, M., & Frieman, J. 2005, `url:http://spiff.rit.edu/richmond/sdss/sn_survey/scan_manual/sn_scan.html`

Cohen, J. 1960, *Educational and Psychological Measurement*, 20, 37

du Buisson, L., Sivanandam, N., & Bassett, B. A., Smith M., 2015, submitted to MNRAS, arXiv:1407.4118 [astro-ph.IM]

Frieman, J. A., Bassett, B., Becker, A., & Choi, C., *et al.* 2008, *AJ*, 135, 338

Graff, P., Feroz, F., Hobson, M. P., & Lasenby, A. 2014, *MNRAS*, 441, 1741

Hanley, J. & McNeil, B. J. 1982, *Radiology*, 143, 29

Kessler, R. 2007, Private Communication

Landis, J. R. & Koch, G. G. 1977, *Biometrics*, 33, 159

Sako, M., Bassett, B., Becker, A. C., & Brown, P. J., *et al.* 1977, *ArXiv e-prints*, `adsurl:http://adsabs.harvard.edu/abs/2014arXiv1401.3317S`

Statistical Challenges in 21st Century Cosmology
Proceedings IAU Symposium No. 306, 2014
A. F. Heavens, J.-L. Starck & A. Krone-Martins, eds.

doi:10.1017/S1743921314013453

Data-mining Based Expert Platform for the Spectral Inspection

Haijun Tian[1,2], Yang Xu[1], Yang Tu[2], Yanxia Zhang[1], Yongheng Zhao[1], Guohong Lei[2], Boliang He[1], Chenzhou Cui[1] and Xuelei Chen[1]

[1]National Astronomical Observatories, Chinese Academy of Sciences, Beijing 100012.

[2]China Three Gorges University, Yichang, 443002. Email: hjtian@lamost.org

Abstract. We propose and preliminarily implement a data-mining based platform to assist experts to inspect the increasing amount of spectra with low signal to noise ratio (SNR) generated by large sky surveys. The platform includes three layers: data-mining layer, data-node layer and expert layer. It is similar to the GalaxyZoo project and it is VO-compatible. The preliminary experiment suggests that this platform can play an effective role in managing the spectra and assisting the experts to inspect a large number of spectra with low SNR.

Keywords. techniques: photometric, methods: data analysis, catalogs

1. Introduction

With the telescopes established and the surveys ongoing, such as the Guoshoujing telescope (LAMOST, Zhao 1999), more and more spectra are collected and released. Unfortunately, except the qualified data, there still exist many spectra unclassified by the automated pipeline. Usually most of these spectra have too low SNR to be classified because of the limiting magnitude of the instruments or other reasons. Some important new discoveries may probably be hidden in these unknown spectra. Therefore, we should not throw away these seemingly useless data, even though they are quite defective. How to handle these unknown spectra is one of the biggest challenges to the modern statistics, data mining techniques as well as eyeball check. In order to ensure the accuracy of the results, we have to motivate users to check these spectra by visual inspection. Owing to huge amount of such spectra generated continuously by the large sky surveys, much time and efforts is needed to check spectra one by one. Consequently, a platform to efficiently manage and coarsely classify these unqualified data is in great requirement for large surveys.

2. Architecture

We propose such a platform, which aims to effectively process and manage a large volume of spectra with low SNR, and mine as much as possible knowledge from the spectra for the scientific research through the integration of machine learning and human inspection. The platform is a three-tier structure including data-node layer, data mining layer and expert layer, as shown in Fig. 1. The core part is data mining. Before the visual inspection, the dataset will be preprocessed with various up-to-date statistical and data-mining techniques. For instance, the dataset can be classified roughly as stars, galaxies or quasars according to the limited information contained in the low SNR spectra. All results from this layer could efficiently help the expert to inspect the spectra. The expert layer provides an interface by which anyone may query and inspect the spectra and save

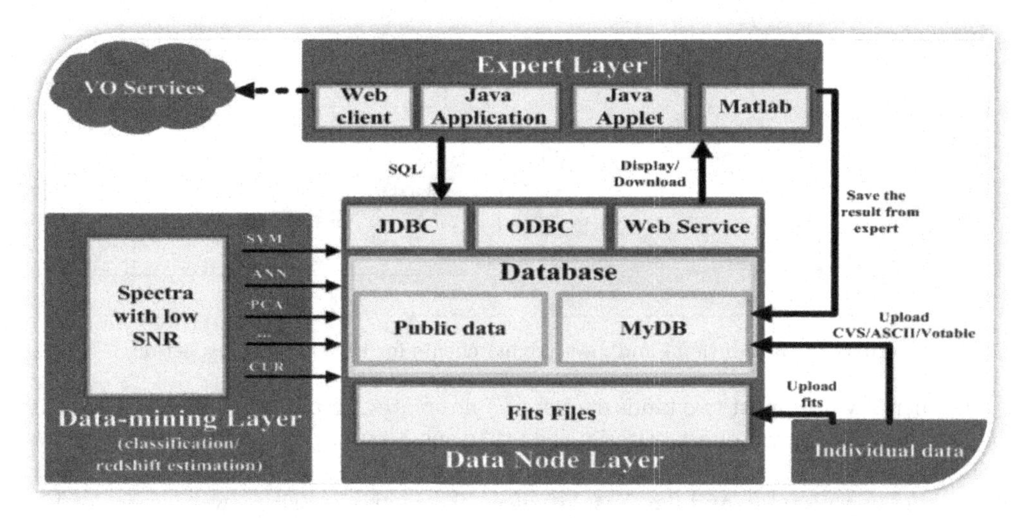

Figure 1. Architecture of expert platform for the spectral inspection

the results in the data node layer. Finally the system can give a more reliable result by efficiently utilizing the experts' input (one spectrum may be inspected by more than one expert). The platform will be VO-compatible and integrated seamlessly with other surveys. Its client will be of diverse forms, such as Matlab-GUI, Java-application, Java-applet, etc.

Several kinds of data-mining algorithms to pre-process the LAMOST spectra have been investigated. For example, Support Vector Machines (SVMs) are first applied to classify the LAMOST spectra, and the classified spectra are saved into the data-node layer as a public dataset. Then the spectra may be queried with a web client and inspected by a Java applet tool, finally the analysis results are returned back into MyDB in the data-node layer.

Data-mining layer. The data-mining layer has two main tasks: classification and redshift or radial velocity estimation for low SNR spectra using various advanced data-mining techniques, such as Support Vector Machines (SVMs, Peng *et al.* 2012), Artificial Neural Networks (ANN, Zhang *et al.* 2009), K-Nearest Neighbors (KNN, Li *et al.* 2008), Principal Component Analysis (PCA, Zhang & Zhao 2003; Yip *et al.* 2004), Wavelet (Machado *et al.* 2014), CUR Matrix Decomposition (Yip *et al.* 2014).

DataNode layer. The data-node layer include two parts: one is based on the database, which manages the catalog generated by the data-mining layer (public dataset available for each user) and the private dataset input by the users (MyDB, visible only for the data owner); the other needs enough storage space to save the spectral fits files, which are also divided into the public and private. The results from the expert layer can also be saved in MyDB as the private file. The data-node layer provides two interfaces, e.g. the database interface (JDBC or ODBC) and the Web Service interface, by which expert layer can query catalogs and retrieve spectra.

Expert layer. The clients with diverse forms, such as Matlab-GUI, Java-application, Java-applet, etc., can be used to display the spectra and assist the experts to inspect the spectra in this layer. These tools provide all kinds of spectral templates and reference lines, which can be easily shaped at expert's will. The redshift can be calculated in real time as the template is moved. The templates will automatically be best-matched based on the least square method, and the optimal redshift can be figured out when the spectrum is best fitted with the template.

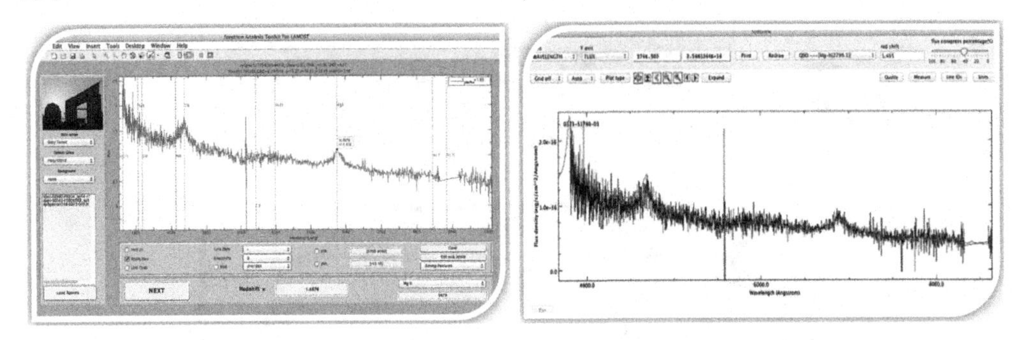

Figure 2. Matlab (left) and Java (right) clients for the spectral inspection.

The clients will support two kinds of queries: one creates simple SQLs through choosing the fields and filling the query criteria; the other one provides a input form, that allows users to compose much more complex SQL statements, and also provides the functions of checking query syntax and database privilege. Through the JDBC, web service or VO interfaces, the clients can access the local or remote databases. Experts with different levels have different weights to contribute the final results.

For the data-node layer, we have finished the database and file systems with public data and MyDB, and released the JDBC and Web Service interfaces for the query functions.

For the expert layer, a Matlab client and a Java client have been preliminary developed, as displayed in the Fig. 2. Actually, the Java client is modified SpecView, an interactive java tool for visualization and analysis of spectral data (Busko 2000), in which we extended some functions including the template fitting, automatic redshift estimation, the interaction with the data-node layer, etc.

3. Conclusions

In order to overcome the difficulty in processing the poor spectra, we propose and develop a data-mining based platform to encourage and assist experts to inspect the low SNR spectra. With the help of the platform, we tentatively process LAMOST spectra. The preliminary experiment indicates that this platform indeed can play an effective role in assisting experts to recognize a large number of spectra with low SNR.

4. Acknowledgments

The authors thank the grants (No. U1231123, U1331202, U1331113, U1231108, 11073024, 11103027, 11303020) from NSFC. THJ thanks the support from LAMOST Fellowship.

References

Busko, I. 2000, *ASPC*, 216, 79B
Li, L. L., Zhang, Y. X., & Zhao, Y. H. 2008, *Science in China*, G, 51(7), 916
Machado, D., Leonard, A., Starck, J. L., & Abdalla, F. 2014, *A&A*, in press
Peng, N., Zhang, Y. X., Zhao, Y. H., & Wu, X. B. 2012, *MNRAS*, 425, 2599
Yip, C. W., Connolly, A. J., Vanden Berk, D. E., *et al.* 2004, *AJ*, 128, 2603Y
Yip, C. W., Mahoney, M. W., Szalay, A. S., Csabai, I., & Budavári, T., *et al.* 2014, *AJ*, 147, 110Y
Zhao, Y. H. 1999, *PYunO*, S, 1Z
Zhang, Y. X., Li, L. L., & Zhao, Y. H. 2009, *MNRAS*, 392, 233
Zhang, Y. & Zhao, Y., 2003, *PASP*, 115, 1006

Statistical Challenges in 21st Century Cosmology
Proceedings IAU Symposium No. 306, 2014
A. F. Heavens, J.-L. Starck & A. Krone-Martins, eds.

doi:10.1017/S1743921314010898

Robust Constraint of Luminosity Function Evolution through MCMC Sampling

Noah Kurinsky and Anna Sajina

Department of Physics and Astronomy, Tufts University,
4 Colby Street, Medford, MA 02155, USA
Email: Noah.Kurinsky@gmail.com

Abstract. We present a new galaxy survey simulation package, which combines the power of Markov Chain Monte Carlo (MCMC) sampling with a robust and adaptable model of galaxy evolution. The aim of this code is to aid in the characterization and study of new and existing galaxy surveys. In this paper we briefly describe the MCMC implementation and the survey simulation methodology and associated tools. A test case of this full suite was to constrain the evolution of the IR Luminosity Function (LF) based on the HerMES (*Herschel* SPIRE) survey of the *Spitzer* First Look Survey field. The initial results are consistent with previous studies, but our more general approach should be of wider benefit to the community.

Keywords. galaxies: luminosity function, galaxies: evolution, galaxies: statistics, methods: analytical, techniques: photometric, infrared: galaxies

1. Motivation

In extragalactic astronomy, we typically conduct surveys which uncover large numbers of high redshift galaxies with sparse SED coverage. Such spectra are insufficient for redshift determination by SED fitting, and spectroscopic follow-up to ascertain the redshifts for all galaxies in a survey is impractical, especially for samples of predominantly optically-faint IR galaxies. Direct measurements of luminosity function evolution are therefore very difficult in this regime. We can use well motivated galaxy spectral colors, however, to characterize cumulative trends in SED and luminosity function evolution in an entire population. By pairing a redshift-evolving SED library and a redshift parameterized luminosity function, we can simulate surveys given knowledge of evolutionary model parameters and local measurements. By sampling the resulting parameter space in a robust manner (as in Marsden *et al.* 2011), one can reconstruct the luminosity function as it evolves with redshift by comparison of the simulated and observed color-color space trends. In this manner, we can not only constrain the luminosity function parameters given observations, but also use the constrained parameters to simulate a high-redshift survey, given characteristics of the instrument performing the survey.

The development of statistically sound fitting methods to accurately constrain the model parameters and generate model populations complicates such an effort, and requires the astronomer have extensive knowledge of a very specific class of statistical and computational methods. We aim to develop methods which make this analysis routine and easily adaptable so that such studies may be more efficient, robust, and accessible to time sensitive studies of new and existing surveys. We are developing a general software package capable of simulating a wide range of surveys with a robust Markov Chain Monte Carlo (MCMC) sampling algorithm, taking into account various LF forms, SED models, and instrumental properties, including noise levels and intensity limits. Our program allows us to characterize a survey as well as highlight the successes and failures of the SED template and the luminosity function form used by the simulation, and their behavior at high redshift, in order to speak to the nature of the data as well as the strength and weakness of the chosen models. We envision this program functioning, in the construction of high-redshift luminosity functions, similarly to how CosmoMC (Lewis & Bridle 2002) and the methods within are applied to constraint of ΛCDM parameters.

2. Current Package

Our simulation approach is centered around the choice of parameterized luminosity function and SED template library. The SED templates, as well as all other data inputs, are stored in FITS files and are easily interchangeable, specified as input parameters to the main program. The data to which parameters are fitted is photometric data in at least three bands, with observations in all bands. For a given set of luminosity function parameters, we follow the following procedure for each redshift-luminosity bin, with luminosity binning determined by the templates and redshift binning defined as a user specified parameter:

• Determine $N(L, z|\mathbf{x})$, the number of galaxies expected in a given luminosity and redshift bin given the luminosity function form and $\mathbf{x}$, the given parameters

• Redshift the SED corresponding to the given $(L, z|\mathbf{x})$ bin to the observed frame, and convolve with instrumental filters

• For each simulated galaxy, add Gaussian noise to each of three generated fluxes with noise determined by the mean of measurement errors found in the observations.

• Determine detectability of the galaxy given survey flux cuts, and add the source to number count calculations if detected.

This procedure across the luminosity-redshift space results in a simulated survey with band-matched fluxes, which is projected into the same color space as the observations. We perform a modified χ^2 test comparing data to observation to determine the quality of the fit.

To constrain the luminosity function parameters, we employ a parallel-chain MCMC simulation approach similar to that employed by Lewis & Bridle (2002), but with updated methods to reflect the more recent findings of Dunkley *et al.* (2005). We employ the common Metropolis sampling algorithm, and the Gelman and Rubin convergence test. The user specifies likely initial parameter values, which parameters to fit, as well as acceptable ranges, and the program uses these as initial conditions. During the burn-in phase, one chain begins at the initial position and the others uniformly distributed throughout the allowable space, and the chains are allowed to freely explore as the temperature is annealed to achieve an optimal acceptance rate. Once the burn-in is complete, the chains are reinitialized closer to the best position, and the simulation runs until the convergence criteria are met. As the simulation progresses, covariance between parameters is continually adjusted to maximize sampling efficiency.

The current package is structured as a script controlled C++ executable, with the main simulation and computationally expensive aspects implemented in C++ and the plotting and user interaction implemented in IDL and python. All data, SED models, and filters are read in as FITS or text files, and we aim to have fully selective and robust options for both of the latter two at first public release, expected to be complete by the end of the year. The luminosity functional form is hard coded in the C++ library, however the computational aspect is contained in a single function which is easily user editable, and the code can easily be recompiled by the standard GNU autotools.

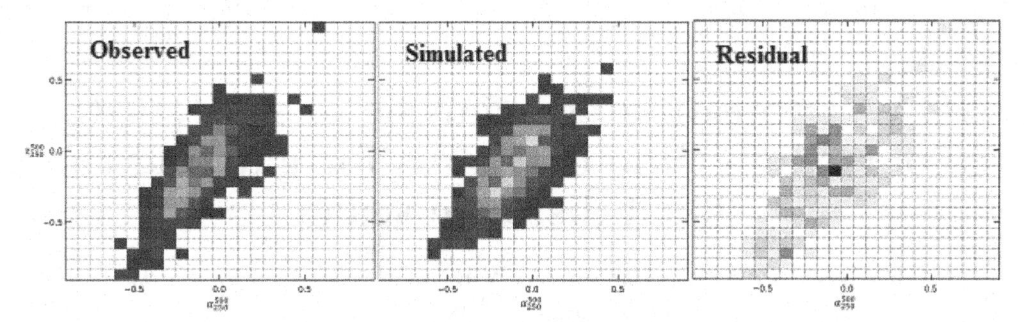

Figure 1. Example of color-color histograms for an observed survey (left) and simulated survey (right), with the difference between the two shown at right. The free parameters are constrained by essentially attempting to minimize the cumulative sum of the right histogram, measured through our modified χ^2 statistic. The discrepancy between these two metrics is on the order of ~ 60 galaxies, compared to an observed sample of ~ 1000 galaxies.

3. Initial Testing and Future Directions

Our initial tests with a ~1000 source HerMES (*Herschel* SPIRE 250, 350, and 500 microns) survey, where only sources with measurements in all three bands were considered, were consistent with the density and luminosity evolution findings of Marsden *et al.* (2011) as well as the SED intrinsic luminosity evolution found by Sajina *et al.* (2012). Using the locally derived SED templates of Rieke *et al.* (2009), we were able to reproduce the survey to within statistical margins. The observed and simulated survey metrics, along with their subtracted residual, can be seen in Fig. 1, where the total number of sources in the residual histogram is ~ 60. We are continuing to refine this package, and aim to publish a paper in Fall 2014 detailing the first public release, and demonstrating test cases in the near and far infrared (Kurinsky *et al.* 2014, in prep). We are also working on testing the code on a wider range of surveys, including WISE (Silva *et al.* 2015, in prep).

Acknowledgements

We are grateful for the support of Tufts University's undergraduate research fund as well as the IAU Student Grant program for making this presentation possible. This work was funded in part by the Tufts Eliopoulos Summer Scholars grant and NSF grant 1313206.

References

Dunkley, J., Bucher, M., Ferreira, P. G., Moodley, K., & Skordis, C. 2005, *MNRAS*, 356, 925
Lewis, A. & Bridle, S. 2002, *Phys. Rev. D.* 66, 10
Kurinsky, N., Sajina, A., & Silva, A. 2014, in. prep.
Marsden, G., Chapin, E. L., Halpern, M., Patanchon, G., Scott, D., Truch, M. D. P., Valiante, E., Viero, M. P., & Wiebe, D. V. 2011, *MNRAS* 417, 1192
Rieke, G. H., Alonso-Herrero, A., Weiner, B. J., Pérez-González, P. G., Blaylock, M., Donley, J. L., & Marcillac, D. 2009, *ApJ* 692, 556
Sajina, A., Yan, L., Fadda, D., Dasyra, K., & Huynh, M. 2012, *ApJ*, 757, 13
Silva, A., Sajina, A., & Kurinsky, N. 2015, in. prep.

Statistical Challenges in 21st Century Cosmology
Proceedings IAU Symposium No. 306, 2014
A. F. Heavens, J.-L. Starck & A. Krone-Martins, eds.

doi:10.1017/S1743921314011077

OCAAT: automated analysis of star cluster colour-magnitude diagrams for gauging the local distance scale

Gabriel I. Perren[1], **Ruben A. Vázquez**[1], **Andrés E. Piatti**[2] **and André Moitinho**[3]

[1]Facultad de Ciencias Astronómicas y Geofísicas (UNLP)
IALP-CONICET, La Plata, Argentina
email: gperren@fcaglp.unlp.edu.ar

[2]Observatorio Astronómico, Universidad Nacional de Córdoba
Córdoba, Argentina

[3] SIM, Faculdade de Ciências da Universidade de Lisboa
Ed. C8, Campo Grande, 1749-016 Lisboa, Portugal

Abstract. Star clusters are among the fundamental astrophysical objects used in setting the local distance scale. Despite its crucial importance, the accurate determination of the distances to the Magellanic Clouds (SMC/LMC) remains a fuzzy step in the cosmological distance ladder. The exquisite astrometry of the recently launched ESA Gaia mission is expected to deliver extremely accurate statistical parallaxes, and thus distances, to the SMC/LMC. However, an independent SMC/LMC distance determination via main sequence fitting of star clusters provides an important validation check point for the Gaia distances. This has been a valuable lesson learnt from the famous Hipparcos Pleiades distance discrepancy problem. Current observations will allow hundreds of LMC/SMC clusters to be analyzed in this light.
Today, the most common approach for star cluster main sequence fitting is still by eye. The process is intrinsically subjective and affected by large uncertainties, especially when applied to poorly populated clusters. It is also, clearly, not an efficient route for addressing the analysis of hundreds, or thousands, of star clusters. These concerns, together with a new attitude towards advanced statistical techniques in astronomy and the availability of powerful computers, have led to the emergence of software packages designed for analyzing star cluster photometry. With a few rare exceptions, those packages are not publicly available.
Here we present OCAAT (Open Cluster Automated Analysis Tool), a suite of publicly available open source tools that fully automatises cluster isochrone fitting. The code will be applied to a large set of hundreds of open clusters observed in the Washington system, located in the Milky Way and the Magellanic Clouds. This will allow us to generate an objective and homogeneous catalog of distances up to ~ 60 kpc along with its associated reddening, ages and metallicities and uncertainty estimates.

Keywords. galaxies: star clusters, methods: data analysis, methods: statistical, cosmology: distance scale

---

1. Introduction

The aim of OCAAT is to bring together many of the algorithms which are usually applied on open clusters (OCs) throughout the literature, in a single easy to use package. OCAAT is able to assign a precise center, calculate a limiting radius, reject stars based on their errors, obtain the cluster's luminosity function and its integrated color, assign a probability of the cluster being a real system and not a random overlapping of field stars, assign membership probabilities for each star within the cluster region and

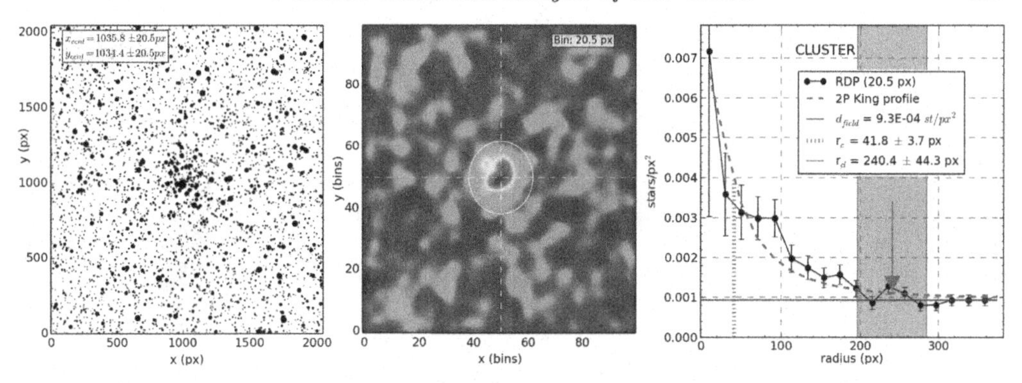

Figure 1. *Left*: Finding chart of synthetic cluster. *Center*: Center determination via a KDE spatial density function. *Right*: radius assignment, obtained as the value where the RDP reaches the field density level.

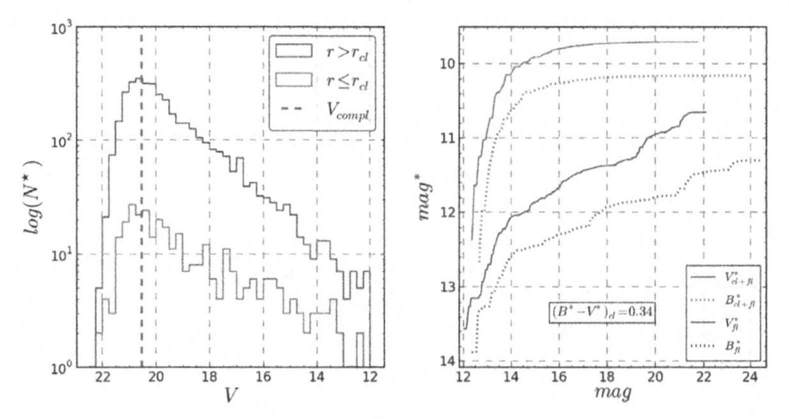

Figure 2. *Left*: Cluster and average field luminosity functions curves in red and blue respectively. *Right*: Integrated B and V magnitudes for the cluster region and averaged field regions.

estimate the cluster's parameters: metallicity, age, distance and extinction along with their uncertainties.

2. Functions description

Structural parameters. Traditionally, centers are inferred by eye after the identification of a visible spatial overdensity, seldom refined applying a simple histogram-based search for the maximum concentration in each axis separately. In OCAAT we employ a 2-dimensional Gaussian kernel density estimate (KDE) function applied to the entire region and the location of its maximum value. Fig. 1 shows to the left a field where a synthetic young cluster is located and to the center its center assignment. The radius is assigned as the value where the radial density profile (RDP) reaches the level of the field density contribution d_{field}, as shown to the left. The error is obtained as the standard deviation of the radius values estimated using different conditions to establish when the RDP has stabilized around d_{field} (grey area in the plot). A 3-parameter King profile is fitted to obtain the core and tidal radius, if the process does not converge then a 2-parameter fit is attempted to derive the core radius only.

Errors, LF and integrated color. A curve composed of a polynomial plus an exponential fit is used as an upper envelope for the error distribution in magnitude. Stars that are

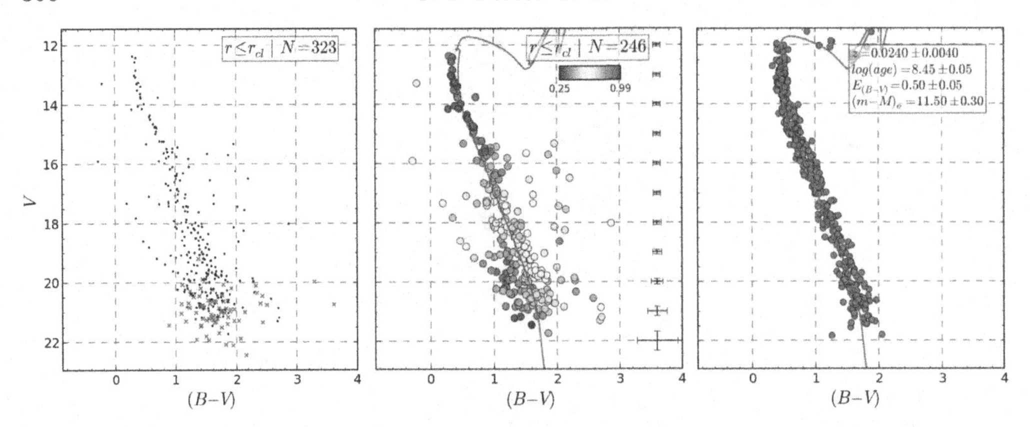

Figure 3. *Left*: CMD of synthetic cluster showing rejected stars (ie: stars above the maximum error limit) as green crosses. *Center*: CMD of the cluster region colored according to the membership probabilities. *Right*: Best fit synthetic cluster created from the red isochrone shown.

located above this curve or above a maximum error threshold in magnitude or color are rejected. The luminosity function curve is obtained for both the cluster region and the field region (scaled to the cluster's area) and the same is done to calculate the integrated color of the stars within the cluster region. Fig. 2 shows these curves for the cluster and field regions.

Cluster KDE probability. A KDE-based algorithm is used to asses the similarity between the arrangement of stars in the cluster region CMD versus CMDs of the field regions, quantified by p-values. The curves formed by these cluster vs field and field vs field region p-values overlap an amount proportional to the probability that the cluster and average field region were drawn from the same distribution. An overlap value close to 1 points to a high probability of the system being a true cluster.

Decontamination algorithm. A Bayesian algorithm inspired by the one presented in Cabrera-Cano & Alfaro (1990) is implemented to assign membership probabilities to stars within the cluster region, based on photometry data alone. Fig. 3 shows to the left all stars inside the defined cluster radius and at the center only those stars below the error cut-off, colored by the membership probabilities assigned by the code. Higher values are usually associated with brighter stars due to the usual field star contamination in the faint region of the CMD.

Synthetic cluster fit. Here is where we obtain the cluster distances. The membership probabilities calculated by the decontamination algorithm are used as weights in a likelihood equation that identifies the best observed vs synthetic cluster fit, based on the one presented in Hernandez & Valls-Gabaud (2008). Each synthetic cluster is generated from a theoretical isochrone accounting for errors, binarity and completeness effects. The search for the synthetic cluster which minimizes the likelihood can be done either via a brute force algorithm or a genetic algorithm (GA). In Fig. 3 (right) the best synthetic cluster found by the GA can be seen, with the underlying theoretical isochrone shown in red (same isochrone shown in gree in the center plot) The uncertainties are estimated applying a bootstrap process.

References

Cabrera-Cano, J. & Alfaro, E. J. 1990, *A&A*, 235, 94

Hernandez, X. & Valls-Gabaud, D. 2008, *MNRAS*, 383, 1603

Statistical Challenges in 21st Century Cosmology
Proceedings IAU Symposium No. 306, 2014
A. F. Heavens, J.-L. Starck & A. Krone-Martins, eds.

doi:10.1017/S1743921314010990

Quantifying correlations between galaxy emission lines and stellar continua using a PCA-based technique

Róbert Beck, László Dobos and István Csabai

Department of Physics of Complex Systems, Eötvös Loránd University,
Pf. 32, H-1518, Budapest, Hungary
E-mail: robert.beck23@gmail.com (RB), dobos@complex.elte.hu (LD),
csabai@complex.elte.hu (ICs)

Abstract. We analyse the correlations between continuum properties and emission line equivalent widths of star-forming and narrow-line active galaxies from SDSS. We show that emission line strengths can be predicted reasonably well from PCA coefficients of the stellar continuum using local multiple linear regression. Since upcoming sky surveys will make broadband observations only, theoretical modelling of spectra will be essential to estimate physical properties of galaxies. Combined with stellar population synthesis models, our technique will help generate more accurate model spectra and mock catalogues of galaxies to be used to fit data from new surveys. We also show that, by combining PCA coefficients from the pure continuum and the emission lines, a plausible distinction can be made between weak AGNs and quiescent star-forming galaxies. Our method uses a support vector machine, and allows a more refined separation of active and star-forming galaxies than the empirical curve of Kauffmann *et al.* (2003).

Keywords. Galaxies: emission lines, starburst, active, classification, Methods: data analysis

---

1. Data sample and preliminary processing

We selected a sample of 13834 galaxies from the SDSS Data Release 7, with both photometric and spectroscopic measurements. The galaxies had to contain 11 given emission lines, and had to have a signal to noise ratio larger than 5. The sample size was limited by allowing only a given range of right ascension.

We performed dereddening on the galaxy spectra, transformed them into the rest frame, and normalised them by dividing by the average of median flux values in given wavelength ranges. Then the spectra were subdivided into stellar continuum and emission line components. The continuum component was given as the linear combination of 10 model spectra, while the emission lines were fitted, and characterized by their equivalent widths.

The continuum components of the spectra were processed further, we performed principal component analysis (PCA) on them via singular value decomposition. We kept the first 5 principal components that explained the largest amount of deviation in the data. Hereafter the continuum properties of the spectra will be characterized by their five singular vector coefficients.

2. Local multiple linear regression

We tried to quantify the connection between the stellar continua and the emission lines. Thus, we performed local multiple linear regression in the PCA space of the continua,

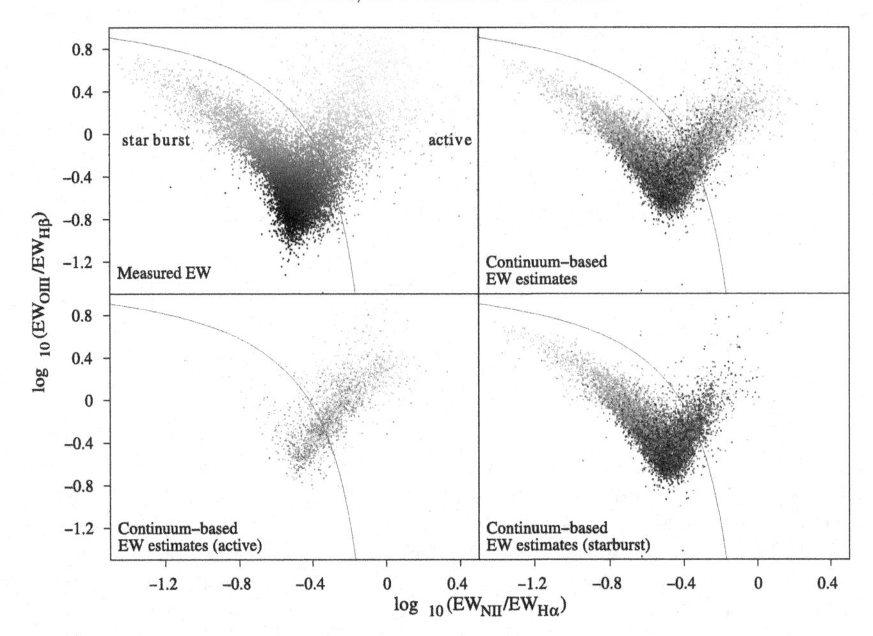

Figure 1. The BPT diagrams of our sample. The colouring corresponds to the galaxies' location on the top left image. The empirical separation line of Kauffmann *et al.* (2003) is shown with a solid line, the classification in the bottom images is based on that. The fitted equivalent widths follow the same V-shaped relationship, but with less deviation around it. It is visible that a significant number of starburst galaxies are estimated to be active based on their stellar continua.

fitting the logarithm of emission line equivalent widths. The 5 singular vector coefficients (that characterize the continua) were the independent variables, and the logarithm of the EWs were the dependent variables in the local regression. In other words, we expressed the line EWs locally as a linear function of continuum PCA coefficients. Locality was determined with a kD-tree, using Euclidean distance, and we took into account only the 30 nearest neighbours. We found that the emission line equivalent widths could be estimated reasonably well from the stellar continua using the method described above.

The lines' measured $\ln(EW)$ values were in the $[-2, 6]$ range, and the RMS values of the estimation (the square root of the mean squared error) were in the $[0.37, 0.78]$ interval, with Pearson's correlation coefficients in the $[0.60, 0.90]$ range. For 13834 data points, this yielded a p-value below floating point accuracy even in the worst case.

3. BPT diagrams

The Baldwin–Phillips–Terlevich diagram (Baldwin *et al.* 1981) is used to quickly characterize a population of galaxies based on emission line properties. The empirical separation line of Kauffmann *et al.* (2003) distuinguishes between active galaxies and star-forming galaxies:

$$\log_{10} \frac{[OIII]}{H\beta} > \frac{0.61}{\log_{10}\left([NII]\,/H\alpha\right) - 0.05} + 1.3$$

If true, a galaxy is classified as an active galaxy. This relation is an empirical refinement of a maximum starburst line by Kewley *et al.* (2001), above which star-forming galaxy models cannot produce examples. However, the intermediate region below the empirical

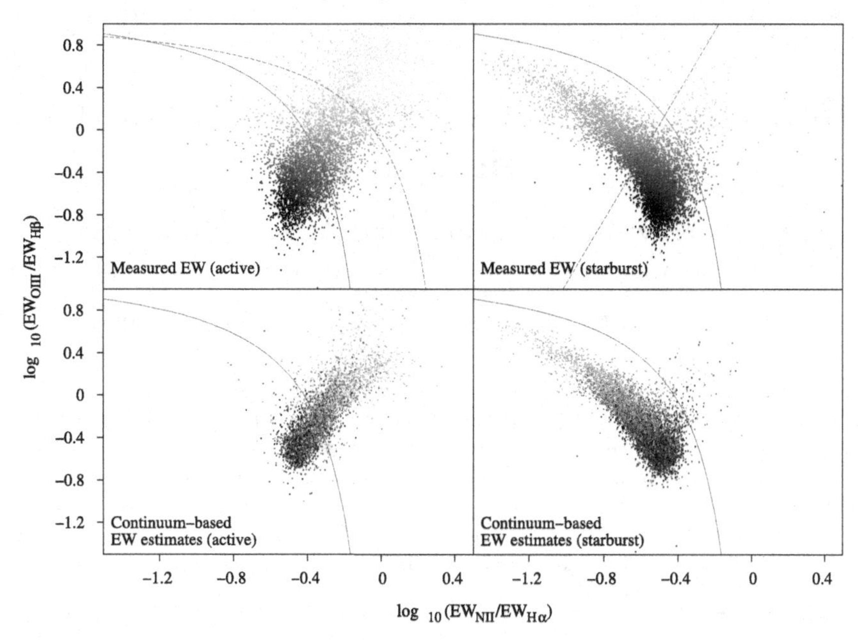

Figure 2. The BPT diagrams of active and star-forming galaxies, as classified by the support vector machine. It is observable that here the stellar continuum-based fitting projects galaxies to belong to the other class only rarely. The training sets for each class lie above the dashed lines on the top two images.

line – where the star-forming and active galaxy branches of the BPT diagram meet – has not yet been described in detail. The BPT diagrams of measured and fitted emission line equivalent widths for our sample are shown on Fig. 1.

4. Support vector machine classification

Based on the results shown on Fig. 1, we attempted to classify galaxies into the active and star-forming categories using the information contained in both their continua and emission lines. We performed principal component analysis on the logarithm of emission line equivalent widths, keeping the first four components. We selected by hand the relevant dimensions in the $5 + 4$ dimensional stellar continuum plus emission line PCA space, then trained a support vector machine on definitive examples. The results of the new classification are plotted on Fig. 2.

5. Acknowledgements

This research was supported by the OTKA-103244 grant of Hungary.

References

Baldwin, J. A., Phillips, M. M., & Terlevich, R. 1981, *PASP*, 93, 5

Kauffmann, G., Heckman, T. M., Tremonti, C., Brinchmann, J., Charlot, S., White, S. D. M.., Ridgway, S., Brinkmann, J., Fukugita, M., Hall, P., Ivezic, Z., Richards, G., & Schneider, D. 2003, *MNRAS*, 346, 1055

Kewley, L. J., Dopita, M. A., Sutherland, R. S., Heisler, C. A., & Trevena, J. 2001, *ApJ*, 556(1), 121

Statistical Challenges in 21st Century Cosmology
Proceedings IAU Symposium No. 306, 2014
A. F. Heavens, J.-L. Starck & A. Krone-Martins, eds.

doi:10.1017/S1743921314010710

Modelling Galaxy Populations in the Era of Big Data

S. G. Murray[1,2]†, C. Power[1,2] and A. S. G. Robotham[1]

[1]ICRAR, University of Western Australia, 35 Stirling Highway, Crawley, Western Australia 6009, Australia
[2]ARC Centre of Excellence for All-Sky Astrophysics (CAASTRO)

Abstract. The coming decade will witness a deluge of data from next generation galaxy surveys such as the Square Kilometre Array and Euclid. How can we optimally and robustly analyse these data to maximise scientific returns from these surveys? Here we discuss recent work in developing both the conceptual and software frameworks for carrying out such analyses and their application to the dark matter halo mass function. We summarise what we have learned about the HMF from the last 10 years of precision CMB data using the open-source `HMFcalc` framework, before discussing how this framework is being extended to the full Halo Model.

---

1. Introduction

The Halo Model has proven successful in predicting the large-scale clustering of dark matter halos and galaxies, and associated properties (Zheng(2004), Zehavi *et al.*(2011), Beutler *et al.*(2013)). With the emerging era of "big data" the Halo Model is set to be one of the key interpretive frameworks in the context of galaxy evolution and spatial statistics. It combines several components, and our aim in this short work is to further understanding of these components in a statistical sense. We also present a seamlessly integrated implementation of the various components, as a platform for next-generation modelling.

2. Uncertainty in the Halo Mass Function

The Halo Mass Function (HMF) is a standard tool in cosmology, used for consistency checking in simulations, fitting of cosmological parameters (especially σ_8) and as a component of other models. However, to properly apply the HMF requires a knowledge of its uncertainty distribution. We thus ask the question: "given our current 'best-bet' cosmology, with associated covariant uncertainty, what is the resulting uncertainty in the predicted mass function?" As a first-order application, this gives a rule-of-thumb estimate to assess whether a given measurement is consistent with current theory, but we show that such an analysis reveals more subtle insights as well (Murray *et al.*(2013a)).

To answer the question, we used the cosmological parameters, with covariant errors, from the past decade of WMAP and Planck results randomly drawing 5000 realisations of the four key parameters $\Omega_b h^2$, $\Omega_c h^2$, σ_8 and n_s from the MCMC chains of four releases. From these parameters we calculated 5000 HMF realisations for each of 12 fitting functions, calculating the resulting 1-σ uncertainties in both the amplitude and slope (shown in Fig. 1).

We find that over the course of the past decade of precision CMB measurements, the uncertainty in the HMF has accordingly decreased, and we give rule-of-thumb figures of 6% (20%) uncertainty at low (high) mass with current Planck results.

† steven.murray@uwa.edu.au

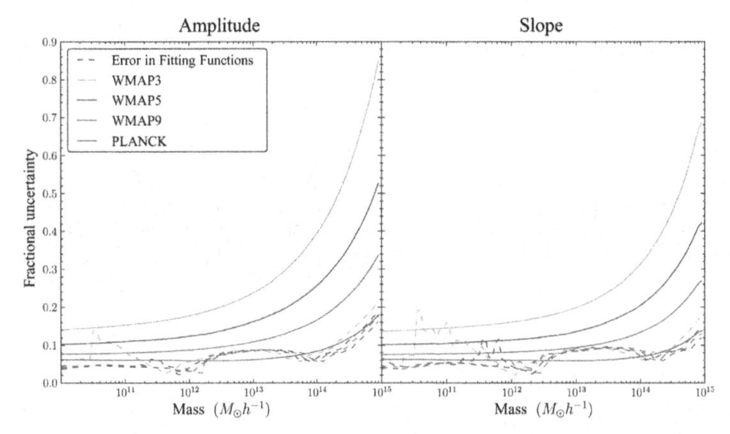

Figure 1. Fractional uncertainty in the amplitude (left) and slope (right) of the HMF for a range of parameter sets. Dashed lines indicate uncertainty due to 'arbitrary' choice of fitting function.

Consistent with previous studies (Vikhlinin *et al.*(2009)), we find that the HMF is most sensitive to the parameters Ω_m and σ_8 (at low-mass and high-mass respectively), which confirms the HMF as a useful tool for constraining these parameters.

Twelve fitting functions from the literature were used at each stage of the analysis which enabled some comparative statistics. Firstly, while the amplitude of each fit is different, the uncertainty behaves in the same way across all fits, allowing the use of one fit representatively. Secondly, we find that the scatter between fits, assuming arbitrary choice of fit, is now comparable to the uncertainty due to cosmology (see Fig. 1), highlighting the need for more robust fitting (cf. Bhattacharya *et al.*(2011)), and an educated choice of fit.

Finally, we calculated the uncertainty at which a viable Warm Dark Matter (WDM) model was more than 1-σ from its CDM counterpart, indicating the level of cosmological uncertainty necessary to detect WDM (see Fig. 2). Linearly extrapolating the trend in uncertainty reduction over the previous decade, we predict another 20 years before we can differentiate CDM and WDM on group mass scales ($\sim 10^{13} M_{\odot}h^{-1}$) using the HMF.

3. Halo Model Implementation

Though the Halo Model is set to be a key framework in the interpretation of the next decade of galaxy survey data, there does not exist an open-source implementation for its calculation. Our new code `halomod`† fills this gap, in the philosophy of our `hmf` code (Murray *et al.*(2013b)).

The `halomod` code was developed with primary goals of flexibility and ease-of-use. To this end, the user may submit a custom fitting function, or one of several other models, to the engine without modifying source code at all – a beneficial feature in a field where new models arise frequently. It also natively supports WDM models, has online documentation, includes most of the models already available in the literature, is simple to install and is algorithmically optimised for iteratively modifying model parameters.

Accompanying our `hmf` code is a web-application interface, `HMFcalc`‡, which exposes all appropriate functionality of `hmf` with a simple web-interface. We are preparing an extension of `HMFcalc` which extends the capabilities to the domain of `halomod`.

† Available at **github.com/steven-murray/halomod**
‡ Found at **hmf.icrar.org**

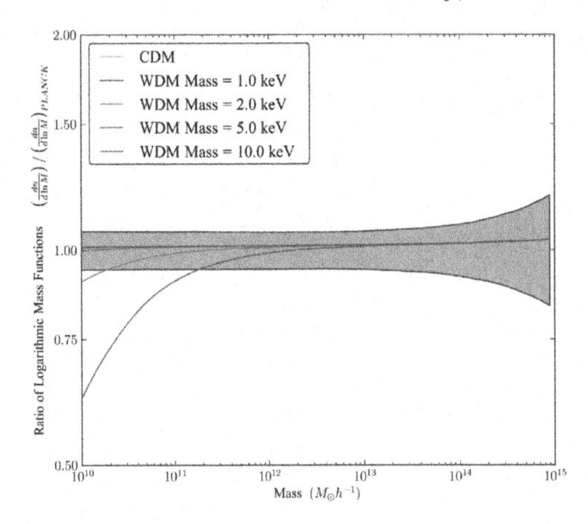

Figure 2. WDM HMF's for specified particle masses against the shaded uncertainty region in Planck CDM. "Boundary Crossing" represents the mass at which a WDM model is detectable for current cosmological uncertainty.

4. Halo Model Fitting

A common application of the Halo Model framework is to estimate the parameters of a Halo Occupation Distribution (HOD) model given an observed projected correlation function (Zehavi *et al.*(2011), Beutler *et al.*(2013)). The HOD parameters and their associated derived parameters are useful tracers of galaxy evolution properties.

Commonly this procedure is performed with a fiducial cosmology while the HOD parameters are fit via a Monte-Carlo routine. We investigate whether fixing the cosmology introduces a bias in the HOD measurement or a misestimation of its uncertainty. This requires leaving the cosmology to vary in the procedure in tandem with the HOD parameters. As a corollary, we determine whether this method can be used to fit certain cosmological parameters.

While this work is preliminary, early results show that several HOD-Cosmology parameter combinations exhibit strong correlations in their posterior distributions, suggesting that they are likely to be biased if the fixed cosmology is biased, and that their uncertainties will have been underestimated. Furthermore, while none of the cosmological parameters are constrained tightly, all marginalised posteriors are bound, suggesting that if appropriate priors are chosen this method could provide an independent and complimentary means of constraining cosmology.

5. Conclusion

We have discussed the application of our new codes `hmf` and `halomod` to the emerging problem of efficient modelling in the era of big data. Specifically, our implementations are simple to use and algorithmically efficient – enabling the seamless modelling of spatial galaxy statistics with current models. We have applied this software to diagnostic statistics for the HMF and two-point galaxy statistics to illustrate their benefit.

References

Bhattacharya, S., Heitmann, K., White, M., *et al.* 2011, *ApJ*, 732, 122
Zehavi, I., Zheng, Z., Weinberg, D. H., *et al.* 2011, *ApJ*, 736, 59
Zheng, Z. 2004, *ApJ*, 610, 61
Beutler, F., Blake, C., Colless, M., *et al.* 2013, *MNRAS*, 429, 3604
Murray, S. G., Power, C., & Robotham, A. S. G. 2013, *MNRAS*, 434, L61
Murray, S. G., Power, C., & Robotham, A. S. G. 2013, *Astronomy and Computing*, 3, 23
Vikhlinin A., *et al.* 2009, *ApJ*, 692, 1060

Statistical Challenges in 21st Century Cosmology
Proceedings IAU Symposium No. 306, 2014
A. F. Heavens, J.-L. Starck & A. Krone-Martins, eds.

doi:10.1017/S1743921314013416

Data-Rich Astronomy: Mining Sky Surveys with PhotoRApToR

Stefano Cavuoti[1], Massimo Brescia[1] and Giuseppe Longo[2]

[1]INAF - Astronomical Observatory of Capodimonte,
I-80131, Napoli, Italy
email: cavuoti@na.astro.it

[2]Dept. of Physics, Naples University,
Box I-80126 Napoli, Italy

Abstract. In the last decade a new generation of telescopes and sensors has allowed the production of a very large amount of data and astronomy has become a data-rich science. New automatic methods largely based on machine learning are needed to cope with such data tsunami. We present some results in the fields of photometric redshifts and galaxy classification, obtained using the MLPQNA algorithm available in the DAMEWARE (Data Mining and Web Application Resource) for the SDSS galaxies (DR9 and DR10). We present PhotoRApToR (Photometric Research Application To Redshift): a Java based desktop application capable to solve regression and classification problems and specialized for photo-z estimation.

Keywords. techniques: photometric, galaxies: distances and redshifts, methods: data analysis, catalogs.

Everyone who has used neural methods to produce photometric redshift evaluation knows that, in order to optimize the results in terms of features, neural network architecture, evaluation of the internal and external errors, many experiments are needed. When coupled with the needs of modern surveys, which require huge data sets to be processed, it clearly emerges the need for a user friendly, fast and scalable application. This application needs to run client-side, since a great part of astronomical data is stored in private archives that are not fully accessible on line, thus preventing the use of remote applications, such as those provided by the DAMEWARE tool (Brescia *et al.* (2014b) and references therein). The code of the application was developed in Java language and runs on top of a standard Java Virtual Machine, while the machine learning model was implemented in C++ language to increase the core execution speed. Therefore different installation packages are provided to support the most common platforms. Moreover, the application includes a wizard, which can easily introduce the user through the various functionalities offered by the tool.

The Fig. 1 shows the main window of the program.

The main features of PhotoRApToR can be summarized as it follows:

• *Data table manipulation.* It allows the user to navigate throughout his/her data sets and related *metadata*, as well as to prepare data tables to be submitted for experiments. It includes several options to perform the editing, ordering, splitting and shuffling of table rows and columns. A special set of options is dedicated to the missing data retrieval and handling, for instance Not-a-Number (NaN) or not calculated/observed parameters in some data samples;

• *Classification experiments.* The user can perform general classification problems, i.e. automatic separation of an ensemble of data by assigning a common label to an arbitrary number of their subsets, each of them grouped on the base of a hidden similarity. The

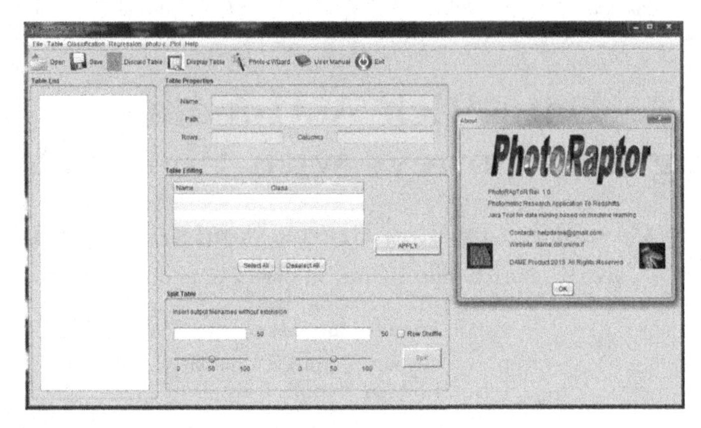

Figure 1. The PhotoRApToR main window.

classification here is intended as *supervised*, in the sense that there must be given a subsample of data for which the right label has been previously assigned, based on the *a priori* knowledge about the treated problem. The application will learn on this known sample to classify all new unknown instances of the problem see for instance Brescia *et al.* (2012), Cavuoti *et al.* (2014);

• *Regression experiments.* The user can perform general regression problems, i.e. automatic learning to find out an embedded and unknown analytical law governing an ensemble of problem data instances (patterns), by correlating the information carried by each element (features or attributes) of the given patterns. Also the regression is here intended in a *supervised* way, i.e. there must be given a subsample of patterns for which the right output is *a priori* known. After training on such Knowledge Base, the program will be able to apply the hidden law to any new pattern of the same problem in the proper way, see for instance Brescia *et al.* (2014a), Cavuoti *et al.* (2012), Brescia *et al.* (2013);

• *Photo-z estimation.* Within the *supervised* regression functionality, the application offers a specialized toolset, specific for photometric redshift estimation. After the training phase, the system will be able to predict the right photo-z value for any new sky object belonging to the same type (in terms of input features) of the Knowledge Base;

• *Data visualization.* The application includes some $2D$ and $3D$ graphics tools, for instance multiple histograms and multiple $2D/3D$ scatter plots. For instance, such tools are often required to visually analyze and explore data distributions and trends;

• *Data statistics.* For both classification and regression experiments, a statistical report is provided about their output. In the first case, the typical confusion matrix (Stehman 1997) is given, including related statistical indicators such as classification efficiency, completeness, purity and contamination for each of the classes defined by the specific problem. For what the regression is concerned, the application offers a dedicated tool, able to provide several statistical relations between two arbitrary data vectors (usually two columns of a table), such as average (bias), standard deviation (σ), Root Mean Square (RMS), Median Absolute Deviation (MAD) and the *Normalized* MAD, NMAD Hoaglin *et al.* (1983), the latter specific for the photo-z quality estimation, together with percentages of *outliers* at the common threshold 0.15 and at different multiples of σ. See Brescia *et al.* (2014a), Ilbert *et al.* (2009).

In Fig. 2 the layout of a general PhotoRApToR experiment workflow is shown. It is valid for either regression and classification cases. The application is available for download from the DAME program web site:

`http://dame.dsf.unina.it/dame_photoz.html#photoraptor`

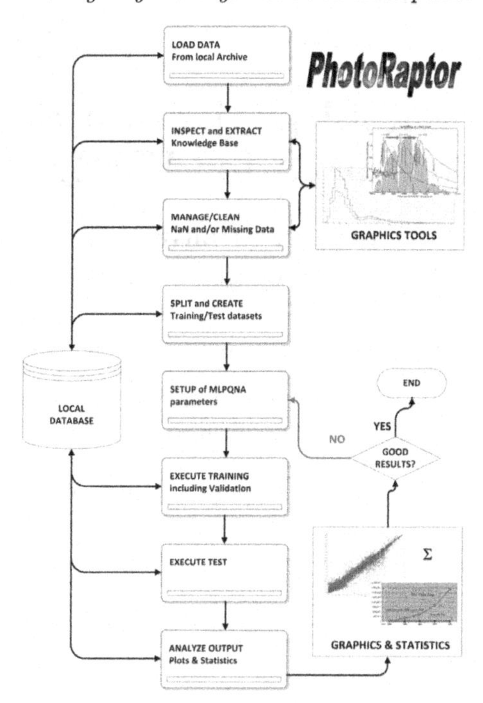

Figure 2. The workflow of a generic experiment performed with PhotoRApToR.

Perspective and conclusions

Nowadays the technological evolution of astronomical detectors and instruments has been so fast to render physically impossible to move the data from their original repositories. This implies that, as it has always been asked for but never implemented, we must be able to move the programs and not the data. Therefore, the future of any data-driven service depends on the capability and possibility of moving the data mining applications to the data centers hosting the data themselves. In this scenario PhotoRApToR represents our test bench of a desktop application prototype, fine tuned to tackle specific astrophysical problems in the big data era.

References

Brescia, M., Cavuoti, S., Paolillo, M., Longo, G., & Puzia, T. 2012, *MNRAS*, 421, 2, 1155

Brescia, M., Cavuoti, S., D'Abrusco, R., Longo, G., & Mercurio, A. 2013. Photometric redshifts for Quasars in multi band Surveys. *ApJ*, 772, 140

Brescia, M., Cavuoti, S., De Stefano, V., Longo, G. 2014a. A catalogue of photometric redshifts for the SDSS-DR9 galaxies, submitted to *A&A*

Brescia, M., Cavuoti, S., Longo, G., *et al.* 2014b. DAMEWARE: a web cyberinfrastructure for astrophysical data mining accepted by *PASP* (in press)

Cavuoti, S., Brescia, M., Longo, G., Mercurio, A. 2012. Photometric redshifts with Quasi Newton Algorithm (MLPQNA). Results in the PHAT1 contest, *A&A*, Vol. 546, A13, 1-8

Cavuoti, S., Brescia, M., D'Abrusco, R., Longo, G., & Paolillo, M. 2014, Photometric classification of emission line galaxies with Machine Learning methods, *Monthly Notices of the Royal Astronomical Society*, Volume 437, Issue 1, p.968-975

Hoaglin, D.C., Mosteller, F., & Tukey, J.W. 1983. *Understanding Robust and Exploratory Data Analysis*, New York: Wiley

Ilbert, O., Capak, P., Salvato, M., *et al.* 2009. Cosmos Photometric Redshifts with 30-bands for $2 - deg^2$. *The Astrophysical Journal* 690, 1236

Stehman, S. V. 1997. Selecting and interpreting measures of thematic classification accuracy. *Remote Sensing of Environment* 62, 1, 77-89

Statistical Challenges in 21st Century Cosmology
Proceedings IAU Symposium No. 306, 2014
A. F. Heavens, J.-L. Starck & A. Krone-Martins, eds.

doi:10.1017/S1743921314013799

Density field projection analysis in search for WHIM

Lauri Juhan Liivamägi

Tartu Observatory, Tõravere, Nõo vald 61602, Estonia

Abstract. We study the possibilities of using line-of-sight projected luminosity density and filament probability fields to find the best locations in the sky to attempt to observe the elusive Warm-Hot Intergalactic Matter (WHIM). We calculate galaxy luminosity fields on a regular grid and then project those to the plane of sky. Column densities are then found by integrating along multiple lines of sight. We also create the projections of the filament probability fields. The results are presented as sky maps of the potentially suitable areas and the corresponding fractions of sky coverage.

Keywords. cosmology: observations, large-scale structure of universe, galaxies: intergalactic medium

1. Introduction

At low redshifts ($z < 2$) all observations of the visible matter sum up to half of the expected cosmological mass density of baryons (Nicastro *et al.* 2008). Large scale structure formation simulations suggest that these missing baryons reside in the form of WHIM in the filamentary cosmic web structure connecting the clusters of galaxies and superclusters (Cen & Ostriker 2006). At the predicted temperatures of 10^5–10^7 K and densities 10^{-6}–10^{-4} cm^{-3} the X-ray emission from WHIM structures is too faint to be detected with current instrumentation. However, the column densities of the highly ionised WHIM metal ions along large-scale structures can reach a level of 10^{15}–10^{16} cm^{-2}, imprinting detectable absorption features on the soft X-ray or UV spectra of background sources. So far, only few tentative observations e.g. at Sculptor Wall (Buote *et al.* 2009) have been made.

Density fields are standard method to study large-scale structures of the universe. Common approach is to assume a Cox model for the galaxy distribution, where the galaxies are distributed in space according to a inhomogeneous point process with the intensity $\rho(\mathbf{r})$ determined by an underlying random field (see, e.g. Martínez & Saar 2003). It means that, although expecting certain biases, we expect galaxies to adequately trace the distribution of mass. As filamets can be a very suitable for WHIM detection we also attempt to use the SDSS filament sample created by Tempel *et al.* (2014).

2. Density fields

Data. We carried out the initial analysis on the SDSS DR8 main galaxy sample (Aihara *et al.* 2011). We used the data from the contiguous 7221 square degree area in the North Galactic Cap (Fig. 1). The sample selection is described in detail in the SDSS DR8 group catalogue paper by Tempel *et al.* (2012). We calculated the absolute magnitudes of galaxies in the SDSS r-band and applied correction for Galactic extinction, k-correction and evolutionary correction. The resulting flux-limited galaxy sample contains 576493

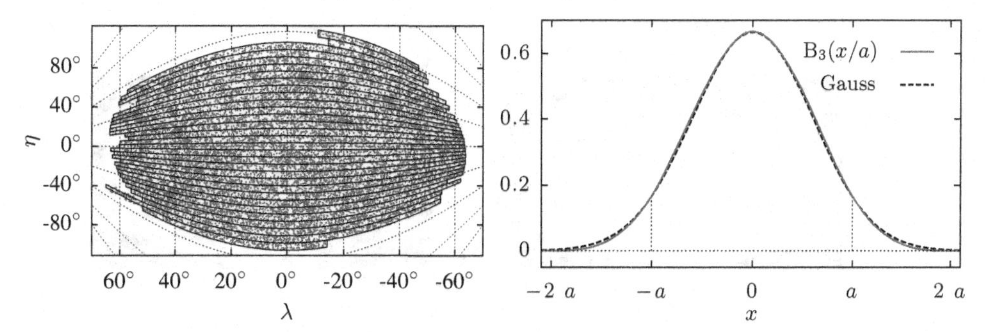

Figure 1. SDSS DR8 main sample footprint.

Figure 2. B_3 and Gaussian ($\sigma = 0.6$) smoothing kernels.

galaxies. In addition, we removed the small-scale finger-of-god redshift distortions to increase the resolution of the density fields (see e.g. Liivamägi *et al.* 2012)

Luminosity and filament probability density. Density fields are calculated on a Cartesian grid using kernel smoothing. We chose B_3 spline function (Fig. 2) as the kernel:

$$B_3(x) = \frac{|x-2|^3 - 4|x-1|^3 + 6|x|^3 - 4|x+1|^3 + |x+2|^3}{12}.$$

The B_3 spline is preferable to e.g. the Gaussian kernel for it is compact in space and therefore does not generate noise in the low density regions as the latter does. We chose smoothing radius to be $a = 1h^{-1}$Mpc as this is the scale we expect to be relevant with respect to the effect. Density and probability field grid cell length is also set to $\Delta = 1h^{-1}$Mpc. We don't use the filament spines presented in the (Tempel *et al.* 2014) catalogue but rather the raw filament probability fields (visit maps) which should offer more continuous representation of the filamentary environment.

We will calibrate the method by performing similar analysis on N-body simulations containing galaxies, dark matter and gas and test our estimates against existing observations. This will be studied in detail in the upcoming paper by Nevalainen *et al.* (in prep).

3. Results

Line of sight analysis. We present the results of the preliminary analysis. The set of limiting parameters used is given in Table 1. We require our structures to have a minimum length to increase the chance of having enough column density. As the WHIM gas temperature has to be in certain interval, it is also reasonable to set both lower and upper limits to the density field values. For the filament probability we require the field value at the vertice to be above 0.05, which corresponds to the 95% probability of being in the filament. The plane of sky is sampled uniformly with lines of sight and following attributes are calculated: the count of continuous structures, the total length covered

Table 1. Parameters for the line of sight analysis.

Field type	Min density	Max density	Minimum length
luminosity density	0.5 mean density[1]	10.0 mean density	3 h^{-1}Mpc
filament probability	5 %	-	3 h^{-1}Mpc

Notes:
[1] Mean luminosity density value for the SDSS DR8 is $1.667 \cdot 10^{-2}$.

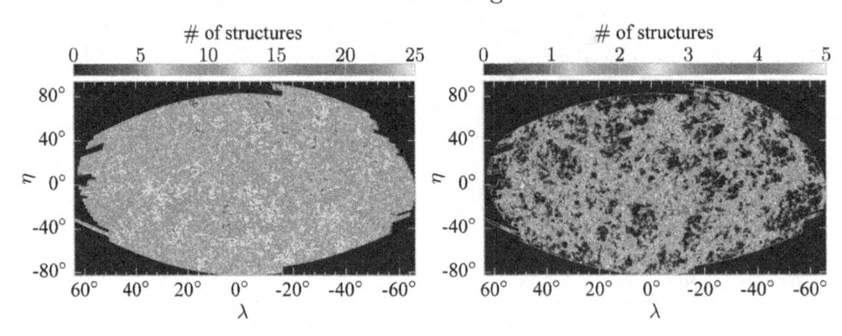

Figure 3. Number of structures along the lines of sight along the line of sight in the luminosity density field (*left*) and in the filament probability field (*right*).

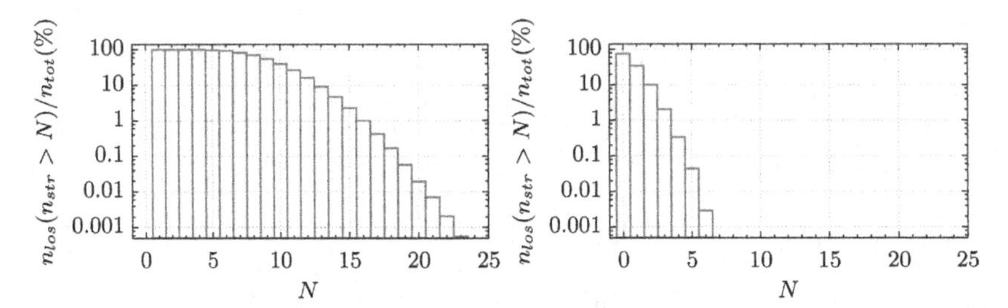

Figure 4. Distribution of the amount of structures along the lines of sight in the luminosity density field (*left*) and in the filament probability field (*right*).

by structures, and in case of luminosity density, the cumulative column density along the line of sight. Due to room limitations, we present here only as example the analysis of number of structures. Fig. 3 shows the sky distribution of the structure counts in both fields. We see that all lines of sight have at least a few structure on them, however there are much less of those with high number. The number of filament structures is significantly smaller and fair amount of directions do not cross any filament. This is also confirmed by the distortion curves on Fig. 4.

Conclusion. We have found that lines of sight with notable number of suitable structures are sparse. Therefore, a careful analysis of foregrounds is necessary e.g. before applying for time for time-consuming observations. Sky maps of projected density are useful tool to be included in the study when estimating the likelyhood for detection of WHIM.

References

Aihara, H. *et al.* 2011, *ApJ*, 29, 193

Buote, D. A., Zappacosta, L., Fang, T., Humphrey, P. J., Gastaldello, F., & Tagliaferri, G. 2009, *Science*, 695, 1351

Cen, R. & Ostriker, J. P. 2006, *Science*, 650, 560

Cui, W., Baldi, M., & Borgani, S. 2012, *MNRAS*, 424, 993

Liivamägi, L. J., Tempel, E., & Saar, E. 2012, *A&A*, 539, A80

Martínez, V. J. & Saar, E. 2003, *Statistics of the Galaxy Distribution* (Chapman & Hall/CRC, Boca Raton)

Nicastro, F., Mathur, S., & Elvis, M. 2008, *Science*, 319, 55

Tempel, E., Tago, E., & Liivamägi, L. J. 2012, *A&A*, 540, A106

Tempel, E., Stoica, R. S., Martínez, V. J., Liivamägi, L. J., Castellan, G., & Saar, E. 2014, *MNRAS*, 438, 3465

Statistical Challenges in 21st Century Cosmology
Proceedings IAU Symposium No. 306, 2014
A. F. Heavens, J.-L. Starck & A. Krone-Martins, eds.

doi:10.1017/S1743921314010874

Application of Statistical Methods on Automated Recognition of Solar Activity Phenomena and Analysis of Their Correlation

Ganghua Lin

Key Laboratory of Solar Activity, Chinese Academy of Sciences
National Astronomical Observatories, Chinese Academy of Sciences
Datun Rd.20A, Chaoyang District, Beijing, China
email: lgh@bao.ac.cn

Abstract. The recent immense volume of observation data has urged us to recognize solar features and to analyze their correlation with automation methods, and many of these automation methods are closely related to statistical methods. Author investigations the extensive literatures to sum up the function of statistical methods and their advantages in recognizing automatically solar features and analyzing their correlation. The review lists the results and summarizes conclusions of the investigations. The results and conclusions will be helpful to promote the further automation processing.

Keywords. solar feature, recognition, automation, methods: data analysis, statistics

1. Introduction

The solar features conclude flare, filament, coronal, and so on. On one hand, the further investigation of these solar features is of particular importance in studies pertaining to space weather. On the other hand, with the improvement of temporal and spatial solution, the volume of observation data has been increased greatly, and that the automation method raise the time efficiency of data process. So in recent years, The study of methods on the automation recognition of the solar features and the analysis of their correlation have been lasted.

2. Review

We do the investigation on the automation process method of solar features. The investigation results are put into table. The table mainly illustrates corresponding relationship between these recognized objects and adopted technology methods, recognition ratio or distinguishing features of these methods. The table also illustrates the application of statistical method in solar features recognition and their correlation. In the table or Figure 1 (Fig. 1), method means the method of recognition, object means recognized solar features. From the table, we learn that these solar features have been recognized automatically, for example, active area, active area properties, quiet area, magnetic field inversion-line, sunspot, sunspot group, facular, flare, flare properties, filament, filament properties, coronal faint variable microwave emission sources, coronal information on transverse and longitudinal wave, coronal Moreton (EIT) waves and propagating EUV dimming, signatures of wave and oscillatory processes in the solar corona, classification of coronal holes and filaments, solar coronal magnetic loops, CME classification, etc.

method	object	Recognition rate
SVM, MLP, RBF. Ming Qu, et al (2003)	flare	96.7%,94.2%,95%
an artificial neural network (ANN), V. V. ZHARKOVA, et al (2005)	Filament	82.5%
Image processing		
enhancement, segmentation, pattern recognition, and mathematical morphology.	Filament, Ming Qu et al, (2005)	Small filament:75%
An intensity filtering threshold value is used for detecting sunspots.	sunspot, faculae, and active-region detections, Colak and Qahwaji, 2008. Colak et al, (2011). Colak and Qahwaji, (2009).	The ASAP tends to detect small sunspots, which can be classified as pores. This is useful especially when grouping and classifying sunspots.
Using edge-detection methods, morphological operations.	Sunspot, Zharkov S. I.,et al, (2005)	95 to 98%
morphological analysis and intensity threshold	Active area, Jie Zhang et al.,(2010)	
A combination of image-processing techniques to determine the boundary of an AR. Using a gaussian kernel to identify potential features.	tracking active area, extracting characterise, Higgins et al.,(2011)	
enhancing their contrast	Classification of Coronal holes and filament, Isabelle F. Scholl, (2007)	
Bayesian spectral analysis of time series and image filtering.	solar atmosphere oscillation, J. Ireland et all,(2010)	
Wavelet Transform	Solar Coronal Magnetic Loops, S. Biskri, (2010). CME classification, D. I. González-Gómez et al, (2010).	it has successfully highlighted loop structures on the TRACE telescope images, without losing an appreciable amount of pixel based loop information.

Figure 1. The application of statistical method in solar feature recognition and their correlation.

In the table or Figure 1 (Fig. 1), the methods are mainly: supervised learning, image processing, Bayesian spectral analysisis, and Wavelet Transform. We learn that among the methods, the method of morphological analysis and intensity threshold is used for detecting small, large features and extracting properties, for example, active area, quiet area, sunspot, flare, filament, coronal dimming, etc. And the method of supervised classification is used for separating main identified object from background and misunderstood objects, for example,sunspot and filament. The wavelet transform is used for structure change, for example, solar coronal magnetic loops, CME. Bayesian spectral analysisis used for process of time series.

References

Biskri, S., Antoine, J.-P., Inhester, B., & Mekideche, F. 2010, *SoPh.*, 373, 385

Colak, T. & Qahwaji, R. 2008, *SoPh.*, 277, 296

Colak, T., Colak, Tufan, Qahwaji, Rami, Ipson, Stan, Ugail, & Hassan 2011, *Advances in Space Research*, 2092, 2104

Croat, T. K., Stadermann, F. J., & Bernatowicz, T. J. 2005, *ApJ.*, 631, 976

Gonzlez-Gmez, D. I., Blanco-Cano, X., & Raga, A. C. 2010, *SoPh.*, 337, 247

Higgins, P. A., Gallagher, P. T., McAteer, R. T. J., & Bloomfield, D. S. 2010, *Advances in Space Research*, 2105, 2117

Ireland, J., Marsh, M. S., Kucera, T. A., & Young, C. A. 2010, *SoPh.*, 403, 431

Isabelle, F. Scholl & Shadia Rifai Habbal 2007, *SoPh.*, 403, 431

Lodders, K. & Fegley, B. 1998, *Meteorit. Planet. Sci.*, 33, 871

Qu, Ming, Shih, Frank Y., Jing, Ju, & Wang, Haimin 2000, *SoPh.*, 157, 172

Qu, Ming, Shih, Frank Y., Jing, Ju, & Wang, Haimin 2005, *SoPh.*, 119, 135

Ott, U., Altmaier, M., Herpers, U., Kuhnhenn, J., Merchel, S., Michel, R., & Mohapatra, R. K. 2005, *Meteorit. Planet. Sci.*, 40, 1635

Zhang Jie, Wang, Yuming, Liu, & Yang 2010, *ApJ.*, 1006, 1018

Zharkova, V. V. & Schetinin, V. 2005, *SoPh.*, 137, 148

Zharkov, S. I., Zharkova, V. V., & Ipson, S. S. 2005, *SoPh.*, 377, 397

Statistical Challenges in 21st Century Cosmology
Proceedings IAU Symposium No. 306, 2014
A. F. Heavens, J.-L. Starck & A. Krone-Martins, eds.

doi:10.1017/S1743921314010849

ANNz2 - Photometric redshift and probability density function estimation using machine-learning

Iftach Sadeh

Astrophysics Group, Department of Physics and Astronomy, University College London, Gower Street, London WC1E 6BT, United Kingdom
email: `i.sadeh@ucl.ac.uk`

Abstract. Large photometric galaxy surveys allow the study of questions at the forefront of science, such as the nature of dark energy. The success of such surveys depends on the ability to measure the photometric redshifts of objects (photo-zs), based on limited spectral data. A new major version of the public photo-z estimation software, `ANNz`, is presented here. The new code incorporates several machine-learning methods, such as artificial neural networks and boosted decision/regression trees, which are all used in concert. The objective of the algorithm is to dynamically optimize the performance of the photo-z estimation, and to properly derive the associated uncertainties. In addition to single-value solutions, the new code also generates full probability density functions in two independent ways.

Keywords. techniques: photometric, galaxies: distances and redshifts

1. Introduction

The different approaches to calculate photometric redshifts (photo-zs) can generally be divided into two categories, template fitting methods and empirical machine-learning (see Hildebrandt *et al.* (2010) for a review of the field). Template fitters employ physically motivated models. On the other hand, Machine-learning methods (MLMs) involve deriving the relationship between the photometric observables and the redshift using a so-called training dataset, which includes both the observables and precise redshift information.

This paper presents `ANNz2`†. (Sadeh *et al.* (2015)), a new implementation of the popular code of Collister & Lahav (2003), which uses artificial neural networks (ANNs) to estimate photometric redshifts. The new code incorporates a variety of machine-learning techniques, such as boosted decision/regression trees (BDTs) in addition to ANNs. The various MLMs are provided by the `TMVA` package (Hoecker *et al.* (2007)). It has been designed to calculate both photometric redshifts and probability distribution functions (PDFs), doing so in several different ways. The introduction of photo-z-PDFs has been shown to improve the accuracy of cosmological measurements (Mandelbaum *et al.* (2008)), and is an important feature of the new version of `ANNz`.

2. Description of the operational modes of `ANNz2`

`ANNz2` uses both regression and classification techniques for estimation of single-value photo-z solutions and PDFs. The different configurations are referred to as *single regression*, *randomized regression* and *binned classification*. A short description follows.

† Code publicly available at https://github.com/IftachSadeh/ANNZ

2.1. *Single regression (single-value photo-z estimator)*

In the simplest configuration of ANNz2, a single regression is performed, using as the output the spectroscopic redshift, denoted hereafter by $z_{\rm spec}$. Consequently, one may derive a per-galaxy photo-z solution. Associated errors are generated using the nearest neighbours error estimation method, discussed by Oyaizu *et al.* (2008).

2.2. *Randomized regression (single-value photo-z and PDF solutions)*

Instead of choosing a single MLM, it is possible to automatically generate an ensemble of regression methods. The *randomized MLMs* differ from each other in several ways. This includes setting unique random seed initializations, as well as changing the configuration parameters of a given algorithm. To give an example, the latter may refer to using various types and numbers of neurons in an ANN, or to arranging neurons in different layouts of hidden layers; for BDTs, the number of trees and the type of boosting algorithm may be changed, etc. Additionally, TMVA provides the option to perform transformations on the input-parameters, including normalization or principal component decomposition. The option is also available to only use a subset of the input parameters, or to train with pre-defined functional combinations of parameters.

Once randomized MLMs are initialized, the various methods are each trained. Subsequently, a distribution of photo-z solutions for each galaxy is generated. A selection procedure is then applied to the ensemble of answers, choosing the subset of methods which achieve optimal performance. The selected methods are used to derive a single photo-z estimator, based on the method with the best performance, denoted by $z_{\rm best}$. The optimized solutions are also used in concert, producing an additional (averaged) single-value solution, referred to as $< z_{\rm best} >$. Finally, the ensemble of estimators are used to derive a complete probability density function, denoted by $\rm PDF_{tmpl}$. This is done by folding a weighted distribution of the solutions with the corresponding uncertainty estimators of the individual MLMs. The weights are derived dynamically, with the purpose of optimizing the quality of the PDF.

2.3. *Binned classification (PDF solution)*

ANNz2 may also be run in classification-mode, employing an algorithm similar to that used by Gerdes *et al.* (2010). The first step of the calculation involves dividing the redshift range of the input samples into many small bins. Within the redshift bounds of a given bin, the *signal sample* is defined as the collection of galaxies for which $z_{\rm spec}$ is within the bin. Similarly, the *background sample* includes all galaxies with $z_{\rm spec}$ outside the confines of the bin.

The algorithm proceeds by training a different classification MLM for each redshift bin. The output of a trained method in a given bin is translated to the probability for a galaxy to have redshift which falls inside that bin. The distribution of probabilities from all of the bins is normalized to unity, accounting for possible varying bin width. It then stands as the photo-z-PDF of the galaxy, denoted in the following as $\rm PDF_{cls}$.

3. Performance of the algorithm

We use data included in data releases 10, DR10, (Ahn *et al.* (2014)) of the Baryon Oscillation Spectroscopic Survey (BOSS) (Dawson *et al.* (2013)), which is part of the Sloan Digital Sky Survey (SDSS), from its current incarnation, SDSS-III (Eisenstein *et al.* (2011)). The data are publicly available as a FITS catalogue, containing flux measurements, which were transformed into magnitudes. Objects in the catalogue were subjected to quality cuts, ensuring that only galaxies with flux measurements in all five bands, u,

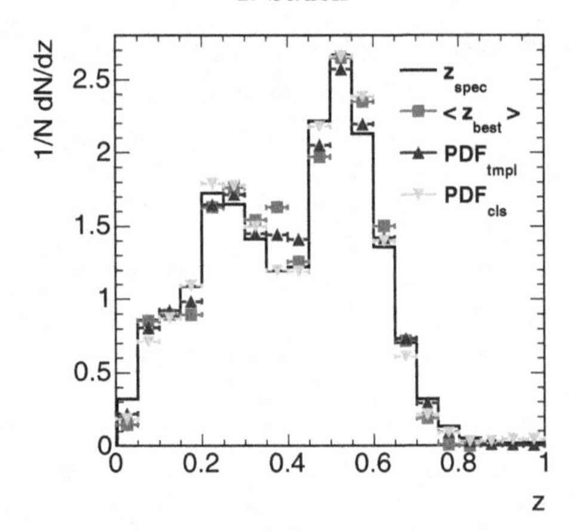

Figure 1. Differential distributions, denoted collectively by z, of the spectroscopic redshift, $z_{\rm spec}$, and of the respective photometric redshift estimated using ANNz2 and described in the text, $< z_{\rm best} >$, $\rm PDF_{tmpl}$ and $\rm PDF_{cls}$, as indicated.

g, r, i and z, which had magnitude errors below 1, were accepted. The five magnitudes served as inputs for training in ANNz2, along with the spectroscopic redshift, which was taken as the true redshift value. The entire sample, consisting of $\sim 150,000$ galaxies, was split into three sub-samples for training, testing and validation purposes; the first sub-sample was used to train the MLMs; the second was used to optimized the performance and derive PDF-weights; using the third sub-sample, the performance was validated by comparing the spectroscopic redshift distribution, with that of the photometric solutions provided by ANNz2.

For this study, ANNz2 was run in randomized regression mode, using 100 ANNs, and in binned classification mode, using collections of BDTs. The results are shown in Fig. 1. One may observe that the PDF solutions provide a better description of the spectroscopic redshift distribution, compared to the single-value photo-zs. Additional tests showing the per-galaxy bias, scatter, outlier fraction and other metrics were performed, and are presented in Sadeh *et al.* (2015).

References

Ahn, C. P., *et al.* 2014, *ApJS*, 211, 17
Collister & Lahav 2003, *Publ.Astron.Soc.Pac.*, 116, 345-351
Dawson, K., *et al.* 2013, *Astron.J.*, 145, 10
Eisenstein, D. J., *et al.* 2011, *Astron.J.*, 142, 72
Gerdes, D., *et al.* 2010, *Astrophys.J.*, 715, 823-832
Hildebrandt, H., *et al.* 2010, *A&A*, 523, A31
Hoecker, A., *et al.* 2007, *PoS*, ACAT, 040
Mandelbaum, R., *et al.* 2008, *Mon.Not.Roy.Astron.Soc.*, 386, 781-806
Oyaizu, H., *et al.* 2008, *Astrophys.J.*, 689, 709-720
Sadeh, I., *et al.* 2015, In preparation

Statistical Challenges in 21st Century Cosmology
Proceedings IAU Symposium No. 306, 2014
A. F. Heavens, J.-L. Starck & A. Krone-Martins, eds.

doi:10.1017/S1743921314013775

Adapting Predictive Models for Cepheid Variable Star Classification Using Linear Regression and Maximum Likelihood

Kinjal Dhar Gupta[1], Ricardo Vilalta[1], Vicken Asadourian[2] and Lucas Macri[3]

[1]Dept. of Computer Science, University of Houston.

[2]Dept. of Mathematics, University of Houston.
4800 Calhoun Road , Houston TX-70004, USA.
email: kinjal13@cs.uh.edu, vilalta@cs.uh.edu, vmasadourian@uh.edu

[3]Dept. of Physics and Astronomy, Texas A&M University.
4242 TAMU , College Station, TX 77843-4242, USA.
email: lmacri@tamu.edu

Abstract. We describe an approach to automate the classification of Cepheid variable stars into two subtypes according to their pulsation mode. Automating such classification is relevant to obtain a precise determination of distances to nearby galaxies, which in addition helps reduce the uncertainty in the current expansion of the universe. One main difficulty lies in the compatibility of models trained using different galaxy datasets; a model trained using a training dataset may be ineffectual on a testing set. A solution to such difficulty is to adapt predictive models across domains; this is necessary when the training and testing sets do not follow the same distribution. The gist of our methodology is to train a predictive model on a nearby galaxy (e.g., Large Magellanic Cloud), followed by a model-adaptation step to make the model operable on other nearby galaxies. We follow a parametric approach to density estimation by modeling the training data (anchor galaxy) using a mixture of linear models. We then use maximum likelihood to compute the right amount of variable displacement, until the testing data closely overlaps the training data. At that point, the model can be directly used in the testing data (target galaxy).

Keywords. (stars: variables:) Cepheids, (galaxies:) Magellanic Clouds, methods: statistical, infrared: stars, methods: data analysis.

---

1. Introduction

Traditional machine learning algorithms assume both training and testing data originate from the same distribution. This comes unwarranted in real-world applications. One approach to handle the discrepancy between source (training) and target (test) domains is called *domain adaptation*, where class-conditional distributions remain equal, though class prior distributions differ Ben-David *et al.* (2006), Storkey (2009), Ben-David *et al.* (2010). Our domain adaptation method learns a model on a source domain without using any information from a target domain; we assume equal class conditional probabilities, but class priors differ by a certain shift across one or more features as explained by Vilalta *et al.* (2013). Different from previous work, we assume a bi-variate Linear Mixture Model with Gaussian noise, and use Maximum Likelihood to find the shift between source and target distributions. The idea is to align the two datasets so that a model learnt on the source domain can be effectively used on the target domain.

We classify a particular type of variable stars named Cepheids into two pulsation modes. We use two features: Apparent Mean Magnitude and Logarithm of Period, and

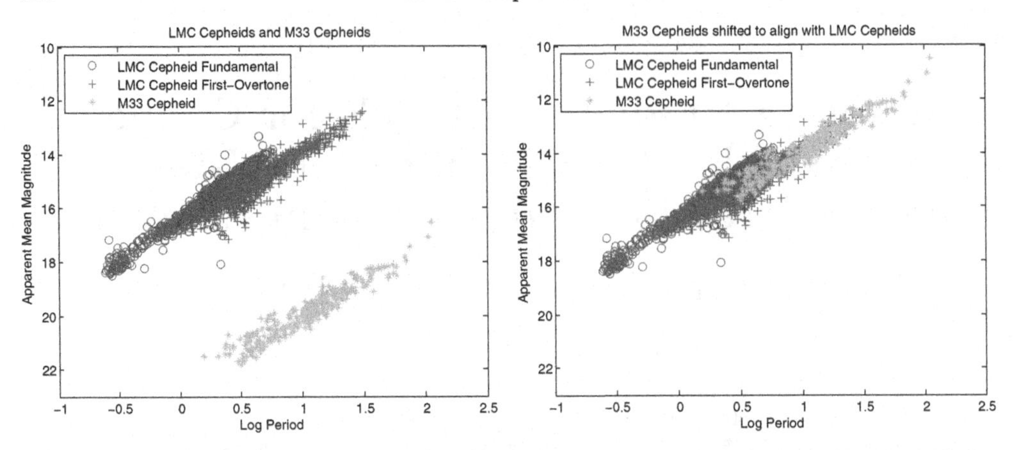

Figure 1. Left. The distribution of Cepheids along the Large Magellanic Cloud LMC (top sample), deviates significantly from M33 (bottom sample). Right. M33 is aligned with LMC by shifting along Mean Magnitude.

two classes: Fundamental and First-Overtone. In our experiments, we take the Large Magellanic Cloud (LMC) (Fig. 1a) as our source domain and M33 (Fig. 1b) as our target domain. All data have been obtained from the Optical Gravitational Lensing Experiment (OGLE) III in the infrared band.

2. Mathematical Formulation

We generate a mixture of linear regression models (as described in Faria & Soromenho (2010)). The log likelihood equation for LMC data for the two features X (Log Period) and Y (Apparent Mean Magnitude) can be written as follows:

$$LogL(\theta|y_1, y_2, ...y_n, x_1, x_2, ...x_n) = \sum_{i=1}^{n} \log\left(\sum_{j=1}^{2} \pi_j \phi_j(y_i|x_i)\right) \quad (2.1)$$

where $y_1, y_2, ...y_n$ are observations of Y, $x_1, x_2, ...x_n$ are observations of X, and each component $\phi_j(y_i|x_i)$ corresponds to a Gaussian distribution $\mathcal{N}(x_i^T\beta_j, \sigma_j)$ with coefficient π_j. Parameter estimates are captured in θ. The equation can be expanded as follows:

$$LogL(\theta|y_1, y_2, ...y_n, x_1, x_2, ...x_n) = \sum_{i=1}^{n} \log\left(\sum_{j=1}^{2} \pi_j \frac{1}{\sigma_j\sqrt{2\pi}} e^{-\frac{(y_i - x_i^T\beta_j)^2}{2\sigma_j^2}}\right) \quad (2.2)$$

where $x_i = [x_{i1}\ x_{i2}]^T$ and $\beta_j = [\beta_{j1}\ \beta_{j2}]^T$. Here x_{i1} and x_{i2} indicate the value of Log Period and the bias variable for the ith observation respectively (x_{i2}=1 for all observations). β_{j1} and β_{j2} are the corresponding regressor variables for the jth component. To align M33 data with LMC data, we assume a shift δ across Mean Magnitude only. The log likelihood equation for M33 data can be written as follows:

$$LogL(\theta, \delta|y_1, y_2, ...y_n, x_1, x_2, ...x_n) = \sum_{i=1}^{n} \log\left(\sum_{j=1}^{2} \pi_j \frac{1}{\sigma_j\sqrt{2\pi}} e^{-\frac{(y_i + \delta - x_i^T\beta_j)^2}{2\sigma_j^2}}\right) \quad (2.3)$$

To find the maximum value of δ we differentiate Eq. (2.3) with respect to δ and equate to zero. After some algebraic manipulation we derive the following equation:

$$\left(\frac{c_{i12}\alpha_{i1}e^{-\gamma_{i1}(\delta)}+c_{i22}\alpha_{i2}e^{-\gamma_{i2}(\delta)}}{\alpha_{i1}e^{-\gamma_{i1}(\delta)}+\alpha_{i2}e^{-\gamma_{i2}(\delta)}}\right)+\left(\frac{c_{i13}\alpha_{i1}e^{-\gamma_{i1}(\delta)}+c_{i23}\alpha_{i2}e^{-\gamma_{i2}(\delta)}}{\alpha_{i1}e^{-\gamma_{i1}(\delta)}+\alpha_{i2}e^{-\gamma_{i2}(\delta)}}\right)=0 \quad (2.4)$$

where

$$c_{ij1}=\frac{(y_i+\delta-x_i^T\beta_j)^2}{2\sigma_j^2}, \qquad c_{ij2}=\frac{(y_i+\delta-x_i^T\beta_j)}{\sigma_j^2}, \qquad c_{ij3}=\frac{1}{2\sigma_j^2},$$

$$\alpha_{ij}=\pi_j\frac{1}{\sigma_j\sqrt{2\pi}}e^{-c_{ij1}}, \qquad \gamma_{ij}(\delta)=c_{ij2}\delta+c_{ij3}\delta^2$$

3. Experiment Results

Fitting a Mixture of Linear Models. We use Eq. (2.2) to fit a mixture of linear models to the source data. The two components refer to the two classes: Fundamental and First-Overtone. We use the EM algorithm (Faria & Soromenho 2010) to do parameter estimation. The EM algorithm stops when the change in Q-value in the E-step falls below 10^{-10}. Initial values and final parameter estimates are shown in Table 1.

Aligning M33 Data with LMC Data. We use the parameter values in Table 1 to find the shift δ between LMC and M33. Solving Eq. (2.6) yields a value of $\delta = -6.06$. After shifting M33 data by δ we obtain the results shown in Table 2. An asterisk denotes a significant difference. Experimental results show how classification accuracy increases significantly after the source and target datasets are aligned.

Table 1. Parameter Estimates

Parameter	Initial Value	Final Value
β_1	$\begin{bmatrix}-3.259\\16.407\end{bmatrix}$	$\begin{bmatrix}-3.209\\16.358\end{bmatrix}$
β_2	$\begin{bmatrix}-2.969\\16.904\end{bmatrix}$	$\begin{bmatrix}-2.929\\16.889\end{bmatrix}$
σ_1	0.1	0.164
σ_2	0.1	0.214
π_1	0.404	0.384
π_2	0.596	0.616

Table 2. Classification Accuracies

Learning Algorithm	Without Shift	With Shift
Neural Networks	92.70(1.19)	97.90(0.89)*
Support Vector Machine with Polynomial Kernel 1	92.70(1.19)	96.27(0.86)*
Support Vector Machine with Polynomial Kernel 2	92.70(1.19)	96.10(0.80)*
Support Vector Machine with Polynomial Kernel 3	92.70(1.19)	96.53(0.88)*
J48 Decision Tree	92.90(1.24)	98.07(0.83)*
Random Forest	92.90(1.24)	98.03(0.69)*

References

Ben-David, S., Blitzer, J., Crammer, K., & Pereira, F. 2006, *Neural Information Processing Systems*, 19, 137-144

Ben-David, S., Blitzer, J., Crammer, K., Pereira, F., & Wortman, J. 2010, *Machine Learning*, 79, 151-175

Faria, S. & Soromenho, G. 2010, *Journal of Statistical Computation and Simulation*, Vol. 80, No. 2, 201-225

Storkey, A. 2009, *Dataset Shift in Machine Learning, MIT Press*, 3-28

Vilalta, R., Dhar Gupta, K., & Macri, L. 2013, *Astronomy and Computing*, 2, 46-53

Statistical Challenges in 21st Century Cosmology
Proceedings IAU Symposium No. 306, 2014
A. F. Heavens, J.-L. Starck & A. Krone-Martins, eds.

doi:10.1017/S1743921314013635

Testing the mutual consistency of different supernovae surveys

N.V. Karpenka[1], F. Feroz[2] and M.P. Hobson[2]

[1]The Oskar Klein Centre for Cosmoparticle Physics, Department of Physics, Stockholm University, AlbaNova, SE-106 91 Stockholm, Sweden
email: nkarp@fysik.su.se
[2]Astrophysics Group, Cavendish Laboratory, J.J. Thomson Avenue, Cambridge CB3 0HE, UK

Abstract. It is now common practice to constrain cosmological parameters using supernovae (SNe) catalogues constructed from several different surveys. Before performing such a joint analysis, however, one should check that parameter constraints derived from the individual SNe surveys that make up the catalogue are mutually consistent. We describe a statistically-robust mutual consistency test, which we calibrate using simulations, and apply it to each pairwise combination of the surveys making up, respectively, the UNION2 catalogue and the very recent JLA compilation by Betoule *et al.* We find no inconsistencies in the latter case, but conclusive evidence for inconsistency between some survey pairs in the UNION2 catalogue.

Keywords. methods: data analysis — methods: statistical — supernovae: general

1. Introduction

The original discovery of the accelerated expansion of the universe (Riess *et al.* 1998; Perlmutter *et al.* 1999), prompted many other teams to start searches for high redshift SNe. This resulted in SNe compilations containing data from different telescopes. Although such analyses result in tighter constraints on parameters, one needs to be careful when using data from different sources, as the presence of unaccounted systematics in any of these data-sets can lead to misleading results. Therefore, it is extremely important to establish whether different data-sets are consistent with each other before performing a joint analysis using them.

A common method to check for consistency between different data-sets, in general, is to perform a χ^2 analysis and compare the best-fit values $\chi^2_{\rm min}$. It is, however, impossible to perform such a test in the standard framework of SNe analysis. Moreover tests based on $\chi^2_{\rm min}$ depend only on the best-fit parameters, and are insensitive to the likelihood over the rest of the parameter space. Thus one must seek an alternative test for testing consistency between SNIa data-sets. We use a method based on the Bayesian consistency test described in Marshall *et al.* (2006).

2. Analysis methodology

Estimation of cosmological parameters from flux measurements is a two-step procedure. First, a SN light-curve fitting algorithms is applied in order to 'standardise' these observations. Second, cosmological parameters are estimated using the output from the light-curve fitting method. In our analysis, this is followed by the application of a consistency test to the constraints derived both on the cosmological parameters and those associated with the SNe population.

Light-curve fitting. They are many different algorithm to perform light-curve fitting, in this paper we employ the SALT-II SN light-curve fitter. When fitting SN light-curves

Parameter	Symbol	Prior	Simulated value
Matter density	$\Omega_{\mathrm{m},0}$	$\mathcal{U}(0,1)$	0.3
Absolute magnitude	M_0	$\mathcal{U}(-20.3,-18.3)$	−19.3
Stretch multiplier	α	$\mathcal{U}(0,1)$	0.12
Colour multiplier	β	$\mathcal{U}(1,5)$	2.6

Table 1. Priors assumed on the parameters, where $U(a,b)$ indicates the uniform distribution in the range $[a,b]$. The final column gives the value assumed in generating the simulated data.

with SALT-II, the outputs are the best-fit values $\hat{m}_B^*$ of the apparent rest frame B-band magnitudes of the SNe at maximum luminosity, the light-curve shape parameter $\hat{x}_1$, the colour $\hat{c}$ in the B-band at maximum luminosity, and the covariance matrix of the uncertainties in the estimated light-curve parameters $\widehat{C}$. Combining these quantities with the estimated redshift $\hat{z}$ of the SN, our basic input data for each SN are thus

$$D_i \equiv \{\hat{z}_i, \hat{m}_{B,i}^*, \hat{x}_{1,i}, \hat{c}_i\} \tag{2.1}$$

for $(i = 1, \ldots, N_{\mathrm{SN}})$. The vector $(\hat{m}_{B,i}^*, \hat{x}_{1,i}, \hat{c}_i)$ for each SN is then assumed to be distributed as a multivariate Gaussian about their respective true values, with covariance matrix $\widehat{C}_i$.

Parameter estimation. The output from the light-curve fitting constitutes the data used to estimate parameters. In this paper we employ the χ^2 method, since it is the most common approach.

In this approach, one defines the χ^2 misfit function to be

$$\chi^2(\mathscr{C}, \alpha, \beta, M_0, \sigma_{\mathrm{int}}) = \sum_{i=1}^{N} \frac{[\mu_i^{\mathrm{obs}}(\alpha, \beta, M_0) - \mu(\hat{z}_i, \mathscr{C})]^2}{(\sigma_{\mu,i}^z)^2 + \sigma_{\mathrm{int}}^2 + \sigma_{\mathrm{fit},i}^2(\alpha, \beta)}, \tag{2.2}$$

where $\mu(\hat{z}_i, \mathscr{C})$ is the predicted distance modulus. In this expression, the 'observed' distance modulus for the ith SN is

$$\mu_i^{\mathrm{obs}} = \hat{m}_{B,i}^* - M_0 + \alpha \hat{x}_{1,i} - \beta \hat{c}_i, \tag{2.3}$$

where M_0 is the (unknown) B-band absolute magnitude of the SNe, and α, β are (unknown) nuisance parameters controlling the stretch and colour corrections. The three dispersion components are: (i) the error $\sigma_{\mu,i}^z$ in the distance modulus; (ii) the intrinsic dispersion σ_{int}, which describes the global variation in the SNIa absolute magnitudes; and (iii) the fitting error $\sigma_{\mathrm{fit},i}(\alpha, \beta)$.

We also assume that dark energy is in the form of a cosmological constant $(w = -1)$ and that the universe is spatially-flat; thus we vary only the parameters $\{\Omega_{\mathrm{m},0}, M_0, \alpha, \beta, \sigma_{\mathrm{int}}\}$.

In keeping with standard practice, we define the 'likelihood' to be simply

$$\mathcal{L}(\Omega_{\mathrm{m},0}, M_0, \alpha, \beta, \sigma_{\mathrm{int}}) = \exp[-\tfrac{1}{2}\chi^2(\Omega_{\mathrm{m},0}, M_0, \alpha, \beta, \sigma_{\mathrm{int}})] \tag{2.4}$$

which is clearly not properly normalized. More importantly, this 'likelihood' is *not* (proportional to) the probability $\Pr(\boldsymbol{D}|\Omega_{\mathrm{m},0}, M_0, \alpha, \beta, \sigma_{\mathrm{int}})$.

Adopting the initial value $\sigma_{\mathrm{int}} = 0.1$, we use the nested sampling algorithm MultiNest (Feroz and Hobson 2008; Feroz *et al.* 2009) to sample from the (unnormalised) 'posterior' $\mathcal{P}(\boldsymbol{\Theta}, \sigma_{\mathrm{int}}) = \mathcal{L}(\boldsymbol{\Theta}, \sigma_{\mathrm{int}})\pi(\boldsymbol{\Theta})$ assuming the separable, uniform priors $\pi(\boldsymbol{\Theta})$ listed in Table 1. The value of σ_{int} is estimated by adjusting it to obtain $\chi^2_{\mathrm{min}}/N_{\mathrm{dof}} \sim 1$.

Test for mutual consistency between data-sets. The standard approach to analysing multiple data-sets jointly is simply to assume that they are mutually consistent. We represent this (null) hypothesis by H_0. It may be the case, however, that the data-sets are inconsistent with one another, resulting in each one favouring a different region of the

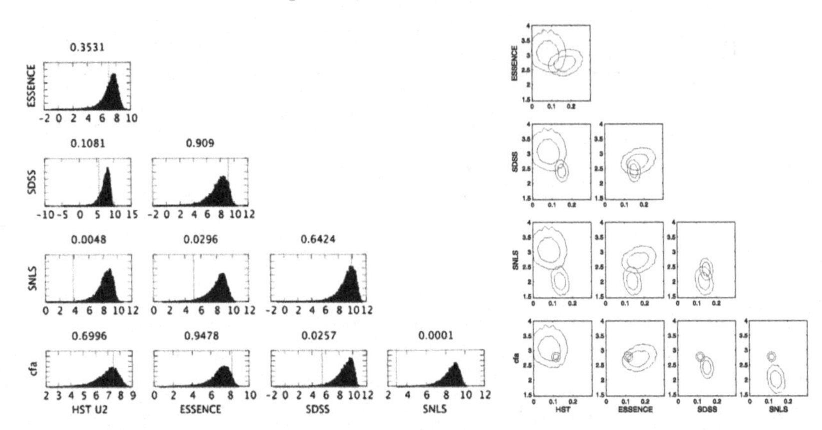

Figure 1. Left panel: Results of the consistency test applied to surveys pairs in the UNION2 compilation. The blue histograms show the distribution of R values obtained from 10^4 consistent simulations of each pair, and the red vertical line indicates the R value obtained from the real data. The corresponding one-sided p-value is given above each panel. Right panel: Two-dimensional marginalised constraints on the parameters (α, β) obtained from the individual constituent surveys contained in the UNION2 catalogue. The red (blue) contours denote the 68 and 95 per cent confidence regions for the survey in that row (column).

model parameter space. We represent this (alternative) hypothesis by H_1. In this case, a joint analysis would lead to completely misleading results (see, for example, Appendix A in Feroz *et al.* 2008 for a demonstration).

In order to determine which one of these hypotheses is favoured by the data, one can perform Bayesian model selection between H_0 and H_1. Assuming that hypothesis H_0 and H_1 are equally likely *apriori*, this can be achieved by calculating the ratio

$$\mathcal{R} = \frac{\Pr(\mathbf{D}|H_0)}{\Pr(\mathbf{D}|H_1)} = \frac{\Pr(\mathbf{D}|H_0)}{\prod_i \Pr(D_i|H_1)}. \tag{2.5}$$

where the probabilities $\Pr(\mathbf{D}|H)$, called Bayesian evidences or marginal likelihoods.

In adapting this Bayesian test to apply it to the standard χ^2 analysis of SNIa data, one replaces $\Pr(\mathbf{D}|\mathbf{\Theta}, H)$ by the 'likelihood' $\mathcal{L}(\mathbf{\Theta}, \sigma_{\rm int})$ given in (2.4) and $\Pr(\mathbf{\Theta}|H)$ by the prior $\pi(\mathbf{\Theta})$ listed in Table 1. In so doing, however, one cannot interpret the terms in (2.5) directly as probabilities, and thus the value of the $\mathcal{R}$ cannot be compared with the normal Jeffrey's scale. Nonetheless, one still expects $\mathcal{R}$ values to be higher for consistent data-sets and lower for inconsistent ones (March *et al.* 2011), so $\mathcal{R}$ may be used as a test statistic that is 'calibrated' with simulations to perform a standard one-sided frequentist hypothesis test.

Thus, we construct the distribution of $\mathcal{R}$ under the null hypothesis H_0 by evaluating it for the simulations in which the individual surveys are mutually consistent. The $\mathcal{R}$ value obtained by analysing the real data can then be used to calculate the p-value as follows:

$$p = \frac{N(\mathcal{R}_{\rm s} < \mathcal{R}_{\rm r})}{N_{\rm tot}}, \tag{2.6}$$

where $\mathcal{R}_{\rm s}$ and $\mathcal{R}_{\rm r}$ are the $\mathcal{R}$ values obtained by analysing simulated and real data-sets respectively, $N(\mathcal{R}_{\rm s} < \mathcal{R}_{\rm r})$ is the number of simulations with $\mathcal{R}$ values less than that obtained by analysing the real data and $N_{\rm tot}$ is the total number of simulations.

3. Data-sets

Real data. In this paper we consider SNe from the UNION2 catalogue (Amanullah *et al.* 2010) and the 'joint light-curve analysis' (JLA) compilation given in Betoule *et al.* (2014). These data-sets consist of SNe observed by different telescopes. Since our aim in this paper is to check for consistency between different SNe surveys/data-sets, we divide these SNe according to the telescope with which they have been observed.

SNANA simulations. In order to calibrate our test statistic $\mathcal{R}$ and calculate the resultant p-values we simulate SNe using the publicly available SNANA package† Our null hypothesis H_0 is that different data-sets are consistent with each other. Thus, we simulate all data-sets with exactly the same parameter values, which are listed in Table 1, and an intrinsic dispersion $\sigma_{\text{int}} = 0.106$.

4. Results and conclusions

Constraints on $\Omega_{\text{m},0}$ and M_0. In analysing compilations of SNe data, it is usual to plot the combined constraints on cosmological parameters obtained from a full joint analysis. Here we instead calculating the constraints imposed by each constituent survey in the UNION2 and JLA compilations. The key observation for our purposes is that the constraints from the constituent surveys are in good agreement, as indicated by the large degree of overlap of the confidence contours for each pairwise comparison for both the UNION2 catalogue and the JLA.

Constraints on α and β. We now consider the constraints on the stretch and colour corrections multipliers α and β, which are not often presented in the analysis of SNe data. The resulting two-dimensional marginalised constraints on the parameters (α, β) are shown in right panel Figs 1 for UNION2. One sees that for several pairings the confidence contours for the two surveys are significantly displaced from one another, and do not overlap at all in some cases. For the JLA compilation for each pairing the confidence contours for the two surveys overlap well, indicating mutual consistency, in sharp contrast to the UNION2 results.

Mutual consistency test between survey pairs. The results of the consistency test for the UNION2 compilation are shown in left panel of Fig. 1, together with the corresponding p-values. One sees that the p-values obtained agree very closely with the degree of overlap between the confidence contours plotted in Fig. 1. In particular, we find very strong evidence, at more than the 99 per cent significance level, for inconsistency between the survey pairs CfA/SNLS and SNLS/HST.

The same analysis for JLA showes that none of the survey pairs shows evidence for inconsistency, even at the 90 per cent significance level, which is in agreement with the good degree of overlap between the confidence contours.

The consistency between the survey pairs in the JLA compilation is in sharp contrast to inconsistencies present the UNION2 compilation. This difference must result from the different SNe selection criteria and calibration techniques used in the construction of the two compilations. The level of inconsistency exhibited by the UNION2 compilation suggests that one must exercise caution when interpreting cosmological constraints derived from it with the usual joint analysis.

References

Amanullah R., *et al.* 2010, *ApJ*, 716, 712
Betoule M., *et al.* 2014, *ArXiv e-prints*
Feroz F., *et al.* 2008, *Journal of High Energy Physics*, 10, 64
Feroz F., Hobson M. P. 2008, *MNRAS*, 384, 449
Feroz F., Hobson M. P., Bridges M. 2009, *MNRAS*, 398, 1601
March M. C., Trotta R., Amendola L., Huterer D. 2011, *MNRAS*, 415, 143
Marshall P., Rajguru N., Slosar A. 2006, *Phys.Rev.D*, 73, 067302
Perlmutter S., *et al.* 1999, *ApJ*, 517, 565
Riess A. G., *et al.* 1998, *AJ*, 116, 1009

† The SNANA package: `http://sdssdp62.fnal.gov/sdsssn/SNANA-PUBLIC/`

Statistical Challenges in 21st Century Cosmology
Proceedings IAU Symposium No. 306, 2014
A. F. Heavens, J.-L. Starck & A. Krone-Martins, eds.

doi:10.1017/S1743921314010928

Improved KPCA for supernova photometric classification

Emille E. O. Ishida[1], Filipe B. Abdalla[2] and Rafael S. de Souza[3,4]

[1]Max-Planck-Institut für Astrophysik,
Karl-Schwarzschild-Str. 1, D-85748 Garching, Germany
email: emille@mpa-garching.mpg.de

[2]Department of Physics and Astronomy, University College London,
London WC1E 6BT, UK
email: fba@star.ucl.ac.uk

[3]Korea Astronomy & Space Science Institute, Daejeon 305-348, Korea
[4]MTA Eötvös University, EIRSA "Lendulet" Astrophysics Research Group,
Budapest 1117, Hungary
email: rafael.2706@gmail.com

Abstract. The problem of supernova photometric identification is still an open issue faced by large photometric surveys. In a previous investigation, we showed how combining Kernel Principal Component Analysis and Nearest Neighbour algorithms enable us to photometrically classify supernovae with a high rate of success. In the present work, we demonstrate that the introduction of Gaussian Process Regression (GPR) in determining each light curve highly improves the efficiency and purity rates. We present detailed comparison with results from the literature, based on the same simulated data set. The method proved to be satisfactorily efficient, providing high purity ($\lesssim 96\%$) rates when compared with standard algorithms, without demanding any information on astrophysical properties of the local environment, host galaxy or redshift.

Keywords. (stars:) supernovae: general, techniques: photometric, methods: statistical

---

1. Introduction

In the late 20th century, type Ia supernovae (SNe Ia) played a central role in the development of the standard cosmological model. Since then, a great effort has been employed towards the construction of a large, reliable SNe Ia sample, which might be able to provide further insights on the nature of the energy component driving the Universe accelerated expansion (dark energy). A major fraction of such efforts is directed to ongoing and planned photometric surveys, which will collect light curve information for thousands of SNe. However, the usefulness of these data to cosmology is limited by our ability to classify them without using spectroscopic information. As an example, the second SN data release from the *Sloan Digital Sky Survey* (Sako *et al.*, 2014) consists of 10258 variable and transient sources from which only 889 were spectroscopically confirmed.

In a previous work (Ishida & de Souza, 2013 - hereafter IdS2013), we demonstrated how the combination of Kernel Principal Component Analysis (kPCA) and Nearest Neighbour (NN) algorithms can be applied to the SN photometric classification problem, leading to competitive results specially when purity is the main feature to be maximise in the final photometric sample. In this work, we show how the introduction of Gaussian Process Regression (GPR) in the light curve fitting process can improve our previous results, allowing for better purity and efficiency rates.

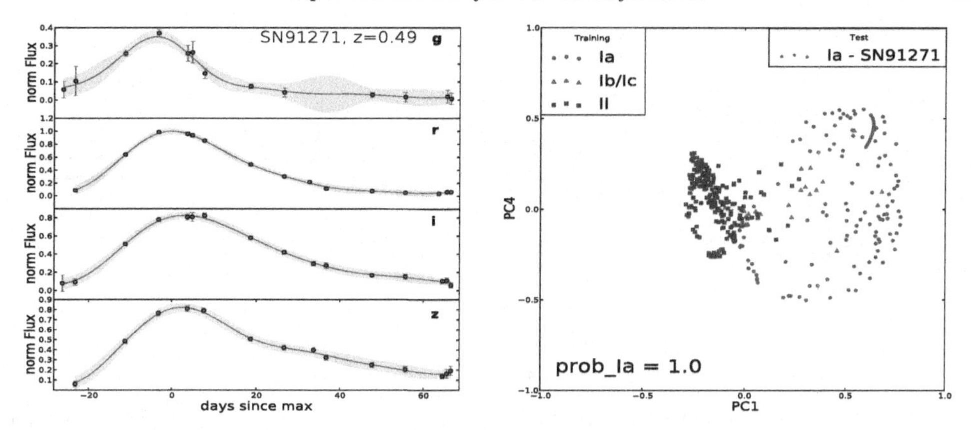

Figure 1. Results from the application of GPR to the pre-processing pipeline of supernovae data. **Left:** Normalized flux measurements as a function of days since maximum brightness. The blue dots denote observed values, the red line corresponds to the mean fitted light curve and the grey region encloses 2 standard deviations at each epoch. **Right:** Projection of supernovae light curves in PC space. The red path denotes the projection of 100 light curves defined within 2σ (grey region in the left panel).

2. Method

The framework presented in IdS2013 consists of applying kPCA to a light curve data set whose types are previously known (called *training set*) and subsequently use the geometrical distribution of SNe in this space to classify a new untyped SN by means of the NN algorithm. kPCA is a generalisation of PCA (see Jolliffe, 2012 for a complete review) built to handle the nonlinear case. It projects the data into a higher dimensional parameter space, where the relations can be linearly described, before determining the directions of each PC (for a deeper discussion see IdS2013 and references therein). After applying kPCA to the training set, the unclassified (or *test*) SN is projected into the PC space and it is assigned the same class as its first NN. As reported in IdS2013, this method achieves extremely high purity rates when applied to simulated data.

Nevertheless, it is important to emphasise that we need flux measurements equally sampled in time for all SNe in order to construct the initial data matrix. In IdS2013, we used a cubic spline interpolation in order to convert the observed flux measurements into a light curve function. Although this procedure proved to be enough to achieve competitive results, it does not enable us to propagate the uncertainty in flux measurements into the classification process. Here, we updated the method by introducing GPR as a light curve fitting technique.

GPR is a Bayesian approach to the linear regression problem, where we consider a prior distribution over the functions likely to describe the behaviour of our data. Before any measurements are taken, we consider a flat prior (e.g. for normalised light curves, $0 \leqslant \text{Flux} \leqslant 1$, $\forall$ epochs). Thus, any smooth function whose image is contained in this flux interval is as likely to describe our light curve. Once we have one measurement $y = F(\text{day}_1)$, it acts like a constraint over the previous prior and now we have a posterior probability distribution, enclosing all smooth functions within [0,1] interval, which pass through the observed point. Applying the same line of thought to successive measurements, we construct a region in the $y \times day$ parameter space, where all functions satisfying the posterior are concentrate (see Rasmussen & Williams, 2006). We used the GPR to determine a light curve function from the observed data points. This allowed

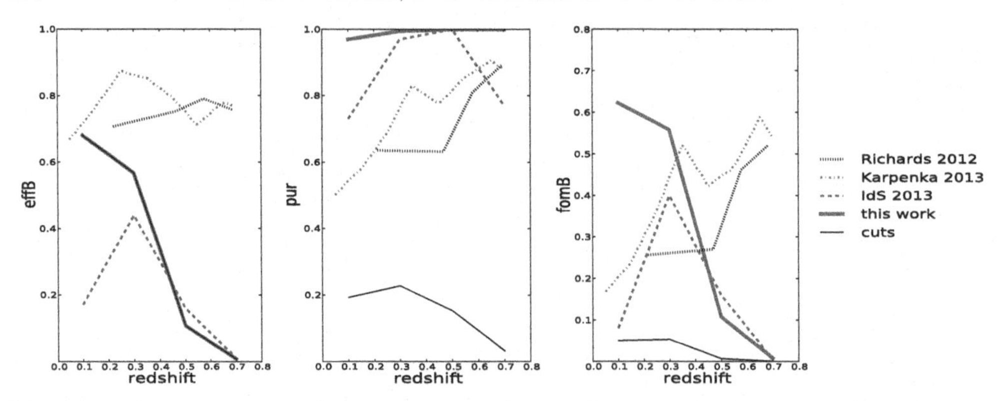

Figure 2. Comparison with literature results from different techniques, applied to the same data set, as a function of redshift. The black (dotted) line correspond to results from Richards *et al.*, 2012 (figure 10, $\mathcal{S}_{m,25}$), the green (dashed) line denotes those presented in Karpenka, Feroz & Hobson, 2013 (figure 3, D_4 without z) and the blue (bold-dashed) line shows our results when using a spline fit, reported in Ishida & de Souza, 2013 (table F2, D_7+SNR5). The red (bold-full) line represents results obtained with GPR, with data from -3 to +24 days, SNR⩾5 and 4-dimensional PC space. The thin line shows the outcomes when all SNe passing selection cuts are classified as Ia.

us to transfer the confidence region coming from the GPR directly into the classification process.

3. Application

We applied the procedure described above to the post-SNPCC data set†, which consists of a simulation corresponding to 5 years of observations from the *Dark Energy Survey*‡. From this initial set, we selected all objects with at least 3 observation epochs with signal to noise ratio, SNR⩾5 in each filter. Moreover, we demanded at least one observation before -3 and one observation after +24 days since maximum brightness. This reduced our sample to 465 and 3495 SNe in the training and test samples, respectively.

The left-hand panel of Figure 1 shows the results of applying GPR to a SN in the test sample, here we used a squared exponential covariance function and marginalized over the hyper-parameters. The right-hand panel illustrates the geometrical distribution of the training set in a 2-dimensional PC space¶ and the corresponding projections of 100 light curves fulfilling the constraints imposed by the posterior. We then applied the NN algorithm to all the 100 points individually, which enabled us to assign a probability of being a SN Ia to this SN. At a first glance, we notice the power of our method in separating Ia and non-Ia SNe. Although there is some degree of contamination, with the most drastic cases corresponding to type Ib/Ic SNe populating the area occupied by SN Ia.

Traditionally, results from such a classification analysis are reported in terms of efficiency (eff), purity (pur) and figure of merit (fom). Mathematically these are defined

† http://sdssdp62.fnal.gov/sdsssn/SIMGEN_PUBLIC/
‡ http://www.darkenergysurvey.org/
¶ Although the figure shows only 2 PC for didactic reasons, our final results were built in a 4-dimension parameter space. The choice of dimensionality and which PC to use was based on a cross-validation procedure as described in IdS2013, section 3.2.

as

$$\text{eff} = \frac{N^{\text{Ia}}_{\text{cc}}}{N^{\text{Ia}}_{\text{tot}}}, \qquad \text{pur} = \frac{N^{\text{Ia}}_{\text{cc}}}{N^{\text{Ia}}_{\text{cc}} + N^{\text{nonIa}}_{\text{wc}}} \qquad \text{and} \qquad \text{fom} = \text{eff} \times \frac{N^{\text{Ia}}_{\text{cc}}}{N^{\text{Ia}}_{\text{tot}}(N^{\text{Ia}}_{\text{cc}} + W N^{\text{nonIa}}_{\text{wc}})}, \tag{3.1}$$

where $N^{\text{Ia}}_{\text{cc}}$, $N^{\text{Ia}}_{\text{tot}}$, $N^{\text{nonIa}}_{\text{wr}}$ are the number of correctly classified SNe Ia, the total number of SNe Ia in the test sample and the number of nonIa's which were wrongly classified as Ia, respectively. We assume a weight factor $W = 3$, which penalises wrong nonIa classifications contaminating the final sample (Kessler *et al.*, 2010). Following the nomenclature of IdS2013, the index "B" used after eff and fom indicates that $N^{\text{Ia}}_{\text{tot}}$ corresponds to the test sample *before* selection cuts. Thus, the effect of applied selection cuts are also taken into account.

Diagnostic results obtained from a GPR will always depend on the probability threshold (the minimum probability required so a SN is considered Ia). Given the statistical level of accuracy of current cosmological analysis, our goal is to maximise purity in our final sample, while keeping fom competitive. In this context, with a probability threshold of 0.95 and SNR$\geqslant$5, our method achieved 98% purity against 91% obtained with the cubic spline fit. The difference is still larger if we consider lower quality data. For SNR$\geqslant$3, GPR obtained purity of 97% while the cubic spline achieved no more than 67%. This proves that GPR is able to provide the same level of purity results even in the presence of lower quality data.

Figure 2 shows how our results compare to those of other methods available in the literature, when applied to the same data set. Here we show the best-case scenario pointed by the authors themselves. Richards *et al.*, 2012 used a combination of diffusion map and random forest algorithms and Karpenka, Feroz & Hobson, 2013 applied a neural network to the post-SNPCC data. Notice that GPR can improve efficiency and purity results over the spline fit even with a smaller light curve coverage (results shown from IdS2013 required data between [-3,+45] days since maximum brightness). Moreover, our method is the only able to achieve purity higher than 96% for the entire redshift range.

4. Conclusions

Our ability to extract cosmological information from future surveys is limit by how well we are able to photometrically classify the available SN samples. Given the current stage of cosmological analysis, where there is no consensus on how the purely photometric data should be used, we believe that our approach is ideal to select a small, high purity SN sample which might help improve the cosmological results.

Acknowledgements

EEOI is partially supported by Brazilian agency CAPES, grant number 9229-13-2. FBA acknowledges the Royal Society for a Royal Society University Research Fellowship.

References

Ishida, E. E. O. & de Souza, R. S. 2013, *MNRAS*, 430, 509
Jolliffe, I. T., 2002, Principal Component Analysis, *Springer-Verlag New York, Inc.*
Karpenka, N. V., Feroz, F., & Hobson, M. P. 2013, *MNRAS*, 429, 1278
Kessler, R., *et al.* 2010, *PASP*, 122, 1415
Rasmussen, C. E., & Williams, C. K. I. 2006, GP for Machine Learning, *MIT Press*
Richards, J. W. *et al.* 2012, *MNRAS*, 419, 1121
Sako, M. *et al.* 2014, arXiv:astro-ph/1401.3317

Statistical Challenges in 21st Century Cosmology
Proceedings IAU Symposium No. 306, 2014
A. F. Heavens, J.-L. Starck & A. Krone-Martins, eds.

doi:10.1017/S1743921314013660

Principcpal Component Analysis of type II supernova V band light-curves

Lluís Galbany[1,2]

[1]Millennium Institute of Astrophysics, Universidad de Chile,
Casilla 36-D, Santiago, Chile

[2]Departamento de Astronomía, Universidad de Chile,
Casilla 36-D, Santiago, Chile
email: `lgalbany@das.uchile.cl`

Abstract. We present a Principal Component Analysis (PCA) of the V band light-curves of a sample of more than 100 nearby Core collapse supernovae (CC SNe) from Anderson *et al.* (2014). We used different reference epochs in order to extract the common properties of these light-curves and searched for correlations to some physical parameters such as the burning of ^{56}Ni, and morphological light-curve parameters such as the length of the plateau, the stretch of the light-curve, and the decrements in brightness after maximum and after the plateau. We also used these similarities to create SNe II light-curve templates that will be used in the future for standardize these objects and determine cosmological distances.

Keywords. techniques: photometric, methods: statistical, (stars:) supernovae: general

---

1. Introduction

SNe II represent the most homogeneous set of CC SNe and, although they are on average ~1.2 mag intrinsically fainter than SNe Ia, their use as independent cosmological distance indicators has been already demonstrated in Hamuy & Pinto (2002). They are also the most common of all SN types, although there is a deficit in the literature as compared to the prioritized SNe Ia, especially at higher redshift. The advantages of SNe II as cosmological probes over SNe Ia reside on the simplicity of their hydrogen dominated atmospheres and low-density surrounding media, and on the current understanding of the explosion mechanism and progenitor system. These advantages together with their high rates and homogeneity, make SNe II promising independent cosmological probes.

2. Analysis

We have used a sample of 116 low-redshift SNe II from Anderson *et al.* (2014) for which several V band light-curve parameters have been determined and measured. This parameters include absolute magnitudes at the peak, at the end of the plateau, and at the beginning of the radioactive decline. There are also determined the explosion epoch and the epochs when the plateau starts and ends, from which the duration of the plateau and of the optically thick phase can be measured. Other parameters such as the amount of ^{56}Ni burned, and the three decrements in magnitude brightness from the peak to the transition, during the plateau, and during the radioactive phase, are also measured from the SN light-curves. The PCA required an optimal selection of the epochs available to both (i) have a wide range that allows a compete standardization of the light curves, and (ii) keep a significant number of SNe with observed data in the range previously selected. The selection of a time reference and a normalization parameter was a relevant

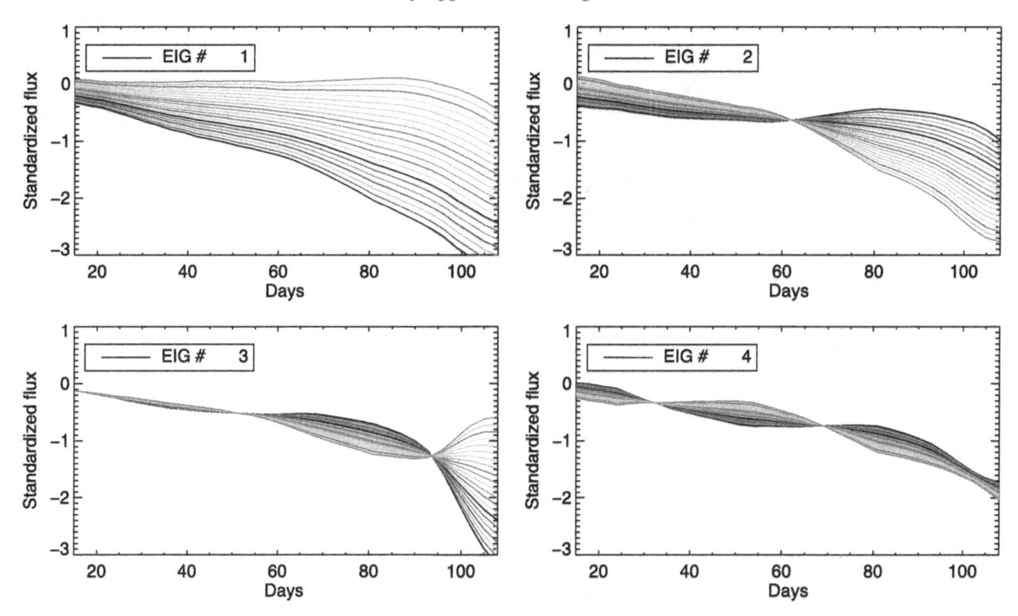

Figure 1. First four eignevectors of the PCA. The variation of the eigenvector when multiplied by different projections is showed in difference colors. The range of projections applied is selected from the 2σ variation from the mean in our sample.

determination as well, since not all the light-curve parameters are available for all the SNe in this sample. After several attempts the best approach was to normalize the light-curves in brightness to the absolute magnitude at the peak and use the explosion date as the reference epoch. Once this was established, we selected all the light-curves with observations before +15 days and after +110 days from the explosion, and applied a linear interpolation in steps of three days in order to have all the light-curves with the same temporal cadence. This resulted in a sample of 29 SNe.

Finally, we have performed a PCA of the 29×32 matrix containing the 32 epochs for each of the 29 SNe. The PCA gave an orthonormal set of 32 eigenvectors (EV) ordered by decreasing variability, and a matrix of 29x32 projections (one projection for SN and EV). In our case, the first four EVs accounted for more than the 99% of the light-curve variation. In Figure 1, one can see the contribution of this four EV to reproduce a SN light-curve by multiplying the 2-sigma variation of all 29 SN projections in each of these 4 EVs.

3. Results

We have correlated each pair of projection and physical or light-curve parameter for all SNe in our sample for which the required information was available. Strong correlations between each projection with different light-curve parameters are found.

- The projections of the first EV correlates to the absolute peak magnitude (M_{max}), and the decline rates from the peak the the plateau ($s1$), and during the plateau ($s2$). In Figure 2 one can see the correlation between these projections and the s2 parameters.
- The second EV correlates with the duration of the plateau (t_{PT}), which is related to the total amount of Hydrogen present in the outer layers of the SNe.
- The projections of EV3 shows the best correlation with the total amount of ^{56}Ni mass synthesized by the SN, which is measured in the radioactive decline phase.

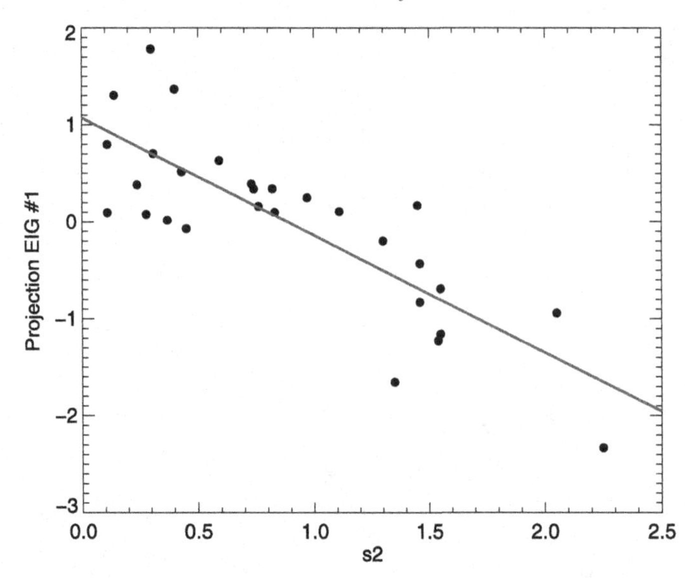

Figure 2. Correlation between the projections of the first eigenvector and the light curve parameter for which the best relation is found. Linear fit in red.

• And the fourth EV with the time elapsed between the explosion and the beginning of the plateau ($t_{trans} - t_{exp}$), which can be understood in terms of the stretch factor for SNe Ia.

Most of these correlations have been previously found in Anderson *et al.* (2014), and here we have been able to reproduce them with a systematic analysis by using a statistical tool. We have also been able to reproduce the original light-curves with only these four EVs. The agreement is remarkable even with the three first components.

4. Summary and outlook

We presented here the results only for the V band. Further analysis will be performed including other passbands. The final goal of this work is to enable the use of SNe II as standard candles only with photometric information, and to produce a Hubble diagram for type II SNe. We have compiled a sample of hundreds of spectroscopically confirmed and photometrically identified type II SNe, up to higher redshift. This sample will be analyzed using the same procedure.

Acknowledgements. Support for LG is provided by the Ministry of Economy, Development, and Tourism's Millennium Science Initiative through grant IC12009, awarded to The Millennium Institute of Astrophysics (MAS), and by CONICYT through FONDECYT grant 3140566.

References

Hamuy, M. & Pinto, P. A 2002, *ApJ*, 566, 63
Anderson, J. P., González-Gaitán, S., Hamuy, M. *et al.* 2014, *ApJ*, 786, 67A

Statistical Challenges in 21st Century Cosmology
Proceedings IAU Symposium No. 306, 2014
A. F. Heavens, J.-L. Starck & A. Krone-Martins, eds.

doi:10.1017/S1743921314010862

Photometric typing of normal and peculiar type Ia supernovae

Santiago González-Gaitán[1] and Filomena Bufano[2]

[1]Departamento de Astronomía, Universidad de Chile
Camino El Observatorio 1515, Las Condes, Santiago, Chile
email: gongsale@gmail.com

[2]Departamento de Ciencias Fisicas, Universidad Andres Bello
Avda. Republica 252, Santiago, Chile

Abstract. We present a new photometric typing technique for normal type Ia supernovae (SNe Ia) and peculiar SNe Ia such as SN 1991bg-like, type Iax and super-Chandrasekhar SNe Ia. This photometric classifier allows the identification of exclusively normal SNe Ia (with a purity above 99% and an efficiency greater than 80%) in current and future large wide field surveys for cosmological studies.

Keywords. Supernovae: general, methods: data analysis

---

1. Introduction

Type Ia supernovae have been fundamental for the discovery of the acceleration of the expansion of the universe driven by a yet unknown dark energy component (Riess *et al.* 1998; Perlmutter *et al.* 1999). With the advent of new powerful instruments, current and future surveys are looking to pin down the nature of dark energy with increasing data and better understanding of systematics. These wide field surveys such as SNLS (Astier *et al.* 2006), DES (Sako *et al.* 2011) and LSST (Ivezic *et al.* 2011) generate hundreds to thousands of transients routinely that cannot be followed-up spectroscopically. Photometric classification techniques need to be able to accurately classifiy these objects in order to prioritize certain of them for follow-up or just make their posterior study possible. For cosmology, one can avoid SN spectroscopy using known spectroscopic redshifts from the host galaxy, provided that the object class is known with a photometric method, and even photometric redshifts have shown potential for cosmology (Campbell *et al.* 2013).

2. Methodology

We use a large dataset of ~500 low-redshift ($z < 0.1$) multi-band light-curves of SNe Ia from the literature, as well as another ~ 200 core-collapse supernovae (CC SNe) to test for contamination. We fit all these light-curves with SiFTO (Conley *et al.* 2008) to two different spectral template series for normal SNe Ia (Hsiao *et al.* 2007) and 91bg-like SNe Ia (Nugent *et al.* 2002). We find that a simple comparison of the resulting fit quality (χ^2_ν) between these two templates is quite successful at separating normal and 91bg-like objects. Figures 1 and 2 show example fits with the two templates for two SNe: SN 2003fa (Hicken *et al.* 2009) and SN 1999by (Ganeshalingam *et al.* 2010), with better normal and 91bg-like template fits, respectively. From these fits, one can see that most of the difference between the two templates stems from the presence/absence of shoulders and secondary maxima in red filters, but also in the different temporal occurence of the primary maxima in all bands.

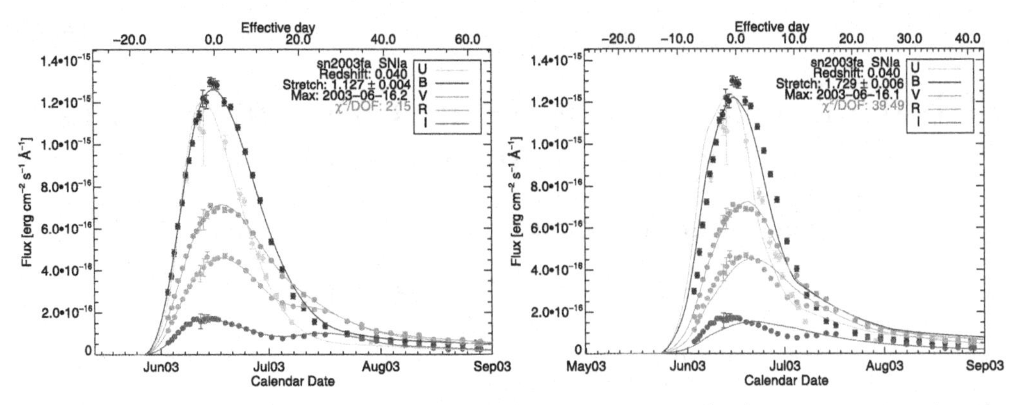

Figure 1. SiFTO fits to SN 2003fa with two templates: normal (left) and a 91bg-like (right) SNIa.

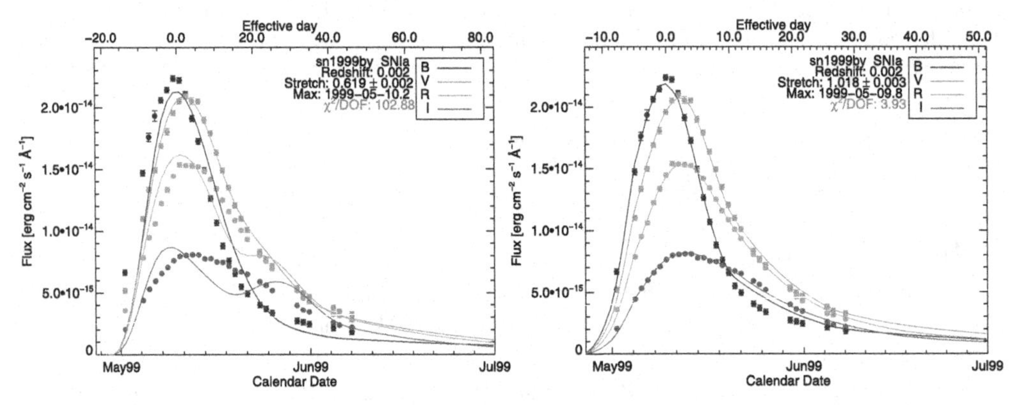

Figure 2. SiFTO fits to SN 1999by with two templates: normal (left) and a 91bg-like (right) SNIa.

We find that this simple classification based on fits to the entire light-curve with two templates is able to differentiate properly between spectroscopic normal and 91bg-like SNe Ia (see figure 3). Additionally, we find that the method consistenly picks up other peculiar SNe Ia such as SN 2002cx-like SNe (Iax) and even super-Chandrasekhar SNe. Taking a closer look at their light-curves, one sees that in fact they also lack secondary maxima in their redder filters, similarly to 91bg-like objects. To disentangle these objects from classical 91bg-like, we perform SiFTO fits with the two templates to only blue (u or U, B and g) and only red (V, r or R and i or I) light-curves. In this way, SNe Iax and super-Chandrasekhar have better 91bg-like template fits in their red but not in their blue light-curves, as opposed to real 91bg-like objects for which the 91bg template fit is always better. This behaviour is interesting and could point towards physical similarities between those sub-groups despite the diversity in other characteristics. More importantly, it opens up the possibility to obtain pure samples of normal SNe Ia for different studies, particularly cosmology.

To study this further, we apply the technique to all low-redshift SNe available, including non-Ia SNe. We perform SIFTO fits with the two templates for the entire light-curves but also for red and blue light-curves. To avoid CC contamination, we additionnally make cuts on the χ^2_ν as shown in figure 4. The final purity we obtain is above 99% for normal SNe Ia (with only one false positive) with an efficiency above 80%. This competes with other classification techniques (e.g Sako *et al.* 2011) with the additional advantage that it

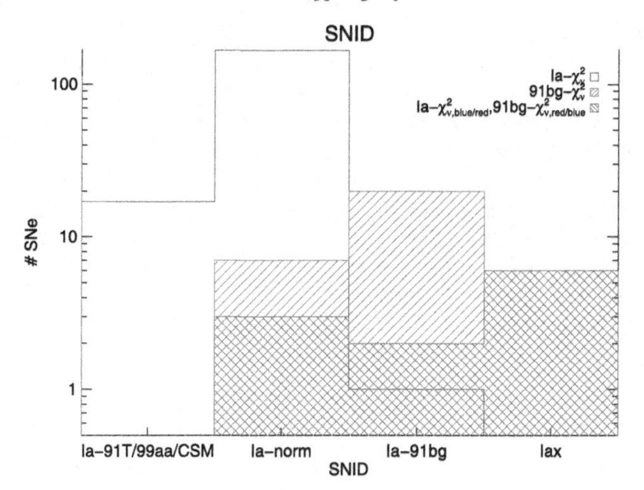

Figure 3. Comparison of our photometric classifier with the spectroscopic classifier SNID (Blondin & Tonry 2007). SNID subclasses are put in different bins in the x-axis. Our photometric normal candidates are shown in blue, 91bg-like according to the overall fit are shown in orange and 91bg-like according to only blue or red band fits are shown in purple.

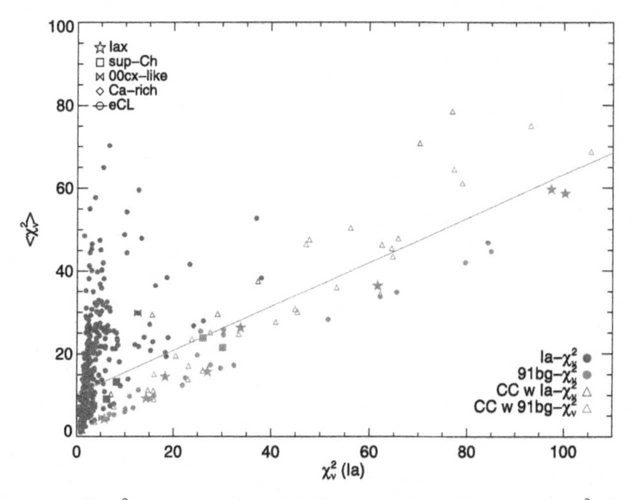

Figure 4. Average overall χ^2_ν of normal and 91bg template fits versus χ^2_ν for the normal template. Circles (normal and 91bg-like), stars (Iax) and squares (super-Chandra) represent SNe Ia whereas triangles are CC SNe. Blue are orange are normal and 91bg-like candidates according to the overall light-curve fits. The dividing line separates SNe Ia from CC contaminants.

is also free of peculiar SNe Ia, which are unfit for cosmology. The method is capable also of selecting peculiar objects such as 91bg-like or SNe Iax, although with a lower purity due to contamination from SNe Ibc.

We have presented a robust technique to identify SNe Ia of different sorts without the need of spectroscopy. Such algrithms are essential for upcoming large wide field surveys and future cosmological work with SNe Ia.

References

Riess, A. G., *et al.* 1998, *AJ*, 116, 1009
Perlmutter, S. *et al.* 1999, *ApJ*, 517, 565

Astier, P. *et al.* 2006, *A&A*, 447, 31
Sako, M. *et al.* 2011, *American Astronomical Society Meetings*, 217, 205
Ivezic, Z. *et al.* 2011, *American Astronomical Society Meetings*, 217, 252
Campbell, H. *et al.* 2011, *ApJ*, 763, 88
Conley, A. *et al.* 2008, *ApJ*, 681, 482
Hsiao, E. Y. *et al.* 2007, *ApJ*, 663, 1187
Nugent, P. *et al.* 2002, *PASP*, 114, 803
Hicken, M. *et al.* 2009, *ApJ*, 700, 331
Ganeshalingam, M. *et al.* 2010, *ApJS*, 190, 418
Blondin, S. & Tonry, J. L. 2007, *ApJ*, 666, 1024
Sako, M. *et al.* 2011, *ApJ*, 738, 162

Statistical Challenges in 21st Century Cosmology
Proceedings IAU Symposium No. 306, 2014
A. F. Heavens, J.-L. Starck & A. Krone-Martins, eds.

doi:10.1017/S1743921314010977

Photometric classification of Supernovae from the SUDARE survey

Giuliano Pignata

Departamento de Ciencias Fisicas, Universidad Andres Bello,
Avenida Republica 252, Santiago, Chile
email: gpignata@unab.cl

Abstract. I present first results from the algorithm we developed to photometrically classify the Supernovae discovered by the Supernova Diversity and Rate Evolution (SUDARE). The classification outcome is compared with that of the public available code SNANA.

Keywords. supernovae: general, methods: statistical, surveys

1. Introduction

SUDARE is a survey which main goal is discover Supernovae (SNe) of all types in the redshift range z = 0.3-0.6, where the rate of Core Collapse (CC) SNe is basically unexplored and the published estimates of the SN Ia rate show some discrepancy. The properties of the SNe discovered will be compared with those of the local sample and linked to the properties of the parent stellar populations. For the latter we choose to monitor the Cosmic Evolution Survey (COSMOS) and Chandra Deep Field South (CDFS) fields. These fields have got an unrivalled coverage from the X-Ray to the Radio. This outstanding data-sets have allowed, among other high-quality data products, a very precise photometric redshifts which are used as a input in our photometric classification algorithm. SUDARE is a rolling-search carried out using the one square degree field of view OMEGACAM camera mounted on the 2.6 meters VLT Survey Telescope (VST) located at ESO Paranal Observatory. To construct a detailed light curve necessary for the photometric classification and to further investigate the SN diversity, we monitor the selected fields in r band with the frequency of one exposure every three days at a limiting magnitude r = 24.5 (S/N = 5.0) and once every week, we are obtaining g and i bands exposures of similar depth. In the images collected on the CDFS and COSMOS field during the first two seasons, we discovered 113 bona-fide SNe.

2. Photometric classification

To classify our discoveries, we developed a simple procedure which assess whether the transient multi-colour light curves match a known SN type and what are the best match sub-type, redshift and extinction. To this aim we selected a sample of SN templates of different types for which both multi-colour light curves and spectral sequence are available. The latter are needed to have an estimate of the K-correction. The templates were selected to represent the well established SN types, namely Ia, Ib, Ic, IIb, II and IIn. We also included some templates of the recently discovered class of very bright SNe (e.g. Gal-Yam (2012)) that, although intrinsically very rare Quimby *et al.* (2013), may easily be detected even at very high redshift.

For all templates we constructed a grid of light curves in apparent magnitudes for each of the survey filters assuming redshifts in the range $0.0 < z < 1.0$ and extinction

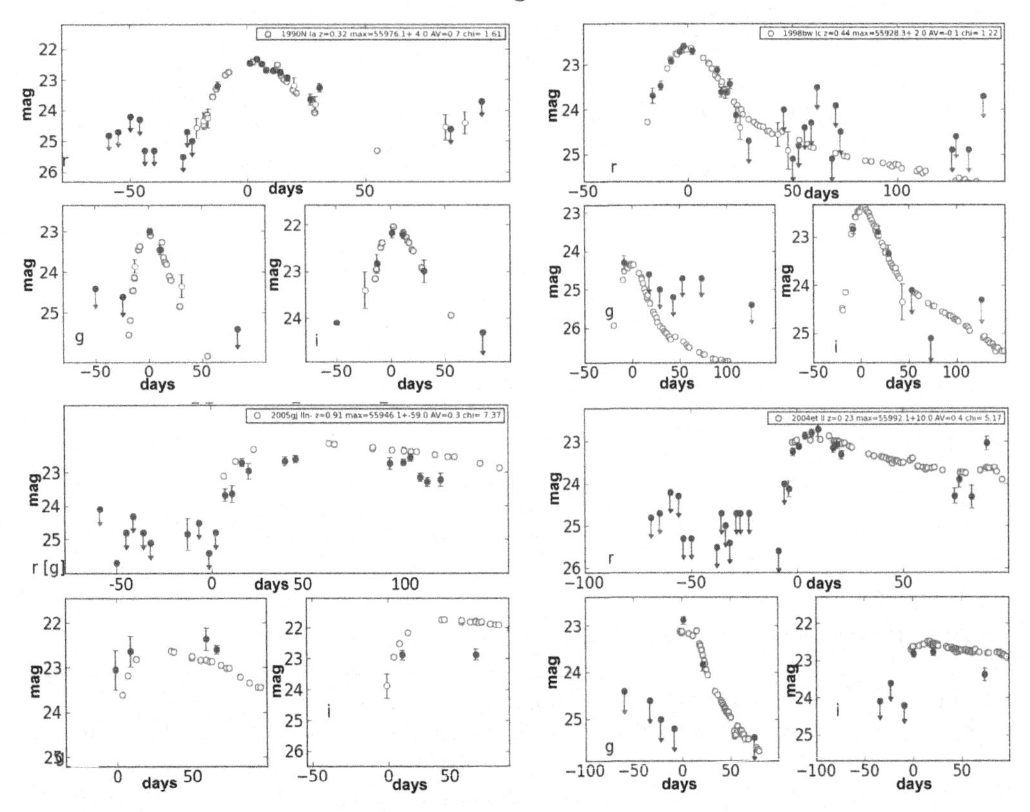

Figure 1. Examples of photometric classifications we obtained using our code. Each quadrant consist of three boxes corresponding to the r band (top), g band (bottom left) and i band (bottom right). Blue points (filled and open) are the SUDARE measured magnitude. Open red points are the template light curves. In the label of the r band box are reported the best match template, the SN type, the redshift, the epoch of the V band maximum light, the extinction in the V band and the value of the reduced χ^2.

$0.5 < A_V < 2.0$ mag. Also the epoch of maximum is allowed to range $\pm$ 100 days from the epoch of the candidate brightest magnitude. We then compare the candidate light curve in each filter with all simulated light curves of the grid. Through χ^2 minimization we derive the SN type, average distance and reddening combining measurement for all filters weighting for the number of observed epochs. In case the host galaxy redshift is known, either from spectroscopic or photometric redshift, the distance range is restricted to the measurement error, making the classification precess much faster. Examples of our classification output are reported in Figure 1. We also photometrically classify our discovery using the public available code SNANA (SuperNova ANAlysis) Kessler *et al.* (2009). The comparison between the results obtained by the two algorithms are shown in the pie chart reported in Figure 2. In this case we collapsed our six SN types in three groups namely: type Ia, Ibc and II. As it is visible there is a good agreement between the output of the two codes. Indeed 86% of the Type Ia SNe classified by our code were confirmed by SNANA. The remaining objects were classified as Type II (5%), Ibc (4%) and for 5% SNANA did not provide a SN classification.

We found a slightly poorer agreement for the type II classification: 74% were confirmed by SNANA, while 11% were classified by the same code as Ia and 4% as Ibc. For the remaining 11% SNANA did not provide a SN classification. Some of these objects could be AGN which nature will be definitely established during the survey progress. Finally, we

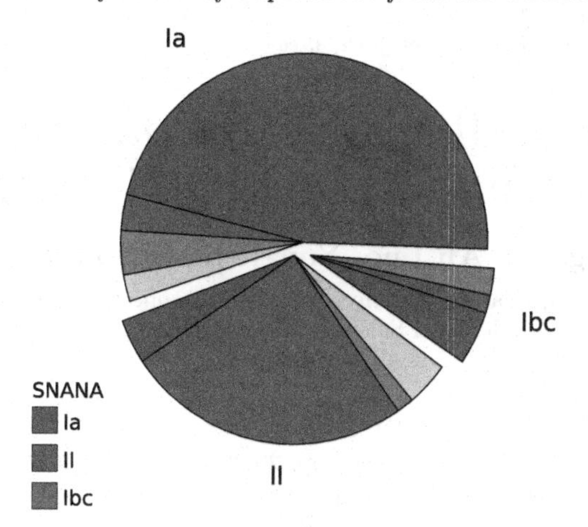

Figure 2. Comparison between the classification obtained using our classifier and SNANA. See text for details.

found the worse agreement in the case of type Ibc SNe were only 50% of our classification was confirmed by SNANA. Considering the reduced numbers of such objects is premature to draw any statistically meaningful conclusion about the latter discrepancy.

3. Future work

For the future we plan to explore more sophisticated techniques for the SN photometric classification of the SUDARE transient starting with the Kernel Principal Component Analysis method (see Ishida & de Souza (2013)) combined with machine learning algorithms.

References

Gal-Yam, A. 2012, *Science*, 337, 927

Kessler, R., Bernstein, J. P., Cinabro, D., Dilday, B., Frieman, J. A., Jha, S., Kuhlmann, S., Miknaitis, G., Sako, M., Taylor, M., & Vanderplas, J. 2009, *PASP*, 121, 883

Ishida, E. E. O. & de Souza R. S. 2013, *MNRAS*, 430, 509

Quimby, R. M., Yuan, F., Akerlof, C., & Wheeler, J. C. 2013, *MNRAS*, 431, 912

Statistical Challenges in 21st Century Cosmology
Proceedings IAU Symposium No. 306, 2014
A. F. Heavens, J.-L. Starck & A. Krone-Martins, eds.

doi:10.1017/S1743921314010825

Automatic stellar spectral parameterization pipeline for LAMOST survey

Yue Wu, Bing Du, Ali Luo, Yongheng Zhao and Hailong Yuan

Key Laboratory of Optical Astronomy, National Astronomical Observatories, Chinese Academy of Sciences, Beijing 100012, China
email: wuyue@bao.ac.cn

Abstract. The Large Sky Area Multi-Object Fiber Spectroscopic Telescope (LAMOST) project performed its five year formal survey since Sep. 2012, already fulfilled the pilot survey and the 1st two years general survey with an output - spectroscopic data archive containing more than 4.1 million observations. One of the scientific objectives of the project is for better understanding the structure and evolution of the Milky Way. Thus, credible derivation of the physical properties of the stars plays a key role for the exploration. We developed and implemented the LAMOST stellar parameter pipeline (LASP) which can automatically determine the fundamental stellar atmospheric parameters (effective temperature $T_{\rm eff}$, surface gravity log g, metallicity [Fe/H], radial velocity $V_{\rm r}$) for late A, FGK type stars observed during the survey. An overview of the LASP, including the strategy, the algorithm and the process is presented in this work.

Keywords. techniques: spectroscopic, methods: data analysis, stars: fundamental parameters

1. Introduction

LAMOST is a unique reflecting Schmidt telescope, with a large aperture and a wide field of view (Cui *et al.* 2012), sited at the Hebei province in China. It can simultaneously observe 4000 celestial objects and obtain the spectra through the multi-fiber system. After the previous two years commissioning survey and one year pilot survey, LAMOST accomplished its 1st two years general survey (Zhao *et al.* 2012) in the period between Sep. 2012 and Jun. 2014. The LAMOST First Data Release - DR1 (Pilot and the 1st year survey data) was published in Aug. 2013, contains more than 2.2 million spectra, the majority of the targets are stars with a quantity of more than 1.9 million, others are Galaxy, QSO and UNKNOWN (low S/N). The whole DR2 (DR1 plus the 2nd year survey data, was published at the end of 2014) sky coverage is demonstrated in the Fig. 1. One of the two main components of the project is the LAMOST Experiment for Galactic Understanding and Exploration (LEGUE) which is focusing on the formation and evolution of the Milky Way (Deng *et al.* 2012). The LAMOST data processing and analysis pipeline is integrated with three modulars: 2D pipeline (raw CCD data reduction, extraction, wavelength calibration), 1D pipeline (classification and the redshift estimation for galaxy and QSO, Luo *et al.* 2012) and the LASP (Wu *et al.* 2011a).

2. Data and Strategy

The LAMOST spectra are with a resolving power of R$\sim$1800 covering 3800 - 9000 $\AA$ wavelength range. The consecutive data processing system of LAMOST is organized with a central versioned main data base (MDB) which is supported by MySQL. The observed data, the conventional observing information as well as all the intermediate and final output of each data reduction and analysis modular are all archived in this MDB, the published spectroscopic data are in the format of fits file. According to the

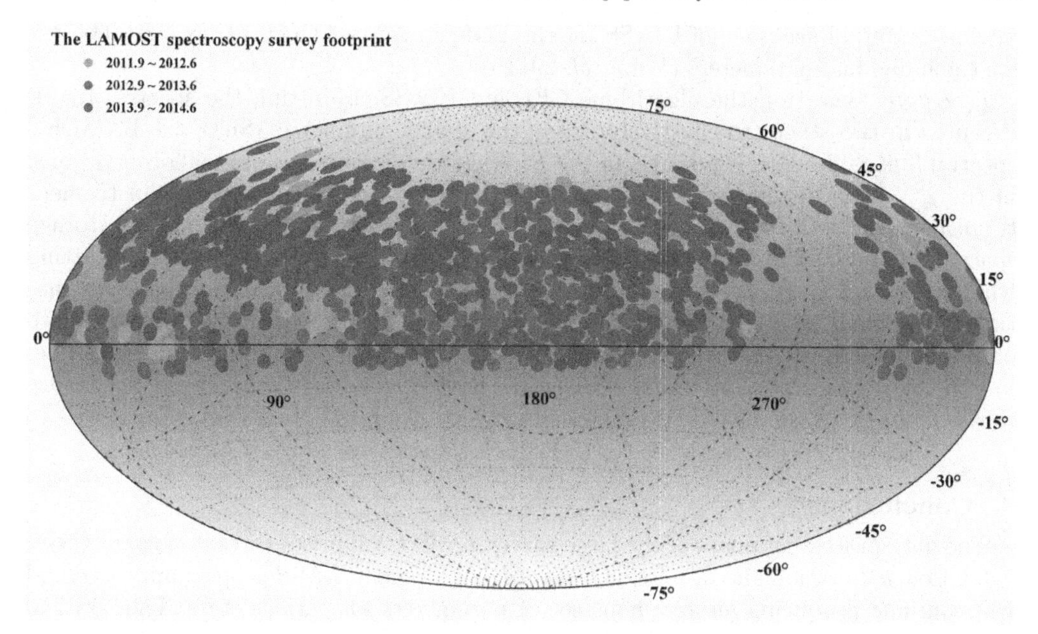

Figure 1. The LAMOST DR2 sky coverage.

moon status, we approximately divided the survey into three parts: dark night, bright night, and instrumental test night. The current selecting criteria for the LASP input are: 'final class' is STAR and 'final subclass' is late A or FGK type, with g band $S/N \geqslant 15$ and $S/N \geqslant 6$ for the bright and dark nights respectively.

3. Methods and Procedure

CFI method. The name CFI is an abbreviation of 'correlation function interpolation' (Du *et al.* 2012). Provided that the observed flux vector is O, and the synthetic model flux vector is S, theoretically, the best-fit pair is that $\cos<O,S> = 1$. So it searchs for the best-fit by maximizing the value of $\cos<O,S>$ as functions of $T_{\rm eff}$, $\log g$ and [Fe/H], where $\cos<O,S> = (O{\cdot}S)/(|O|\times|S|)$ which is referred to as correlation coefficient. For the selection of the synthetic spectral girds, CFI employed Kurucz spectrum synthesis code based on the ATLAS9 stellar atmosphere model (Castelli & Kurucz 2003).

ULySS method. ULySS (Koleva *et al.* 2009, Wu *et al.* 2011b) determines the atmospheric parameters via minimizing the χ^2 between the observation and the model spectra which are generated by an interpolator built based on the ELODIE stellar library (Prugniel & Soubiran 2001,2007). The model is:

$\mathrm{Obs}(\lambda) = \mathrm{P_n}(\lambda) \times \mathrm{G}(v,\sigma) \otimes \mathrm{TGM}(\mathrm{T_{eff}}, \log \mathrm{g}, [\mathrm{Fe/H}], \lambda)$, where $\mathrm{Obs}(\lambda)$ is the observed spectrum sampled in $\log\lambda$, $\mathrm{P_n}(\lambda)$ a series of Legendre polynomials of degree n, and $\mathrm{G}(v,\sigma)$ a Gaussian broadening function parameterized by the residual velocity v, and the dispersion σ. The multiplicative polynomial is meant to absorb errors in the flux calibration, Galactic extinction or any other source affecting the shape of the spectrum.

The TGM function is an interpolator of the ELODIE library, it consists of polynomial expansions of each wavelength element in powers of $\log(T_{\rm eff})$, $\log g$, [Fe/H] and $f(\sigma)$ (a function of the rotational broadening parameterized by σ). Three sets of polynomials are defined for three temperature ranges (roughly matching OBA, FGK, and M types) with important overlap between each other where they are linearly interpolated. The derived intrinsic external accuracy of ULySS for the FGK stars are 43 K, 0.13 dex, and 0.05 dex for the individual parameters (Wu *et al.* 2011b).

Procedure. Based on the algorithms CFI and ULySS, by fitting the spectra, LASP executed in two stages to effectively derive the stellar parameters. Since the LAMOST spectral flux calibration is relative, in the 1st stage we measure the original observations, in the 2nd stage, we measure the normalized spectra. Considering the low instrumental response in both edges, as well as for optimizing the computing time and the storage space, CFI and ULySS respectively selects [3850, 5500] $\AA$ and [4100, 5700] $\AA$ as the fitting window. In each stage, firstly, we utilize CFI to quickly get a set of initial coarse estimations, after that, by using CFI results as a guessed starting point, we adopt ULySS to obtain more accurate and credible measurements as the final output. CFI can rapidly select the appropriate stellar template fitting range, this advantage helps effectively reduce the computational quantity and time for ULySS by using less guessed starting grid.

4. Conclusions

The data processing phase of the large survey project comprises three important tasks: validation, calibration and in-mission software development. LASP is being improved and will continue producing an enormous set of parameters for various stars. Thus critical assessment of its measurements is significant for the realization of the project's scientific goal. In one of the independent external validations, by comparing the common hundreds of stars between LAMOST DR1 and the PASTEL Catalog (Soubiran *et al.* 2010), we obtain precisions of 110 K, 0.19 dex, 0.11 dex and 4.91 km/s for $T_{\rm eff}$, log g, [Fe/H] and $V_{\rm r}$ respectively in the specified temperature range (Gao *et al.* 2014). A systemic comprehensive validation and calibration works according to different spectral types and SNRs have already been carried out, the results will be published in a separate work.

With the LAMOST DR1, a catalogue of stellar parameters which contains more than 1 million stars was simultaneous published, it has been characterized as the largest one in the world so far. For the spectra acquired during the 2nd year survey, LASP already determined nearly one million stars' parameters. Considering the five years operation, further LAMOST survey will yield more than 6 million stars covering in a wider sky area and depth, this will broaden the expected science especially for the characterization of the Galaxy with more various evolutionary stages of the stars collected.

Acknowledgements. The authors thank the grants (Nos. 11403056, 11103031, 11273026, 11178021) from NSFC and the National Key Basic Research of China 2014CB845700.

References

Castelli, F. & Kurucz, R. L. 2003, *IAUS*, 210, A20
Cui, X. Q., Zhao, Y. H., Chu, Y. Q., *et al.* 2012, *RAA*, 12, 1197
Deng, L. C., Newberg, H. J., Liu, C., *et al.* 2012, *RAA*, 12, 735
Du, B., Luo, A. L., Zhang, J. N., Wu, Y., *et al.* 2012, *SPIE*, 8451, 37
Gao, H., Zhang, H. W., Liu, X. W., *et al.* 2014, *RAA*, in press
Koleva, M., Prugniel, P., Bouchard, A., & Wu, Y. 2009, *A&A*, 501, 1269
Luo, A. L., Zhang, H. T., Zhao, Y. H., *et al.* 2012, *RAA*, 12, 1243
Prugniel, P. & Soubiran, C. 2001, *A&A*, 369, 1048
Prugniel, P. & Soubiran, C. 2007, arXiv:astro-ph/0703658
Soubiran, C., Le Campion, J. F., Cayrel de Strobel, G., *et al.* 2010, *A&A*, 515, 111
Wu, Y., Luo, A. L., Li, H. N., *et al.* 2011a, *RAA*, 11, 924
Wu, Y., Singh, H. P., Prugniel, P., *et al.* 2011b, *A&A*, 525, 71
Zhao, G., Zhao, Y. H., Chu, Y. Q., *et al.* 2012, *RAA*, 12, 723

Statistical Challenges in 21st Century Cosmology
Proceedings IAU Symposium No. 306, 2014
A. F. Heavens, J.-L. Starck & A. Krone-Martins, eds.

doi:10.1017/S1743921314011144

New Challenges in Cosmology Posed by the Sloan Digital Sky Survey Quasar Data

Adrija Banerjee and Arnab Kumar Pal

Dept.: Physics and Mathematics Unit, Indian Statistical Institute
203, Barrackpore Trunk Road, Kolkata, West Bengal- 700108, India
email: adibanhere@gmail.com, arnabandstats@gmail.com

Abstract. For SDSS quasar data (2005) we have truncated data structure whereas for the survey of 2007 the data is no longer truncated. This calls for development or use of completely different statistical methodology to study the data for the evolution of the same objects like quasars. These different methodologies suggest different interpretation for a particular phenomenon in nature. This leads to the issue of validation of the data. More intriguing and challenging issue crops up as, given all of the data, what can be said about the laws of physics that have been operating over the universe? Over here we have used the concept of Neural Network to model the relationship between redshift and apparent magnitude.

Keywords. galaxies: distances and redshifts

1. Introduction

According to the Hubble Law, under the assumption of the universe being homogeneous and isotropic, the galaxies appear to be receding with a velocity v proportional to their distance d from the observer: $v = H_o d$, where H_o is called the Hubble constant. There are two luminosity functions, absolute luminosity and apparent luminosity. Normally astronomers work with absolute magnitude M and apparent magnitude m, where $d \approx m - M$ is known as distance modulus and is related to the luminosity distance. Therefore, the relation between the distance modulus and $\log z$ is considered to test Hubble law. Roy *et al.* studied the bivariate distribution of redshift and apparent magnitude observed in Sloan Digital Sky Survey(SDSS) for quasar redshift (2005) in which the data was truncated. However, the redshift data in SDSS quasar survey of 2007 was no longer truncated. Mukhopadhayay *et al.* (2009) discussed a general framework to study data for redshift and apparent magnitude in quasar sample and found a nonlinearity of the fitting curve for Hubble like relation. Now the intriguing issue is, for SDSS quasar data (2005) we have truncated data structure whereas for the survey of 2007 the data is no longer truncated. This calls for development or use of completely different statistical methodology to study the data for the evolution of the same objects like quasars. These two different methodologies suggest different interpretation for a particular phenomenon in nature. This leads to the issue of validation of the data. In order to take a closer look into the debate we have taken up the Tenth Data Release of Sloan Digital Sky Survey on Quasars and investigate the validation of Hubble Law on it.

2. Data

Our massive cosmological data set, consists of 96307 data points on logarithm of red-shift (z) and apparent magnitude (m) for Qsasars (qsasi-stellar objects) collected from SDSS data. We have cleaned the data by removing outliers and those values of for which ln z turns out to be undefined.

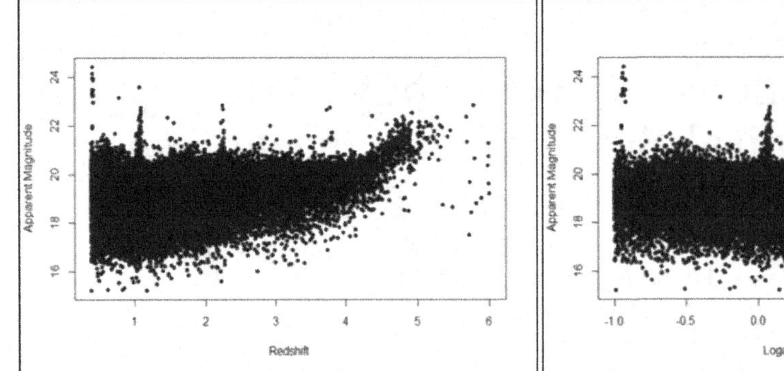

Figure 1. Scatter plot of z vs m

Figure 2. Scatter plot of inz vs m

From the Scatter Plot of z vs m (Fig 1) it is clear that the data is non-truncated.

3. Methodology

The huge and rich in information data as the SDSS Quasar data require cutting edge statistical technologies for their analysis. Thus over here we have used the concept of Neural Network Model Fitting. A Neural Network Model is very similar to non- linear regression model, with the exception that the former can handle an incredibly large amount of model parameters. For this reason, Neural Network Models are said to have the ability to approximate any continuous function.

At first to understand the overall pattern of the data we have binned the data, i.e., we have divided the whole support of lnz into 558 equal classes of width 0.005 and have taken the mean of corresponding variables. Fig 3 illustrates the scatter plot of the above data after binning. Thereafter on the binned data we have carried out Neural Networking and obtained the predicted values (See Fig 4). In order to deduce the upper and lower prediction lines we have taken the respective maximum and minimum points to each point of the binned data (See Fig 5) and fit Neural Networking on them in similar fashion (See Fig 6). Next in Fig 7 we have constructed the prediction curves and modified data after binning with the data as a whole.

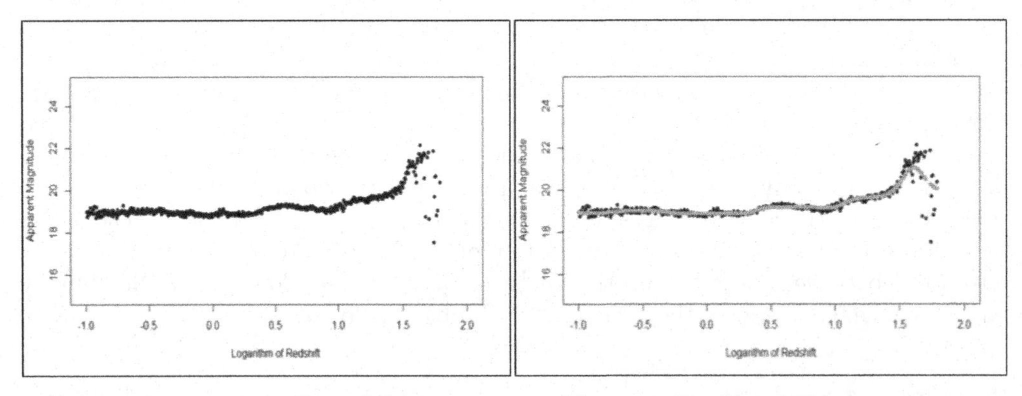

Figure 3. Scatter plot of Binned Data

Figure 4. Predicated Values

Fig 4 illustrates the scatter plot of the above data after binning. Thereafter on the binned data we have carried out Neural Networking and obtained the predicted values (See Fig 5).

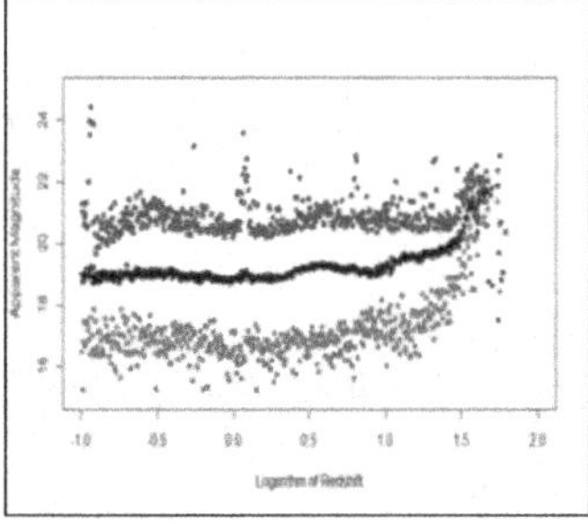

Figure 5. Maximum and Minimum Points Corresponding to Each of the Binned Data

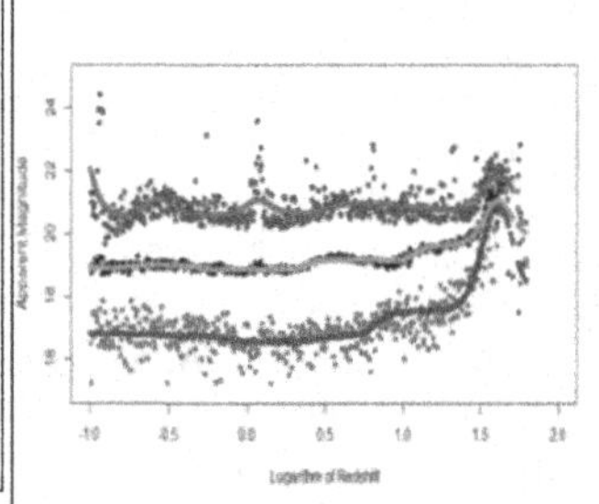

Figure 6. Fitted Upper and Lower Prediction Lines and Modified Data after Binning

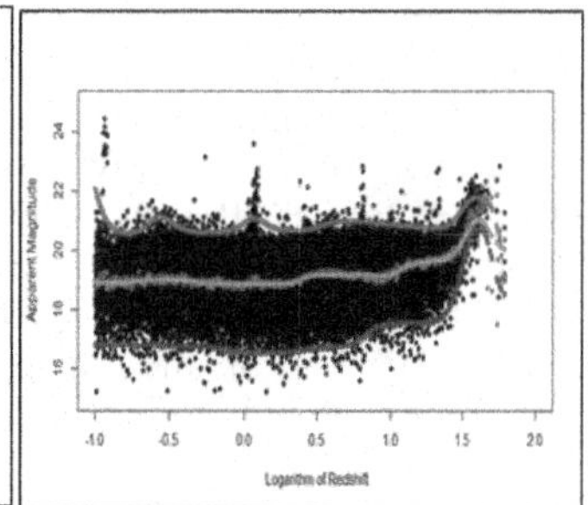

Figure 7. Prediction Curves Prediction Lines and Modified With Data as a Whole

4. Results

It is evident from the fitted values that for $lnz < 1$ the relationship between lnz and m is almost linear (Refer Fig 7). Thus, In order to validate Hubbles Law, we have tested for linearity when $lnz < 1, i.e., z \in [0, 2.718281]$. On constructing the test for linearity for the Upper, Mean and Lower Data Clouds we get the following results:

For Upper Boundary Data Cloud				For Lower Boundary Data Cloud				For Mean Constructed Data Cloud			
Coefficients	Est.	Std. Error	p-value	Coefficients	Est.	Std. Error	p-value	Coefficients	Est.	Std. Error	p-value
Intercept	20.840	0.0006	<2E-16	Intercept	16.700	0.0007	<2E-16	Intercept	19.010	0.0004	<2E-16
β	0.064	0.0012	<2E-16	β	0.119	0.0014	<2E-16	β	0.198	0.0008	<2E-16
Adjusted R^2		0.032		Adjusted R^2		0.087		Adjusted R^2		0.467	

5. Possible Implication

It is found from our analysis of the Tenth Data Release of Sloan Digital Sky Survey on Quasars that the Hubble relation between m and log z is valid for small values of redshift (z) , i.e., for the range: $z \in [0, 2.718281]$ For higher values of z, Hubble Law breaks down because we get a non-linear curve. Now the deviation from Hubble relation may be due to some other mechanism for redshift. It may be pointed out that the environmental effects for the Quasars may be taken into consideration to explain the deviation. This kind of environmental effect has been modeled in Doopler like mechanism. The deviation may also be a hint at the presence of dark energy or vacuum energy.

6. Acknowledgment

We would like to acknowledge and thank Prof. Sisir Roy of Indian Statistical Institute for his precious guidance and support.

References

Abell, G. O.(1958) *Astrophys.* J.Suppl. 3, 211-288

Coles, P. & Lucchin, F., (2002) *Cosmology : The Origin and Evolution of Cosmic Structure.* 2nd Edition, John Wiley & Sons, Ltd. p.77

Datta, S., Roy, S., Roy, M., & Moles, M. (1998), *Phys.Rev.A*, 58,720

Efron, B. & Petrosian, V. (1992) *Astrophys. J.*, 399, 345-352

Hoessel, J. G., Gunn, J. E. & Thuan, T.X. (1980) *Astrophys.*, 241,486-492

Korany, D. M. & Strauss, M. A. (1996) *Testing the Hubble Law with the IRAS 1.2 Jy redshift Survey.* http://xxx.lanl.gov/astro-ph/9610034

McLaren, C., Wagstaff, M., Brittegram, G., & Jacobs, A. (1991) *Biometrics* 47, 607-708

Roy, S., Datta, S., Roy, M., & Kafatos, M. (2006) *Statistical Analysis of Quasar Data and Hubble Law*, arXiv:astro-ph/0605356v1

Roy, S., Kafatos, M., & Datta, S. (1999), *Phys. Rev. A*, 60, 273

Segal, I. E. & Nicoll, J. F. (1992) *Proc. Nat. Acad. Sci.* (USA) 89,11669-11672

Veron-Cetty, M. P. & Veron, P. (2001) *A&A*, 374, 92

Wolf, E. & James, D. F. V. (1996), *Rep. Prog. Phys.*, 59, 771

Statistical Challenges in 21st Century Cosmology
Proceedings IAU Symposium No. 306, 2014
A. F. Heavens, J.-L. Starck & A. Krone-Martins, eds.

doi:10.1017/S1743921314010771

Semi-analytical study on the generic degeneracy for galaxy clustering measurements

Alejandro Guarnizo[1], Luca Amendola[1], Martin Kunz[2] and Adrian Vollmer[1]

[1]Institut Für Theoretische Physik, Ruprecht-Karls-Universität Heidelberg, Philosophenweg 16, 69120 Heidelberg, Germany

[2]Département de Physique Théorique and Center for Astroparticle Physics, Université de Genève, Quai E. Ansermet 24, CH-1211 Genève 4, Switzerland

Abstract. From the galaxy power spectrum in redshift space, we derive semi-analytical results on the generic degeneracy of galaxy clustering measurements. Defining the observables $\bar{A} = Gb\sigma_8$ and $\bar{R} = Gf\sigma_8$, (being G the growth function, b the bias, f the growth rate, and σ_8 the amplitude of the power spectrum), we perform a Fisher matrix formalism to forecast the expected precision of these quantities for a Euclid-like survey. Among the results we found that galaxy surveys have generically a slightly negative correlation between $\bar{A}$ and $\bar{R}$, and they can always measure $\bar{R}$ about 3.7 to 4.7 times better than $\bar{A}$.

Keywords. cosmology: observations, cosmology: theory, methods: statistical, surveys

1. Introduction

Future galaxy surveys will provide new opportunities to verify the current standard cosmological model, and also to constrain modified gravity theories, invoked to explain the present accelerated expansion of the universe. Before studying general parametrizations of dark energy, its however important to understand first which quantities can be really observed From this direction recently Amendola *et al.* (2013) shown that cosmological measurements can determine, in addition to the expansion rate $H(z)$, only three additional variables R, A and L, given by

$$A = Gb\delta_{\mathrm{m},0}\,, \qquad R = Gf\delta_{\mathrm{m},0}\,, \qquad L = \Omega_{\mathrm{m}0}GY(1+\eta)\delta_{\mathrm{m},0}\,. \tag{1.1}$$

with G is the growth function, b is the galaxy bias with respect to the dark matter density contrast, and $\delta_{\mathrm{m},0}$ is the dark matter density contrast today. The functions η (the anisotropic stress $\eta = -\frac{\Phi}{\Psi}$) and Y (The clustering of dark energy $Y \equiv -\frac{2k^2\Psi}{3\Omega_{\mathrm{m}}\delta_{\mathrm{m}}}$), describe the impact of the dark energy on cosmological perturbations. In Amendola *et al.* (2014), a Fisher analysis was made using galaxy clustering, weak lensing and supernovae probes, in order to find the expected accuracy with which an Euclid like survey can measure the anisotropic stress η, in a model-indepndent way.

In this work we want to obtain some reults on the intrinsic degeneracy on galaxy clustering measurements, using the quantities A and R. We use a flat ΛCDM fiducial model, with $\Omega_{\mathrm{m},0}h^2 = 0.134$, $\Omega_{b,0}h^2 = 0.022$, $n_s = 0.96$, $\tau = 0.085$, $h = 0.694$, $\Omega_k = 0$, Euclid like survey specifications are used Amendola *et al.* (2013): we divided the redshift range $[0.5, 2.0]$ in 5 bins of width $\Delta z = 0.2$ and one of width $\Delta z = 0.4$; an spectroscopic error $\delta z = 0.001(1+z)$, a fraction of sky $f_{\mathrm{sky}} = 0.375$ and area of 15 000 deg^2. Later on, the bias b is assumed to be unity.

2. Fisher matrix for Galaxy Clustering

Observations of the growth rate f from large scale structures using Redshift Space Distortions (RSD), give a direct way to test different dark energy models, Song & Percival (2009), Percival & White (2009), Racanelli *et al.* (2013). Lets consider now the galaxy power spectrum in redshift space

$$P(k,\mu) = (A + R\mu^2)^2 = (\bar{A} + \bar{R}\mu^2)^2 \delta_{\mathrm{t},0}^2(k), \tag{2.1}$$

whit $\bar{A} = Gb\sigma_8, \bar{R} = Gf\sigma_8$, and we explicitly use $\delta_{\mathrm{m},0} = \sigma_8\delta_{\mathrm{t},0}$. The Fisher matrix is in general

$$F_{\alpha\beta} = \frac{1}{8\pi^2}\int_{-1}^{1} d\mu \int_{k_{\min}}^{k_{\max}} k^2 V_{\mathrm{eff}} D_\alpha D_\beta \,\mathrm{d}k, \tag{2.2}$$

where $D_\alpha \equiv \frac{d\log P}{dp_\alpha}$, and V_{eff} is the effective volume of the survey

$$V_{\mathrm{eff}} = \left(\frac{\bar{n}P(k,\mu)}{\bar{n}P(k,\mu)+1}\right)^2 V_{\mathrm{survey}}, \tag{2.3}$$

being $\bar{n}$ the galaxy number density in each bin. We want to study the dependence on the angular integration in the Fisher matrix for the set of parameters $p_\alpha = \{\bar{A}(z_\alpha), \bar{R}(z_\alpha)\}$. The derivatives of the power spectrum are

$$D_\alpha = \frac{2}{\bar{A}+\bar{R}\mu^2}\{1,\mu^2\}. \tag{2.4}$$

we consider two cases depending on the behavior of V_{eff}, equation (2.3):

(*a*) "Enough data" $\bar{n}P(k,\mu) \gg 1$, then we have $V_{\mathrm{eff}} \approx V_{\mathrm{survey}}$ and the Fisher matrix could be written as

$$F_{\alpha\beta} \approx \frac{1}{2\pi^2}\int_{k_{\min}}^{k_{\max}} k^2 V_{\mathrm{survey}} M_{\alpha\beta}\,\mathrm{d}k, \tag{2.5}$$

where

$$M_{\alpha\beta} = 4\begin{pmatrix} \dfrac{\sqrt{\bar{S}} + (\bar{A}+\bar{R})\tan^{-1}\sqrt{P_1}}{\bar{A}^{3/2}\sqrt{\bar{R}}(\bar{A}+\bar{R})} & \dfrac{-\sqrt{\bar{S}} + (\bar{A}+\bar{R})\tan^{-1}\sqrt{P_1}}{\bar{R}^{3/2}\sqrt{\bar{A}}(\bar{A}+\bar{R})} \\[2ex] \dfrac{-\sqrt{\bar{S}} + (\bar{A}+\bar{R})\tan^{-1}\sqrt{P_1}}{\bar{R}^{3/2}\sqrt{\bar{A}}(\bar{A}+\bar{R})} & \dfrac{\bar{R}(3\bar{A}+2\bar{R}) - 3(\bar{A}+\bar{R})\sqrt{\bar{S}}\tan^{-1}\sqrt{P_1}}{\bar{R}^3(\bar{A}+\bar{R})} \end{pmatrix} \tag{2.6}$$

being $\bar{S} = \bar{A}\bar{R}$.

(*b*) Shot-noise dominated $\bar{n}P(k,\mu) \ll 1$, and then $V_{\mathrm{eff}} \approx (\bar{n}P(k,\mu))^2 V_{\mathrm{survey}}$ and since we are interesting only in the μ dependence we can write $V_{\mathrm{eff}} \approx P(k,\mu)^2$. Then the Fisher matrix becomes

$$F_{\alpha\beta} \approx \frac{1}{2\pi^2}\int_{k_{\min}}^{k_{\max}} k^2 \delta_{\mathrm{t},0}^4(k) N_{\alpha\beta}\,\mathrm{d}k, \tag{2.7}$$

with

$$N_{\alpha\beta} = 8\begin{pmatrix} \bar{A}^2 + \dfrac{2\bar{A}\bar{R}}{3} + \dfrac{\bar{R}^2}{5} & \dfrac{\bar{A}^2}{3} + \dfrac{2\bar{A}\bar{R}}{5} + \dfrac{\bar{R}^2}{7} \\[2ex] \dfrac{\bar{A}^2}{3} + \dfrac{2\bar{A}\bar{R}}{5} + \dfrac{\bar{R}^2}{7} & \dfrac{\bar{A}^2}{5} + \dfrac{2\bar{A}\bar{R}}{7} + \dfrac{\bar{R}^2}{9} \end{pmatrix}. \tag{2.8}$$

We notice that in the two limiting cases above, we can move the matrices $M_{\alpha\beta}$ and $N_{\alpha\beta}$ outside of the integral, as for the fiducial model $\bar{A}$ and $\bar{R}$ do not depend on k. This

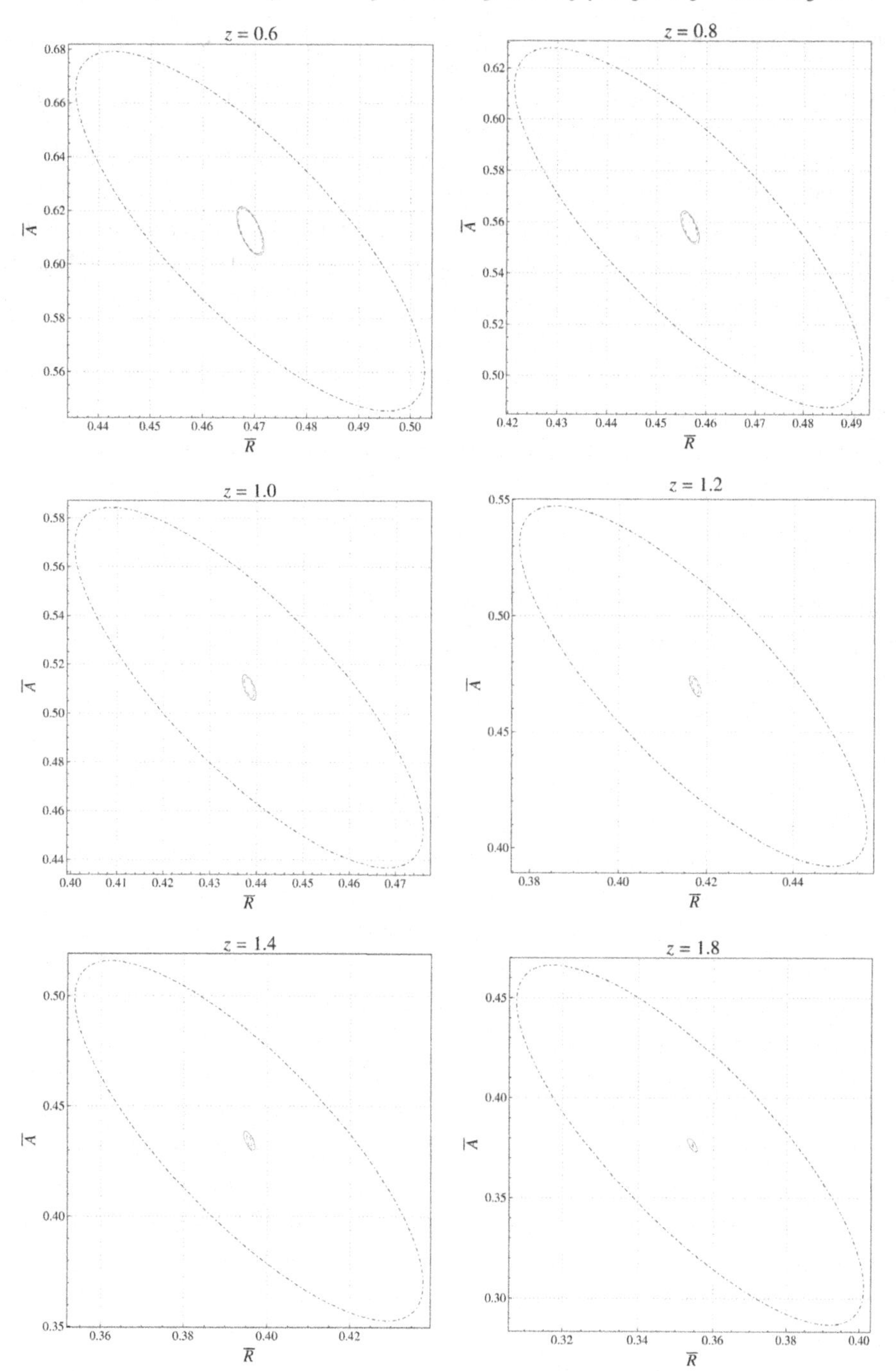

Figure 1. Confidence contours for $\bar{A}$ and $\bar{R}$ in the three cases: Solid line $V_{\rm eff}$, dot dashed line $V_{\rm eff} \approx V_{\rm survey}$, and dashed line $V_{\rm eff} \approx P(k,\mu)^2$..

means that, although the absolute size of the error ellipse depends on the integral, the relative size and orientation do not. In other words, we can obtain 'generic expectations' for the shape of the degeneracy between $\bar{A}$ and $\bar{R}$ from galaxy clustering surveys. These results are quite representative for the full range of $\bar{A}$ and $\bar{R}$, i.e. galaxy surveys have generically a slightly negative correlation between $\bar{A}$ and $\bar{R}$, and they can always measure $\bar{R}$ about 3.7 to 4.7 times better than $\bar{A}$, see Figure 1. In comparisson to the results of Song & Kayo (2010), we remove the dependence on $\delta_{t,0}$, eq. (1.1), which is a quantity that depends on inflation or other primordial effects.

Acknowledgements

L.A., A.G. and A.V. acknowledge support from DFG through the project TRR33 "The Dark Universe", A.G. also acknowledges support from DAAD through program "Forschungsstipendium für Doktoranden und Nachwuchswissenschaftler". M.K. acknowledges financial support from the Swiss NSF.

References

Amendola, L., Kunz, M., Motta, M., Saltas, I. D., & Sawicki, I. 2013, *Phys. Rev. D*, 87, 023501
Amendola, L., Fogli, S., Guarnizo, A., Kunz, M., & Vollmer, A. 2014, *Phys. Rev. D*, 89, 063538
Amendola, L., *et al.* (Euclid Theory Working Group) 2013, *Living Rev. Relativity*, 16 , 6
Percival, W. J. & White, M. 2009, *MNRAS*, 393, 297-308
Raccanelli, A., Bertacca, D., Pietrobon, D., Schmidt, F., Samushia, L., Bartolo, N., Dor, O., Matarrese, S., & Percival, W. J. 2013, *MNRAS*, 436, 89-100
Song Y.-S. & Kayo I. 2009, *MNRAS*, 407, 1123-1127
Song Y.-S. & Percival W. J. 2009, *J. Cosmol. Astropart. Phys.*, 0910, 004

Statistical Challenges in 21st Century Cosmology
Proceedings IAU Symposium No. 306, 2014
A. F. Heavens, J.-L. Starck & A. Krone-Martins, eds.

doi:10.1017/S174392131401076X

Distribution of Maximal Luminosity of Galaxies in the Sloan Digital Sky Survey

E. Regős, A. Szalay, Z. Rácz, M. Taghizadeh and K. Ozogany
[1]CERN (Geneva), [2]Johns Hopkins U. (Baltimore), [3]ELTE (Budapest)

Abstract. Extreme value statistics (EVS) is applied to the pixelized distribution of galaxy luminosities in the Sloan Digital Sky Survey (SDSS). We analyze the DR8 Main Galaxy Sample (MGS) as well as the Luminous Red Galaxy Sample (LRGS). A non-parametric comparison of the EVS of the luminosities with the Fisher-Tippett-Gumbel distribution (limit distribution for independent variables distributed by the Press-Schechter law) indicates a good agreement provided uncertainties arising both from the finite size of the samples and from the sample size distribution are accounted for. This effectively rules out the possibility of having a finite maximum cutoff luminosity.

Keywords. Galaxies: formation, Galaxies: luminosity function, Cosmology: large-scale structure

1. Introduction

Extreme value statistics analyzes the behavior of the tails of distributions. The distribution of extreme values for i.i.d. (independent, identically distributed) variables converge to a few limiting distributions depending on the tail behavior of the parent population (Fisher-Tippet-Gumbell, Weibull, Fisher-Tippet-Frechet). The onset of the scaling behavior is quite slow, therefore requires very large samples. The galaxy samples in the SDSS redshift survey may be just large enough to attempt such an analysis, and we present here a study of the distribution of maximal luminosities.

We minimize the correlations between luminosities and positions by selecting the maximal luminosities from the batches of galaxies in elongated conical regions (defined by the footprints of HEALPix cells on the sky). We also explore random sampling.

The shape of the galaxy luminosity function is important for the EVS. This function is well described by the Schechter function, functionally similar (and motivated by) the theoretically derived Press-Schechter formula, with a power law distribution and an exponentially falling tail. Such a tail would imply a Fisher-Tippett-Gumbel (FTG) EVS distribution, with corrections for the finite sample sizes depending on the power law at low luminosities. In this analysis we will show that there is an excellent agreement with these expectations, implying that the Schechter function extends to very high luminosities, i.e. there is no indication for a sharp cutoff at a high but finite luminosity.

In order to arrive at the above results, it was important to notice that the galaxies can be divided into two types with significantly distinct luminosities and spatial distributions.

Previous studies have examined whether Brightest Cluster Galaxies (BCG) are the extremes of a red early-type galaxy luminosity function or they are different. Tremaine and Richstone (1977) investigated the statistics of the luminosity gap between the first and second brightest galaxies in clusters of galaxies and found that it is substantially larger than what can be explained with an exponentially decaying luminosity function. BCGs in high-luminosity clusters are not drawn from the luminosity distribution (LD) of all red cluster galaxies, while BCGs in less luminous clusters are the statistical extreme. For a volume-limited sample of Luminous Red Galaxies (many of them BCGs) we find that they are the extremes of a LD with an exponentially decaying tail.

Even though the SDSS sample is large, the residual from the FTG distribution can be explained only when we consider the corrections due to both the finite size of the samples and the distribution present in the sample sizes (the number of galaxies in a cone is finite and varies from cone to cone). The distributions of the galaxy luminosities and the galaxy counts in pencil beams are constructed in Section 2.

2. Sample Creation

We use data from SDSS-DR8, available in a MS-SQL Server database that can be queried online via CasJobs (`http://casjobs.sdss.org`). The spectroscopic survey renders a complicated geometry defined by `sectors`.

We explore 2 different galaxy samples: the Luminous Red Galaxies (LRGs) (Eisenstein 2001) and the Main Galaxy Sample (MGS) (Strauss 2002). The LRGs (magnitude limit 19.2) are slowly evolving and intrinsically luminous red galaxies composed of old stellar populations, selected for tracing the structure at a higher redshift [0.20, 0.38] than MGSs. For selecting the MGS we impose a redshift interval of [0.065, 0.22] and magnitude [13.5,17.65] in r-band Petrosian magnitude limit of 17.77 . A total of $N_{\rm T} = 348975$ MGS and $N_{\rm T} = 52579$ LRG galaxies are obtained in a survey's comoving volume of $V_{\rm S} = 0.56\,{\rm Gpc}^3$ and $V_{\rm S} = 2.18\,{\rm Gpc}^3$ respectively.

2.1. *Footprint and HEALPix based pencil beams, Distributions of galaxy counts in a HEALPix cell*

In order to have close-to-i.i.d realizations (batches) from which to draw the maximal luminosities, we tessellate the sphere into regions defined by individual HEALPix cells (Gorski 2005), all of which have the same area. This creates 3-dimensional pencil-like beams that sample the galaxy populations across different redshifts.

We create the entire SDSS-DR8 spectroscopic footprint with resolution $N_{side} = 512$ ($\sqrt{\Omega_{pix}} \simeq 6.87'$). We degrade the footprint to 3 lower resolution maps defined by $N_{side} = 16, 32$ and 64, creating thus the cells that define the pencil beams. We use only the group of cells which satisfy that their fractional area occupancy inside the footprint $f \geqslant 0.97$.

The other approach we use is the random sampling of the luminosity parent distribution in batches of fixed size.

We construct the luminosity functions of the different galaxy samples and fit the MGS sample with Schechter, generalized gamma distribution (see also Figures 2 and 4).

The LRG sample was made to include the brightest early types. Therefore, they are naturally fitted - their luminosity function itself - with an extreme value distribution or its special case the Gumbel.

3. Theory of Extreme Value Statistics

Extreme value statistics (EVS) is concerned with the probability of the largest value in a batch of N measurements. For us, they are galaxy luminosities in a given solid angle of the sky and N is the number of galaxies in the given angle.

The results of the EVS are simple for i.i.d. variables. The limit distribution belongs to one of three types and the determining factor is the large-argument tail of the parent distribution. Frechet type distribution emerges if the parent distribution f decays as a power law, Fisher-Tippett-Gumbel (FTG) distribution is generated by fs which decay faster than any power law and parent distributions with finite cutoff and power law behavior around the cutoff yield the Weibull distribution. All the above cases can be unified as a generalized EVS whose integrated distribution $F_N(v)$ is given in the $N \to \infty$ limit by $F(v) = \exp\left[-(1+\xi v)^{-1/\xi}\right]$ where $\xi > 0, = 0, < 0$ correspond to the Frechet, FTG, Weibull classes, respectively, with the parameter ξ being the exponent of the power law behavior.

The parent distribution for galaxy luminosities is known, it is the Gamma-Schechter distribution. For this the limit distribution of extremal luminosities belongs to the FTG class ($\xi \to 0$) $P(v) = \frac{dF(v)}{dv} = \frac{1}{b}\exp\left[-\frac{v-a}{b} - \exp(-\frac{v-a}{b})\right]$ where the parameters can be fixed by setting $\langle v\rangle = 0$ and $\sigma = \sqrt{\langle v^2\rangle - \langle v\rangle^2} = 1$. This choice ($a = \pi/\sqrt{6}$ and $b = \gamma_E \approx 0.577$) leads to a parameter-free comparison with the experiments provided the histogram of the maximal luminosities $P(v)$ is plotted in terms of the variable $x = (v - \langle v\rangle_N)/\sigma_N$ where $\langle v\rangle_N$ is the average of

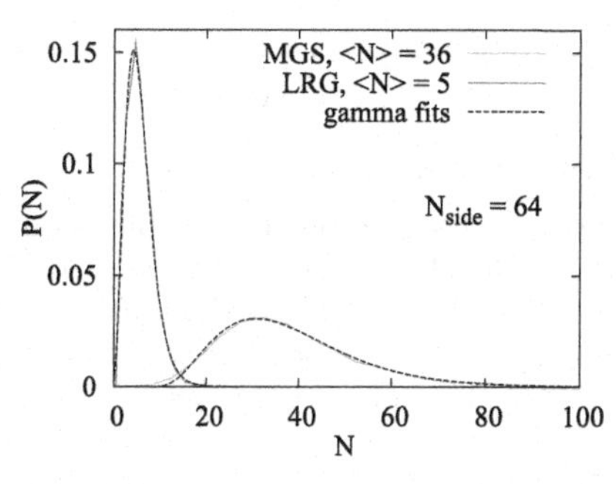

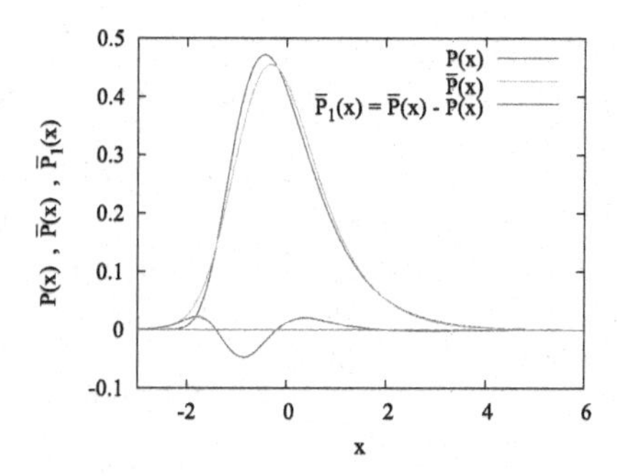

Figure 1. (a) Distribution of the number of galaxies N in the pencil beams for $N_{side} = 64$, for the MGS and the LRG samples. Fits to the empirical data are also shown. (b) Comparison of FTG with the count-averaged EVS, and their difference.

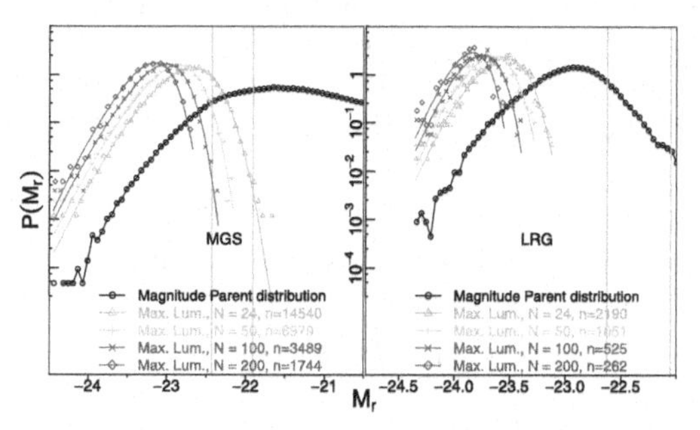

Figure 2. Distribution of r-band absolute magnitude M_r and distribution of maximal luminosity for the full samples according to the fixed batch size random sampling.

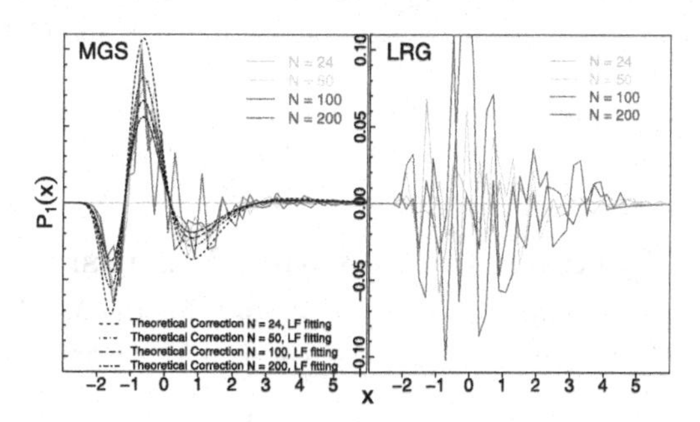

Figure 3. Empirical finite-size corrections from batches of fixed size N for random sampling.

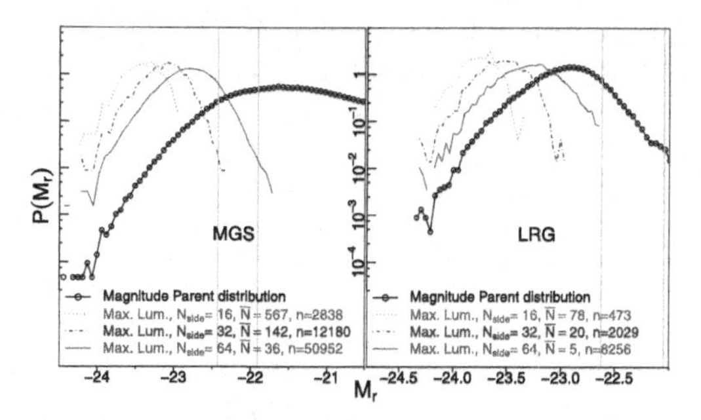

Figure 4. Parent distribution of r-band absolute magnitude M_r and distribution of maximal luminosity for the full samples according to various HEALPix resolutions.

the maximal luminosity while $\sigma_N = \sqrt{\langle v^2\rangle_N - \langle v\rangle_N^2}$ is its standard deviation. The resulting scaling function is the universal function in the limit $N \to \infty$: $\Phi_N(x) = \sigma_N P_N(\sigma_N x + \langle v\rangle_N) \to \Phi(x)$.

3.1. *Deviations from the ideal case*

In addition to the assumption of i.i.d. variables, there are two additional problems with comparing data with theory. A notorious aspect of EVS is the slow the convergence to the limit distribution. Second, the batch size N (the number of galaxies in a given solid angle) varies with the direction of the angle. Thus the histogram of the maximal luminosities $P_N(v)$ is built from a distribution of Ns. Both effects introduce corrections to the limit distribution.

3.2. *Finite size corrections*

Finite size corrections to first order can be written as $\Phi_N(x) \approx \Phi(x) + q(N)\Phi_1(x)$, where $q(N \to \infty) \to 0$ and the shape correction $\Phi_1(x)$ is universal. Both the amplitude q and the shape correction Φ_1 are known for Press-Schechter type parent distributions. The convergence is slow since $q(N) = -\theta/\ln^2 N$. The value of θ is roughly 1 thus for characteristic range of $N \approx 10 - 200$ one can expect a 20-4% deviation coming from finite-size effects.

The finite-size shape correction is $\Phi_1(x) = [M_1(x)]'$ where $M_1(x) = \Phi(x)\left[\frac{ax^2}{2} - \frac{\zeta(3)x}{a^2} - \frac{a}{2}\right]$.

3.3. *Variable batch size*

If the normalized distribution $F(N)$ of N is known, the count-averaged EVS is

$$\overline{P}(v) = \lim_{\overline{N}\to\infty} \int_0^\infty \mathcal{F}(N) P_N(v) dN \tag{3.1}$$

and the difference $\overline{P}_1(x) \equiv P(x) - \overline{P}(x)$ provides us an estimate of corrections coming from the variable sample size.

We carried out simulations for the following: $F(N)$ of the sample is fitted to $(aN+b)^3 \exp(-(aN+b))$, and $F(N)$ is just the exact empirical distribution for the MGS and LRG samples (see Figures 1 and 3).

Combining the finite-size effects with the effects coming from the variable batch-size can produce the features observed in the deviations from the FTG limit.

4. Distribution of Maximal Luminosities, Discussion

Figure 2 shows the distribution of maximal luminosity for the MGS and LRG samples according to the fixed batch size random sampling. Figure 4 is the same but according to cones of various HEALPix resolutions.

The distributions obtained are FTG distributions. This effectively rules out the possibility of having a finite maximum cutoff luminosity.

We can claim that at large distances r the luminosity-luminosity correlations should decay with an exponent $C(r) \sim r^{-\sigma}$ with $\sigma > 2$. The FTG-s found suggest weak correlations, and the noise and correlated features of the noise can be explained by the combination of finite-size and variable batch-size effects.

We find that the LRGs – many of them BCGs – are the extremes of a luminosity distribution with exponential decay.

References

Taghizadeh-Popp, M. *et al.* 2012, *ApJ*, 759, 100
Fisher, R. A. & Tippet, L. H. C. 1928. *Proc. Camb. Phil. Soc.* 24,180
Gnedenko, B. V. 1943. *Ann. Math.* 44, 423-453
Górski, K. M. *et al.* 2005, *ApJ*, 622:759 , 109, 301

Statistical Challenges in 21st Century Cosmology
Proceedings IAU Symposium No. 306, 2014
A. F. Heavens, J.-L. Starck & A. Krone-Martins, eds.

doi:10.1017/S1743921314010758

Reconstructing light curves from HXMT imaging observations

Zhuo-Xi Huo[1], Juan Zhang[2], Yi-Ming Li[1] and Jian-Feng Zhou[1]

[1]Tsinghua University, Beijing, 100084, China
email: huozx@tsinghua.edu.cn

[2]Institute of High Energy Physics, Chinese Academy of Sciences,
Beijing, 100049, China
email: zhangjuan@ihep.ac.cn

Abstract. The Hard X-ray Modulation Telescope (HXMT) is a Chinese space telescope mission. It is scheduled for launch in 2015. The telescope will perform an all-sky survey in hard X-ray band (1 - 250 keV), a series of deep imaging observations of small sky regions as well as pointed observations. In this work we present a conceptual method to reconstruct light curves from HXMT imaging observation directly, in order to monitor time-varying objects such as GRB, AXP and SGR in hard X-ray band with HXMT imaging observations.

Keywords. method: data analysis, techniques: image processing, X-rays: time-varying objects.

1. Introduction

1.1. *The HXMT mission*

Hard X-ray Modulation Telescope (HXMT) is a planned Chinese space telescope mission. This is a telescope dedicated to X-ray astronomy in 1 - 250 keV. There are three individual telescopes, i.e., the high-energy telescope (HE), the medium-energy telescope (ME) and the low-energy telescope (LE) on board of HXMT.

HE detects hard X-ray photons from 20 keV to 250 keV with 18 scintillator detectors. The field of view (FoV) of each detector is constrained by an individual collimator. Thus we have $1.1^\circ \times 5.7^\circ$, $5.7^\circ \times 5.7^\circ$ (in FWHM) and covered FoVs separately. HE detector can resolve photon-arrival events between 25 μs. The energy resolution of HE is 19% at 60 keV. Its effective area is 5 400 cm^2.

ME covers the energy range from 5 keV to 30 keV with 54 collimated Si-PIN detectors. It also has two different FoVs, $1^\circ \times 4^\circ$ and $4^\circ \times 4^\circ$ as well. 24 collimated CCD detectors are included in LE, which covers the soft X-rays from 1 keV to 15 keV. LE has two different FoVs too. One is $1.6^\circ \times 6^\circ$ while the other is $4^\circ \times 6^\circ$ (in FWHM). The effective areas of ME as well as LE are 952 cm^2 and 384 cm^2 respectively.

There are also a set of anti-coincidence detectors (ACDs) installed around HE detectors (Wu *et al.* 2004) and a space environment monitor (Shen *et al.* 2008) on board of the satellite to help screen background events and to measure the spectra and time-varying properties of ambient protons and electrons alongside the covered HE detector.

To achieve the scientific objectives of this mission the telescope will operate in three observing modes by its current design. First, it will perform an all-sky imaging survey in 1 - 250 keV. An all sky map in this energy range will be reconstructed from the observed data, and a catalog of detected objects will be compiled. At least 25% available observing time of this telescope will be allocated to the all-sky imaging survey. Second, it will carry out a series of deep imaging observations of small sky regions. Maps of these regions

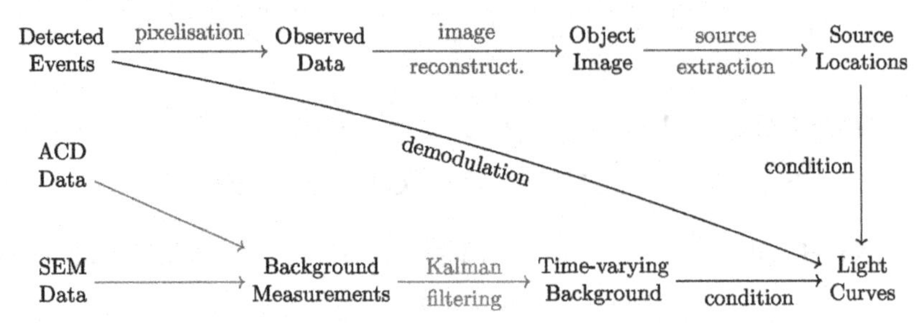

Figure 1. Data analysis pipeline

will be reconstructed from the observed data. Finally a considerable percentage of its observing time will be consigned to pointed observations of various targets to obtain their energy spectra as well as light curves (Lu 2012).

1.2. *Time-varying objects in hard X-rays*

There are various interesting time-varying objects visible in hard X-rays. Gamma-ray bursts (GRB) and their X-ray afterglows both contribute to the their emissions in hard X-rays (O'Brien *et al.* 2006). Because of instability of the accretion disc most compact X-ray sources are variable (Matsuoka *et al.* 2009). Short-term variability of soft Gamma-ray repeaters (SGRs) and anomalous X-ray pulsars (AXPs) have been detected in hard X-ray. (Mereghetti 2008).

Given its effective area and its FoVs, the geometric factor of HXMT/HE is competitive in hard X-ray telescopes, which makes it suitable for monitoring all-sky objects. It has temporal-resolution as good as 25 μs so it is adequate to monitor time variability of hard X-ray objects. However, by its current design HXMT do not distribute burst alert in real time or perform follow-up observations triggered by other telescopes. In this work we have present a conceptual pipeline for reconstructing light curves of time-varying objects solely from HXMT imaging observations. In this way at least 25% of its total observing time could contribute to monitoring time-varying objects.

2. Data analysis pipeline

2.1. *Temporal-spatial modulation in HXMT imaging observation*

HXMT imaging observation through its collimated detectors is described by the following temporal-spatial modulation equation

$$d(\boldsymbol{\omega}) = \int_t^{t+\Delta t} \int f(t,\boldsymbol{x}) p(\boldsymbol{\omega}(t),\boldsymbol{x}) \mathrm{d}t\boldsymbol{x} \approx \Delta t \int f(t,\boldsymbol{x}) p(\boldsymbol{\omega},\boldsymbol{x}) \mathrm{d}\boldsymbol{x}, \tag{2.1}$$

where $\boldsymbol{\omega}(t)$ is the status vector of the telescope as a function of time, $\boldsymbol{x}$ is coordinate vector in celestial coordinate system, $d(\boldsymbol{\omega})$ is the observed data as a function of $\boldsymbol{\omega}$, $f(t,\boldsymbol{x})$ is the temporal-spatial distribution of the object intensity, and $p(\boldsymbol{\omega},\boldsymbol{x})$ is the response of the telescope to a unit source at $\boldsymbol{x}$ while the status of the telescope being $\boldsymbol{\omega}$.

2.2. *High-temporal-resolution light curve estimation*

To solve Eq. 2.1 we present in this work a three-pass pipeline. The first pass goes through blue paths, while the second and third passes go through red and black paths in Fig. 1.

First, bright sources are located through image reconstruction and source extraction. The approximation in Eq. 2.1 is valid for time-varying object only in short time scales. Hence in each image reconstruction we only use data observed along the same scanning path. Therefore the reconstruction problem is of obvious illness, where two-dimensional image is to be reconstructed from one-dimensional data. We take advantage of the parallel observations with different collimators and positions to add on the dimensionality of the observed data. However it imposes a shift-variant deconvolution difficulty on the image reconstruction procedure. The difficulty can be tackled by a clustering-based approximation (Huo & Zhou 2013).

Second, the time-varying background is estimated by Kalman filtering technique. The detected events in average originate from background emissions, including diffusive X-rays, cosmic rays, protons trapped in South Atlantic Anomaly (SAA), albedo photons, etc. The diffusive X-rays consist of CXB and Galactic diffuse X-rays. Cosmic rays have complicated interactions with instruments, which result in a significant background. SAA-trapped protons are the dominant factor of HE background. In addition, this factor will further excite part of the instrument, and induce a large amount of background after leaving SAA region. Background induced by albedo photons depends on the elevation and altitude of the satellite. Therefore the total background can be summarized as:

$$B_{\text{total}} = B_{\text{CR}} + B_{\text{A}}(\boldsymbol{\omega}(t)) + B_{\text{SAA}}(\boldsymbol{\omega}(t)) + B_{\text{G}} + B_{\text{C}} + B_{\text{I}} \tag{2.2}$$

where B_{CR}, B_{G}, B_{C} and B_{I} denote backgrounds from CR, Galactic diffusive X-ray, CXB and the instrument, which are considered constant by the first-order approximation, while $B_{\text{A}}(\boldsymbol{\omega}(t))$ and $B_{\text{SAA}}(\boldsymbol{\omega}(t))$ represent backgrounds from albedo photons and SAA-trapped protons, which are time-varying.

Finally, light curves of the objects are demodulated in time domain from the observed data directly with their locations fixed and the time-varying background estimated previously.

3. Summary

Light curves of the time scales in 10 - 100 s or in several days can be reconstructed from HXMT imaging observations. We can monitor X-ray time-varying objects bright enough with the conceptual method suggested in this work, which requires no extra observations. HXMT is planned to launch in 2015. Comprehensive feasibility study as well as reliable implementation are in urgent need amongst all ground-based software researches and developments.

References

Wu, YP, Ren DH & You Z 2004, *Adv. Sp. Res.* 34, 12

O'Brien, P., Willingale, R., Osborne, J., Goad, M. R., Page, K. L., *et al.* 2006, *ApJ* 647, 2

Mereghetti, Sandro 2008, *A&AR* 15, 4

Shen, Guo-Hong, Wang, Shi-Jin, Zhu, Guang-Wu, Zhang, Sheng-Yi & Wang, Yue 2008, in: Jian-Cheng Fang, Zhong-Yu Wang (eds.), *Seventh International Symposium on Instrumentation and Control Technology: Optoelectronic Technology and Instruments, Control Theory and Automation, and Space Exploration*, Proc. of SPIE Vol. 7129, p. 71292O

Li, Gang, *et al.* 2008, *Chinese Journal of Space Science*, 28, 6

Matsuoka, M., Kawasaki, K., Ueno, S. *et al.* 2009, *PASJ* 61, 5

Jin, Jing, *et al.* 2010, *Chinese Physics C*, 34, 1

Lu, Fang-Jun 2012, in: Tadayuki Takahashi, Stephen S. Murray, Jan-Willem A. den Herder (eds.), *Space Telescope and Instrumentation 2012: Ultraviolet to Gamma Ray*, Proc. of SPIE Vol. 8443, p. 84431

Huo, Zhuo-Xi & Zhou, Jian-Feng 2013, *Research in Astronomy and Astrophysics* 13, 8

Statistical Challenges in 21st Century Cosmology
Proceedings IAU Symposium No. 306, 2014
A. F. Heavens, J. -L. Starck & A. Krone-Martins, eds.

doi:10.1017/S1743921314010746

Statistical Challenges in the Photometric Calibration for 21st Century Cosmology: The J-PAS case.

Jesús Varela[1], David Cristóbal-Hornillos[1], Javier Cenarro[1], Alessadro Ederoclite[1], David Muniesa[1], Héctor Vázquez Ramió[1], Nicolas Gruel[2] & Mariano Moles[1]

[1]Centro de Estudios de Física del Cosmos de Aragón, Plaza San Juan 1, Planta 2
E-44001, Teruel, Spain N
email: `jvarela@cefca.es`

[2]University of Sheffield, Department of Physics and Astronomy, Hicks Building, Hounsfield Road, Sheffield, S3 7RH, United Kingdom

Abstract. The success of many cosmological surveys in the near future is highly grounded on the quality of their photometry. The Javalambre-PAU Astrophysical Survey (J-PAS) will image more than 8500 deg^2 of the Northern Sky Hemisphere in 54 narrow + 2 medium/broad optical bands plus Sloan u, g and r bands. The main goal of J-PAS is to provide the best constrains on the cosmological parameters before the arrival of projects like Euclid or LSST. To achieve this goal the uncertainty in photo-z cannot be larger than 0.3% for several millions of galaxies and this is highly dependent on the photometric accuracy.

The photometric calibration of J-PAS will imply the intensive use of huge amounts of data and the use of statistical tools is unavoidable. Here, we present some of the key steps in the photometric calibration of J-PAS that will demand a suitable statistical approach.

Keywords. Cosmology: observations, techniques: photometric, methods: data analysis

---

1. Introduction

The Javalambre-PAU (Physics of the Accelerating Universe) Astrophysical Survey (J-PAS, Benítez *et al.* 2014) is an astrophysical survey that will image more than 8500 deg^2 of the Northern Sky Hemisphere in a particular filter system made of 54 narrow ($FWHM \sim 130\AA$) contiguous optical bands (from $\sim 3700\AA$ to $\sim 9150\AA$) and two additional wider bands in the blue and red extremes. The filter system is complemented with a special u filter and the standard Sloan g and r.

The J-PAS filter system has been designed to be optimal in the determination of photometric redshifts with a goal of $\delta_z/(1+z) \sim 0.3\%$ for more than 100 million galaxies, an accuracy required to improve the constrains on the cosmological parameters.

2. The Photometric Calibration Statistical Challenges

The photometric calibration of J-PAS will be done as follows (Varela *et al.* 2013):

• A previous survey (Javalambre-Photometric Local Universe Survey, J-PLUS) of the same area of the sky will be carried out with a large field of view 83cm telescope using a special set of 12 filters designed to better estimate the spectral type of the stars.

• J-PLUS images will be calibrated using the extinction measurements obtained by an extinction monitor, and the observation of Primary Spectrophotometric Standard Stars.

• We will identify the stars with S/N>50 in the 12 bands (Secondary Standard Stars, SSSs).

• The photometry of the SSSs will be fitted using libraries of stellar spectra.

• From the best fitted stellar spectra, we will compute the synthetic J-PAS photometry of the SSSs.

• On the J-PAS images, we will identify the SSSs, and comparing the observed instrumental magnitudes with the synthetic magnitudes we will provide a preliminary photometric calibration of the J-PAS images.

• Taking advantage of the high degree of overlapping between different exposures of the same filters and using the spectra of many spectrophotometric stars in the area of J-PAS, we will perform a spectro-übercalibration, that will help to improve not only the relative calibration for a single band but also among the different bands of the survey.

The large amount of the data that we will manage and the accuracy level that we require for the final photometry implies the constant use of statistical tools.

Following we present some of the "statistical challenges" that we need to face in the quest to achieve the best photometry and the best photometric redshifts required by the 21st Century Cosmology.

2.1. *Challenge 1. Extinction coefficients*

The atmospheric extinction for J-PAS will be measured with a dedicated extinction monitor. This will observed 10 optical bands that will allow to derive an atmospheric extinction curve (see Challenge 2). The measurement of the extinction in each single band will be done following the traditional procedure of observing the same star or group of stars at different airmasses. Given a star s with measurements $\{m_{s,i}\}$ at airmasses x_i, the magnitude outside the atmosphere m_s^o is given by the expression $m_s^o = m_{s,i} + k_n x_i$, where k_n is the extinction coefficient in the given band for the night n.

In the case of a dedicated telescope, it is possible to relate the observations in different nights if there are stars that are observed in common between different nights. In that case, and assuming the stability of the optical system (ie. that the zero point doesn't change), m_s^o can be considered constant and it is possible to perform a fit using data from more than one night obtaining more robust values of k_n and m_s^o.

2.2. *Challenge 2. Atmospheric Extinction Curve*

Since the extinction monitor is observing in bands different from those used in J-PLUS and J-PAS, we need to construct an atmospheric extinction curve from which estimate the atmospheric extinction in any optical band.

This will be done assuming a model of the atmosphere of 3 components (Hayes & Latham 1975): Rayleigh scattering, Ozone absorption and aerosol scattering. The general form of such model would be $\kappa(\lambda) = A_R\lambda^{-\alpha_R} + A_O\kappa_O(\lambda) + A_{ae}\lambda^{-\alpha_{ae}}$, where $\kappa(\lambda)$ is the atmospheric extinction at wavelength λ, $\kappa_O(\lambda)$ is a normalized atmospheric extinction by ozone, α_R and α_{ea} provide the wavelength dependence of the Rayleigh and aerosol scattering, respectively, and, finally, A_R, A_O and A_{ea} give the amplitude of the extinction for each component. While the normalized wavelength dependence of the absorption of each component (ie., $\lambda^{-\alpha_R}$, $\kappa_O(\lambda)$ and $\lambda^{-\alpha_{ae}}$) can be considered constant given a location, their amplitudes depend on the atmospheric conditions and are the values that we are interested on estimating.

The estimation of the amplitudes is done performing a fitting of the observed extinction coefficients obtained in Challenge 1, with the values of the model given by $\kappa(\lambda)$, convolved with the transmission of the system in each band.

2.3. *Challenge 3. PSF Homogenization*

The PSF homogenization needed to obtain consistent photometry between bands and to combine single exposures obtained in different seeing conditions will be performed using a combination of the PSFEx software (Bertin 2011) and software developed by our group.

This procedure will be performed in $\sim 10^7$ single exposures to produce ~ 34000 pseudo-IFU cubes of 56 images.

2.4. *Challenge 4. Stellar Spectral Fitting*

The transportation of the calibration from J-PLUS (12 bands) to J-PAS (56 bands) will be done using high signal-to-noise stars whose SEDs will be fitted with spectra from libraries of stellar spectra (either theoretical or empirical). Then, the best fit stellar models will be convolved with the transmission curves of J-PAS to obtained the calibrated magnitudes of the stars in the J-PAS filter system. With several dozens of stars in each CCD of JPCam†, it will be possible to compute the photometric zero points for each exposure of J-PAS.

2.5. *Challenge 5. Spectro-übercalibration*

The survey strategy devised for J-PAS will ensure a high degree of overlapping between contiguous exposures. Following the ideas of Padmanabhan *et al.* (2008), we will used the information of the overlapping areas to improve the relative photometry in each band. However, we can push this idea further and using the synthetic photometry of many objects with flux calibrated spectra in the J-PAS area, we can tie the photometry between the different bands of J-PAS in a kind of spectro-übercalibration.

The photometric model that we will solved will contain 10^7 zero points (one for each single exposure) and $\sim 10^4$ δZP, that will model the dependence of the ZP with the position in the CCD, the filter and the possible temporal variation of these coefficients, and the amount of data in the system will be of the order of 10^9 measurements of stars.

2.6. *Challenge 6. Photometric Redshift*

With the calibrated photometry in all the J-PAS bands, the last step will be the computation of the phometric redshift. This will be done using the Bayesian approach implemented in BPZ (Benítez 2000), updated in BPZ2.0. It has been shown that the analysis of the residuals of the photo-z procedure in the different bands can help to detect offsets in the original photometry, that applied to all the objects can improve the final photometric redshifts (Molino *et al.* 2014).

References

Benítez, N. 2000, *ApJ*, 536, 571

Benítez, N., Dupke, R., Moles, M., *et al.* 2014, arXiv:1403.5237

Bertin, E. 2011, *Astronomical Data Analysis Software and Systems XX*, 442, 435

Hayes, D. S. & Latham, D. W. 1975, *ApJ*, 197, 593

Molino, A., Benítez, N., Moles, M., *et al.* 2014, *MNRAS*, 441, 2891

Padmanabhan, N., Schlegel, D. J., Finkbeiner, D. P., *et al.* 2008, *ApJ*, 674, 1217

Taylor, K., Marín-Franch, A., Laporte, R., *et al.* 2014, *Journal of Astronomical Instrumentation*, 3, 50010

Varela, J., Gruel, N., Cristóbal-Hornillos, D., *et al.* 2013, *Highlights of Spanish Astrophysics VII*, 957

† Javalambre Panoramic Camera. This is the camera that will be used to carried out J–PAS (Taylor *et al.* 2014).

Statistical Challenges in 21st Century Cosmology
Proceedings IAU Symposium No. 306, 2014
A. F. Heavens, J.-L. Starck & A. Krone-Martins, eds.

doi:10.1017/S1743921314010734

Statistical assessment of the relation between the inferred morphological type and the emission-line activity type of a large sample of galaxies

R. A. Ortega-Minakata[1], J. P. Torres-Papaqui[1], H. Andernach[1] and J. M. Islas-Islas[1,2]

[1]Departamento de Astronomá, Universidad de Guanajuato, 36000, Guanajuato, Mexico
[2]Universidad Tecnológica de Tulancingo, 43642, Hidalgo, Mexico

Abstract. We quantify the statistical evidence of the relation between the inferred morphology and the emission-line activity type of galaxies for a large sample of galaxies. We compare the distribution of the inferred morphologies of galaxies of different dominant activity types, showing that the difference in the median morphological type between the samples of different activity types is significant. We also test the significance of the difference in the mean morphological type between all the activity-type samples using an ANOVA model with a modified Tukey test that takes into account heteroscedasticity and the unequal sample sizes. We show this test in the form of simultaneous confidence intervals for all pairwise comparisons of the mean morphological types of the samples. Using this test, scarcely applied in astronomy, we conclude that there are statistically significant differences in the inferred morphologies of galaxies of different dominant activity types.

Keywords. methods: statistical, galaxies: statistics, galaxies: active

---

1. Introduction and sample selection and classification

There is anecdotical evidence of galaxies with different morphologies having different dominant emission-line activity types (Heckman 1980; Ho, Filippenko & Sargent 1997; Sabater *et al.* 2012).

We inferred the morphological type of 800,685 galaxies of the spectroscopic sample of the SDSS-DR7 (Abazajian *et al.* 2009) using the relations between the broad-band colours u-g, g-r, r-i and i-z in the SDSS system and a concentration index of their brightness profile, $R_{90}(r)/R_{50}(r)$ (where $R_{90}(r)$ and $R_{50}(r)$ are the Petrosian radii containing 90% and 50% of the Petrosian flux in the r band, respectively), of subsamples of these galaxies and their morphology classified by eye (Shimasaku *et al.*, 2001; Fukugita *et al.*, 2007). The numerical scale of our inferred morphology (1–6) is that of Fukugita *et al.*'s (2007) visual classification, so galaxies of earlier types have an inferred morphology closer to 0, while later types have one closer to 6. Fig. 1 (left) shows the distribution of our inferred morphology.

Also, using a standard diagnostic diagram based on emission-line ratios (Baldwin, Phillips, & Terlevich 1981, BPT) we classified our large sample of galaxies according to their dominant activity type (Fig. 1, right). We used the STARLIGHT code (Cid Fernandes *et al.* 2005, `www.starlight.ufsc.br`) to measure the emission-line flux of all the lines seen in the spectra of the entire sample of galaxies. For our analysis, we only considered 216,510 galaxies with redshifts 0.03–0.30, with good quality spectra according to relevant flags in the spectroscopic catalogue of the SDSS-DR7, and with the lines

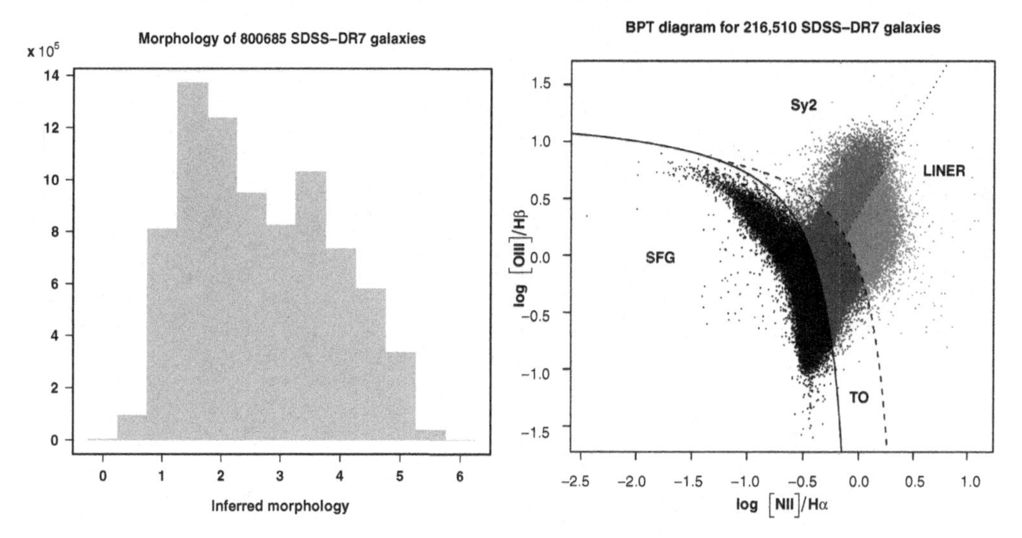

Figure 1. *Left:* Distribution of our inferred morphology for 800,685 galaxies. *Right:* BPT diagnostic diagram for 216,510 galaxies and samples defined from it.

relevant to the diagnostic diagram having $S/N > 3$. The criterion to separate star-forming galaxies (SFGs) from AGN-dominated galaxies is that of Kauffmann *et al.* (2003); to separate transition objects (TOs) from AGN-dominated galaxies we use the criterion proposed by Kewley *et al.* (2001), and to distinguish between Sy2s and LINERs we use the method explained by Torres-Papaqui *et al.* (2012).

2. Analysis and conclusions

Using the R suite of statistical analysis (`www.r-project.org/`), we compared the distribution of our inferred morphology for samples of galaxies with different dominant emission-line activity (Fig. 2, left). The samples are those obtained from the BPT diagram (see previous section) and a sample of 16,733 passive galaxies, i.e. those in which none of the emission lines relevant for the BPT diagram are detected, also in the 0.03–0.30 redshift range. The vertical extent of each box is the inter-quartile range (IQR), while the whiskers length is 1.5 times the IQR from the extremes of the box in each direction. The median of each sample is the thick, black bar, while the mean is the black dot.

The error of the median inferred morphology of each activity type is shown by the width of the notch (a double-cone-shaped region drawn around the median), which is proportional to $\mathrm{IQR}/\sqrt{\mathrm{N}}$, where N is the number of galaxies in each sample of different activity type (shown below each). Comparing any two samples, no overlapping notches imply significantly different medians. Note that a large number of galaxies or a small dispersion within each sample can make the notches small, sometimes barely visible.

The figure shows that there are significant differences in the median morphology for different dominant activity types, going from later to earlier morphologies along a sequence of SFGs, TOs, Sy2s, LINERs and passive galaxies. This suggests a strong relationship between the dominant activity type and the morphology of galaxies, with SFGs being predominantly later types, and LINERs and passive galaxies being mostly earlier types.

Fig. 2 (right) shows a test of the significance of the difference in the mean morphological type between all pairs of activity-type samples. This test uses an ANOVA model with a modified Tukey test. It compares all possible pairs of means simultaneously; the null

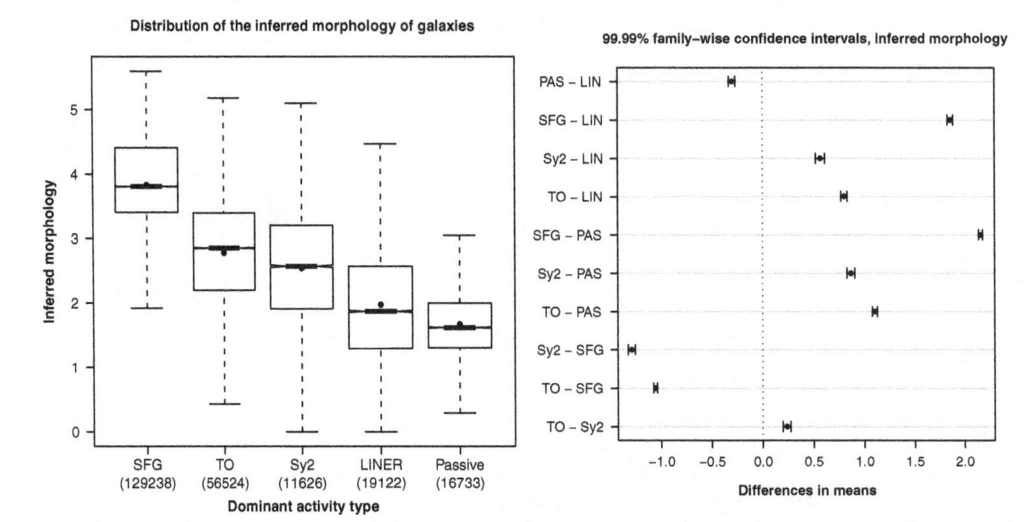

Figure 2. *Left:* Distribution of our inferred morphology for samples of galaxies with different dominant emission-line activity. Sample sizes are given in parentheses. *Right:* Test of the significance of the difference in the mean morphological type between all the activity-type samples.

hypothesis is that all means being compared are equal. Simultaneous confidence intervals for all pairwise comparisons are formed by adding or subtracting the standard error multiplied by the value of a t-distribution at the level of confidence considered (here 99.99%). For this test, the standard error is calculated considering the unequal sample sizes and their heteroscedasticity (different variance for different samples). Details of this test are explained in Herberich, Sikorski & Hothorn (2010) and Hothorn, Bretz & Westfall (2008).

The figure shows these confidence intervals for all the differences of the mean morphological types of the samples defined using the BPT diagram and the sample of passive galaxies. Any confidence intervals including zero would indicate no significant difference for that pair of means. For all our activity-type samples, the differences in the mean inferred morphology are significant at least at the 99.99% confidence level.

J. P. Torres-Papaqui acknowledges DAIP-UG for support grant 432/14. R. A. Ortega-Minakata thanks DAIP-UG for support for travel expenses to attend the Symposium.

References

Abazajian, K. N., Adelman-McCarthy, J. K., Agüeros, M. A., *et al.* 2009, *ApJS*, 182, 543
Baldwin, J. A., Phillips, M. M., & Terlevich, R. 1981, *PASP*, 93, 5
Cid Fernandes, R., Mateus, A., Sodré Jr., L., *et al.* 2005, *MNRAS*, 358, 363
Fukugita, M., Nakamura, O., Okamura, S., *et al.* 2007, *AJ*, 134, 579
Heckman, T. M. 1980, *A&A* 87, 152
Herberich, E., Sikorski, J., & Hothorn, T. 2010, *PLoS ONE*, 5, 1 (Plos ONE 5(3): e9788.doi:10.1371/journal.pone.0009788)
Ho, L. C., Filippenko, A. V., & Sargent, W. L. W. 1997, *ApJ*, 487, 568
Hothorn, T., Bretz, F., & Westfall, P. 2008, *Biom. J.*, 50, 346
Kauffmann, G., Heckman, T. M., Tremonti, C., *et al.* 2003, *MNRAS*, 346, 1055
Kewley, L. J., Dopita, M. A., Sutherland, R. S., *et al.* 2001, *ApJ*, 556, 121
Sabater, J., Verdes-Montenegro, L., Leon, S., *et al.* 2012, *A&A*, 545, A15
Shimasaku, K., Fukugita, M., Doi, M., *et al.* 2001, *AJ*, 122, 1238
Torres-Papaqui, J. P., Coziol, R., Andernach, A., *et al.* 2012, *Rev. Mexicana AyA*, 48, 275

Statistical Challenges in 21st Century Cosmology
Proceedings IAU Symposium No. 306, 2014
A. F. Heavens, J. -L. Starck & A. Krone-Martins, eds.

doi:10.1017/S1743921314013805

Low/High Redshift Classification of Emission Line Galaxies in the HETDEX survey

Viviana Acquaviva[1], Eric Gawiser[2], Andrew S. Leung[2] and Mario R. Martin[1]

[1]Physics Department, New York City College of Technology, Brooklyn, NY 11201
email: vacquaviva@citytech.cuny.edu
[2]Department of Physics and Astronomy, Rutgers, The State University of New Jersey, Piscataway, NJ 08554

Abstract. We discuss different methods to separate high- from low-redshift galaxies based on a combination of spectroscopic and photometric observations. Our baseline scenario is the Hobby-Eberly Telescope Dark Energy eXperiment (HETDEX) survey, which will observe several hundred thousand Lyman Alpha Emitting (LAE) galaxies at 1.9 < z < 3.5, and for which the main source of contamination is [OII]-emitting galaxies at z < 0.5. Additional information useful for the separation comes from empirical knowledge of LAE and [OII] luminosity functions and equivalent width distributions as a function of redshift. We consider three separation techniques: a simple cut in equivalent width, a Bayesian separation method, and machine learning algorithms, including support vector machines. These methods can be easily applied to other surveys and used on simulated data in the framework of survey planning.

Keywords. galaxies: distances and redshifts, methods: statistical, Machine Learning

---

1. The HETDEX survey

The spectroscopic galaxy survey HETDEX (Hobby-Eberly Telescope Dark Energy eXperiment, http://hetdex.org/) is going to investigate the largely unexplored redshift range of $1.9 < z < 3.5$. By accurately mapping the 3D positions of galaxies, HETDEX will pinpoint the expansion rate of the Universe with a precision of 1% and detect dark energy, if it exists, at this early epoch. To reach the exciting objective of measuring the expansion rate of the Universe far back in the past, HETDEX will find and catalog several hundred thousand Lyman Alpha Emitting galaxies (LAEs) in the aforementioned redshift range. LAE galaxies are identified through the Lyman$-\alpha$ emission line, a transition between the $n = 2$ and $n = 1$ energy levels of hydrogen in which a photon of wavelength $\lambda = 1216\mathring{A}$ is emitted. Since hydrogen is by far the most common element in the Universe, this high-luminosity line acts as a lamppost even for galaxies very far away. The main source of contamination in HETDEX is an emission line from singly ionized oxygen, [OII], whose rest-frame wavelength is 3727 $\mathring{A}$. In the detection range of the HET spectrographs (3500-5500$\mathring{A}$), [OII] emitters at redshift $z < 0.56$ might be mistaken for Ly-α emitters, as illustrated in Fig. 1. Therefore, a crucial issue involved in maximizing the science impact of the survey is **to make sure that we classify high-redshift LAEs and low-redshift [OII] emitters as accurately as possible.** Towards this end, HETDEX plans a broadband imaging survey in addition to IFU spectroscopy.

2. The name of the game: contamination and incompleteness

If we were only armed with the measurement of an emission line flux, we would have a hard time distinguishing between Lyman-α and [OII] emitters. Fortunately, there are

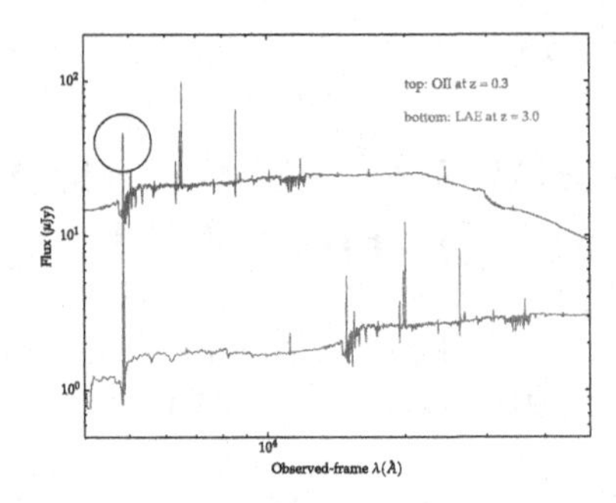

Figure 1: The Lyman-α line from a $z = 3$ galaxy is observed at exactly at the same wavelength as the [OII] line from a $z = 0.305$ galaxy. The emission line flux (shown by the circle) is the same for these two galaxies in this example, but the intensity of the continuum (and therefore the equivalent width, EW) is different. On average, LAEs have much higher EWs than [OII]s.

additional differences between the low-redshift and high-redshift galaxies. The first and most intuitive one is the fact that [OII] emitters, being much closer to us, are on average much brighter. This is usually quantified by the *continuum*, the density of detected photons with wavelengths near the wavelength of the emission line. The key ratio of (emission line flux)/(continuum flux density) is called "equivalent width" (EW):

$$\mathrm{EW} = \frac{f_{\mathrm{EL}}}{f_{\lambda,\mathrm{cont}}} \tag{2.1}$$

LAEs have high emission line fluxes and dim continua (because they are farther away), so they have on average much higher equivalent widths. The traditional requirement that LAE have rest-frame EW > 20 $\mathring{A}$ (e.g., Gronwall *et al.* 2007, Adams *et al.* 2011), is effective in limiting the misclassification of [OII] emitting galaxies as LAEs, at the expense of having a highly incomplete sample. Simulations show that this criterion induces a contamination fraction of over 4% at low-redshift.

To describe our challenge more quantitatively, we can define our scientific objective as to minimize the error $\sigma(D_A)$ with which the angular diameter distance to the median redshift of the survey can be measured. This uncertainty depends on two factors: the number of [OII] emitters that are erroneously classified as LAE (contamination), and the number of LAEs that are "lost" because they are classified as [OII] emitters (incompleteness). Through analysis of errors in the power spectrum reconstruction (E. Komatsu, private comm.) we found that the error in the angular diameter distance can be written as

$$\sigma^2_{D_A} \propto \left(\frac{f_{\mathrm{cont}}}{0.025}\right)^2 + \frac{270,000}{N_{\mathrm{LAE}}} \tag{2.2}$$

where f_{cont} is the contamination fraction, and N_{LAE} is the number of recovered LAEs in the survey. This is the quantity that we shall try to minimize in our quest.

3. Bayesian method

We developed a technique to separate low- and high- redshift galaxies based on a more comprehensive analysis of the information available in the literature about LAE and [OII] galaxies. The main idea is the following:

- Start from the luminosity functions and equivalent width distributions for LAE and [OII] galaxies published in Ciardullo *et al.* (2012) and Ciardullo *et al.* (2013);

• Interpolate those above to obtain a fitting formula valid in the range of redshift of interest;

• Use these as priors to write a probability P(LAE) (P([OII])) that a galaxy of given observed line flux and continuum flux is an LAE ([OII] emitter), modulo a normalization factor;

• Consider the ratio P(LAE)/P([OII]) as the main indicator of the classification;

• Optimize the method by tuning the above ratio as a function of redshift in order to achieve the smallest possible error from Eq. (2.2).

This technique has been shown to reduce both contamination and incompleteness with respect to the "$EW > 20\mathring{A}$" criterion. The details will be discussed in an upcoming publication (Leung *et al.*2014, in preparation).

4. Machine Learning

Another approach that we propose here is to classify LAE/[OII] galaxies using supervised *machine learning*. Machine learning is the concept of having a computer program learn about the properties of a dataset by looking for correlations within the different objects in the dataset (called instances). We plan to use ancillary data existing on some areas of the sky overlapping with the HETDEX survey to identify a subset of galaxies for which a label (LAE or [OII]) can be assigned with a high degree of confidence. This subset is divided into a "training sample", which is then used to "learn" what is the best rule to classify an object as an LAE or an [OII] emitter, and a "validation" sample, which is used to evaluate the performance of the classification procedure and make adjustments as needed. Contrary to the previous methods, no knowledge of the detailed physical model of galaxies is needed; the algorithm learns the classification rule exclusively by comparison with previous cases for which the correct answer was known. In principle, this method can perform better than any of those described above because there is no "assumption" involved in estimating whether an object is an LAE or an [OII] emitter, and there is no formal limit to how complicated the classification rule can be. On the other hand, the ability of the algorithm to learn the rule correctly depends crucially on the quality and size of the training sample, which needs to be a fair representation of the overall sample of galaxies in order for the method to succeed.

Support Vector Machines: We chose to use the Support Vector Machine (SVM) classification algorithm (e.g. Boser *et al.* 1992) to decide whether a galaxy is a Lyman Alpha or [OII] emitter. We assume that labels are available for a fraction (up to 1/10th) of the entire dataset. Each galaxy is an instance, and each instance consists of three attributes, which summarize our anticipated knowledge of the galaxies from the observations, and a label (LAE/[OII]). The attributes are the observed wavelength of the emission line, the emission line flux, and the continuum flux density.

How SVM Works: SVM works by finding a hyperplane within the dataset that separates the two classes (see Fig. 2). In order to find this hyperplane in a non-linearly-separable dataset, the algorithm must transform the data into higher dimensions. This is done through different mappings, often called kernels. The kernel that we have found to be most successful is the radial basis function (rbf) kernel. This kernel is Gaussian, and has several parameters which can be adjusted:

(*a*) The amplitude parameter γ, which defines the complexity of the decision boundaries.

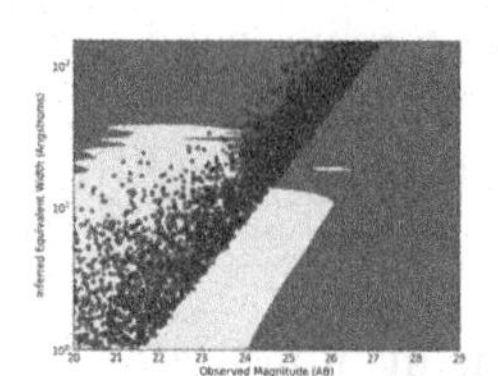

Results; γ = 0.1; class weight = 15		
	Soft Margin C	
	1	10
contamination		
unMADened data	2.9%	7.5%
EW instead of cont. flux	0.5%	0.5%
MADened data	0.5%	0.6%
incompleteness		
unMADened data	26.5%	15.7%
EW instead of cont. flux	23.5%	23.1%
MADened data	21.0%	20.8%
$\sigma(D_A)$		
unMADened data	1.652	3.195
EW instead of cont. flux	1.158	1.156
MADened data	1.142	1.149

Figure 2: *Left:* A graphical representation of our data. Cyan dots represent LAEs and yellow ones represent [OII] emitters. The separation between colors is the decision boundary. Any objects that falls into the blue area is classified by the SVM as an LAE. Note: this is a simplified version with a 2D decision boundary for illustration purposes. *Right:* A summary of the results obtained by optimizing the SVM algorithm. The numbers for $\sigma(D_A)$ are *proportional to* the actual measurement error.

(*b*) The soft margin constant C. Margins are defined by maximizing the distance between the boundary hyperplane and the closest objects in the dataset (also called support vectors). A soft margin, as is used in the rbf kernel, allows for some error. For example, if one instance is beyond the margin, one can allow for that instance to be misclassified if this overall increases the performance of the algorithm. A large value of C corresponds to a small tolerance for these misclassification.

(*c*) The class weight. Adjusting this parameter corresponds to assigning a higher penalty to misclassifications of one type of object (in practice, a different value of C according to the class).

Results: Through a grid search of all possible values of C, γ, and class weight, we found the most efficient hyperplane, as shown in the Table of Fig. 2. The results for soft margin values of C = 1 and C = 10 are both shown since they are close to optimal. Our best values of γ and class weight were 0.1 and 15 respectively. Other steps in the optimization process consisted of "standardizing" data (or more precisely, "MADening" the data, by rescaling them to all have the same median and median-based-deviation), and checking which combination of attributes delivers the best results.

5. Work in progress

Both the Bayesian method and SVMs showed substantial improvement with respect to the EW = $20\mathring{A}$ cut. We plan to keep optimizing both techniques by: 1. including information about the color in the SEDs; 2. extending the simulations to include other emission lines that can help the classification of [OII] emitters (such as the other emission lines in the Balmer series, and [NeIII]); 3. considering other machine learning algorithms, such as boosted decision trees and random forests, in addition to SVMs.

References

Adams, J. J., *et al. ApJS* 192, 5 (2011).
Ciardullo, R., *et al. ApJ* 744, 110 (2012).
Ciardullo, R., *et al. ApJ* 769, 83 (2013).
Gronwall, C., *et al. ApJ* 667, 79 (2007).
Leung, A. S., *et al.*, in preparation (2014).
Boser, E. B., Guyon, I. B., & Vapnik, V. N., *5th Annual ACM Workshop on COLT*, 144 (1992).

Statistical Challenges in 21st Century Cosmology
Proceedings IAU Symposium No. 306, 2014
A. F. Heavens, J.-L. Starck & A. Krone-Martins, eds.

doi:10.1017/S1743921314013659

A high-dimensional look at VIPERS galaxies

Benjamin R. Granett[1] and the VIPERS Team[2]

[1]INAF OA Brera,
Via E. Bianchi 46, 23807 Merate, Italy
email: ben.granett@brera.inaf.it

[2]The VIPERS Team: U. Abbas, C. Adami, S. Arnouts, J. Bel, M. Bolzonella, D. Bottini, E. Branchini, A. Burden, A. Cappi, J. Coupon, O. Cucciati, I. Davidzon, G. De Lucia, S. de la Torre, C. Di Porto, P. Franzetti, A. Fritz, M. Fumana, B. Garilli, L. Guzzo, P. Hudelot, O. Ilbert, A. Iovino, J. Krywult, V. Le Brun, O. Le Fèvre, D. Maccagni, K. Małek, A. Marchetti, C. Marinoni, F. Marulli, H. J. McCracken, Y. Mellier, L. Moscardini, R. C. Nichol, L. Paioro, J. A. Peacock, W. J. Percival, M. Polletta, A. Pollo, M. Scodeggio, L. A. M. Tasca, R. Tojeiro, D. Vergani, G. Zamorani and A. Zanichelli
Web address: http://vipers.inaf.it

Abstract. We investigate how galaxies in VIPERS (the VIMOS Public Extragalactic Redshift Survey) inhabit the cosmological density field by examining the correlations across the observable parameter space of galaxy properties and clustering strength. The high-dimensional analysis is made manageable by the use of group-finding and regression tools. We find that the major trends in galaxy properties can be explained by a single parameter related to stellar mass. After subtracting this trend, residual correlations remain between galaxy properties and the local environment pointing to complex formation dependencies. As a specific application of this work we build subsamples of galaxies with specific clustering properties for use in cosmological tests.

Keywords. Galaxy surveys, statistical methods, galaxy properties

---

1. Parameter space of VIPERS galaxies

The VIMOS Public Extragalactic Redshift Survey (VIPERS) (Guzzo *et al.*, 2014; Garilli *et al.*, 2014) is targeting 100k galaxies with spectroscopy at redshifts z=0.5 to 1.2. With complementary photometric coverage from UV to IR and morphological measures from CFHTLS, this unique dataset is providing unprecedented detail on the properties of galaxies and the formation of structure at intermediate redshifts.

We examine the correlations across the parameter space of observable and derived properties with special focus on spatial clustering. We adopt a volume limited sample at $0.5 < z < 1$ with luminosity cut $M_B < -19.5 - z$ giving 15 494 galaxies. We consider the following set of 12 parameters: (1) $\underline{M_\star}$: stellar mass; (2) $\underline{U - V}$: rest-frame spectral slope in blue; (3) $\underline{M_V}$: rest-frame V magnitude; (4) $\underline{\phi}$: PCA 1 spectral index; (5) $\underline{D4000}$: Balmer break depth; (6) $\underline{OIIEW}$: OII line equivalent width; (7) $\underline{OIIFlux}$: OII line apparent flux; (8) $\underline{R - K}$: rest-frame spectral slope in red; (9) $\underline{\theta}$: PCA 2 spectral index; (10) $\underline{sersic}$: Sersic morphological index; (11) $\underline{\gamma}$: Clustering slope at $1 < r < 30\mathrm{Mpc}/h$; (12) $\underline{A}$: Clustering amplitude at $5 < r < 30\mathrm{Mpc}/h$.

The spectral principal component analysis is described in Marchetti *et al.* (2013); the derivations of the rest-frame galaxy properties and stellar mass are given in Davidzon *et al.* (2013) and Fritz *et al.* (2014); spectral line measurements are described in Garilli *et al.* (2014); the spatial clustering properties are described in de la Torre *et al.* (2013) and Marulli *et al.* (2013); the galaxy morphologies are measured in Krywult *et al.* (in preparation). These parameters represent diverse quantities with different units. To

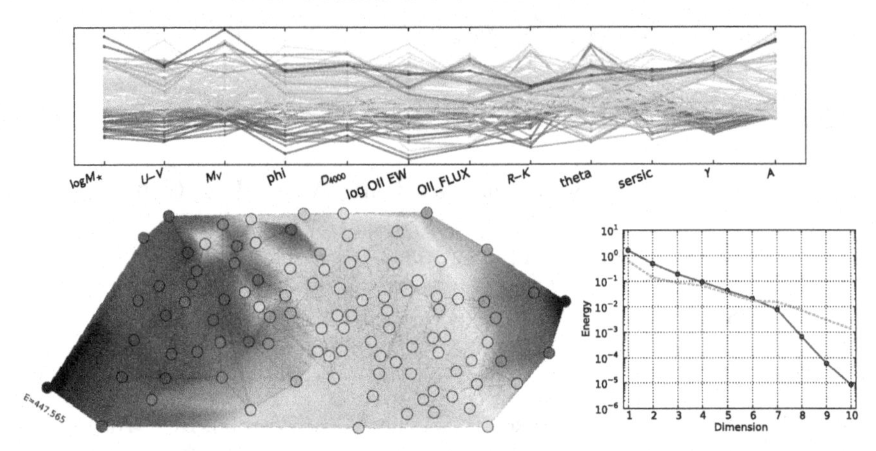

Figure 1. *Top:* A parallel-coordinates plot of group centres in the 12D space coloured according to $M_\star$. The order of colours is mostly preserved across the plot indicating strong correlations. *Left:* A graph of the connections between galaxies coloured by $M_\star$. Neighbouring group centres are connected by an edge with length given by their Gaussianised separation. The graph is flattened onto the page allowing edges to stretch and compress while similar groups remain near to each other. The uniform colour gradient shows that sorting by a single parameter, $M_\star$, can organise the groups into a one-dimensional sequence. *Right:* The solid line shows the graph energy with dimension D. The energy follows a power-law trend; the knee at $D = 7$ points to the underlying dimension. For comparison, the eigenvalues are overplotted (orange dashed line).

measure separations between points in parameter space, we first apply a Gaussianisation transform to the distributions.

We group the galaxies by properties to further reduce the dimensionality of the dataset. We apply the k-means group-finding algorithm to determine 100 group centres. This algorithm partitions the space into subgroups with approximately equal size. The projected correlation function of each subgroup is measured to derive the γ and A parameters. The γ and A values are then assigned to individual galaxies by extrapolating between the group centres. What is the underlying dimensionality of the dataset? We show in Fig. 1 that a single parameter such as stellar mass can explain the primary trends.

2. Correlations in parameter space

We further examine the correlations between parameters. Fig. 2 shows the joint distributions between all pairs of parameters. The bimodal nature of galaxies is reflected in most of the properties considered. Most striking is the clustering amplitude A that characterises how galaxies are distributed over the density field. Considering the two locii of galaxies, the reddest, luminous and massive group is also highly clustered signifying dense environments, while the bluer galaxies are more uniformly distributed and show weaker clustering slopes γ.

To investigate the factors that influence the galaxy properties we subtract the median trend with $M_\star$ from each parameter (not shown here). Once removed, the correlations are greatly reduced. In particular, correlations with M_V are reduced since M_V is degenerate with stellar mass. However, we find that colour parameters $U - V$, ϕ and $D4000$ remain strongly correlated and still relate to clustering.

Stellar mass is a major driver of galaxy properties, but it cannot explain everything. After subtracting trends with stellar mass, residual correlations exist reinforcing the conclusion that environment must play an important role. This is highlighted by the dependencies seen in the correlation function since the clustering properties may be

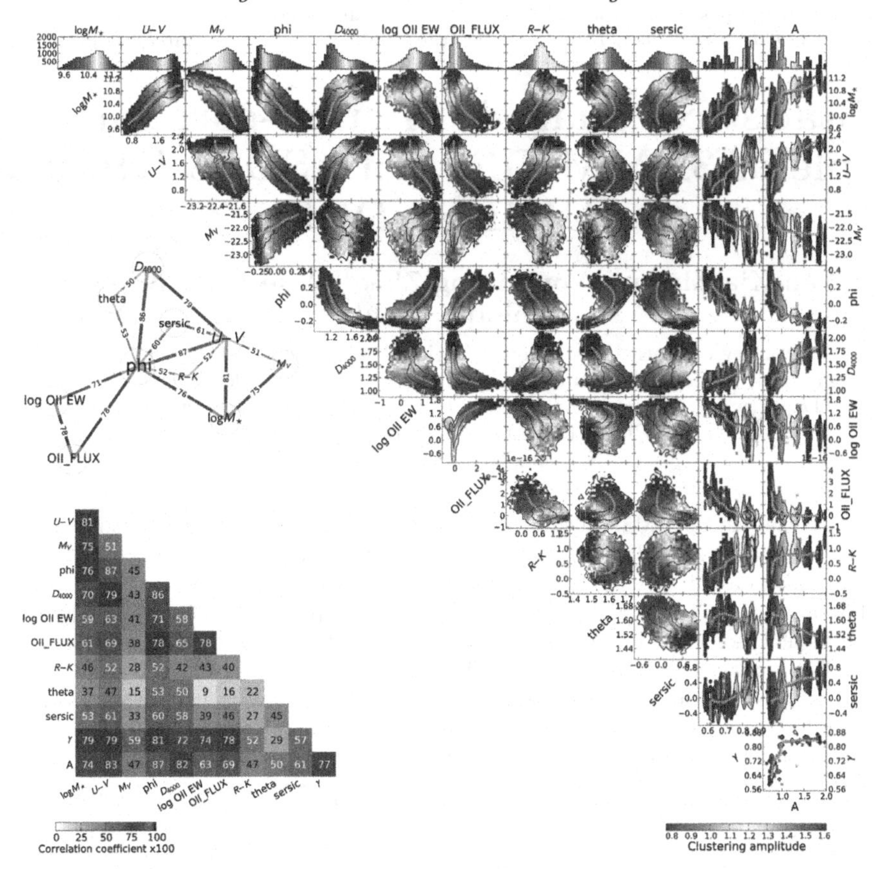

Figure 2. Connections between parameters quantified by Spearman's rank correlation coefficient. Stellar mass, luminosity and D4000 are connected but separated from line strength. The PCA ϕ parameter emerges as one of the most informative parameters. The panels show the joint distributions of parameters coloured by clustering amplitude A. The contours contain 50% and 90% of galaxies. Orange dashed line gives the trend with stellar mass while blue dashed line gives the trend with M_V showing that mass and luminosity can account for the primary trends.

related directly to how galaxies are distributed within underlying dark matter clumps. We see that the dependencies on galaxy properties are complex as they encode the subtle mechanisms relating the formation of galaxies to the large-scale density field.

We acknowledge the ESO staff for the management of service observations. We are grateful to M. Hilker for his constant help and support. BRG acknowledges support of the European Research Council through the Darklight ERC Advanced Research Grant (# 291521). A complete list of the funding agencies that support VIPERS may be found at http://vipers.inaf.it.

References

Davidzon, I., *et al.* 2013, *A&A*, 558, A23

Fritz, A., *et al.* 2014, *A&A*, 563, A92

Garilli, B., *et al.* 2014, *A&A*, 562, A23

Guzzo, L., *et al.* 2014, *A&A*, 566, A108

Marchetti, A., *et al.* 2013, *MNRS*, 428, 1424

Marulli, F., E., *et al.* 2013, *A&A*, 557, A17

de la Torre, S., *et al.* 2013, *A&A*, 557, A54

Statistical Challenges in 21st Century Cosmology
Proceedings IAU Symposium No. 306, 2014
A. F. Heavens, J.-L. Starck & A. Krone-Martins, eds.

doi:10.1017/S1743921315000022

Statistical analysis of cross-correlation sample of 3XMM-DR4 with SDSS-DR10 and UKIDSS-DR9

Yan-Xia Zhang[1], Yong-Heng Zhao[1], Xue-Bing Wu[2] and Hai-Jun Tian[1,3]

[1]Key Laboratory of Optical Astronomy, National Astronomical Observatories, Chinese Academy of Sciences, 20A Datun Road, Chaoyang District, 100012, Beijing, P.R.China
[2]Department of Astronomy, Peking University 100871, Beijing, P.R.China
[3]College of Science, China Three Gorges University, 443002 Yichang, P.R.China
email:zyx@bao.ac.cn

Abstract. We match the XMM-Newton 3XMMi-DR4 catalog with the Sloan Digital Sky Survey (SDSS) Data Release 10 and the United Kingdom Infrared Deep Sky Survey (UKIDSS) Data Release 9. Based on this X-ray/optical/infrared catalog, we probe the distribution of various types of X-ray emitters in the multidimensional parameter space. It is found that quasars, galaxies and stars have some kind distribution rule, especially for stars. The result shows that only using the X-ray/optical features, stars are difficult to discriminate from galaxies and quasars, the added information from infrared band is very helpful to improve the classification result of any classifier. Comparing the classification accuracy of random forests with that of rotation forests, rotation forests show better performance.

Keywords. catalogs - surveys - X-rays: diffuse background - X-rays: galaxies - X-rays: stars

---

1. Introduction

Various large sky surveys from different bands provide abundant resources of studying multiwavelength properties of objects. X-ray satellites are helpful to explore some types of objects, notably active galaxies (AGN), clusters of galaxies, interacting compact binaries and active stellar coronae. Zhang *et al.* (2013) cross-correlated the XMM-Newton 2XMMi-DR3 catalog with the Sloan Digital Sky Survey (SDSS) Data Release 8, investigated the distribution of various classes of X-ray emitters in the multidimensional photometric parameter space, then applied random forests on this sample and found that the X-ray emitting stars had poor classification accuracy compared to galaxies and quasars.

2. Data

3XMM-DR4 is generated from the European Space Agency's (ESA) XMM-Newton observatory. The catalogue contains 531,261 X-ray source detections above the processing likelihood threshold of 6. These X-ray source detections relate to 372,728 unique X-ray sources.

The Sloan Digital Sky Survey (SDSS) experiences over eight years of operations (SDSS-I, 2000-2005; SDSS-II, 2005-2008), now SDSS enters SDSS-III. Data Release 10 (DR10) is the first release of the spectra from the SDSS-III's Apache Point Observatory Galactic Evolution Experiment (APOGEE). DR10 also includes hundreds of thousands of new galaxy and quasar spectra from the Baryon Oscillation Spectroscopic Survey (BOSS), in addition to all imaging and spectra from prior SDSS data releases.

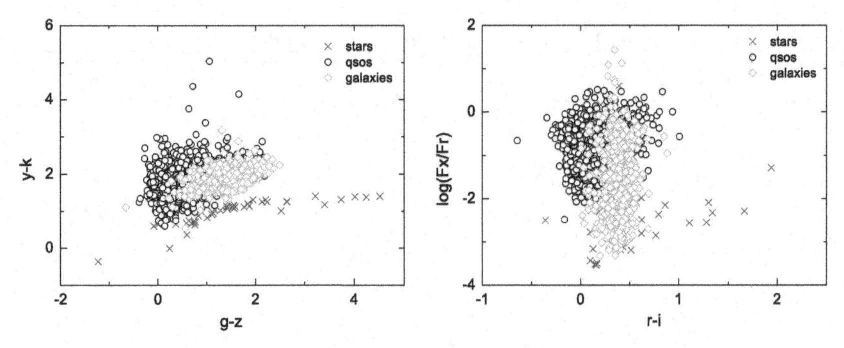

Figure 1. Left: Distribution of sources in $y-k$ vs. $g-z$ diagram. Right: Distribution of sources in the $\log(f_{\mathrm{x}}/f_{\mathrm{r}})$ vs. $r-i$ diagram. Stars are represented as crosses, galaxies as open diamonds, quasars as circles.

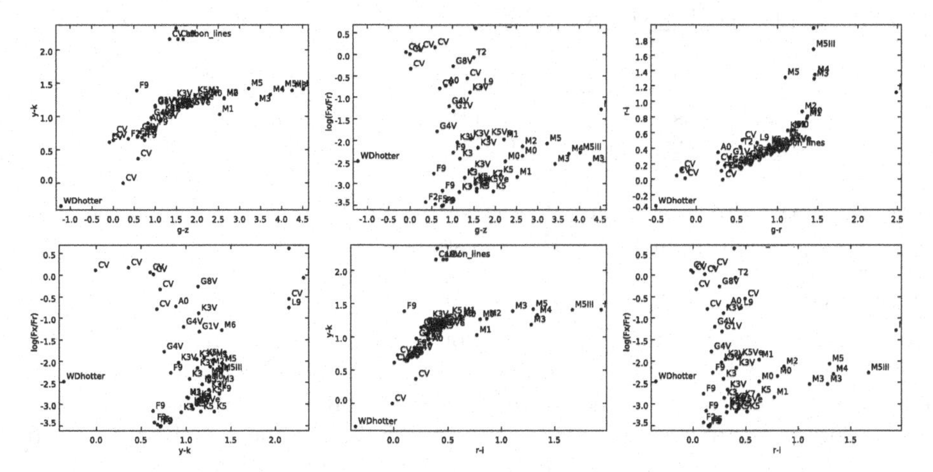

Figure 2. Distributions of stars in different parameter spaces.

The UKIRT InfraRed Deep Sky Survey (UKIDSS) consists of five surveys: Large Area Survey (LAS), Galactic Plane Survey (GPS), Galactic Clusters Survey, Deep Extragalactic Survey (DXS), Ultra Deep Survey (UDS). As for LAS, the number of objects observed is 212,234,393, and each band depth is Y=20.2, J=19.6, H=18.8, K=18.2, respectively. Here we use the public data release 9.

We cross-match 3XMM-DR4 with SDSS DR10 and UKIDSS DR9, and obtain the sample with information from X-ray, optical and infrared bands. In order to study statistical properties of this sample, Figures 1-2 show the distribution of quasars, galaxies and various stars in the multidimensional space.

3. Methods and Results

Random forests, developed by Leo Breiman and Adele Cutler, are one kind of decision tree methods by constructing a lot of decision trees and outputting the class that is the mode of the classes provided by individual trees. They have wide applications in astronomy for classification and regression.

Rotation forests (Rodriguez *et al.* 2006) are a method for creating classifier ensembles, based on feature extraction by Principal Component Analysis (PCA). Decision trees are adopted here. Rotation forests are firstly used in astronomy.

Using the open source software WEKA (Hall *et al.* 2009) and adopting the default setting, random forests and rotation forests are performed on the cross-matched sample.

Table 1. Accuracy with different input patterns

Input pattern	Random Forest Accuracy	Rotation Forest Accuracy
$g-z, y-k$	87.00%	88.70%
$i, g-z, y-k$	88.34%	88.83%
$i, g-z, y-k, log(f_X/f_r)$	90.27%	91.03%
$hr1, hr2, hr3, hr4, i, g-z, y-k, log(f_X/f_r)$	91.12%	91.12%
$hr1, hr2, hr3, hr4, SC\_EXTENT, i, g-z, y-k, log(f_X/f_r)$	90.31%	91.21%
$u-g, g-r, r-i, i-z$	88.30%	89.60%
$i, u-g, g-r, r-i, i-z$	89.64%	91.30%
$r, u-g, g-r, r-i, i-z$	90.27%	91.21%
$y-j, j-h, h-k$	85.38%	87.94%
$y, y-j, j-h, h-k$	90.36%	91.79%
$u-g, g-r, r-i, i-z, y-j, j-h, h-k$	91.39%	92.42%
$i, u-g, g-r, r-i, i-z, y-j, j-h, h-k$	92.12%	**93.50%**
$i, u-g, g-r, r-i, i-z, y, y-j, j-h, h-k$	92.47%	93.05%
$u-g, g-r, r-i, i-z, g-z, y-j, j-h, h-k, y-k$	92.11%	93.00%
$i, u-g, g-r, r-i, i-z, g-z, y-j, j-h, h-k, y-k$	92.15%	93.27%
$i, u-g, g-r, r-i, i-z, log(f_X/f_r), y-j, j-h, h-k$	**92.65%**	93.41%
$SC\_EXTENT, i, u-g, g-r, r-i, i-z, log(f_X/f_r), y-j, j-h, h-k$	92.15%	93.18%
$hr1, hr2, hr3, hr4, i, u-g, g-r, r-i, i-z, log(f_X/f_r), y-j, j-h, h-k$	91.57%	92.78%

Table 2. The detailed accuracy for the best classification result by rotation forests

	TP Rate	FP Rate	Precision	Recall	F-Measure	ROC Area
GALAXY	0.866	0.032	0.909	0.866	0.887	0.971
QSO	0.968	0.142	0.943	0.968	0.956	0.971
STAR	0.712	0.000	0.974	0.712	0.822	0.965

The accuracy with different input patterns is listed in Table 1. The detailed accuracy for the best classification result by rotation forests is shown in Table 2.

4. Conclusions

We compare the performance of random forests and rotation forests on the classification of all types of objects. Obviously, rotation forests are superior to random forests. The accuracy of classification improves when infrared information added, especially that of stars. Moreover all kinds of stars are easily discriminated and they have a clear distribution rule. In some parameter space, the special stars have different distribution from the most of stars. Therefore it is easy to choose special X-ray emitting star candidates from the huge photometric survey data depending on the information from optical and infrared bands. Firstly the rotation forest classifier is applied to select star candidates, then the special star candidates are picked out from the y-k vs. g-z diagram.

Acknowledgements

This paper is funded by National Key Basic Research Program of China 2014CB845700, National Natural Science Foundation of China under grants Nos.11178021, 11033001 and NSFC-Texas A&M University Joint Research Program No.11411120219. We acknowledge the SDSS, UKIDSS and XMM databases.

References

Hall, M., Frank, E., Holmes, G., Pfahringer, B., Reutemann, P., & Witten, I. H. 2009, *SIGKDD Explorations*, 11(1), 10

Rodriguez, J. J., Escuela Politecnica Superior, Burgos Univ., Kuncheva, L. I., & Alonso, C. J. 2006, *Pattern Analysis and Machine Intelligence*, 28(10), 1619

Zhang, Y., Zhou, X., Zhao, Y., & Wu, X. 2013, *AJ*, 145, 42

Statistical Challenges in 21st Century Cosmology
Proceedings IAU Symposium No. 306, 2014
A. F. Heavens, J.-L. Starck & A. Krone-Martins, eds.

doi:10.1017/S1743921314011089

Euclid space mission: a cosmological challenge for the next 15 years

R. Scaramella[1], Y. Mellier, J. Amiaux, C. Burigana, C.S. Carvalho, J.C. Cuillandre, A. da Silva, J. Dinis, A. Derosa, E. Maiorano, P. Franzetti, B. Garilli, M. Maris, M. Meneghetti, I. Tereno, S. Wachter, L. Amendola, M. Cropper, V. Cardone, R. Massey, S. Niemi, H. Hoekstra, T. Kitching, L. Miller, T. Schrabback, E. Semboloni, A. Taylor, M. Viola, T. Maciaszek, A. Ealet, L. Guzzo, K. Jahnke, W. Percival, F. Pasian, M. Sauvage and the Euclid Collaboration

[1]Osservatorio di Roma
email: kosmobob@oa-roma.inaf.it

Abstract. Euclid is the next ESA mission devoted to cosmology. It aims at observing most of the extragalactic sky, studying both gravitational lensing and clustering over ∼15,000 square degrees. The mission is expected to be launched in year 2020 and to last six years. The sheer amount of data of different kinds, the variety of (un)known systematic effects and the complexity of measures require efforts both in sophisticated simulations and techniques of data analysis. We review the mission main characteristics, some aspects of the the survey and highlight some of the areas of interest to this meeting.

Keywords. cosmology: cosmological parameters, gravitational lensing, dark matter, large-scale structure of universe

---

1. Overview

Science. Euclid is an ESA Cosmic Vision mission (Laureijs *et al.* 2011) mainly devoted to the study of cosmology, namely precise measurements of the expansion factor $a(t)$ and of the growth of density perturbations, $\delta\rho/\rho$, whose power spectrum is $P(k)$. This will be achieved via two main probes: (i) gravitational lensing [WL] via exquisite shape measures + photoz of over a billion of galaxies and (ii) the clustering of galaxies via accurate redshift mostly obtained by emission lines, measured through a slitless spectrograph.

An important aspect of Euclid is the contemporary use of additional cosmological probes (to WL and clustering add clusters of galaxies – counts, density profiles and mass function –, the Integrated Sachs-Wolfe, and possibly high-z SNe) and their overall complementarity. Moreover, it is possible to study not only the geometry but also the detailed evolution of density perturbations: the information on the growth factor (usually parametrised as $d\ln\delta/d\ln a \propto \Omega_m^\gamma$, where for dark matter dominated models is $\gamma \sim 4/7$) allows one to better differentiate among competing classes of models, such as dark energy vs. modified gravity (see Fig. 1, and Amendola *et al.* 2013).

Moreover, both the variety of data and the large sky coverage ($\sim 15,000$ sq. deg. in the wide survey) will yield a wealth of information which will be crucial for several areas in astronomy, yielding a large legacy value to Euclid databases. Also the deep fields (~ 40 sq. deg. observed repeaditly to reach two magnitudes fainter than the wide survey) will be unique for these purposes (Laureijs *et al.* 2011).

Instruments and observations. In Euclid direct contributions to ESA are provided by an international consortium [EC], lead by Y. Mellier, which comprises most of european

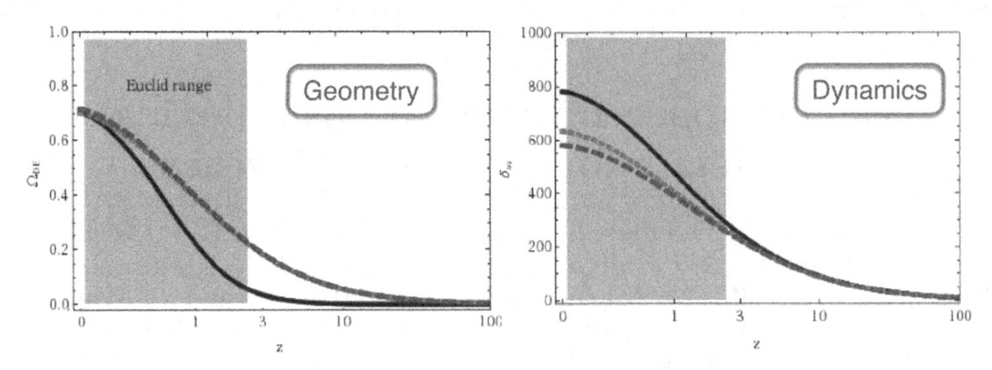

Figure 1: Effect of dark energy on the evolution of the Universe. Left: Fraction of the density of the Universe in the form of dark energy as a function of redshift, z, for a model with a cosmological constant ($w = -1$, black solid line), and two models with identical background expansion: quintessence with a different equation of state (blue dashed line), and a modified gravity model -DGP- (red dotted line, with $w_0 = -0.77$ and $w_a = 0.29$, with $w = w_0 + w_a(1+z)/z$). In all cases, dark energy becomes dominant in the low redshift Universe era probed by Euclid. Right: Growth factor of cosmic structures for the same three models. Only by adding at low redshifts the measure of the growth of structure (right panel) to measure of the geometry (left panel) a modification of dark energy can be distinguished from that of gravity because of the different predicted growth rate. Weak lensing and clustering measure both effects.

nations (as of today $\sim$ 1200 participants from over $\sim$150 institutes) and provides the two instruments VIS and NISP. VIS (Cropper *et al.* 2012) is an optical imager well matched to the telescope sampling (see Fig. 3), while NISP (Prieto *et al.* 2012) can take either slitless spectra or NIR photometry in the Y, J, H bands. EC is also providing the Ground Segment data analysis, which will be carried out in several data centers. Also NASA contributes to the mission with the NIR arrays and therefore a selected number of US scientists participate as full members to the EC.

The Euclid spacecraft has stringent requirements on operations in order to minimise systematics and maintain the superb optical quality over a large field of view. This in turn translates in the need of observing always orthogonally to the sun and in a step

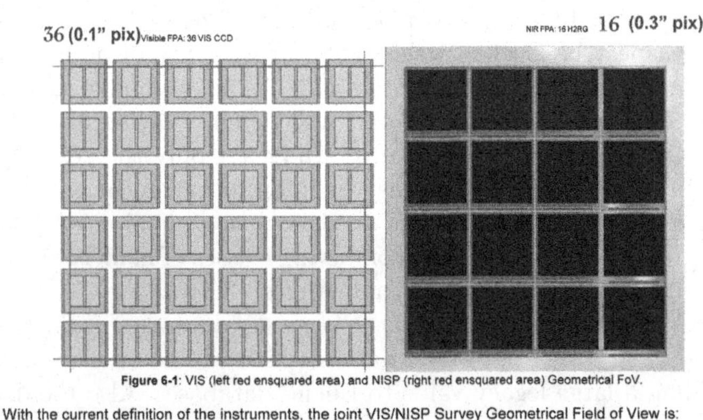

Figure 2: The focal plane of the two instruments VIS and NISP. The joint FoV is $\sim$0.5 sq. degs.

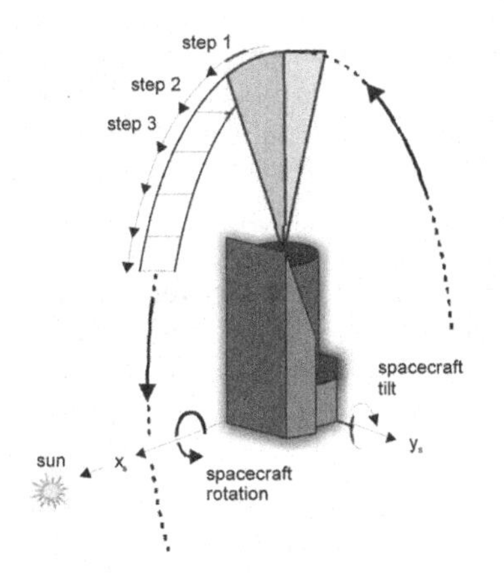

Figure 3: The telescope will operate in a step and stare fashion. Precise thermal stability requires to observe orthogonally to the sun, limiting time visibility of areas at low ecliptic latitudes.

and stare scan strategy (see Fig. 2) which approximately covers $\sim$ 20 fields, i.e. $\sim$ 10 square degrees, per day. The sun-angle constraints imply only few days of visibility for low ecliptic latitudes and that perennial visibility (needed to satisfy specific needs of cadence for calibration purposes) is achieved only over two circles of 3 deg radius centered on the ecliptic poles. This forces to set the position of the deep fields as close as possible to the ecliptic poles. The Southern one cannot be centered because of the location of the Large Magellanic Cloud. Details are given in Amiaux *et al.* (2012) and Tereno *et al.* (2014).

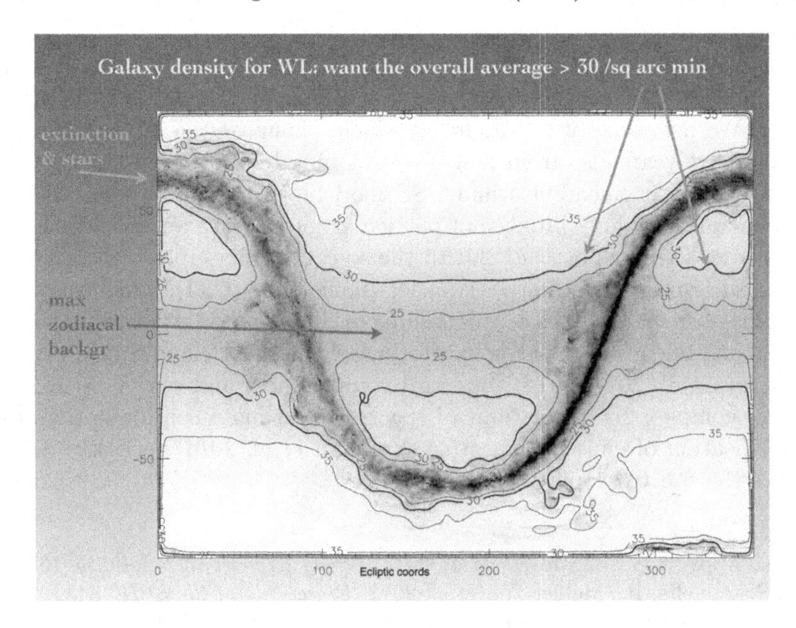

Figure 4: Euclid expected galaxy density on the sky [counts/arc min^2] g as a function of average zodiacal background and extinction. These galaxies have S/N>10 with a limit of M_{ab} = 24.5 (wide optical filter) and minimum size of FWHM[gal]>1.25 FWHM[PSF]

Each field is observed in four identical sequences but for small dithers (Laureijs *et al.* 2011; Amiaux *et al.* 2012). In each sequence first the longest exposures for spectra (NISP) and morphology (VIS) are taken at the same time, then these are followed by a sequence of three photometry exposures in Y, J and H band (change in filters would perturb the VIS exposure, so no data are taken during this phase). The four slitless exposures will be taken with a red grism ($\sim$ 1.25–1.85 μm) but with three different main orientations for the dispersion direction, to take care of possible line contamination in spectra caused by nearby objects. Also a blue grism ($\sim$ 0.92–1.3 μm) will be present on board to be used on the deep fields. This will extend the wavelength spectral coverage in a way which is very important for legacy purposes (e.g. line ratios; Laureijs *et al.* 2011).

2. Data models and systematic effects

The large number of sampled objects (see Fig. 4) and the wide area coverage (see Fig. 5) translate into a good sampling of all scales of interest such that a very good modelling and control of biases is needed for Euclid so to reach its primary science goals. Therefore a lot of work continues to be done on these aspects, taking also advantage of

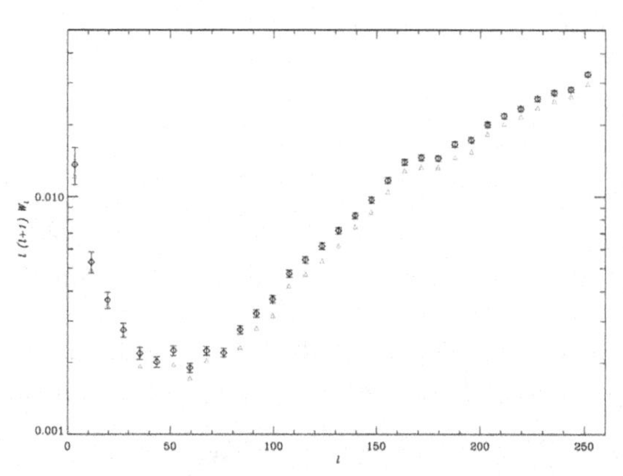

Figure 5: Window function (binned) coefficients (we plot the customary $\ell(\ell + 1)W_\ell$) obtained via Healpix routines Gorski *et al.* 2013 for two different sky coverages (Amiaux *et al.* 2012, Tereno *et al.* 2014).

experience on real data from ground based surveys or in similar situations (e.g. HST, also slitless). We list some of these efforts which, though often advanced, in part will benefit in the next years also from new or developing techniques in statistical analysis. Important aspects are: Shear bias and PSF modelling (Cropper *et al.* 2013); intrinsic color gradients across galaxy profiles, of particular importance for Euclid because of the wide optical filter (Semboloni *et al.* 2013); the so-called "noise bias" (Viola *et al.* 2014). Another crucial issue is the Charge Transfer Inefficiency (CTI), which originates from radiation damage on CCDs. This causes trailing of charges and therefore shape distortions in a complex, non-linear way. Strenuous efforts went into modelling it in HST, GAIA and Euclid (Massey *et al.* 2013, 2014). Spectral completeness and purity estimates require extensive and complex 2D simulations of spectra to estimate contamination, extraction and misclassification of emission lines (see Zoubian *et al.* 2014) to assess completeness and purity of the spectroscopic sample.

References

Amendola, L., Appleby, S., Bacon, D., *et al.* 2013, *Living Reviews in Relativity*, 16, 6
Amiaux, J., Scaramella, R., Mellier, Y., *et al.* 2012, *Proceeding of the SPIE*, 8442
Cropper, M., Cole, R., James, A., *et al.* 2012, *Proceeding of the SPIE*, 8442
Cropper, M., Hoekstra, H., Kitching, T. D., *et al.* 2013, *MNRAS*, 431, 3103
Gorski, K. M., Wandelt, B., Hivon, E., *et al.* *http://healpix.sourceforge.net*
Laureijs, R., *et al.* 2011, *Euclid Definition Study report*, ESA/SRE(2011)1, arXiv:1110.3193
Massey R., Hoekstra, H., Kitiching, T. D., *et al.* 2013, *MNRAS*, 429, 661
Massey, R., Schrabback, T., Cordes, O., *et al.* 2014, *arXiv e-print*, arXiv:1401.1151v1
Prieto, E., Amiaux, J., Augueres, J.-L., *et al.* 2012, *Proceeding of the SPIE*, 8442
Semboloni, E., Hoekstra, H., Huang, Z., *et al.* 2013, *MNRAS* 432, 2385
Tereno, I., Carvalho, C. S., Dinis, J., *et al.* 2014, *This volume*
Viola, M., Joachimi, B., & Kitching, T. D. 2014, *MNRAS*, 439, 1909
Zoubian, J., Kmmel, M., Kermiche, S., *et al.* 2014, *ASP conference series*, 485, 509

Statistical Challenges in 21st Century Cosmology
Proceedings IAU Symposium No. 306, 2014
A. F. Heavens, J. -L. Starck & A. Krone-Martins, eds.

doi:10.1017/S174392131401093X

Euclid Space Mission: building the sky survey

I. Tereno[1,2], C.S. Carvalho[1], J. Dinis[1,2], R. Scaramella[3], J. Amiaux[4], C. Burigana[5], J.C. Cuillandre[4], A. da Silva[1,2], A. Derosa[5], E. Maiorano[5], M. Maris[6], D. Oliveira[7], P. Franzetti[8], B. Garilli[8], P. Gomez-Alvarez[9], M. Meneghetti[10], S. Wachter[11] and the Euclid Collaboration

[1]Instituto de Astrofísica e Ciências do Espaço, Tapada da Ajuda, 1349-018 Lisboa, Portugal,
email: `tereno@fc.ul.pt` ,
[2]Faculdade de Ciências da Universidade de Lisboa, Campo Grande, 1749-016 Lisboa, Portugal
[3]INAF - Osservatorio Astronomico di Roma, Via Frascati 33, 00040 MontePorzio Catone, Italy
[4]Commissariat a l'Energie Atomique, Orme des Merisiers, 91191 Gif sur Yvette, France
[5]INAF - IASF Bologna, via Piero Gobetti 101, 40129 Bologna, Italy
[6]INAF - Osservatorio Astronomico di Trieste Via G. B. Tiepolo 11, 34131 Trieste, Italy
[7]Max Planck Institut fuer Astrophysik, Karl Schwarzschild Str.1., 85740 Garching, Germany
[8]INAF - IASF Milano, Via E Bassini 15, 20133 Milano, Italy
[9]European Space Agency, ESAC, POB 78, 28691 Villanueva de la Canada, Madrid, Spain
[10]INAF - Osservatorio Astronomico di Bologna, Via Ranzani 1, 40127 Bologna, Italy
[11]Max Planck Institut fuer Astronomie, Konigstuhl 17, 69117 Heidelberg, Germany

Abstract. The Euclid space mission proposes to survey 15000 square degrees of the extragalactic sky during 6 years, with a step-and-stare technique. The scheduling of observation sequences is driven by the primary scientific objectives, spacecraft constraints, calibration requirements and physical properties of the sky. We present the current reference implementation of the Euclid survey and on-going work on survey optimization.

Keywords. Euclid space mission, cosmology: observations, survey, operation

1. Introduction

The Euclid space mission (Laureijs *et al.*2011), part of the European Space Agency Cosmic Vision Program, is currently under construction by a large international consortium of research centres together with the ESA and aerospace industry partners. The mission addresses the big scientific questions related to fundamental physics and cosmology on the nature and properties of dark energy, dark matter and gravity, as well as on the physics of the early universe and the initial conditions that seed the formation of cosmic structure. For this purpose, the collaboration is building a space telescope with associated visible imager, near infrared photometer and slitless spectrograph, scheduled to be launched on 2020 and to stay in operation for 6 years. Euclid will map the sky with a step-and-stare strategy imaging an average of 30 galaxies per square arc-minute, with mean redshift around 1, which means a total of around 2 billion galaxies over 15000 square degrees of the extragalactic sky, and obtaining the spectra of an average of 3500 galaxies for square degree. The large catalogue will be used to detect signatures of the expansion rate of the Universe and the growth of cosmic structures using two main probes: weak gravitational lensing effects on galaxies and the properties of galaxy clustering.

The mission is defined as a survey that aims at tilling the useful sky with the reference elementary observation pattern in the most efficient way. The elementary observation pattern includes a dithering pattern of 4 frames and the exposure time (including exposure time per instrument and readout overheads). The survey consists of a Wide Survey that ensures the performance of the primary science and a Deep Survey that allows for legacy science. A large volume of calibration data, amounting to 20% of the total observing time, is needed for characterisation of the on-board instruments and for sample characterization in order to control systematic effects. These data are obtained in various time scales, using both survey data and dedicated observations on specific targets.

2. The Reference survey

The limits of the extragalactic sky that meet the area and depth requirements needed for the two main probes to reach the science objectives, are set by considering the main contributors to galaxy counts. These are zodiacal background and extinction, which impact the signal-to-noise ratio (SNR), and bright stars, which impact the number of useful objects. The optimal region excludes low galactic latitudes and low ecliptic latitudes.

The Wide survey is built on the optimal region at the time slots not occupied by calibrations. Indeed, given the needs for specific cadences and periodic observations, calibration observations are defined first, fixing a time skeleton. The result of the galaxy counts analysis also helps in defining the strategy. The survey is built by starting observations at the region of higher density of galaxies, starting from the ecliptic poles where the zodiacal background is minimum, so to have best SNR in the early stages of the mission. The Deep survey is built on fields located close to the ecliptic poles to maximize visibility from space. The basic survey strategy is implemented using the Euclid Sky Survey Planning Tool and described in Amiaux *et al.*2012.

When building the survey, stringent constraints on the spacecraft attitude must be followed. Indeed, in order to ensure that the thermo-mechanical stability of the Payload Module is sufficient to preserve the weak lensing measurement accuracy, the spacecraft will be operated with its sunshield normal to the sun direction, with only small variations around this attitude being authorized. The telescope is only allowed to depoint from the orthogonal direction to the centre of the solar disk by a solar aspect angle of -3 to 10 degrees. If observing with a small depointing angle, the field-of-view on the sky would be tilted. To minimize gaps in the sky coverage, this tilt may be compensated by observing with an equivalent tilt in the telescope. The authorized range of this tilt is severely constrained by the structure of the solar shield. Other important limits to take into account are the total authorized number of telescope operations and the maximum allowed number of large slews between non spatial consecutive pointings.

Given all these science and technical constraints, we have produced a survey, which is the current reference to be used in all relevant Euclid applications, and is shown in Fig. 1. This reference survey achieves a sky coverage of 15000 square degrees in 5.5 years, with over 40000 pointings all within the depointing constraints and 85% with small tilts (less than 1 deg). Calibrations are made in time blocks of up to one week, and the number of large slews is kept under 10 per month on average.

3. Optimizing the Euclid Survey

We aim to produce various alternative versions of compliant surveys, which calls for a larger degree of automation. On the other hand, the current reference survey has room for optimization. To meet these needs, we present two complementary approaches: a deterministic one, where the sequence is constructed step by step, and a stochastic one,

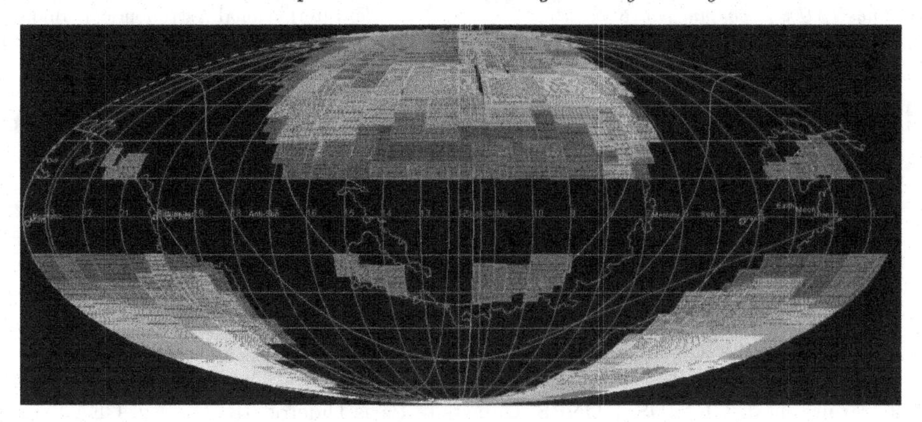

Figure 1. The present Euclid reference survey fulfilling all mission specifications, shown in a Mollweide projection of the entire sky in ecliptic coordinates with the ecliptic North pole up. Jagged line is an iso-contour of extinction. Different colours indicate different years of the survey.

where we seek for a global solution. In both cases we start by pre-tessellating the sky with around 50000 identical fields (corresponding to the telescope field-of-view, FoV).

Computing the Euclid Survey from a Set of Conditions. At a given time, only a small part of the sky is available for observation due to the depointing constraints, defining a spatial time window. We start by considering all FoVs within the time window and build one sequence of FoVs for the time slot between two consecutive calibration blocks. From a starting field at high latitude, we build a progression mostly up and down ecliptic meridians, within a limited latitude range producing thus horizontal bands. The progression should leave no gaps, minimize the number of large slews and prioritize higher quality fields. There are fixed conditions to determine the latitude range of the horizontal bands, to recover missing fields in a compact region (left during calibration times) and to establish when and where to shift between hemispheres. The conditions may also be used as variables in a stochastic approach.

Computing the Euclid Survey with Simulated Annealing. We consider the Euclid survey as a combinatorial optimization problem, whose solution is a sequence that visits all the fields without repetitions, in the shortest time, and fulfils all constraints. In this approach, the space of configurations (all possible sequences) will be searched in a controlled manner. We again consider the time windows, but this time they are defined taking into account both depointing and tilting constraints. For each time slot between two consecutive calibration blocks, we define a number of patches grouping neighbouring fields and tile them in an optimal way. The tiled patches for all time windows form an initial sequence. We generate neighbouring sequences from a current one based on random sets of changes. We will use the simulated annealing technique (Kirkpatrick *et al.*1983) to evaluate the sequences. A sequence with lower cost or with a slight increase in cost (to escape from local minima) is kept, in a converging process. Gradually the acceptance threshold (the so-called temperature parameter) is lowered, restricting the acceptance of new sequences and ensuring convergence to a sequence with minimal idle time.

References

Amiaux, J., Scaramella, R., Mellier, Y., *et al.* 2012, *Proceeding of the SPIE*, 8442
Kirkpatrick, S., Gelatt, C. D., & Vecchi, M. P. 1983, *Science*, 220,671
Laureijs, R. *et al.* 2011, arXiv e-print, arXiv:1110.3193

Statistical Challenges in 21st Century Cosmology
Proceedings IAU Symposium No. 306, 2014
A. F. Heavens, J. -L. Starck & A. Krone-Martins, eds.

doi:10.1017/S1743921314010953

Testing the Equivalence Principle in space with MICROSCOPE: the data analysis challenge

Joel Bergé[1], Quentin Baghi[1] and Sandrine Pires[2]

[1]ONERA, the French Aerospace Lab,
29 avenue de la Division Leclerc, 92320 Châtillon, France
email: joel.berge@onera.fr; quentin.baghi@onera.fr

[2]Laboratoire AIM, CEA/DSM–CNRS, Université Paris Diderot, IRFU/SAp, CEA Saclay,
Orme des Merisiers, 91191 Gif-sur-Yvette, France
email: sandrine.pires@cea.fr

Abstract. Theories beyond the Standard Model and General Relativity predict a violation of the Weak Equivalence Principle (WEP) just below the current best experimental upper limits. MICROSCOPE (Micro-Satellite à traînée Compensée pour l'Observation du Principe d'Equivalence) will allow us to lower them by two orders of magnitude, and maybe to detect a WEP violation. However, analyzing the MICROSCOPE data will be challenging, mostly because of missing data and a colored noise burrying the signal of interest. In this communication, we apply an *inpainting* technique to simulated MICROSCOPE data and show that *inpainting* will help detect a WEP violation signal.

Keywords. Gravitation, Methods: data analysis, Methods: statistical

1. Introduction

Theories beyond the Standard Model and General Relativity (GR) are under development to solve the various challenges fundamental physics and cosmology are faced with. Among them, we can cite string theory and loop quantum theory, which aim to unify GR and quantum physics, or modified gravity theories such as scalar-tensor theories that aim to explain the accelerated expansion of the Universe with a modification of gravity on cosmological scales. Most of those theories predict a violation of the Weak Equivalence Principle (WEP).

The WEP states the equivalence of the gravitational mass m_g and of the inertial mass m_i. It is a cornerstone of GR: any detection of a WEP violation will be a smoking gun for new physics, with significant implications for cosmology. Various experiments have been conducted over a century on the ground, bringing more and more stringent limits on possible WEP violations, by comparing how two bodies "A" and "B" made of different composition behave in the same gravitational field: the Eötvös ratio

$$\eta = 2\frac{(m_g/m_i)_A - (m_g/m_i)_B}{(m_g/m_i)_A + (m_g/m_i)_B} \tag{1.1}$$

quantifies a WEP violation. The best upper limit is currently $\eta \leqslant 10^{-13}$. As new gravity theories predict a WEP violation at the level $10^{-18} \leqslant \eta \leqslant 10^{-14}$, it is important to better test the WEP. Since on-ground experiments are now reaching their limits, it becomes important to bring experiments to space. MICROSCOPE is a French Space Agency funded mission which aims to test the WEP in space at the level $\eta = 10^{-15}$, expected for launch in 2016.

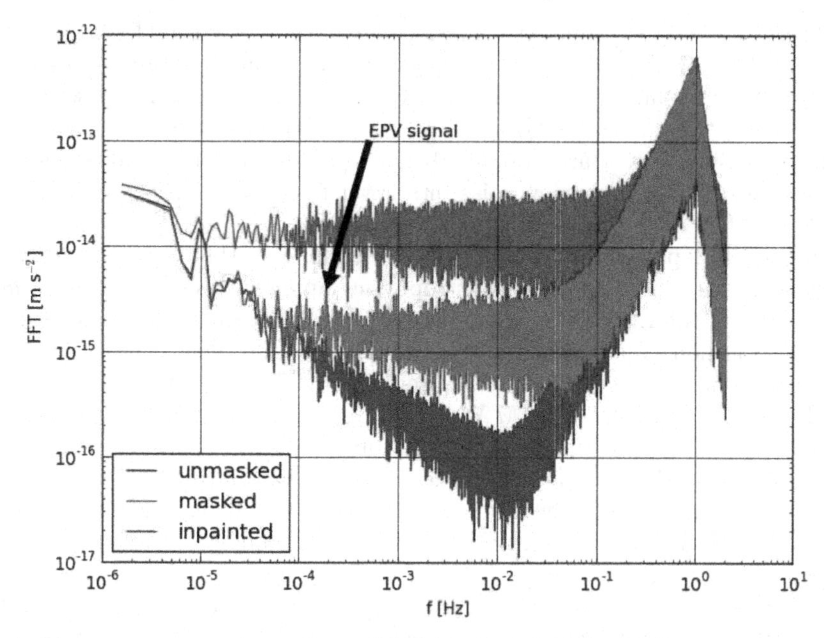

Figure 1. Linear power spectrum of simplified, simulated MICROSCOPE data. In blue, the original data; in green, the data with gaps; in red, the data with gaps filled in by *inpainting*. The simulation includes instrumental noise and a pure sine signal representing the WEP violation signal we are looking for, shown by the black arrow.

In this communication, we first present the MICROSCOPE mission, before emphasizing on its measurement challenge: detecting a very small signal burried in a colored noise and plagued by missing data.

2. MICROSCOPE

MICROSCOPE (Touboul 2009, Touboul *et al.* 2012) is a drag-free microsatellite which aims to test the WEP by comparing the acceleration experienced by two free-falling test masses in the Earth gravity field. It embarks two ultrasensitive electrostatic differential accelerometers. Each accelerometer is made up of two coaxial cylindrical proof masses whose motion is electrostatically constrained. In one (reference) accelerometer, the proof masses are made of the same material to demonstrate the experiment accuracy; they are made of different materials in the second accelerometer, which is used to test the WEP. The electric potential applied to keep the masses in equilibrium is a measure of the difference in the proof masses motion; hence, a non-zero applied potential is a measure of a WEP violation.

3. Data analysis

The MICROSCOPE data will consist of time series from which we will have to extract the signal of interest at a well known frequency that depends on the satellite's orbit and attitude. However, this signal will be burried in a colored stochastic noise, contaminated by various deterministic signals, and plagued by holes due to missing or invalid data. Therefore, its precise extraction and estimation require sophisticated methods. The blue curve in Fig. 1 shows the linear power spectrum of a simplified expected simulated signal: the instrumental noise increases both at low and high frequency, where it largely exceeds the amplitude of the expected WEP violation signal, shown by the "EPV signal" arrow. For the sake of simplicity, we did not simulate any contaminating deterministic signal.

Gaps in time series and invalid data are expected to occur because of several causes, such as telemetry loss, micrometeorite impacts, saturation due to mechanical crackles, and result in missing data (Hardy *et al.* 2013). They affect the measured power spectrum by a leakage of the high frequency noise into lower frequencies, where a WEP violation signal is looked for, as shown by the green curve in Fig. 1.

Here, we fill in gaps thanks to an *inpainting* technique, that has already been used to deal with missing data in several astrophysics applications (e.g. Pires *et al.* 2009, 2014): the missing information is extrapolated from the available data using a sparsity prior on the solution (Elad *et al.* 2005). Let $X(t)$ be the ideal complete time series, $Y(t)$ the measured incomplete time series and $M(t)$ the binary mask (1 where we have data, 0 elsewhere), such that $Y = MX$. Inpainting consists of recovering $X(t)$ knowing $Y(t)$ and $M(t)$. Among all the possible solutions, we search for the sparsest solution in a given representation ϕ^T (i.e. the time series X that can be represented with the fewest coefficients $\alpha = \phi^T X$). The solution is obtained by solving min $||\phi^T X||_0$ under the constraint $||Y - MX||^2 \leqslant \sigma^2$ where the l_0 norm is the number of non-zero coefficients, $||z||^2 = \sum_k z_k^2$ is the classical l_2 norm and σ is the noise rms. If the time series X is sparse enough in the representation ϕ^T (i.e. few strong coefficients can represent the data), the l_0 pseudo-norm can be replaced by the convex l_1 norm, i.e. $||z||_1 = \sum_k |z_k|$ (Donoho & Huo 2001), and threfore its global minimum can be reached by descent techniques.

The red curve in Fig. 1 shows the result of using *inpainting* on our simulated MICROSCOPE data. The noise level (due to missing data) is clearly decreased, especially at low frequency, where we search for the WEP violation signal. From the figure, we see that after inpainting, it becomes possible to detect the peak due to the signal.

A weighted least square method can then be used to detect and characterize the WEP violation signal.

Acknowledgements

We thank Pierre Touboul, Gilles Métris, Emilie Hardy, Manuel Rodrigues and Bruno Christophe for useful discussions. This work makes use of technical data from the CNES-ESA-ONERA-CNRS-OCA Microscope mission, and has received financial support from ONERA and CNES. We acknowledge the financial support of the UnivEarthS Labex program at Sorbonne Paris Cité (ANR-10-LABX-0023 and ANR-11-IDEX-0005-02).

References

Donoho, D. & Huo, X. 2001, *IEEE Trans. Inf. Theory*, 47, 2845

Elad, M., Starck, J.-L., Querre, P., & Donoho, D. 2005, *J. Appl. Comp. Harmon. Anal.*, 19, 340

Hardy, É., Levy, A., Métris, G., Rodrigues, M., & Touboul, P. 2013, *Space Sci. Rev.*, 180, 177

Pires, S., Starck, J.-L., Amara, A., *et al.* 2009, *MNRAS*, 395, 1265

Pires, S., *et al.* submitted

Touboul, P. 2009, *Space Sci. Rev.*, 148, 455

Touboul, P., Métris, G., Lebat, V., & Robert, A. 2012, *Classical Quant. Grav.*, 29, 184010

Statistical Challenges in 21st Century Cosmology
Proceedings IAU Symposium No. 306, 2014
A. F. Heavens, J.-L. Starck & A. Krone-Martins, eds.

doi:10.1017/S1743921314013441

Fundamental Cosmology with the E-ELT

C. J. A. P. Martins[1]†, A. C. O. Leite[1,2] and P. O. J. Pedrosa[1,2]

[1]Centro de Astrofísica, Universidade do Porto, Rua das Estrelas, 4150-762 Porto, Portugal
email: Carlos.Martins@astro.up.pt

[2]Faculdade de Ciências, U. Porto, Rua do Campo Alegre, 4150-007 Porto, Portugal

Abstract. The evidence for the acceleration of the universe shows that canonical theories of cosmology and particle physics are incomplete, and that new physics is out there, waiting to be discovered. Forthcoming high-resolution ultra-stable spectrographs will play a key role in this quest for new physics. Here we focus on astrophysical tests of the stability of nature's fundamental couplings, and by taking existing VLT data as a starting point we discuss how forthcoming improvements (in particular with the E-ELT) will impact on fundamental cosmology.

Keywords. Cosmology: Observations, Cosmology: theory, Techniques: spectroscopic, Methods: data analysis, Instrumentation: spectrographs

---

1. Introduction

While ΛCDM provides the simplest viable cosmological model, fine-tuning arguments suggest that alternatives involving scalar fields may be more likely. Astrophysical measurements of nature's fundamental couplings can be used to study the properties of dark energy, either by themselves or in combination with other cosmological datasets. They complement other methods due to the large redshift lever arm. In Amendola *et al.* (2012) we extended Principal Component Analysis (PCA) methods and studied the feasibility of applying them to astrophysical measurements of varying couplings—whether they are detections of variations or null results.

Recent results of Webb *et al.* (2011) suggest that the fine-structure constant has varied over the last ten billion years, the relative variation being at the level of a few parts per million. Efforts to test this claim are ongoing, but a detailed answer may have to wait for the next generation of facilities. It is clear that observation time on these will scarce, and optimized observational strategies are essential. In Leite *et al.* (2014) we take some steps towards quantifying the potentialities of this method. We use current α measurements from VLT/UVES as a benchmark that can be extrapolated into future (simulated) datasets whose impact for dark energy characterization can be studied, specifically focusing on ESPRESSO (for the VLT), and especially in the E-ELT's high-resolution spectrograph (ELT-HIRES). We summarize these results here.

Our formalism is described in Amendola *et al.* (2012). We will consider quintessence-type models coupled to electromagnetism, though $\mathcal{L}_{\phi F} = -1/4 B_F(\phi) F_{\mu\nu} F^{\mu\nu}$ where the gauge kinetic function $B_F(\phi)$ is assumed linear. Then, the evolution of α is linearly proportional to the field displacement, the proportionality parameter being a dimensionless coupling parameter ζ. From this one can calculate the Fisher matrix using standard techniques. We consider three fiducial forms for the equation of state parameter, respectively $w_c(z) = -0.9$, $w_s(z) = -0.5 + 0.5\tanh(z - 1.5)$, and $w_b(z) =$

† We acknowledge the financial support of grant PTDC/FIS/111725/2009 from FCT (Portugal). C.J.M. is also supported by an FCT Research Professorship, contract reference IF/00064/2012, funded by FCT/MCTES (Portugal) and POPH/FSE (EC).

Table 1. The coefficients A and B in the fitting formula 1.1, assuming 20 or 30 PCA bins in the redshift range $0 < z < 4$ and uncertainties σ_α in parts per million.

Model	A ($N_{\rm bin}$ = 20)	B ($N_{\rm bin}$ = 20)	A ($N_{\rm bin}$ = 30)	B ($N_{\rm bin}$ = 30)
Constant	1.14	0.52	1.39	0.63
Step	2.10	0.96	2.53	1.16
Bump	1.65	0.75	2.00	0.91

$-0.9 + 1.3 \exp\left(-(z-1.5)^2/0.1\right)$. Phenomenologically, these describe the three qualitatively different interesting scenarios: an equation of state that remains close to a cosmological constant throughout the probed redshift range, one that evolves towards a matter-like behaviour by the highest redshifts probed, and one that has non-trivial features over a limited redshift range. In what follows we will refer to these three cases as the *constant*, *step* and *bump* fiducial models.

In order to systematically study possible observational strategies, it's of interest to find an analytic expression for the behaviour of the uncertainties of the best determined PCA modes described above. By exploring numbers of measurements N_α between 20 and 200, uniformly distributed in redshift up to $z = 4$, and individual measurement uncertainties between 10^{-5} and 10^{-8} we find the following fitting formula for the uncertainty σ_n for the n-th best determined PCA mode

$$\sigma_n = A\frac{\sigma_\alpha}{N_\alpha^{0.5}}[1 + B(n-1)]\,. \tag{1.1}$$

The coefficients A and B will depend on the choice of fiducial model, and also on the number of PCA bins assumed for the redshift range under consideration. Table 1 lists these coefficients for choices of 20 and 30 bins. Notice that it's useful to provide the uncertainly σ_α in the fitting formula in parts per million, since in that case the coefficients A and B are of order unity.

2. Calibrating with VLT data

A time normalisation can be derived from the present VLT performances. We can assume a simple (idealised) observational formula, $\sigma^2_{sample} = C/T$, where C is a constant, T is the time of observation necessary to acquire a sample of N measurements and σ_{sample} is the uncertainty in $\Delta\alpha/\alpha$ for the whole sample. This is expected to hold for a uniform sample (ie, one in which one has N_α identical objects, each of which produces a measurement with the same uncertainty σ_α in a given observation time). Clearly any real-data sample will not be uniform, so there will be corrections to this behaviour. The uncertainty of the sample will be given by $\sigma^2_{sample} = 1/\sum_{i=1}^N \sigma_i^{-2}$, and for the above simulated case with N measurements all with the same α uncertainty we trivially have $\sigma^2_{sample} = \sigma^2_\alpha/N$. We have used the UVES data from King (2011), complemented by observation time data provided by Michael Murphy, to build a sample to calibrate the observational formula.

The analysis of Leite *et al.* (2014) shows strong correlation between the number of transitions used to make one measurement (N_λ) and the uncertainty corresponding to it. A simple parametrisation shows the following approximate relation

$$\sigma_{\Delta\alpha/\alpha} = 1,39 \times 10^{-4} N_\lambda^{-1,11}\,. \tag{2.1}$$

One consequence of these properties is that the simple relation above will not strictly hold. Nevertheless, there is a simple way to correct it, which consists of allowing the

Table 2. Number of nights needed to achieve an uncertainty $\sigma_1 = 1$ in the best-determined mode, as a function of the fiducial model and the spectrograph used.

Model	Baseline	ESPRESSO	ELT-HIRES
Constant	6.9	0.7	0.02
Step	23.5	2.5	0.07
Bump	14.5	1.5	0.05

former constant C to itself depend on the number of sources. This is easy to understand: in a small sample one typically will have the best available sources; by increasing our sample we'll be adding sources which are not as good as the previous ones, and therefore the overall uncertainty in the α measurement will improve more slowly than in the ideal case (or alternatively one will need more telescope time.

Using standard Monte Carlo techniques we find that a good fit is provided by the linear relation

$$C(N_\alpha) = 0.31\, N_\alpha + 5.02\,. \tag{2.2}$$

Here the constant has been normalised such that σ_{sample} is given in parts per million and T is in nights. As a simple check, for the UVES Large Program for Testing Fundamental Physics, with about 40 nights and 16 sources, we infer from the fitting formula a value of 0.5 parts per million, consistent with the expectations of the collaboration.

3. Future observational strategies

We thus obtain a UVES-calibrated PCA formula

$$\sigma_n = A[1 + B(n-1)] \left[\frac{C(N_\alpha)}{T}\right]^{1/2}\,, \tag{3.1}$$

where the UVES $C(N)$ formula is given by Eq. 2.2. The most striking feature of this result is the explicit (and strong) dependence on the number of sources. Future improvements will come from a better sample selection and optimised acquisition/calibration methods and both of these are expected to significantly reduce this dependence, even eliminating if for moderately sized samples of absorbers. In the case of the ELT-HIRES, a further improvement will come from the larger collecting power.

With simple but reasonable extrapolations, Leite *et al.* (2014) forecast the expected changes to the UVES formula, and from this carry out an assessment of the impact of these measurements for constraining dark energy. These are summarized in Table 2. The gains achieved in going from the current baseline to ESPRESSO and ELT-HIRES are quite clear. We note that a uniform redshift cover is important in obtaining these results, and there are mild dependencies on the fiducial model and other effects. A detailed study of possible observational strategies will be presented elsewhere.

References

Amendola, L., Leite, A. C. O.., Martins, C. J. A. P., Nunes, N. J., Pedrosa, P. O. J.., & Seganti, A. 2012, *Phys. Rev. D*, 86, 063515

King, J. A. 2011, UNSW PhD thesis, *Searching for variations in the fine-structure constant and the proton-to-electron mass ratio using quasar absorption lines* (online e-Print: arXiv:1202.6365)

Leite, A. C. O., Martins, C. J. A. P., Pedrosa, P. O. J., & Nunes, N. J., 2014, *Phys. Rev.* D 90, 063519

Webb, J. K., King, J. A., Murphy, M. T., Flambaum, V. V., Carswell, R. F., & Bainbridge, M. B. 2011, *Phys. Rev. Lett.*, 107, 191101

Statistical Challenges in 21st Century Cosmology
Proceedings IAU Symposium No. 306, 2014
A. F. Heavens, J. -L. Starck & A. Krone-Martins, eds.

doi:10.1017/S1743921314010941

Entropy in universes evolving from initial to final de Sitter eras

José P. Mimoso[1] and Diego Pavón[2]

Dept. Física, Fac. Ciências, Universidade de Lisboa & CAAUL
Campo Grande, Edifcio C8 - P-1749-016 Lisbon, Portugal
email: jpmimoso@fc.ul.pt
[2]Dept. Física, Universidad Autónoma de Barcelona, 08193 Bellaterra (Barcelona), Spain
email: Diego.Pavon@uab.es

Abstract. This work studies the behavior of entropy in recent cosmological models that start with an initial de Sitter expansion phase, go through the conventional radiation and matter dominated eras to be followed by a final de Sitter epoch. In spite of their seemingly similarities (observationally they are close to the Λ-CDM model), different models deeply differ in their physics. The second law of thermodynamics encapsulates the underlying microscopic, statistical description, and hence we investigate it in the present work. Our study reveals that the entropy of the apparent horizon plus that of matter and radiation inside it, increases and is a concave function of the scale factor. Thus thermodynamic equilibrium is approached in the last de Sitter era, and this class of models is thermodynamically correct. Cosmological models that do not approach equilibrium appear in conflict with the second law of thermodynamics. (Based on Mimoso & Pavon 2013)

Keywords. equation of state, cosmology: theory, miscellaneous

---

1. Introduction

Macroscopic systems tend spontaneously to thermodynamic equilibrium. This constitutes the empirical basis of the second law of thermodynamics: The entropy, S, of isolated systems never decreases, $S' \geqslant 0$, and it is concave, $S'' < 0$, at least in the last leg of approaching of the equilibrium (see, e.g. Callen 1960). The second law of thermodynamics encapsulates the underlying microscopic, and statistical description, and hence is an important tool to investigate the consistency of cosmological models.

We have compelling reasons to believe that there was a primordial stage of inflation, and observations revealed another late stage of accelerated expansion. Models interpolating between two end stages dominated by a cosmological constant are therefore of manifest interest. The idea that the universe may have started from an instability of a de Sitter stage can be traced back to Barrow (1986)'s deflationary universe, to Prigogine *et al.* (1989), as well as to Carvalho *et al.* (1992) and Lima & Maia (1994) phenomenological models of irreversible particle production, or equivalently of dissipative effects as discussed by Lima & Germano (1992). The initial de Sitter space exists for the decay time of its constituents. Subsequently, the universe transits its normal radiation and matter dominated eras, to become finally attracted to a new, late de Sitter stage. The details of these Λ decaying models and transitions have been the endeavor of various proposals, e.g. Carneiro (2006), Carneiro and Tavakol (2009), Basilakos *et al.* (2012) and references therein. Recently two models starting and ending in de Sitter eras were put forward in Lima *et al.* (2012) and in Lima *et al.* (2009). We denote these latter models I and II, respectively. Both assume a spatially flat Friedmann-Robertson-Walker metric, and from the observational viewpoint they are very close to the conventional ΛCDM model. The physics behind the models is deeply different. While model I rests on the production of particles induced by the gravitational field (and dispenses altogether with dark energy), model II assumes dark energy in the form of a cosmological constant that in reality varies with the Hubble factor in a manner prescribed by quantum field theory.

Here we report on our findings of Mimoso & Pavon (2013), where have addressed the consistency from the perspective of the second-law of thermodynamics of the latter models promoting a cosmological transition from de Sitter initial and final stages. The laws of thermodynamics set macroscopic criteria that should be met by sound cosmological models.

2. The entropy

The entropy S of the universe is the entropy of the apparent horizon, $S_h = k_B\,\mathcal{A}/(4\,\ell_{pl}^2)$, plus the entropy of the radiation, S_γ, and/or pressureless matter, S_m, inside it, where $\mathcal{A}$ and ℓ_{pl} denote the area of the horizon and Planck's length, respectively (see Radicella & Pavon 2012 and references therein). The area of the apparent horizon

$$\mathcal{A} = 4\pi\tilde{r}_{\mathcal{A}}^2\,, \tag{2.1}$$

where $\tilde{r}_{\mathcal{A}} = (\sqrt{H^2 + ka^{-2}})^{-1}$ is the radius of the horizon. Accordingly, for a flat model, the entropy of the apparent horizon is $S_h = k_B\pi/(\ell_{pl}\,H)^2$, as the universe transits from de Sitter, $H = H_I$, to a radiation dominated expansion. In turn, the evolution of the entropy of the radiation fluid inside the horizon can be determined with the help of Gibbs equation (Callen 1960)

$$T_\gamma\,dS_\gamma = d\left(\rho_\gamma\,\frac{4\pi}{3}\tilde{r}_{\mathcal{A}}^3\right) + p_\gamma\,d\left(\frac{4\pi}{3}\,\tilde{r}_{\mathcal{A}}^3\right)\,, \tag{2.2}$$

On the other hand, for the entropy of dust matter, it suffices to realize that every single particle contributes to the entropy inside the horizon by a constant bit, say k_B. Then,

$$S_m = k_B\frac{4\pi}{3}\tilde{r}_{\mathcal{A}}^3\,n\,, \tag{2.3}$$

where the number density of dust particles obeys the conservation equation

$$n' = (n/(aH))[\Gamma_{dm} - 3H] < 0 \tag{2.4}$$

with the decay into matter characterized by the rate $\Gamma_{dm} = 3H_0^2\,\tilde{\Omega}_\Lambda/H > 0$.

In model I the creation rate of massless, radiation particles is given by $p_c = -(1 + w)\rho\Gamma_r/(3H)$. One derives $H = H(a)$, and subsequently $S(a)$, $S'(a)$, $S''(a)$, and $T(a)$ from the above expressions for S_h, S_γ and S_m We find that in the phenomenological model of Lima *et al.* (2012) the universe behaves as an ordinary macroscopic system (Radicella & Pavon 2012); i.e., it eventually tends to thermodynamic equilibrium characterized by a never ending de Sitter expansion era with $H_\infty = H_0\,\sqrt{\tilde{\Omega}_\Lambda} < H_0$.

In the model II (Lima *et al.* 2009) it is assumed that in quantum field theory in curved spacetime the cosmological constant is a parameter that runs with the Hubble rate in a specified manner (see Parker & Toms 2009 and Solá 2011):

$$\Lambda(H) = c_0 + 3\nu H^2 + 3\alpha\frac{H^4}{H_I^2}\,, \tag{2.5}$$

where c_0, α and ν are are constant parameters of the model. The absolute value of the latter is constrained by observation as $|\nu| \sim 10^{-3}$. At early times the last term dominates, and at late times ($H \ll H_I$) it becomes negligible whereby (2.5) reduces to $\Lambda(H) = \Lambda_0 + 3\nu(H^2 - H_0^2)$ with $\Lambda_0 = c_0 + 3\nu H_0^2$.

Proceeding to equate the relevant quantities in terms of the scale factor, we find that, as in the previous model, $S_h' > 0$ and $S_h'' < 0$. However at variance with it, the matter entropy, $S_m = k_B \frac{4\pi}{3}\tilde{r}_{\mathcal{A}}^3 n \propto H^{-3} n$, decreases with expansion and is convex. This is so because, in this case, the rate of particle production, Γ_{dm}, goes down and cannot compensate for the rate of dilution caused by cosmic expansion. Nevertheless, S_h' and S_h'' dominate over S_m' and S_m'', respectively, as $a \to \infty$. Thus, as in model I, the total entropy results a growing and concave function of the scale factor, in the far future stage. Hence, the universe gets asymptotically closer and closer to thermodynamic equilibrium.

3. Conclusions

In both models considered, the entropy, as a function of the scale factor, never decreases and is concave at least at the last stage of evolution, signaling that the universe is finally approaching thermodynamic equilibrium. So, we conclude that models I and II show consistency with thermodynamics, and that their overall behavior can be most easily understood from the thermodynamic perspective. Further, these results remain valid also if quantum corrections to Bekenstein-Hawking entropy law are incorporated.

Acknowledgements

This research was partially funded by FCT through the projects CERN/FP/123618/2011 and CERN/FP/123615/2011, by the "Ministerio Español de Economía y Competitividad" under Grant number FIS2012-32099, and by the "Direcció de Recerca de la Generalitat" under Grant No. 2009SGR-00164.

References

Barrow, J. D. 1986, *Phys. Lett. B*, 180, 335

Basilakos, S., Polarski, D., & Sola, J. 2012, *Phys. Rev. D*, 86, 043010, arXiv:1204.4806

Callen, H. 1960, *Thermodynamics*, J. Wiley, N.Y.

Carneiro, S. 2006, *Int. J. Mod. Phys. D*, 15, 2241, arXiv:gr-qc/0605133

Carneiro, S. & Tavakol, R. 2009, *Gen. Rel. Grav.*, 41, 2287; *Int. J. Mod. Phys. D*, 18, 2343, arXiv:0905.3131

Carvalho, J. C., Lima, J. A. S., & Waga, I. 1992, *Phys. Rev. D*, 46, 2404

Lima, J. A. S., Basilakos, S., & Solá, J. 2012, arXiv:1209.2802

Lima, J. A. S., Basilakos, S., & Costa, F. E. M. 2012, *Phys. Rev. D*, 86, 103534

Lima, J. A. S. & Germano, A. S. M. 1992, *Phys. Lett. A*, 170, 373

Lima, J. A. S. & Maia, J. M. F. 1994, *Phys. Rev. D*, 49, 5597

Mimoso, J. P. & Pavón, D. 2013, *Phys. Rev. D*, 87, 047302

Parker, L. E. & Toms, D. J. 2009, *Quantum Field Theory in Curved Spacetime: quantized fields and gravity*, Cambridge University Press.

Prigogine, I., Geheniau, J., Gunzig, E., & Nardone, P. 1989, *Gen. Rel. Grav.*, 21, 767

Prigogine, I. 1989, *Int. J. Theor. Phys.*, 28, 927

Radicella, N. & Pavón, D. 2012, *Gen. Relativ. Grav.*, 44, 685

Shapiro, I. L. & Sola, J. 2009, *Phys. Lett. B*, 682, 105

Solá, J. 2011, *J. Phys. Conf. Ser.*, 283, 012033

Statistical Challenges in 21st Century Cosmology
Proceedings IAU Symposium No. 306, 2014
A. F. Heavens, J.-L. Starck & A. Krone-Martins, eds.

doi:10.1017/S1743921314010850

The Stochastic Gravitational Wave Background Generated by Cosmic String Networks

L. Sousa[1] and P. P. Avelino[1 2]

[1]Centro de Astrofísica da Universidade do Porto, Rua das Estrelas, 4150-762 Porto, Portugal
[2]Departamento de Física e Astronomia da Faculdade de Ciências da Universidade do Porto, Rua do Campo Alegre 687, 4169-007 Porto, Portugal
emails: Lara.Sousa@astro.up.pt, pedro.avelino@astro.up.pt

Abstract. Cosmic string interactions often result in the formation of cosmic string loops that detach from the long string network and radiate their energy in the form of gravitational waves. Loop production occurs copiously throughout the cosmological evolution of a cosmic string network and the superimposition of their emissions gives rise to a stochastic gravitational In this essay, we briefly review our recent work on the stochastic gravitational wave background generated by cosmic string networks and introduce a set of numerical and analytical tools for the computation of this background.

Keywords. cosmology: theory, early universe, gravitational waves

1. Introduction

The production of cosmic string networks as remnants of symmetry-breaking phase transitions in the early universe is predicted in a wide variety of grand unified scenarios. These cosmic string networks, despite being formed in the early universe, may survive throughout the cosmological history, potentially leaving behind a plethora of observational signatures. One such signature would be a stochastic gravitational wave background (SGWB) generated as a result of the radiative emissions of cosmic string loops (Vilenkin (1981)). These loops are created in significant numbers as a result of cosmic string collisions and interactions (for instance, when a string self-intersects or when two strings intersect simultaneously at more than one point). Once these loops are created, they detach from the long string network and start to oscillate relativistically under the effect of their tension and decay radiatively by emitting gravitational waves.

If one assumes that the cosmic string network is statistically homogeneous on sufficiently large scales, two variables are sufficient to describe its large-scale dynamics: the characteristic lengthscale — $L \equiv (\mu/\rho)^{1/2}$, where μ is the string tension, and ρ is the average energy density of long strings — and its root-mean-square velocity, $\bar{v}$. Under this assumption, the rate of loop production (per comoving volume) is given by

$$\frac{dn_c}{dt} = \frac{\tilde{c}}{\alpha}\frac{\bar{v}}{L^4}\,, \tag{1.1}$$

where α is a parameter that characterizes the size of loops, $l_c = \alpha L(t_c)$, at the moment of creation, t_c, and $\tilde{c}$ is a phenomenological parameter that characterises the efficiency of the loop-chopping mechanism. The rate of loop production is, then, strongly dependent on the large-scale properties of the network and, therefore, its accurate characterisation requires an accurate description of cosmic string network dynamics.

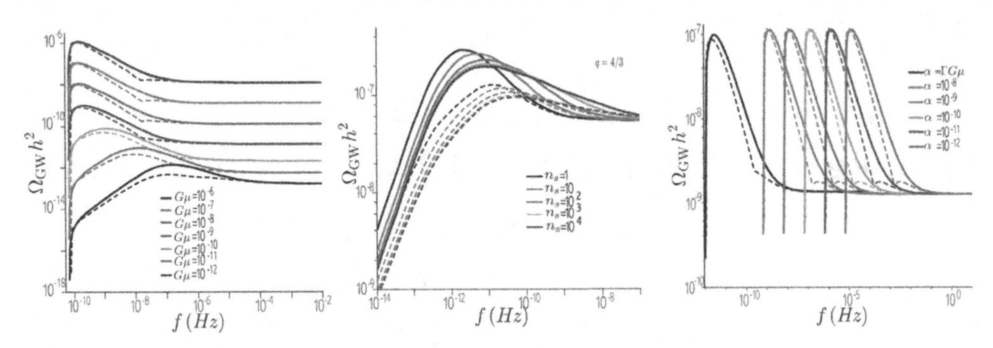

Figure 1. Same examples of SGWB power spectra. *From left to right:* the SGWB spectrum for 1. various values of cosmic string tension (and fixed $\alpha = 10^{-7}$); 2. large cuspy loops ($\alpha = 10^{-1}$), considering n_s harmonic modes of emission; 3. small loops (and fixed $G\mu = 10^{-7}$). Solid lines represent the SGWB spectra computed numerically using the VOS model to describe cosmic string dynamics, while dashed lines represent the spectra of networks experiencing scale-invariant evolution.

2. SGWB: Numerical results

The Stochastic gravitational wave background is often quantified using the spectral density of gravitational waves per logarithmic frequency (f) interval, in units of critical density (ρ_c),

$$\Omega_{\rm GW} = \frac{1}{\rho_c}\frac{d\rho_{\rm GW}}{d\log f}, \qquad (2.1)$$

where $\rho_{\rm GW}$ is the gravitational radiation energy density. This quantity is strongly dependent on the rate of loop production as a function of time, and, consequently, on the large-scale cosmic string parameters. In this context, it is generally assumed that cosmic string networks experience scale-invariant evolution throughout cosmological history, with the characteristic lengthscale growing linearly with time ($L \propto t$) and constant $\bar{v}$. However, this is not an accurate depiction of realistic cosmic string network dynamics: once the radiation-matter transition is triggered, the network cannot maintain scale-invariant evolution. In order to improve current estimations of SGWB spectrum, we have developed, in Sousa & Avelino (2013), an algorithm to compute the spectral density of gravitational waves generated by realistic cosmic string networks. This algorithm relaxes the assumption of scale-invariant cosmic string evolution, and uses the Velocity-Dependent One-Scale (VOS) model (Martins & Shellard (1996)) — which was previously shown to provide an accurate description of the large-scale dynamics of cosmic string networks — to describe their dynamics.

We have demonstrated that the simplifying assumptions about network dynamics used in the majority of SGWB computations lead to an underestimation of the number of loops created during most of the matter era and that this causes a significant underestimation of the amplitude and broadness of the peak of the spectrum (see Fig. 1). As a result, the total gravitational wave energy density at the present time obtained using the VOS model may be 25 – 70% larger (depending on the size of cosmic string loops) than that obtained under the scale-invariant assumption. The use of simplified models of string evolution may, therefore, lead to inaccuracies in the computation of the observational constraints on cosmic string tension using data from direct or indirect gravitational wave detection experiments.

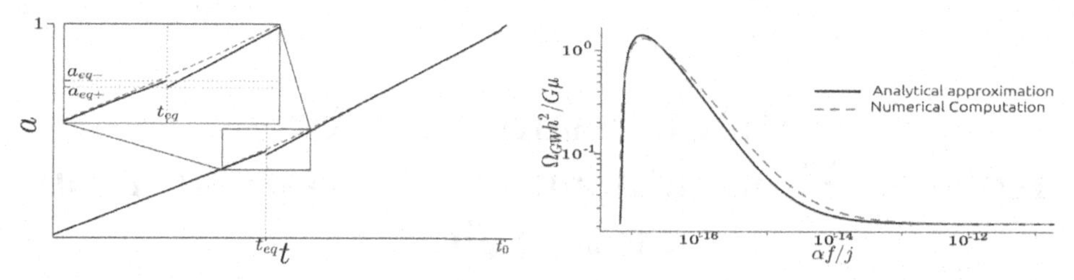

Figure 2. *Left Panel:* Fitting function for the cosmological scale factor. *Right Panel:* Analytical approximation to the SGWB power spectrum in the small loop regime.

3. SGWB: Analytical Approximation for the small-loop regime

If loops are small — smaller than the gravitational backreaction scale — they have a short lifespan, surviving less than a Hubble time. In this limit, it is reasonable to assume that loops radiate their energy effectively immediately in the cosmological timescale. Using this simplification, we have developed in Sousa & Avelino (2014) an alternative method to compute the SGWB spectrum in the small-loop regime that has the advantage of requiring significantly less computation time.

This method also allowed us to derive an analytical approximation to the realistic SGWB generated by small cosmic string loops:

$$\Omega_{\rm gw}^j h^2(f) = \mathcal{K}\left\{\frac{\bar{v}_r}{\xi_r^3}\frac{\alpha}{jt_{eq}^2}\left(\frac{a_{eq-}}{a_0}\right)^4 + \frac{3}{f}\frac{\bar{v}_m}{\xi_m^4 t_\Lambda^3}\left(\frac{a_{\Lambda-}}{a_0}\right)^5\left[1-\frac{2j}{\alpha\xi_m t_\Lambda f}\left(\frac{a_{\Lambda-}}{a_0}\right)\right]\right\}, \quad (3.1)$$

for $f \geqslant 2ja_{\Lambda-}/t_\Lambda$. Here we have introduced the constants $\mathcal{K} = (16j\pi G\mu h^2\tilde{c})/(3H_0^2\alpha)$, and $a_{\Lambda-} = a_{eq+}(t_\Lambda/t_{eq})^{2/3}$, used $j = 1, 2, ...$ to label the harmonic mode of loop emission, and $t_{\rm eq}$ and t_Λ, respectively, for the time of radiation-matter equality and of matter-cosmological-constant equality. Moreover, a_0 and H_0 represent, respectively, the value of the scale factor and the Hubble parameter at the present time. In deriving this expression, we have assumed that the network undergoes scale-invariant evolution throughout cosmological history — characterized by $\bar{v} = \bar{v}_r$ ($\bar{v} = \bar{v}_m$) and $L = \xi_r t$ ($L = \xi_m t$) during the radiation (matter) era. We have also assumed that the dynamics of the universe is determined, at any given time, by the dominant component of the its energy density, in order to construct an analytical fitting function for the scale factor (see the left panel of Fig. 2). This analytical approximation to the SGWB spectrum provides an excellent fit to that of realistic cosmic string networks (see the right panel of Fig. 2). It constitutes a useful tool for a first estimation of the SGWB spectrum, thus allowing for simple estimates of the associated observational constraints in the small-loop regime.

Acknowledgements

L.S. is supported by FCT and POPH/FSE through the grant SFRH/BPD/76324/2011. P.P.A. is supported by FCT and POPH/FSE through contract IF/00863/2012. This work was also partially supported by grant PTDC/FIS/111725/2009 (FCT).

References

Vilenkin, A. 1981, *Phys.Lett.* B107, 47
Sousa, L. & Avelino, P. P. 2013, *Phys.Rev.* 2 D88, 023516
Sousa, L. & Avelino, P. P. 2014, *Phys.Rev.* D89, 083503
Martins, C. J. A. P. & Shellard , E. P. S 1996, *Phys.Rev.* D54, 2535-2556

Statistical Challenges in 21st Century Cosmology
Proceedings IAU Symposium No. 306, 2014
A. F. Heavens, J.-L. Starck & A. Krone-Martins, eds.

doi:10.1017/S1743921314013519

Shape estimation for Košice, Almahata Sitta and Bassikounou meteoroids

Vladimir Vinnikov[1], **Maria Gritsevich**[1,2,3] **and Leonid Turchak**[1]

[1]Dorodnicyn Computing Centre of the Russian Academy of Sciences, Department of Computational Physics, Vavilova ul. 40, 119333, Moscow, Russia
email: vvinnikov@list.ru

[2]Finnish Geodetic Institute, P.O. Box 15, FI-02431 Masala, Finland;
email: maria.gritsevich@fgi.fi

[3]Institute of Physics and Technology, Ural Federal University, Mira ul. 21, 620002, Ekaterinburg, Russia

Abstract. This paper is concerned with a meteoroid shape estimation technique based on statistical laws of distribution for fragment masses. The idea is derived from the experiments that show that brittle fracturing produces multiple fragments of size lesser than or equal to the least dimension of the body. The number of fragments depends on fragment masses as a power law with exponential cutoff. The scaling exponent essentially indicates the initial form of the fragmented body. We apply the technique of scaling analysis to the empirical data on the mass distributions for Košice, Almahata Sitta and Bassikounou meteorites.

Keywords. Methods: data analysis, methods: statistical, Shape estimation, scaling law, meteoroid fragmentation, mass distribution.

Introduction

The estimation of initial meteoroid shape is crucial, since the meteoroid pre-entry mass, terminal meteorite mass, and fireball luminosity are proportional to the pre-entry shape factor of the meteoroid to the power of 3 (Gritsevich & Koschny 2011). However, its reconstruction is not straightforward. Unlike a classical puzzle, the meteorite fragments found within one fall do not match each other since not all body splinters reach the Earth's surface, and those that do suffer from melting and ablation, which smooth distinct crack surfaces and round sharp corners. Fortunately, in many cases the ablation is considered to be weak after the major fragmentation occurred (Spurny *et al.* 2003; Gritsevich 2008) and thus the resulting mass distribution does not change significantly (Oddershede *et al.* 1998). Moreover, the brittle shattering has fractal properties similar to many other natural phenomena (e.g. Lang & Franaszczuk 1986; Nagahama 1993). This self-similarity for scaling mass sequences is described by power law statistical expressions (Oddershede *et al.* 1993; Amitrano 2012). Simply speaking, it means that any observed limited subset of small splinter masses does not give any information about their relative location in the overall distribution. This subset can be placed anywhere on the probability density curve, resulting only in scale change of the total mass. The same feature is exhibited by more classical fractals, such as clouds or mountain rocks. It is theoretically impossible to deduce the absolute size of fractal objects while watching their parts without comparison to surroundings.

1. Shape estimation

The experiments on brittle fracturing demonstrate that the resulting multiple fragments have the size lesser than or equal to the least dimension of the body (Oddershede *et al.* 1993). The shape-estimation technique is based on the idea that such pieces obey the power law with scaling exponent β_0 and exponential cutoff: $F_c(m) = C\cdot m^{-\beta_0}\cdot \exp(-m/m_U)$, where $C = (\beta_0 - 1)\cdot m_L^{\beta_0 - 1}$ is a normalization constant, $m_U > m_L$ is an upper cutoff fragment mass and m_L is an arbitrary lower mass limit, acting as an additional constraint for undersampled tiny unrecoverable fragments. These mass constraints are also among the sought parameters. The meteorite sample consists of N fragments ranging in mass from m_0 to m_{N-1} presented in ascending order. If two or more fragments yield equal tabular masses, we add a small value corresponding to the fragment mass measurements error (for example, 0.001 g) to one of them, and move the masses apart. Since the density distribution $n(m)$ of fragment masses $\{m_i\}_{i=1}^{N}$ is discrete, we convert it to the piecewise complementary cumulative distribution function (CCDF) $N^*(m)$ as suggested in (Oddershede *et al.* 1998): $N^*(m) = \frac{1}{m}\int_m^\infty n(x)\,dx$, where $m \geqslant m_L$, $n(m) = \sum_{i=1}^{N}\delta(m - m_i)$ and δ is the Dirac delta function.

In order to match $F_c(m)$ to $N^*(m)$ we conduct a normalization procedure, equaling their values at the point m_L. For simplicity, we also assume that the lower bound m_L coincides with the mass of the fragments m_j, i.e. $n_L = j$, $m_L = mj$. Thus the normalization constant C can be expressed as follows: $C = (m_j)^{\beta_0 - 1}\cdot \exp(-m_j/m_U)\cdot(N - j)$. The CCDF becomes: $F_c(m) = \frac{N-j}{m_j}\cdot(m/m_j)^{-\beta_0}\cdot\exp(-(m - m_j)/m_U)$. We note here that the scaling exponent β_0 is non-dimensional and has no dependence on the dimensions of $N^*(m)$ and $F_c(m)$.

Next, the least-squares method is applied to minimize the expression: $S(\beta_0, j.m_U) = \sum_{i=j}^{N}[F_c(m_i) - (N - j)/m_j]^2$. We use the Newtons method to get β_0 and m_U for the valid range of m_L. Once β_0 is obtained, we can estimate dimensionless shape parameter d and size proportions a_x, a_y, a_z from the empirical equations $0.13d^2 - 0.21d + (1.1 - \beta_0) = 0$ and $d = 1 + 2(a_x\cdot a_y + a_y\cdot a_x + a_z\cdot a_x)\left(a_x^2 + a_y^2 + a_z^2\right)^{-1}$. We emphasize that these equations and relations do have the area of applicability limited to homogeneous brittle solids that are not prestrained. It is known that prestrained materials (e.g. tempered glass) while being brittle do not follow power law mass distribution under fragmentation.

2. Results

We use the described technique on the set of mass distributions, corresponding to well known meteorite showers with large number of recovered fragments. Here we present results based on 3 cases, namely: Košice, Bassikounou and Almahata Sitta. The respective data can be found in (Buhl & Baermann 2007; Shaddad *et al.* 2010; Gritsevich *et al.* 2014). All mass distributions exhibit common features of undersampling for small fragments and exponential cutoff for a finite size effect. The estimated values of scaling exponent β_0, lower and upper constraints m_L, m_U and dimensionless shape parameter d are presented in the table 1. The corresponding plots are provided in figures 1, 2, 3.

The most interesting mass distribution is related to the Almahata Sitta meteorite shower (Shaddad *et al.* 2010). The plot (Fig.3) demonstrates two distinctive linear subranges. The first subrange doubtfully can be attributed only to the undersampling effect. The number of fragments within it is large, and their masses obey the scaling law with β_0 equal to 1.0. It is possible to construct the composite power law which fits both subranges. However the coefficients of the empirical equation imply that the scaling exponent β_0 should be greater than 1.1, if the roots d are expected to be real value. We anticipate that the Almahata Sitta distribution is caused by the independently fragmented meteoroids with different properties. This problem is a subject of the following study.

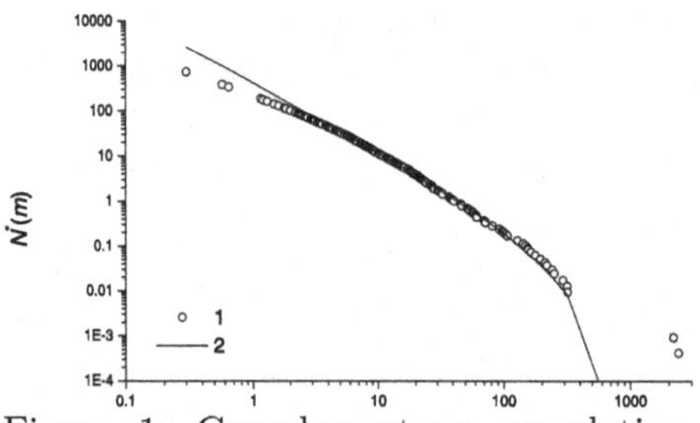

Figure 1: Complementary cumulative number of fragments $N^*(m)$ vs m. 1 – Observed data, 2 – Power law distribution with exponential cutoff. Košice meteorite.

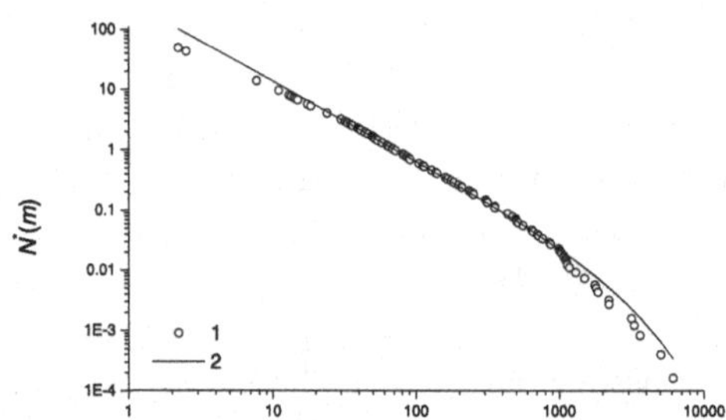

Figure 2: Complementary cumulative number of fragments $N^*(m)$ vs m. 1 – Observed data, 2 – Power law distribution with exponential cutoff. Bassikounou meteorite.

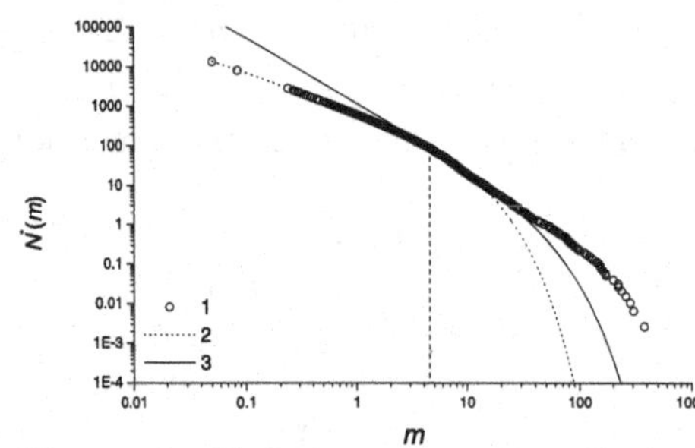

Figure 3: Complementary cumulative number of fragments $N^*(m)$ vs m. 1 – Observed data, 2,3 – Power law distributions with exponential cutoff. Almahata Sitta meteorite.

Table 1: The derived shape parameters for the selected meteoroids

Meteoroid	β_0	m_L,[g]	m_U,[g]	d
Košice	1.53	5.64	155.17	2.8
Bassikounou	1.32	29.9	2839.42	2.34
Almahata Sitta	1.0	0.6	7.84	–
	1.64	4.51	31.97	3.0

References

Amitrano, D 2012, *Eur. Phys. J. Spec. Top.*, 205, 199

Buhl, S. & Baermann, M. 2007, *Niger Meteorite Recon*, www.niger-meteorite-recon.de

Gritsevich, M. 2008, *Sol. Syst. Res.*, 42, 372

Gritsevich, M. & Koschny, D. 2011, *Icarus*, 212, 877

Gritsevich, M., Vinnikov, V., Kohout, T., Tóth, J., Peltoniemi, J., Turchak, L., & Virtanen, J. 2014, *Meteorit. Planet. Sci.*, 49, 328

Lang, B. & Franaszczuk, K. 1986, *Meteoritics*, 21, 425

Nagahama, H. 1993, *Int. J. Rock Mech. Min. Sci. & Geomech. Abstr.*, 30, 469

Oddershede, L., Dimon, P., & Bohr, J. 1993, *Phys. Rev. Lett.*, 71, 3107

Oddershede, L., Meibom, A., & Bohr, J. 1998, *Europhys. Lett*, 43, 598

Shaddad, M. H., Jenniskens, P., Numan, D., Kudoda, A. M., Elsir, S., Riyad, I. F., Ali, A. E., Alameen, M., Alameen, N. M., Eid, O., Osman, A. T., Abubaker, M. I., Yousif, M., Chesley, S. R., Chodas, P. W., Albers J., Edwards, W. N., Brown, P. G., Kuiper, J., & Friedrich, J. M. 2010, *Meteorit. Planet. Sci.*, 45, 1557

Spurny, P., Oberst, J., & Heinlein, D. 2003, *Nature*, 423, 151

Statistical Challenges in 21st Century Cosmology
Proceedings IAU Symposium No. 306, 2014
A. F. Heavens, J. -L. Starck & A. Krone-Martins, eds.

doi:10.1017/S1743921314010655

X-ray cross-correlation analysis of the low-mass X-ray binary 4U 1636-53

Ya-Juan Lei

Key Laboratory of Optical Astronomy, National Astronomical Observatories, Chinese Academy of Sciences, Beijing 100012, China
email: leiyjcwmy@163.com

Abstract. We analyze the cross-correlation function of the soft and hard X-rays of the atoll source 4U 1636-53 with *RXTE* data. The results show that the cross-correlations evolve along the different branches of the color-color diagram. At the lower left banana states, we have both positive and ambiguous correlations, and positive correlations are dominant for the lower banana and the upper banana states. The anti-correlation is detected at the top of the upper banana states. The cross-correlations of two atoll sources 4U 1735-44 and 4U 1608-52 have been studied in previous work, and the anti-correlations are detected at the lower left banana or the top of the upper banana states. Our results show that, in the 4U 1636-53, the distribution of the cross-correlations in the color-color diagram is similar to those of 4U 1735-44 and 4U 1608-52, and confirm further that the distribution of cross-correlations in color-color diagram could be correlated with the luminosity of the source.

Keywords. X-rays: binary, accretion

1. Introduction

Low-mass X-ray binaries (LMXBs) with weakly magnetized neutron stars (NSs) can be classified into two subclasses, Z sources with high luminosity and atoll sources with low luminosity, according to their X-ray spectral and timing properties (Hasinger & van der Klis 1989). In an X-ray color-color diagram (CCD), the Z source traces out a Z track within the shorter timescales (hours to days), whereas the typical atoll source shows a C-shaped track within days to weeks or so. It is suggested that most timing and spectral characteristics on the position along the Z or atoll track are related to the mass accretion rate (van der Klis 2000). With the mass accretion rate increasing, the atoll sources evolve from the island state (IS) to the banana branch, which is usually subdivided into lower left banana (LLB), lower banana (LB), and upper banana (UB) states.

Both the spectral and timing analyses are usually used to explore physical changes in the source emission. The cross-correlation function between different X-ray energy bands is also a useful means for studying the accretion disk. The anti-correlated lags of a few hundred seconds are found in some black hole X-ray binaries, Z and atoll sources. It is suggested that the large timescale of lags could correspond to the viscous timescales of the inner accretion disk (Sriram *et al.* 2012). In earlier work, we firstly detect the anti-correlations in two atoll sources 4U 1735-44 and 4U 1608-52, which occur in the LLB or the top of the UB. The cross-correlated results suggest that the observed anti-correlation may be relevant to the transition between the hard and soft states (Lei *et al.* 2013, 2014). It is obvious that more observations are needed to consolidate the inference. In the work, we study the cross-correlations of 4U 1636-53 with *RXTE* data. 4U 1636-53 is one of the most interesting LMXBs of the atoll source, and it contains a companion star mass $\sim 0.4\ M_{\odot}$ with an orbital period of 3.8 hr (van Paradijs *et al.* 1990). We describe the observations in section 2, show the results and discussion in section 3.

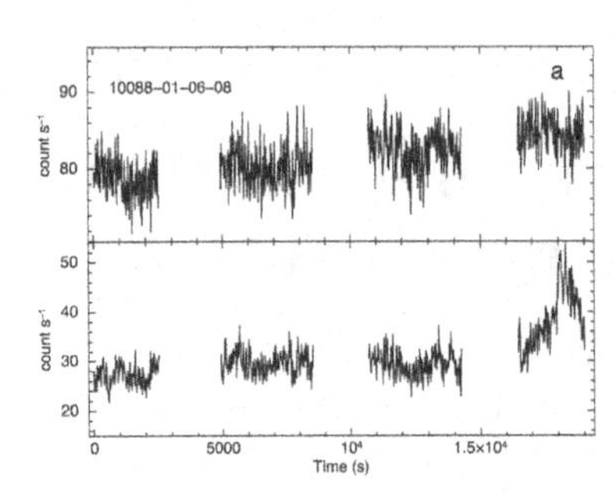

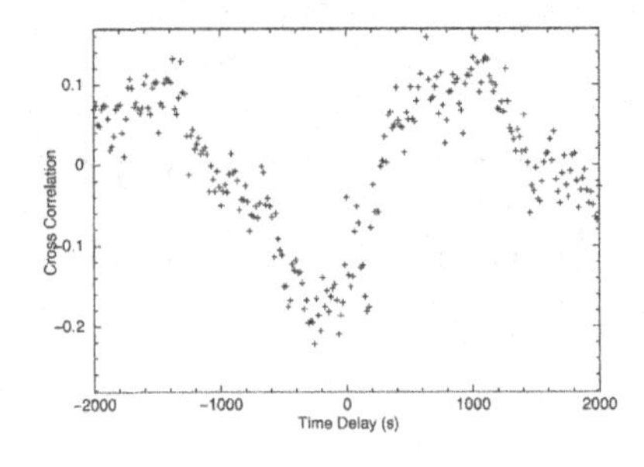

Figure 1. Light curves with bin size of 16 s and cross-correlation of the segment with anti-correlation detected. Symbol " a " corresponds to the individual segment for which anti-correlation is detected.

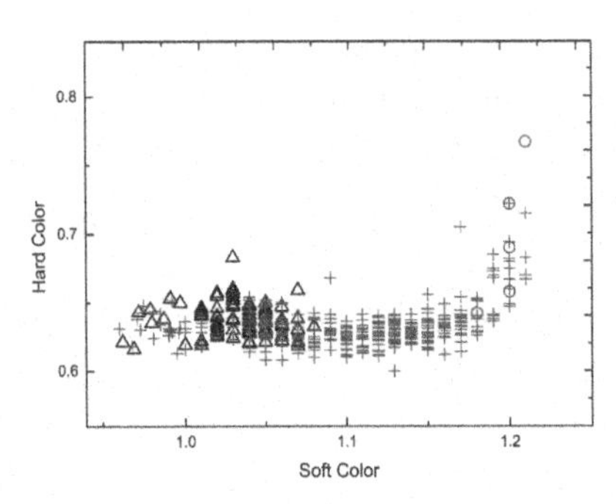

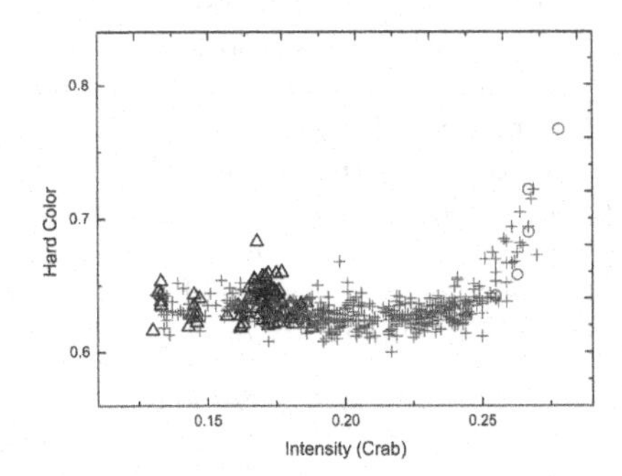

Figure 2. CCD and HID of 4U 1636-53 of the observation P10088. The cross stands for the positive-correlation, the triangle corresponds to the ambiguous and the circle indicates the anti-correlation. Each point is on average of 512 s.

2. Observations

The observations of 4U 1636-53 analyzed in this paper are from the Proportional Counter Array (PCA) on board the *RXTE* satellite during April 1996 to February 1997 (P10088). PCA consists of 5 non-imaging, co-aligned Xe multiwire proportional counter units (PCUs). In this work, only PCU2 data are adopted. The light curves are extracted using standard2 data mode, with bin size of 16 s. The XRONOS tool "*crosscor*" is used for estimating the cross-correlation between the soft X-rays (2.0-3.5 keV) and the hard X-rays (12-30 keV) for the observations whose the exposure time lasting longer than 2000 s. The data of type I X-ray bursts are excluded. For CCD analysis, the soft and the hard colors are defined as the count-rate ratios 3.5-6.0 keV/2.0-3.5 keV and 9.7-16 keV/6.0-9.7 keV, respectively. In the hardness intensity diagrams (HID), the intensity in the energy band 2.0-16 keV is calculated and is normalized to the Crab count rate.

In the same way as the previous work, the cross-correlation results are divided into three groups: positive, ambiguous and anti-correlated classes (Lei *et al.* 2008). An anti-correlation between the soft and hard X-rays refers to a negative cross-correlation coefficient, and vice versa for a positive correlation. Fig. 1 shows the background subtracted light curves, the anti-correlation between soft and hard X-rays for the observation with anti-correlation detected. The distribution of cross-correlations in the CCD of the analyzed observations is shown in Fig. 2.

3. Results and Discussion

We have studied the cross-correlations between the soft and hard X-rays of 4U 1636-53 of the observation P10088 of *RXTE* and find that it is evolutional along the different branches of the CCD. Fig. 2 shows the distribution of cross-correlations in the CCD. One sees from Fig. 2 that, both positive and ambiguous correlations are detected at the LLB in similar proportion, but the positive correlations dominant for the LB and the UB. The anti-correlation of the soft and hard X-rays is detected at the top of the UB (ObsID 10088-01-06-08), where the time lag is about two hundred seconds (Fig. 1). Anti-correlated X-ray time lags of a few hundred seconds seen in some LMXBs suggest that the accretion disk could be truncated (Sriram *et al.* 2012). Therefore at the top of the UB of 4U 1636-53 where the accretion rate is most highest, the accretion disk could be truncated in a radius very close to the NS.

For LMXBs, it is suggested that the cold photons come from the accretion disk, and the high energy photons from the hot corona. In the LLB, the lower accretion rate leads to the accretion disk boundary be away from the hot corona, and which creates the ambiguous and weakly positive correlations of the soft and hard X-rays. In the LB and the UB, the accretion rate increases, and the accretion disk moves inward and is close to the hot corona, so the soft seed photons increase and can be Comptonized effectively. The increasing of both the soft and hard X-rays results in the positive correlation. However, when the luminosity is beyond a critical luminosity at the top of the UB, the Comptonizing cloud shrinks. The soft seed photons increase and cool the Compton cloud, which results in the decreasing of the hard photons, and hence the anti-correlation.

The distribution of cross-correlation in the CCD of 4U 1636-53 is similar to those of atoll sources 4U 1735-44 and 4U 1608-52, where anti-correlations are detected at the LLB or the top of the UB. The spectral analyses of 4U 1735-44 and 4U 1608-52 suggest that both the LLB and the top of the UB could be corresponding to the transition between the hard and soft states. The cross-correlated results of 4U 1636-53 are also consistent with those of black hole X-ray binaries whose anti-correlations are also detected in the transition state.

The studies of 4U 1735-44 and 4U 1608-52 suggest that the position of the anti-correlation in the CCD can be related to the luminosity of the source, i.e., the anti-correlation is located at the top of the UB when the luminosity of a atoll source is lower, and occurs in the LLB when the luminosity is higher (Lei *et al.* 2013, 2014). For the observation P10088 of 4U 1636-53, the luminosity is lower and the anti-correlation is detected at the top of the UB, which is consistent with those of 4U 1735-44 and 4U 1608-52. In addition, the study of XTE J1701-462 also implies that the position of anti-correlations in the CCD might depend on the luminosity of the source (Wang *et al.* 2014). Further work is needed in the future for exploring the relation.

References

Hasinger, G. & van der Klis, M. 1989, *A&A*, 225, 79
Lei, Y. J., Qu, J. L., Song, L. M., *et al.* 2008, *ApJ*, 677, 461
Lei, Y. -J., Zhang, H. -T., Zhang, C. -M., *et al.* 2013, *AJ*, 146, 60
Lei, Y. -J., Zhang, S., Qu, J. -L., *et al.* 2014, *AJ*, 147, 67
Sriram, K., Choi, C. S., & Rao, A. R. 2012, *ApJS*, 200, 16
van der Klis, M. 2000, *ARAA*, 38, 717
van Paradijs, J., van der Klis, M., van Amerongen, S., *et al.* 1990, *A&A*, 234, 181
Wang, Y. N., Lei, Y. J., Ding, G. Q., *et al.* 2014, *MNRAS*, 440, 3726

Statistical Challenges in 21st Century Cosmology
Proceedings IAU Symposium No. 306, 2014
A. F. Heavens, J.-L. Starck & A. Krone-Martins, eds.

doi:10.1017/S1743921314013507

The Role of Statistics and Statisticians in the Future of Astrostatistics

Joseph M. Hilbe

T. Denney Sanford School of SFD, Arizona State University,
P.O. BOX 873701,Tempe, AZ 85287-3701
email: `hilbe@asu.edu`; `j.m.hilbe@gmail.com`

Solar System Ambassador Program, Jet Propulsion Laboratory,
4800 Oak Grove Dr., Pasadena, CA 91109

Abstract. Given the generic definition of statistics, it is clear that astronomers have engaged in statistical analysis of some variety since astronomy first emerged as a science. However, from the early nineteenth century until the beginning of the twenty-first the two disciplines have been somewhat estranged — there was no formal relationship between the two. This has now changed, as is evidenced by the recent creation of the International Astrostatistics Association (IAA), the ISI astrostatistics committee, astrostatistics working groups authorized by the IAU and AAS, and this Symposium. The challenge for us to come is in establishing how statisticians and astronomers relate in developing the discipline of astrostatistics. I shall propose a direction for how the discipline can progress in both the short term and well as for future generations of astrostatisticians.

Keywords. methods: statistical

---

1. Introduction

When I was first invited to present this Closing Address from my perspective as a statistician, I thought of a host of items that could be mentioned and discussed. However, when it came time to actually preparing my remarks for this occasion, I decided that it was best to focus on a single theme. This theme relates to the role that statistics and statisticians have in the future of astrostatistics. However, dual sub-themes are entailed in this general theme. I shall provide an outline of the recent events leading to the emergence of astrostatistics as a discipline in its own right, and second, I should like to propose a vision of the future of astrostatistics. Once I have proposed how I envision this future, I hope that the role statistician's play in it will be apparent.

2. Astrostatistics as a discipline

Astrostatistics may in general be defined as the discipline related to the statistical analysis of astronomical data. Some may incorporate astroinformatics as a component of astrostatistics, and a few years ago it was even argued that astrostatistics is a component of astroinformatics.

Astrostatistics focuses on statistically evaluating astronomical data; astroinformatics focuses on the problems and techniques of collecting and managing astronomical data. These are definitely related disciplines, and astrostatistics largely depends on, and will continue to rely on, the advances made in astroinformatics. However, the fields have essentially different goals, and I will separate the two for my purpose here.

Is astrostatistics a subdivision of the astronomical sciences? I tend to think that most astrophysicists believe this to be the case. Many statisticians, however, appear to think of astrostatistics as a branch of statistics, as they analogically do for biostatistics, geostatistics, environmetrics, chemometrics, psychometrics, econometrics, and epidemiology, to name a few such joint statistical disciplines.

Those in the biological and medical sciences with little exception think of biostatistics and biometrics as separate disciplines to biology. Geologists, though, I believe consider geostatistics as both a subdivision of geology and as a discipline in its own right. I envisage astrostatistics to be more in line with geostatistics in this regard, but the analogy to biostatistics may characterize the astrostatistics of the future.

Until recently astrostatistics has differed from these other joint statistical disciplines in that each of them have had an international association devoted to their discipline for a number of years. There are also either MS or PhD degree programs in these joint disciplines, and grad students can earn a degree with a clear concentration in the joint discipline.

A few examples of organizations that have been formed to support joint statistical disciplines include:

the *International Environmetrics Society*,
the *International Association for Mathematical Geology*,
the *International Biometric Society*,
the *International Society for Computational Biology*,
the *International Epidemiology Association.*

A few examples of such organizations based within the social sciences include

the *Econometric Society for the statistical analysis of economic data*,
the *International Association for Business and Industrial Statistics*,
the *International Association for Official Statistics*, and
the *International Association of Survey Statisticians*, to name a few.

There are similar international associations for most every recognized joint statistical discipline I can think of. At the very minimum about every international scientific association I researched had a section or committee related to the statistical analysis of that discipline's data. Until four years ago astrostatistics had no international association, and only in August, 2012 — twenty-one months ago — did the IAA come into existence. Moreover, there are currently still no MS or PhD degree programs in astrostatistics of which I am aware, although a few universities are moving in that direction.

Astrostatistics, as a discipline, is still in its infancy. This is not to say that some type of statistical analysis, albeit primarily descriptive analyses, has not been applied to astronomical data for over two thousand years. It has. But except for a relatively few occasions, astronomy progressed perfectly well without the need for inferential statistical techniques beyond linear regression – until recently.

3. The recent development of astrostatistics

It was not until computers were powerful enough in terms of RAM, processor speed, and storage space, that more complex non-linear and non-parametric statistical procedures were developed. The advent of personal computers made interactive statistical analysis a possibility. With them came the rapid development of more complex statistical procedures, and as computing power increased, statistical capabilities increased. As more complex statistical procedures were developed, the more they could be of use to astronomers who desired to statistically analyze their data.

Although there were a few conferences between statisticians and astronomers prior to 1991, it was not until then that astronomers became more interested in statistics. The first of the quint-annual *Statistical Challenges in Modern Astronomy* conferences was developed by Pennsylvania State Professors Eric Feigelson in astronomy, and Jogesh Babu, in statistics. Sponsored by the Center for Astrostatistics at Penn State, these conferences have been the mainstay of the discipline until just recently.

Table 1. Recent Milestones in Astrostatistics (selected)

Jul 2008	ISI Astrostatistics Interest Group formed (Hilbe)
Mar 2009	Astro2010 State of the Profession Paper: Astroinformatics: A 21st Century Approach to Astronomy (Kirk Borne, lead author)
Aug 2009	Meeting of ISI Astrostatistics Interest Group, ISI World Statistics Congress, Durban, South Africa (Hilbe)
Dec 2009	ISI Astrostatistics Committee approved (Hilbe)
Jan 2010	ISI International Astrostatistics Network formed (Hilbe)
Aug 2011	2 Invited Papers and 2 Special Topics in astrostatistics sessions: (Sen; Hilbe) World Statistics Congress, Dublin, Ire – 1st invited sessions at WSC
Sep 2011	Statistical Challenges in Modern Astronomy V (Penn. State- Feigelson, Babu); Springer Series in Astrostatistics formed at SCMAV conference (Hilbe)
Feb 2012	ASAIP started (Feigelson & Hilbe editors)
Jun 2012	AAS Astrostatistics & Astroinformatics Working Group approved (Ivezic)
Aug 2012	IAU Astrostatistics & Astroinformatics Working Group approved (Feigelson)
Aug 2012	IAA officially approved 30 Aug 2012 (Hilbe)
Jan 2013	1st meeting of the AAS A&A WG (at Winter meeting of the AAS), Long Beach, CA. (Ivezic)
Aug 2013	Invited and Special topics sessions in astrostatistics at ISI World Statistics Congress, Hong Kong, China (Sen; Hilbe; van Dyk)
Mar 2014	American Statistical Assoc. astrostatistics interest group approved (Cisewski)
Apr 2014	IAA Working Group Cosmostatistics formed (de Souza)
May 2014	Ist IAU Astrostatistics Symposium - Statistical Challenges in 21st Century Astronomy, IAUS306, Lisbon, Portugal (Heavens; Starck)

Several collaborations of astronomers and statisticians were also developed during the 1990s and into the 21st century. But there were no committees or even working or interest groups in astrostatistics under the auspices of any astronomical or statistical organization or society until 2008.

Although I have a varied background, to be sure, including being an Emeritus Professor at the University of Hawaii and now adjunct professor of statistics at Arizona State University, I came to astrostatistics as a biostatistician having rather substantial experience in handling large national-level health-outcomes data. I also had spent considerable time programming statistical software procedures, a number of which became components of commercial statistical software.

As a youth, though, I had planned on becoming as astronomer, and only changed majors half-way through my undergraduate career in the mid-1960s. However, I still retained my interest in astronomy over the years, and as an avocation, collected meteorites. I have well over 100 meteorites of all varieties including Martian and Lunar stones and every classification of carbonaceous chondrite. In addition, I have been a member of the international Meteoritical Society for a number of years, which is mostly comprised of planetary scientists and geologists. In 2007 I became associated with the Jet Propulsion Laboratory, representing them in giving presentations on meteorites to schools, community groups, astronomy clubs, and the like. It was that association which resulted in my being here today.

In 2008 I was on one of my bi-weekly conference calls with the Principal Investigator of a JPL research study. I don't recall the nature of the project or mission now, or who the PI conducting the call happened to be, but I clearly recollect that he complained that he was not sure of how best to statistically evaluate study results. I had heard this same complaint before on several previous conference calls with other principal investigators, but this time the words stuck.

Not being aware at the time of the already ongoing collaborations in astrostatistics, I got off the call thinking that there should be a way to get astronomers and statisticians who have an interest in astronomical research together, from the very outset of a project, in order to help affect a proper statistical analysis of the study data. I considered that astrophysicists have training and expertise on the physics side of such a project, but were likely not trained in statistical analysis — in particular only a relatively few are likely current with recent advances in statistics, but most would value having competent statistical advice. On the other hand, I thought that there would be a number of statisticians who would love to collaborate with astrophysicists on various astronomical research projects. It could be a win-win for both astronomers and statisticians, with better science as the result. I have amended my view a bit since 2008, but in large part I still maintain this position.

The call spurred me to form an interest group in astrostatistics under the auspices of the International Statistics Institute, or ISI, which is to the discipline of statistics as the IAU is to astronomy. I already chaired an ISI standing committee, and was acquainted with ISI administration and the procedures of becoming an interest group. Moreover, I soon learned that several very competent groups of statisticians and astronomers had been gathering for collaboration and conferences for a decade and a half, but I believed that more collaboration between astronomers and statisticians could exist if astrostatistics as a discipline was perhaps formally recognized by the ISI and IAU. Remember, I first conceived of this idea a result of hearing complaints about the need for statistical support from principal investigators who were associated with major research organizations.

The ISI Interest Group was formed for the purpose of gathering interested statisticians together at the 2009 ISI World Statistics Congress in Durban, South Africa to discuss astrostatistics and the role statisticians can play in the analysis of astronomical data. I convened the meeting, which was attended by some 50 astronomers and statisticians from around the world. The ISI president-elect also attended. At the end we voted to seek full committee status under the ISI. A full standing committee on astrostatistics was approved at the 2009 winter conference of the ISI executive committee — the first such committee under any astronomical or statistical organization.

I contacted those who I believed to be recognized at the time as the leaders in the discipline to join the committee. The response was thoroughly positive. The ISI, though, limited committee sizes to 15, and we had many more researchers interested in being involved. I therefore initiated the ISI Astrostatistics Network in January 2010, a few weeks after the committee had been approved.

In February of 2012 Eric Feigelson, also a member of the Network executive committee, was able to secure support from his department at Pennsylvania State University for the creation and maintenance of an Astrostatistics & Astroinformatics Portal, or ASAIP. He kindly invited me to be co-editor, a position we jointly share, but he has been the mainstay of the effort.

The ASAIP serves as a means for anyone interested in astrostatistics or astroinformatics to obtain information about forthcoming astrostatistically-related conferences, to receive announcements from other members, to learn about journal articles and books related to astrostatistics, and to even serve as a resource for job openings and job searches. There is also the opportunity for Portal members to join blogs on various topics.

For example, the ASAIP has 8 general purpose Sections of Blogs hosted by Portal members as well as three sections of Blogs that are primarily hosted by outside sites - 8 in astronomy, 15 in statistics, and 17 in information sciences. In addition to Blogs, the Portal provides information on locating tutorials and classes on subjects related to statistical modeling, software languages, and the like. The Portal had some 150 members by April 2012 and now has over 700 members, just two years later.

In June of 2012 the *American Astronomical Society* Council approved a new Working Group in Astrostatistics and Astroinformatics - the first such Working Group authorized by an astronomical society. Shortly afterwards, in August 2012, Commission 5 of the IAU approved the creation of a similar Working Group, with Feigelson as its inaugural chair. Later that month, on the 30th of August, the executive committee of the ISI

Astrostatistics Network voted to be renamed the *International Astrostatistics Association*, or IAA. It is now an independent professional association for astrostatisticians with ties to the ISI through the ISI astrostatistics committee, and to the IAU through its relationship to the IAU Working Group in Astrostatistics & Astroinformatics.

Currently the IAA has some 462 members from 47 nations, with membership sections of regular, PostDoc and grad student. Regular members are also partitioned as being astronomers, statisticians, information scientists or astroinformaticists, or other. Last month the IAA inaugurated a Working Group on Cosmostatistics aimed at promoting the development of new statistical tools for cosmological data exploration. Rafael de Souza of the Korea Astronomy and Space Science Institute initiated and heads the working group. As of May 15th, there were 31 members participating in the Working Group, with the four project groups displayed on the screen already having been formed. A fifth project group was formed this past week.

General Relationships into Astronomical Datasets,
Multiple Survey Analysis,
CosmoR : R oriented packages for cosmostatistics,
Generalized Linear Models for Astro/Cosmo-Statistics.

4. Astrostatistics and extended generalized linear models

With respect to the last mentioned Project Group—on Generalized Linear Models or GLMs–I believe that astronomers have underutilized this class of models and their extensions. Basic GLMs include normal or Gaussian regression, gamma and inverse Gaussian models, and the discrete response binomial, Poisson and negative binomial models.

Table 2. Range of Generalized Linear Models

Range of Generalized Linear Models
Continuous Response Models
Normal or Gaussian, Log-normal
Gamma
Inverse Gaussian
Binomial Response Models
Logistic - probit - complementary loglog - loglog
Count Response Models
Poisson (POI)
Negative Binomial (NB)

Extended Generalized Linear Models include the addition of scale parameters for continuous response GLM models, as well as a wide variety of discrete response regression models. Many astrostatistical data situations have discrete data to be modeled, but are instead modeled as if the variable of interest to be modeled is continuous. Or, if they model count data, many astronomers employ a Poisson model, somehow forgetting about the Poisson distributional assumption of the equality of the mean and variance —called equi-dispersion. This is a terribly restrictive assumption, and is rarely found in real data.

From looking at the literature in the field, it appears that most astronomers are not aware of the explosion of discrete response models that have occurred recently — many of which I believe can be applied to astronomical data from both frequency-based and Bayesian perspectives. Project Groups like this — comprised of both astronomers and statisticians — have the prospect of bringing more modeling examples, and software, to the attention of astronomers seeking to properly evaluate their study data.

Table 3. Selected Extensions to the Basic GLM Algorithm

Binomial	
Beta	Generalized maximum entropy logistic
Generalized binomial	
Nested logit	Fractional polynomial binomial
Proportional odds models	Partial proportional odds models
Multinomial models	Heterogeneous multinomial models
Beta-binomial	
Count	
Heterogeneous negative binomial	Poisson-lognormal
Poisson Inverse Gaussian (PIG)	3-parameter generalized NB-F
Generalized Poisson (GP)	3-parameter generalized NB-W
Generalized negative binomial P	Finite mixture models
Double Poisson	Quantile count models
Double hurdle	
2-part Hurdle POI, NB, PIG, GP, NBP, NBF & Logit, Probit, POI, etc	
Zero-inflated POI, NB, PIG, GP, NBP, NBF & Logit, Probit, cloglog, loglog	
Truncated POI, NB, PIG, GP, NBP, NBF (right, left, interval, 3+levels)	
Censored POI, NB, PIG, GP, NBP, NBF (right, left, interval, 3+levels)	
Hierarchical and mixed standard and Bayesian models for the above	

5. The future of astrostatistics

I hope that we can encourage more participation in the IAA *Working Group in Cosmostatistics*, with other astrostatisticians proposing or becoming members of similar project groups. New working groups can be developed as well. To my mind this represents a true advance in the discipline, and our discipline will thrive as this program expands.

Regarding my themes — the role of statistics and of statisticians in the future of astrostatistics: I have argued that both statisticians and astronomers play essential roles in the discipline. Statisticians will continue to develop ever more complex statistical methods, many of which will be applicable to astronomical science. In fact, recent advances in environmental and ecological statistics, which entail work in spatial analysis, geostatistics, and discrete response modeling, and advances made by econometricians in time series, can be of particular use to astrostatisticians. A number of other newly developed statistical methods can as well.

Statisticians who are working in these areas, as well as competent statisticians who have a particular interest in astronomy, can be of considerable help to astronomers when analyzing data.

Of course, there may be some astronomers who love to engage in statistics and perhaps thoroughly enjoy developing statistical models and software appropriate for astronomical analysis. Those attending this Symposium give evidence of this type of astrophysicist. You have the capability of developing statistical procedures and software that not only is applicable to astronomy, but which can also favorably impact research in other disciplines.

Table 4. Proposals for the Future

- Joint Astrostatistics MS & PhD Degree Programs
- Having more researchers propose and develop IAA Working and Project Groups
- Structure IAA to be like similar international subject area associations
  - § Add committees to IAA
    - Committee for PostDocs
    - Committee for Graduate Students
    - Membership committee
    - Awards-Recognition committee
      - For outstanding publication award
      - Young researcher awards
      - IAA Fellow
      - Signature award for contributions to the discipline
    - Meetings committee
      - Coordinate and advise members of the various meetings and conferences related to astrostatistics & astroinformatics being held throughout the year
    - Finance committee
      - I have sought to have external funding support the activities of the IAA. I hope we can maintain this philosophy. We need more than one person working on this task.
  - § Other functions ???

I also propose that we encourage and participate in the development of interdisciplinary degrees in astrostatistics, from the MS to the PhD level. These should be joint projects by University Astronomy/Astrophysics and statistics departments. Such programs will entail the need for participating statistics departments to hire professors who specialize in astrostatistics. This is exactly what occurs in other joint disciplines, and will in astrostatistics as well. This is good for the discipline, and will help increase job opportunities for astrostatisticians.

Finally, I believe it is important to develop and support the IAA, as well as the IAU and AAS Working Groups. But if our discipline is to advance and flourish, and be thought of as are other successful joint disciplines, maintaining and supporting a strong international astrostatistics association is vital. It has been for other joint disciplines, and it will be for us as well.

As a statistician who has had a lifelong interest in astronomy and cosmology, I have been pleased to participate in this IAU Astrostatistics symposium and to serve on its scientific organizing committee. The outstanding success of this Symposium should signal the IAU that astrostatistics is a valuable component of professional astronomy, as well as an important discipline in its own right.

It's now up to us, both individually and collectively, as astronomers, statisticians and information scientists, to decide the future direction of astrostatistics, and how we wish to shape its future.

Reference

Hilbe, J. M. 2014, *Modeling Count Data*, Cambridge: Cambridge University Press

Statistical Challenges in 21st Century Cosmology
Proceedings IAU Symposium No. 306, 2014
A. F. Heavens, J. -L. Starck & A. Krone-Martins, eds.

doi:10.1017/S1743921314011090

Concluding Remarks from a Cosmologist

Andrew H. Jaffe
Astrophysics, Blackett Laboratory
Prince Consort Road
Imperial College London
UK
email: a.jaffe@imperial.ac.uk

Abstract. I conclude the SCCC21 conference highlighting some of the contrasts we heard about, some specific topics (the statistics of random fields), and discuss how some of these play out in the analysis of cosmic microwave background data. I conclude with a hopeful look at the efficacy of blind analyses in the CMB and elsewhere in cosmology.

Keywords. cosmology: cosmological parameters, methods: statistical

1. (False?) Dichotomies

Throughout this meeting, we saw contrasts between different groups of researchers, different philosophies, and different techniques to attack the same problem. Some of these distinctions are real, but we often find that they are not as sharp as we might initially think.

Cosmologists & Statisticians. By their very presence, pretty much everyone here straddles this line, no matter what our formal training might be.

Inference & Algorithms. Perhaps this is a better distinction than the previous: do you come with science questions to answer, looking for the right tool? Or do you start with the tools, and look for applications?

Theorists & Observers. Astronomy has always been a discipline straddling observation and theory. Although there have always been observers who put a lot of theory into the interpretations of their data, astronomy has traditionally been led by data, from the purely observational beginnings of the Hertzsprung-Russell diagram and the Morgan-Keenan-Kellman system of stellar classification, both of which only latterly gained theoretical underpinnings. Nowadays, that tendency has only got stronger, with the rise of a group of theoretically trained astrophysicists who spend much of their time wondering about how to analyze data. They (we) are perhaps the equivalent of the *phenomenologists* of particle physics, charged with making predictions for what will be seen by experimenters.

Bayesians & Frequentists. Historically, the interpretation of probability has been a contentious point. This remains so in the statistics community as well as amongst practicing scientists. There is no need to rehearse the controversies here; rather, it's important to point out that we practicing scientists cannot really afford idealogical purity (see also Section 3) — it is more important to get some sort of answer out even if lacking a principled statistical basis. It is also important to remind ourselves that "Bayesian" and "frequentist" are themselves catch-all descriptions which can hide vast philosophical and practical differences — see, for example, Good (1971) and his "46,656 varieties of Bayesians".

Laggards & Leaders. See astronomer-turned-statistician Ewan Cameron's talk and his contribution to these proceedings, "What we talk about when we talk about fields,"

Cameron (2014) in which he gently berates the cosmology community for lagging behind the statisticians.

Humans & Machines; Supervised & Unsupervised. Many algorithms discussed here require human intervention: we come with particular parameterizations of the phenomena we are studying. But in the coming "big data" era, that may not always be possible: we want the data to teach us what's in it. Of course, these dichotomies are particularly false: any unsupervised algorithms is written by humans, able to make certain kinds of classifications more easily than other.

Real & Idealized. It is heartening to see even the statisticians working with real data, eschewing the caricature of simplifying our real scientific problems to ones that are mathematically tractable but unrecognizable and scientifically uninteresting. (When I first starting learning more formal statistics than I had managed as an undergraduate, I was amused to learn the polysyllabic term "heteroscedastic", so complicated it was clearly seen as a special case — we just call them "errors".)

Statistics & Systematics. If we think we can describe the distribution of some quantity contributing to our data, we call it a "statistical error"; otherwise, we call it a "systematic". As CMB experimentalist Paul Richards from UC Berkeley once said, systematic errors are what can make a good experiment get worse instead of better when you get more data.

Sparse & Dense. The first thing a physicist will do with almost data she can get her hands on is to Fourier transform it. We do this because in many cases the properties of the data are simpler — more sparse — in the harmonic domain. In recent years, the idea of sparsity has taken hold: can we find general methods for finding transformations of the data into a sparse basis?

Blind & Open. At a higher methodological level, cosmologists are beginning to realize that we are not without our own unacknowledged prejudices: sometime we find the answer we are looking for. To combat this, blind analysis, a tool that has a long tradition in the particle physics community, is finding its place in cosmology (see Section 3.1).

2. Random Fields

Although pioneering cosmologists like Peebles (1980) had clearly been thinking of the distribution of matter in the Universe as realizations of a random field, it took the work of Adler (1981) and its subsequent use in a cosmological context by the immense and much-cited (if less often read) Bardeen *et al.* (1986) to bring this formulation into broader use, specifically for the calculation of the distribution of extrema, as proxies for the location of galaxies and clusters as well as voids.

Those works were largely concerned with Gaussian fields — luckily for us, this seems like a good description of the distribution of matter on large scales. We may well ask ourselves why this is the case. There are two somewhat obvious, but not necessarily equivalent, routes. The first is through the *central limit theorem*: if we can describe the distribution of matter by a suitable combination of individual processes, we can expect the result to be well-approximated by a Gaussian distribution (at least far from the tails...). But in fact we know more about the distribution than this: it seems to be an isotropic field, described by an approximately scale-invariant Harrison-Zeldovich-Peebles-Yu power spectrum (Peebles & Yu 1970; Harrison 1970; Zeldovich 1972). To get this, we need more physics. A free (non-interacting) quantum field will have an isotropic Gaussian correlation function, and cosmic inflation is one way to embed such a field in an expanding Universe and get a (nearly) scale-invariant power spectrum (see, e.g., Planck Collaboration XXII 2014, and references therein).

Furthermore, we know that the Gaussian description isn't enough: late-time effects such as nonlinear gravitational evolution and lensing certainly modify the statistical properties of the field, as do any departures from a free quantum field. The former effects have already been observed (Planck Collaboration XXIV 2014); the latter, which would give us a hint about the nature of the interactions experienced by the field, were the subject of much discussion at the conference, and the target of ongoing observations.

The statistical challenge lies in what infinitesimal subset of the ways in which a distribution may depart from an isotropic Gaussian are of cosmological interest. Some of those may be driven by particular theoretical ideas, such as the weak non-Gaussianity induced by the coupling of the inflationary potential (Planck Collaboration XXIV 2014), while others may just be convenient phenomenological descriptions of non-Gaussian distributions (Planck Collaboration XXIII 2014). During this meeting, we heard about many different versions of these:

- simple phenomenological parameterizations, driven by even simpler models ($f_{\rm NL}$);
- going beyond point estimators to various ansätze for the distribution $Pr(f_{\rm NL}|\text{data})$;
- going beyond single numbers to full n-point polyspectra;
- Minkowski functionals and the beautiful Gaussian Kinematic Formula (Taylor 2006);
- physical reconstruction vs nonlinear transformations and clipping; and
- going beyond the CMB — today: QSOs, tomorrow: Euclid, 21cm, and the coming age of photo-z.

3. Case Study: The CMB

Almost all of these topics show up in the study of the Cosmic Microwave Background. Indeed, several of the first theorists to investigate the details of the distribution of the CMB on the sky (Bond & Efstathiou 1984) went on to pioneer the analysis of CMB data from real experiments (Bond *et al.* 1998; Efstathiou 2006).

A simple flowchart for CMB data analysis is shown in Fig. 1: we start with raw time-ordered data, average these into a map of the sky, evaluate the angular power spectrum, C_ℓ of this map, and determine the set of cosmological parameters responsible for this spectrum. Each step represents significant data compression. For *Planck*, there are trillions of initial samples, tens of millions of pixels in a map, 2500 multipoles in the power spectrum which is finally represented by six parameters (Planck Collaboration XVI 2014). At each step, the intermediate result (i.e., the map or the power spectrum) can be thought of as a *sufficient statistic* — in a Bayesian sense we don't really need to even specify a prior for them as they are just convenient representations of our original data. For the map, we only need to assume that the signal is fixed on the sky (so any time evolution violates the model); for the power spectrum, we need to further assume isotropy and Gaussianity.

Even without further complications, this is a computationally daunting task. If the noise is not white, the time for the mapmaking step (see, e.g., Cantalupo *et al.* 2010) naively scales as the cube of the number of pixels, as does the power spectrum step (Bond *et al.* 1998). Both of these can be simplified by clever numerics (Ashdown *et al.* 2007) or by using frequentist methods (Hivon *et al.* 2002) but these result in approximate or incomplete solutions. For mapmaking, it is impossible to calculate the full noise covariance of the resulting Gaussian distribution, while for power spectra even the form of the distribution is not well determined.

And even this picture is quite a bit simpler than what is needed in practice. A raft of "systematic effects" need to be accounted for, usually only in an approximate or parameterized way. The most important is probably the effect of astrophysical foregrounds which contaminate the primordial signal. Much of this can be taken care of by

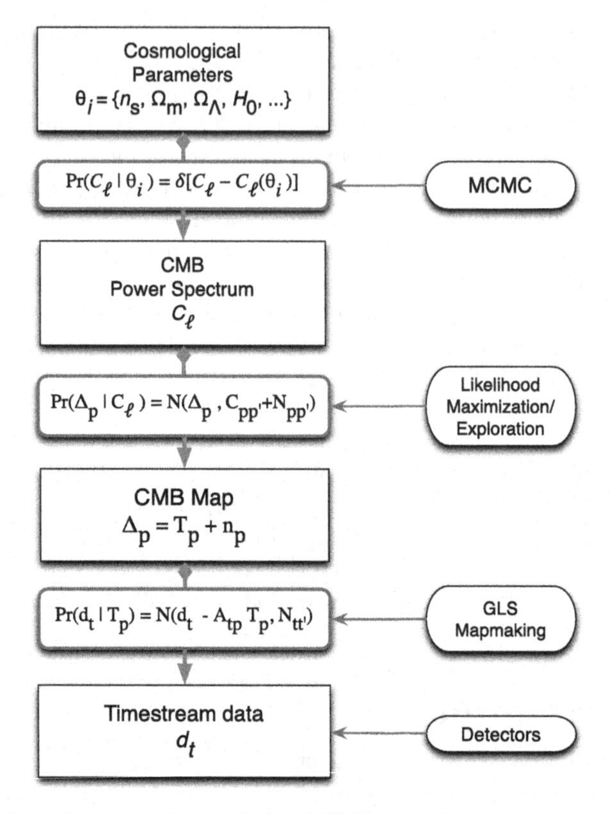

Figure 1. The flow of data in the analysis of CMB experiments — the simplified version.

conservative masking of the sky but the low-level remnants of foreground emission remain throughout the sky and must be accounted for (e.g., Planck Collaboration XII 2014). Similarly, the data-flow described above assumes axisymmetric beam-shapes (i.e., the point-spread function in the usual astronomical parlance) and dense, uniform sampling of the sky in the time domain. Except in special cases (Armitage-Caplan & Wandelt 2004), this cannot be accounted for exactly.

In practice, then, we cannot afford idealogical purity — the cosmological community is largely Bayesian in its outlook, but in fact many of our algorithms are essentially frequentist, assuming that the asymptotic correspondence between Bayesian and frequentist results holds at least at small scales (high ℓ). We further remove at least some systematic effects by measuring some appropriate set of instrumental parameters and fixing their values, or in some cases use forward Monte-Carlo modeling to approximately parameterize the posterior distribution (e.g., Planck Collaboration VII 2014).

3.1. *Blind Analyses*

We never have enough information to give a complete statistical description of a realistic experiment: there are *always* systematic effects for which we have not accounted. When do we stop looking? At this meeting, in talks from Hiranya Peiris and others, we were admonished to do something other than just stopping when we thought we had the right answer. One way to ensure this, used very often in the particle physics community, is to use a blind analysis: perfect the tools on simulations and on subsets or combinations of the data which are expected to have none of the signal being sought after.

We attempted such an analysis for the recent publications from Polarbear (The Polarbear Collaboration 2014, 2013), an experiment measuring the polarization of the CMB, in particular searching for the so-called B-mode pattern which results from a background of inflationary gravitational waves and/or lensing of polarized CMB photons along the line of sight. The Polarbear Collaboration (2014) gives the first direct detection of the B-mode power spectrum due to the latter lensing effect, analyzed using largely blind techniques. These are especially useful during the initial discovery phase during which it would be very easy to confuse an instrumental or foreground systematic effect for the sought-after signal.

References

Adler, R. J. 1981, The Geometry of Random Fields (Society for Industrial and Applied Mathematics (SIAM, 3600 Market Street, Floor 6, Philadelphia, PA 19104))
Armitage-Caplan, C. & Wandelt, B. D. 2004, *PRD*, 70, 123007
Ashdown, M. A. J., Balbi, A., Bartlett, J. G., *et al.* 2007, *A&A*, 467, 761
Bardeen, J. M., Bond, J. R., Kaiser, N., & Szalay, A. S. 1986
Bond, J. R. & Efstathiou, G. 1984, 285, L45
Bond, J. R., Jaffe, A. H., & Knox, L. E. 1998, *PRD*, 57, 2117
Cameron, E. 2014, arXiv:1406.6371v1, 6371
Cantalupo, C. M., Borrill, J. D., Jaffe, A. H., Kisner, T. S., & Stompor, R. 2010, *ApJ*, 187, 212
Efstathiou, G. 2006, *MNRAS*,370, 343
Good, I. J. 1971, *American Statistician*, 25, 62
Harrison, E. 1970, *PRD*, 1, 2726
Hivon, E., Górski, K. M., Netterfield, C. B., *et al.* 2002, *ApJ*, 567, 2
Peebles, P. J. E. 1980, The large-scale structure of the universe
Peebles, P. J. E. & Yu, J. T. 1970, *ApJ*, 162, 815
Planck Collaboration VII. 2014, *A&A*, in press, arXiv:1303.5068
Planck Collaboration XII. 2014, *A&A*, in press, arXiv:1303.5072
Planck Collaboration XVI. 2014, *A&A*, in press, arXiv:1303.5076
Planck Collaboration XXII. 2014, *A&A*, in press, arXiv:1303.5082
Planck Collaboration XXIII. 2014, *A&A*, in press, arXiv:1303.5083
Planck Collaboration XXIV. 2014, *A&A*, in press, arXiv:1303.5084
Taylor, J. 2006, *The Annals of Probability*, 34, 122
The Polarbear Collaboration. 2013, arXiv:1312.6645
The Polarbear Collaboration. 2014, arxiv:1403.2369
Zeldovich, Y. B. 1972, *MNRAS*,160, 1P

Author Index

Printed in the United States
by Baker & Taylor Publisher Services